Oxygen Production and Reduction in Artificial and Natural Systems

Oxygen Production and Reduction in Artificial and Natural Systems

Edited by

James Barber
Imperial College London, UK

Alexander V Ruban
Queeen Mary University of London, UK

Peter J Nixon
Imperial College London, UK

World Scientific

NEW JERSEY • LONDON • SINGAPORE • BEIJING • SHANGHAI • HONG KONG • TAIPEI • CHENNAI • TOKYO

Published by

World Scientific Publishing Co. Pte. Ltd.
5 Toh Tuck Link, Singapore 596224
USA office: 27 Warren Street, Suite 401-402, Hackensack, NJ 07601
UK office: 57 Shelton Street, Covent Garden, London WC2H 9HE

Library of Congress Cataloging-in-Publication Data
Names: Barber, J. (James), 1940– editor.
Title: Oxygen production and reduction in artificial and natural systems /
edited by James Barber, Imperial College London, UK, Alexander V. Ruban,
Queen Mary University of London, UK, Peter J. Nixon, Imperial College London, UK.
Description: New Jersey : World Scientific, [2019] | Includes bibliographical references.
Identifiers: LCCN 2019006784 | ISBN 9789813276918 (hardcover : alk. paper)
Subjects: LCSH: Bioenergetics. | Energy metabolism.
Classification: LCC QH510 .O98 2019 | DDC 572/.43--dc23
LC record available at https://lccn.loc.gov/2019006784

British Library Cataloguing-in-Publication Data
A catalogue record for this book is available from the British Library.

For any available supplementary material, please visit
https://www.worldscientific.com/worldscibooks/10.1142/11173#t=suppl

Typeset by Stallion Press
Email: enquiries@stallionpress.com

Contents

Foreword by James Barber FRS

This book is the outcome of a Bioenergetics workshop held at Nanyang Technological University (NTU), Singapore in April 2018 to honour Professor Bertil Andersson for his outstanding contributions to scientific research and administration, particularly his very successful 11 years at NTU as Provost (2007–2011) and President (2011–2018) and to celebrate his 70th birthday.

During his PhD studies at Lund University, Bertil introduced the technique of polymer 2-phase partition for the subfractionation of plant photosynthetic membranes [Andersson, 1978]. This new approach developed by his PhD supervisor, Professor Per Albertsson (see photograph below) allowed him to separate inside-out and right-side out chloroplast thylakoid membrane vesicles. He concluded that the thylakoid membranes of higher plants are highly heterogeneous with photosystem II (PSII) and photosystem I (PSI) being located in different membrane regions (PSII in the stacked membranes of the grana and PSI in the exposed stomal membranes). It was also discovered that there was a partitioning of other components such as the ATP-synthase and light-harvesting complexes [Andersson and Akerlund, 1978]. At the time this was a highly controversial since it went against the prevailing models of the organisation of photosynthetic supercomplexes with PSI and PSII directly linked. Nevertheless, the lateral heterogeneity model that he and Jan Anderson [Andersson and Anderson, 1980] published almost 40 years ago (shown on the cover) has stood the test of time.

Bertil with his PhD supervisor and mentor, Per Albertsson who passed away during the preparation of this book.

Bertil was also involved in demonstrating that the lateral heterogeneity of the thylakoids is highly dynamic in response to the physiological state of the plant. Thus, it was discovered that protein phosphorylation of the light-harvesting complex II (LHC II) leads to its migration from the photosystem II rich membrane regions to the photosystem I membrane regions [Andersson *et al.*, 1982]. This dynamic and reversible mechanism was of great significance for understanding the so-called state transition mechanism which controls the excitation distribution between the PSI and PSII in response to changes in the external light-regime. In so doing, the efficiency of photosynthesis is optimised. Later he also showed that the activation of the kinase was under redox control exerted *via* the occupancy of plastoquinol at the Q_0-site of the cytochrome *b*/*f* complex [Vener *et al.*, 1997].

The isolation of inside-out thylakoids opened up new experimental approaches to study PSII and the oxygen evolving complex (OEC). He and his colleagues identified the OEC protein components 33, 23 and 16 kDa out of which the latter two had not been described before [Akerlund *et al.*, 1982]. This was followed up by the identification of the genes, protein sequences and leader peptide compositions of the 23 and 16 kDa proteins (Jansen *et al.*, 1987).

Bertil was also a co-discoverer of the Psb-S protein that is a unique analogue of the family of light-harvesting proteins, but does not possess a direct light-harvesting function [Wedel *et al.*, 1992]. It has more recently been suggested that this protein could play a role in the so-called Q_E-quenching, a crucial molecular mechanism for regulation of light harvesting.

For many years, Bertil was the leader in the elucidation of the molecular basis of photoinhibition with several important contributions. For example he and his co-workers were the first to identify specific proteolytic enzymes involved in the processing of the D1 protein such as FtsH [Lindahl *et al.*, 2000] and the Deg P2 type [Haussuhl *et al.*, 2002]. And, of course, the reviews on photoinhibition that he co-authored have been very influential and exceptionally highly cited [Aro *et al.*, 1993; Barber and Andersson, 1992].

In addition to his many research achievements, Bertil has a remarkable record in promoting science in general. For about 20 years he was involved in the selection of Nobel Prize winners, particularly in chemistry and was chairman of the Chemistry Nobel Committee in 1997. In the period 2004–2007, Bertil was the Head of the European Science Foundation (ESF) in Strasbourg. In his capacity as Director of ESF and Vice-President of the European Commission Research Advisory Board (EURAB), he was one of the driving forces behind the establishment of the European Research Council (ERC). Prior to this he was a very effective and successful Rector of Linkoping University, Sweden (1999–2004).

In 2007, Bertil joined the academic staff of NTU in Singapore, first as Provost (2007–2011) and then as President (2011–2018). During Andersson's leadership, NTU developed into a research-intensive university and is today one of the fastest developing universities on the international scene. Bertil Andersson is a truly exceptional person. His many successes have been motivated by his love of scientific research and the quest for new knowledge. We have been close friends for almost 40 years and I know he will not sit back in retirement but continue to use his experience and talents in constructive ways to support science and science policies.

References

Andersson, B. (1978). Separation of spinach chloroplast lamellae fragments by phase partition: Including the isolation of inside-out thylakoids. *PhD Thesis*, Lund University, Sweden.

Albertsson, P.Å., Andersson, B., Larsson, C. and Åkerlund, H.E., (1982). Phase partition — a method for purification and analysis of cell organelles and membrane vesicles. *Meth. Biochem. Anal.*, 28, 115–150.

Andersson, B. and Åkerlund, H.E. (1978). Inside-out membrane vesicles isolated from spinach thylakoids. *Biochim. Biophys. Acta Bioenerg.*, 503, 462–472.

Andersson, B. and Anderson, J.M. (1980). Lateral heterogeneity in the distribution of chlorophyll-protein complexes of the thylakoid membranes of spinach chloroplasts. *Biochim. Biophys. Acta Bioenerg.*, 593, 427–440.

Andersson, B., Åkerlund, H.E., Jergil, B. and Larsson, C. (1982). Differential phosphorylation of the light-harvesting chlorophyll — protein complex in appressed and non-appressed regions of the thylakoid membrane. *FEBS Lett.*, 149, 181–185.

Aro, E.M., Virgin, I. and Andersson, B. (1993). Photoinhibition of photosystem II. Inactivation, protein damage and turnover. *Biochim. Biophys. Acta Bioenerg.*, 1143, 113–134.

Åkerlund, H.E., Jansson, C. and Andersson, B. (1982). Reconstitution of photosynthetic water splitting in inside-out thylakoid vesicles and identification of a participating polypeptide. *Biochim. Biophys. Acta Bioenerg.*, 681, 1–10.

Barber, J. and Andersson, B. (1992). Too much of a good thing: light can be bad for photosynthesis. *Trends Biochem. Sci.*, 17, 61–66.

Haussuhl, K., Andersson, B. and Adamska, I. (2001). A chloroplast DegP2 protease performs the primary cleavage of the photodamaged D_1 protein in plant photosystem II. *EMBO J.*, 20, 713–722.

Jansen, T., Rother, C., Steppuhn, J., Reinke, H., Beyreuther, K., Jansson, C., Andersson, B. and Herrmann, R.G. (1987). Nucleotide sequence of cDNA clones encoding the complete '23 kDa' and '16 kDa' precursor proteins associated with the photosynthetic oxygen-evolving complex from spinach. *FEBS Lett.*, 216, 234–240.

Lindahl, M., Spetea, C., Hundal, T., Oppenheim, A.B., Adam, Z. and Andersson, B. (2000). The thylakoid FtsH protease plays a role in the light-induced turnover of the photosystem II D1 protein. *Plant Cell,* 12, 419–431.

Vener, A.V., Van Kan, P.J., Rich, P.R., Ohad, I. and Andersson, B. (1997). Plastoquinol at the quinol oxidation site of reduced cytochrome *bf* mediates signal transduction between light and protein phosphorylation: thylakoid protein kinase deactivation by a single-turnover flash. *Proc. Natl. Acad. Sci.*, 94, 1585–1590.

Wedell, N., Klein, R., Ljungberg, U., Andersson, B. and Herrmann, R.G. (1992). The single-copy gene psbS codes for a phylogenetically intriguing 22 kDa polypeptide of photosystem II. *FEBS Lett.*, 314, 61–66.

Preface

This book covers some of the most recent advances in understanding oxygen recycling in the Biosphere. The processes of oxygen production and oxygen reduction lie at the very heart of the biological energy transformation events that sustain almost all life on our planet. The Earth, or the "blue dot" as described by James Barber in Chapter 1, is a very special place in the universe that is now under serious threat from human activity. The very strong message of this chapter — that we must develop alternative sources of renewable energy based on solar energy — is a clarion call not only to politicians but also to scientists. Other contributors to this book are the two outstanding Nobel Laureates — Rudolph Marcus (Chapter 2) and John Walker (Chapter 6). Marcus formulated a theory to describe electron transfer between molecules that provides a kinetic and thermodynamic description of electron transfer in both natural and artificial systems. Walker solved the atomic structure of the ATP synthase — a molecular motor that uses the electrochemical energy of the proton gradient across the membrane to drive ATP synthesis. Not surprisingly, the theme of electron and proton transport reactions in the key redox enzymatic complexes of mitochondria and chloroplasts is covered in the majority of chapters of this book. Hence, Chapters 3 (Kaila), 4 (Siegbahn) and 17 (Guo) are dedicated to fundamental theoretical aspects of electron transport reactions and the proteins involved. The state-of-the-art reviews on events and molecules involved in the proton energy transformation are described in Chapters 2 (Marcus), 6 (Walker), 7 (Wikström), 8 (Rich), 10 (Ishikita) and 15 (Dobbek). Processes involved in oxygen production via the water-splitting system of Photosystem II and its damage, assembly and repair are described in Chapters 5 (Vass), 11 (Llobet), 13 (Agapie), 14 (Brudvig), 16 (Nixon) and 18 (Dau). Moreover, the applied aspects of the science of bioenergetics are clearly visible as the first seedlings of a new generation of applied multidisciplinary research aiming at enhancing photosynthesis in crop plants and developing alternative sources of clean energy. Indeed, Chapter 19 (Jansson) is dedicated to improving biomass production and Chapters 9 (Zhang) and 12 (Wong) focus on artificial water-splitting systems.

We recommend this book to specialists in Bioenergetics and Photosynthesis, researchers and postgraduate students who are working and studying various aspects of respiratory and photosynthetic electron transport chains, proton energy transformation events and application of principles of biological energy transformation for the creation of alternative, renewable energy sources for humankind.

We would like to acknowledge the Institute of Advanced Studies at Nanyang Technological University for supporting and hosting the workshop in March 2018 and encouraging us to produce this book.

J. Barber, P.J. Nixon and A.V. Ruban
23 November 2018

Chapter 1

How Biology Solved Its Energy Problem and Implications for the Future of Humankind

James Barber

Imperial College London, South Kensington Campus SW7 2AZ, UK
j.barber@imperial.ac.uk

The appearance of life on our planet occurred about four billion years ago following five hundred million years of geological and chemical evolution. It took the form of single cell prokaryotic chemotrophs obtaining their energy from chemicals which could act as electron donors. After 1.5 billion years there were a wide variety of anaerobic prokaryotes which fell into two groups, archaea and bacteria. Darwinian principles had allowed these prokaryotes to be widely distributed in many different types of environments and evolve complex systems to capture and utilise energy for the reductive process of cellular biochemistry. Sooner or later, however, the energy sources (oxidisable substrates) available to these chemotrophic prokaryotes became limiting and some developed the capacity to supplement their metabolic energy requirements by absorbing solar radiation. This heralded the "Big Bang of Evolution" since this strategy gave rise to phototrophic organisms which could use sunlight to totally drive their bioenergetics. This was achieved by using solar energy to extract "hydrogen" (in the form of high energy electrons and protons) from water and release oxygen into the atmosphere. From then on, life prospered and diversified on an enormous scale driven by Darwinian evolution. It was from this success that humans evolved as a very special species with abilities far greater than any other animal that has ever lived on planet Earth. Yet we are on a very dangerous course which some believe could result in the elimination of our species in a matter of a few hundred years if we carry on the way we do today, that is, continue to use fossil fuel as a major energy source. Biology solved its energy problem by using sunlight to split water and extract the stored energy by recombining the "hydrogen" with oxygen *via* respiration. We must do the same by using energy from the sun captured directly *via* photovoltaic cells and indirectly by hydro-, wind- and wave-power to give electricity and hydrogen as the two main energy carriers. The latter complements batteries for renewable energy storage and must come from the electrolysis of water. There is an urgent need to progress with expanding and developing the appropriate technologies and putting in place the infrastructure, particularly for the generation of hydrogen and its storage, transport and use.

1. Evolution of Bioenergetics

1.1. *Under anaerobic conditions*

According to the Big Bang Theory, the Universe came into being about 13.7 billion years ago, leading to the formation of our Planet and Solar System about 9.2 billion years later. As depicted in Figure 1, it took half a billion years for the Earth to cool and establish a chemical environment for life to evolve. Primordial Earth must have been a very hostile environment. Just how the complex reactions that define life came about remains a mystery. It is usually assumed that it took place in a soup of organic and inorganic molecules in a very specific aqueous environment [Sarchet, 2018], possibly around a hydrothermal vent [Sojo *et al.*, 2016] or in a shallow pond exposed to intense UV light [Mulkidjanian *et al.*, 2012].

Present day knowledge of molecular genetics and cell biology indicates it was a unique event and all life can be traced back to a single starting point leading to

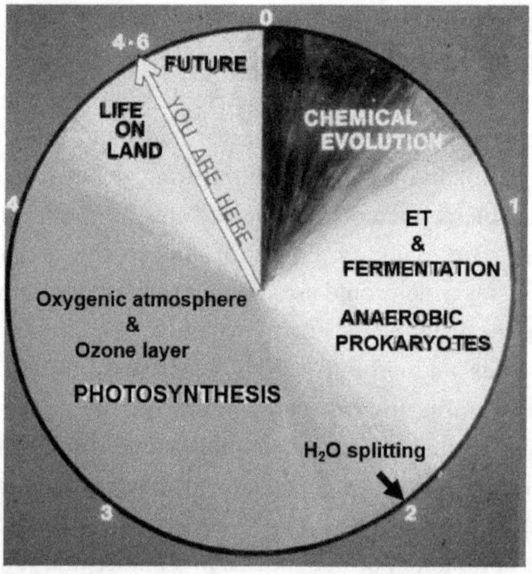

Figure 1. A five billion year clock of evolution of life on our planet. ET = electron transport.

the Last Universal Common Ancestor (LUCA). Within the primeval soup polymers of nucleic acids would need to form which could act as templates for replication. It is not so difficult to imagine this chemistry occurring given the properties of polynucleotides and rate of chemical mutations that could occur over very long periods of time. Ultimately these chains of nuclear acids could act as templates for amino acids also thought to have been present in the primeval soup. It seems almost incomprehensible that from these initial chemical events that life would evolve to the complexity of anaerobic prokaryotic organisms similar to those found on our planet today. However to rationalise this incredible achievement it has to be recognised that it occurred over an almost inconceivable period of time, in the region of about 1.5 billion years (Figure 1). For this reason alone we should respect and marvel at the evolution of life on our planet.

The early forms of anaerobic life had to establish a myriad of enzymes necessary for the prokaryote cell to grow and reproduce and be able to adapt to different environments. But this massive achievement required a continuous input of energy which would have been supplied from the stock of high energy organic and inorganic molecules created during the period of chemical evolution. Sooner or later the reductive chemistry would have become more reliant on sources such H_2, CO and CH_4 as electron donors and evolve the ability to generate electrochemical potential gradients across the cell membrane and convert this energy source to adenosine triphosphate (ATP). This is rather like that found in oxygenic organisms but with no O_2 available as a terminal electron acceptor [Olsen, 2006]. Instead they could use a range of electron acceptors such SO_4, NO and S. As the biochemistry of these early stages of bioenergetics advanced, fermentation pathways must also have evolved to obtain ATP without the need for an electron transport chain. For all this to occur, a cell membrane had to exist which almost certainly started off as vesicles consisting of a double layer of amphiphilic lipids (liposome) which form readily when these molecules are agitated in an aquatic environment, for example by the bursting of bubbles in the water close to hydrothermal vents or at the water surface by wave action. The formation of these vesicles could encapsulate organic molecules and inorganic ions present in the primeval soup. At that time the CO_2 level in the environment was high and several biochemical pathways seemed to emerge to aid its reduction [Apel *et al.*, 2013]. Coupled with this would be the requirement to maintain osmotic stability so that energy driven pumping mechanisms across the cell membrane would have to be established. It seems much of this evolved during LUCA and thus before the diversion to archaea and bacteria around 3.8 billion years ago [Ouzounis *et al.*, 2000; Weiss *et al.*, 2016].

Under some circumstance an energy crisis almost certainly would have developed when chemically-based reductants were not sufficient. Amazingly some organisms developed a way to supplement their metabolism utilising the energy of sunlight. Certain archaea developed their own "photovoltaic cells" as represented today by Halobacteriaceae. When there is a shortage of chemical substrates they produce a pigmented cell membrane containing bacteriorhodopsin [Sharma *et al.*, 2006]. This pigment undergoes a conformational change on absorbing light and in so doing, drives the accumulation of energised protons across the cell membrane which can supplement the metabolic energy requirement by powering the production of ATP. But far more important for future evolution of life was the appearance of anaerobic photosynthetic bacteria about 3.5 billion years ago [Olsen and Blankenship, 2005; Hohmann-Marriott and Blankenship, 2011]. In this case chlorophyll was the pigment of choice. Unlike bacteriorhodopsin, the absorption of light energy by bacteriochlorophyll results in charge separation with the electron elevated to a high energy level. The energy rich electrons could then be used to supplement the energy demands of these photosynthetic bacteria. The positive hole left in the chlorophyll is neutralised by an electron donor. The donors could now be less reducing and a range of organic and inorganic species were likely to be available including SO_4, NO_3, H_2S, NH_3, and various organic acids. Of particular importance would have been the availability of multivalent reduced metal ions as electron donors, particularly Fe^{2+}, but of considerable significance would be the availability of Mn as an electron donor which would ultimately be the basis of the water splitting catalytic centre. Mn would have been very abundant and widely distributed as it is today and available as Mn^{2+} in solution or in a mineralised form [Sauer and Yachandra, 2002; Najafpour, 2011]. This heralded a major step in the evolution of life on our planet.

Before this occurred anaerobic prokaryotic life had been highly successful and Darwinian principles would have driven this success. It is reasonable to assume that those organisms which utilised light energy had a distinct advantage over those who could not. This would mean that the aquatic environment would be coloured. Goldsworthy [1987] has made the point that the absorption spectrum of bacteriorhodopsin is in the green part of the solar spectrum while that of bacteriochlorophyll is in the near UV/blue and far red/near IR regions. This would allow both types of organisms to live side-by-side. However halobacteria have the disadvantage of requiring a supply of energy rich substrates to drive their reductive metabolism while photosynthetic bacteria could exist on more abundant less reducing electron donors. On the other hand halobacteria are archaea rather than bacteria, many of which can grow in extreme environmental conditions and presumably in such cases survived where photosynthetic bacteria could not. This is the case today. Based on the existence of modern anaerobic photosynthetic bacteria we can assume there were two classes differing in their photosynthetic reaction centres (RC); with Type 1 RC where the terminal electron acceptors are ferredoxins (*e.g.* heliobacteria, green sulphur bacteria, green non-sulphur bacteria) and Type II RC where quinones act as terminal electron acceptors (*e.g.* purple bacteria such as *Rhodobacter sphaeroides*, *Rhodopseudomonas viridis*). Both types of RCs consist of bacteriochlorophyll absorbing in the near far red (850 nm to 960 nm) and they share structural features which indicate that they are derived from a common ancestor.

It is often said that before the water splitting mechanism evolved, anaerobic prokaryotes were simple forms of life. This is far from the truth. They were already very advanced in their cellular metabolism and bioenergetics much of which was established during LUCA and therefore before the separation into archaea and bacteria. They were able to adapt to many different, and often extreme environmental niches. But inevitably their future development was limited by the resources available to them, particularly energy. However the complex machinery of anoxygenic bacterial photosynthesis and the associated respiratory complexes had evolved and established a framework for aerobic bioenergetics.

1.2. *Under aerobic conditions*

Clearly having the ability to use light energy to supplement or fully drive cellular metabolism has a distinct evolutionary advantage [Nelson, 2011]. Even so, ultimately the biomass of anaerobic prokaryotes would be limited by the availability of appropriate electron donors. As emphasised above, this energy problem was solved when organisms evolved the ability to split water using sunlight. Using water as an electron donor, coupled with the production of O_2 as a by-product, set the scene for the "Big Bang of Evolution." The release of this gas about 2.6 billion years ago had dramatic consequences for the development of life, since it created an oxygenic atmosphere and in time allowed the ozone layer to form [Canfield *et al.*, 2005; Fischer *et al.*, 2016]. With oxygen available, the efficiency of cellular metabolism increased dramatically, since for a given amount of substrate, aerobic respiration provides in the region of 20 times more energy than anaerobic respiration. It was probably this improved efficiency of cellular bioenergetics which paved the way for the subsequent evolution of eukaryotic cells and multicellular organisms. The establishment of the ozone layer provided a shield against harmful UV radiation allowing organisms to exploit the terrestrial, as well as the aquatic environments. Consequently life has been able to prosper and diversify on an enormous scale as testified by the fossil records and by the extent and variety of living organisms on our planet today. For this to happen several major steps had to be made:

(i) Incorporation and electronic coupling of Type 1 and Type 2 RCs into the same organism so that the energy of two photons of light could span the energy gap between the oxidation of water and the reduction of ferredoxin and also power ATP formation (photophosphorylation).

(ii) Modification of the synthesis of chlorophyll so that its spectral properties did not compete with those of bacteriochlorophyll and also be able to use a more energy rich part of the long wavelength visible spectrum (600–700 nm).

(iii) Modification of Type I RC to photosystem 1 (PS1) by introducing P700 as the primary donor and creating binding sites for electron donation from plasocyanin/cytochrome *c* and for the reduction of bound low potential iron-sulphur proteins.

(iv) Modification of Type II RC to photosystem 2 (PSII) by introducing P680 as primary donor and creating binding sites for electron donation from the oxygen evolving catalytic site and for the reduction of plastoquinone. In contrast rather than those used by anaerobic photosynthetic bacteria such as ubiquinone and menoquinone.

(v) Redox coupling of PSII to PSI with the cytochrome b_6f complex such that protons are actively pumped across a membrane and power the conversion of ADP to ATP as achieved by the cytochrome *b/c* complex of bacteria and mitochondria.

(vi) Setting the redox potential of the P680 reaction centre chlorophyll of PSII to 1.2 eV and adjusting the redox potentials of the electron transport chain, including PSI accordingly.

(vii) Introduction of protection mechanisms, particularly in PSII, to reduce oxidative damage (especially at high light intensities) and create a repair system.

(viii) Introduction of new types of light harvesting systems based on different forms of chlorophyll and phycobilins.

It is simply amazing that these changes, together with other requirements for oxygenic photosynthesis evolved from photosynthethetic bacteria. However, many of the building blocks were already established [Baymann *et al.*, 2003]. Of course, it occurred over a very long period of time and unfortunately there is no fossil evidence to establish the journey taken. Clearly the evolutionary pressure to use water as an electron donor was very high and the levels of UV radiation at that time would enhance the rate of mutations as it would have also done during the anaerobic period. The adoption of Mn by PSII as the metal for the oxygen evolving complex (OEC) [Debus, 1992; Barber, 2003, 2009] suggests that Mn^{2+} was a major electron donor prior to the appearance of the water splitting reaction itself. Indeed, the assembly of this catalytic centre in modern PSII is achieved by a process called photoactivation which involves a successive light induced oxidation of Mn^{2+} to form a cluster of four Mn ions in its most reduced functional state ($3Mn^{3+}/Mn^{4+}$) [Ananyev and Dismukes, 1996]. Associated with this light induced uptake of four oxidised Mn ions is the binding of Ca^{2+} which is required as a cofactor for the water splitting reaction [Debus, 1992].

It is estimated that the OEC must have emerged about three billion years ago when cyanobacteria became the dominant prokaryotic organisms, able to eradicate competition from anaerobic prokaryotes by oxygen poisoning. Since this would have included halobacteria then the removal of bacteriorhodopsin as a major pigment would allow ancestral cynanobacteria to introduce phycobilins as an additional light harvesting system to complement the light absorping properties of chlorophyll. Cyanobacteria would now have dominated the aquatic environment, so much so, that over a period of 600 million years they had produced enough O_2 to oxidise vast amounts of minerals, particularly Fe^{2+} to Fe^{3+} as well as a host of organic molecules. When the oxygen sinks were saturated the level of O_2 in the atmosphere and oceans rose to levels which were intolerable to anaerobic organisms except those in special ecological niches. During this dramatic extinction event, the level of atmospheric O_2 is debatable with some claims that it could have been comparable with that of today [Lyons *et al.*, 2014]. This would mean that the global cyanobacterial water splitting activity would have reached a level equal to that of all types of photosynthetic organisms existing at present. This is unlikely and perhaps an oxygen level of 5% is nearer to the truth [Kump, 2008]. As stated above it is generally agreed that the Great Oxygen Event (GOE) happened 600 million years later. During this time there were opportunities to adapt to a wide range of environmental conditions as evidenced by some present day cyanobacteria which can flourish under extreme conditions such as high (80°C) and low temperatures (below freezing) and very acidic (pH 1) and alkaline conditions (pH 10) [Blankenship, 2001].

Sometime following this period, a form of cyanobacterium became incorporated into cells of non-photosynthetic prokaryotes by endosymbiosis to ultimately form chloroplasts of eukaryotic photosynthetic algae. The formation of the extensive folding thylakoid membranes in chloroplasts of green algae (and later plants) provided light scattering properties which would make a phycobilin based light harvesting system of cyanabacteria redundant. However today red algae maintain phycobilins for light harvesting and have a different membrane organisation. To complement this endosymbiotic event another type of non-photosynthic prokayotic bacterium having a respiratory electron transport chain also became symbiotic with a prokaryote which ultimately evolved into a mitochondrion. With both symbionts in the same cell gene transfer occurred to form the cell nucleus, although some genes remained in the respective organelles. Eukaryotic cells could now efficiently utilise the high energy compounds derived from light driven reductive chemistry using O_2 as the terminal electron acceptor. With the birth of eukayotic cells the door was open for the evolution of complex multicelluar organisms.

2. Water Splitting by Photosystem II

The importance of this fundamental chemical reaction, which is the entry point for the energy cycle of essentially all life on our planet, cannot be overstated. The water molecule is very stable and splitting it into its elemental constituents is thermodynamically and chemically demanding especially when achieved in a delicate biological environment and powered by the relatively low energy content of four photons (4 hν) of long wavelength visible light. As emphasised above, in so doing it provided biology with an unlimited supply of reducing equivalents needed to convert carbon dioxide into the organic molecules of life while at the same time producing oxygen to transform our planetary atmosphere from an anaerobic to an aerobic state.

The water splitting reaction of PSII requires four quanta of light to produce one O_2 molecule and involves five intermediate states (S_0 to S_4). Four oxidising equivalents of about the same potential are sequentially accumulated at the catalytic site with absorption of each photon (see Figure 2) [Blankenship, 2014; Barber, 2016a]. The catalytic centre consists of a $Mn_3Ca^{2+}O_x$ having cubane geometry with a fourth Mn

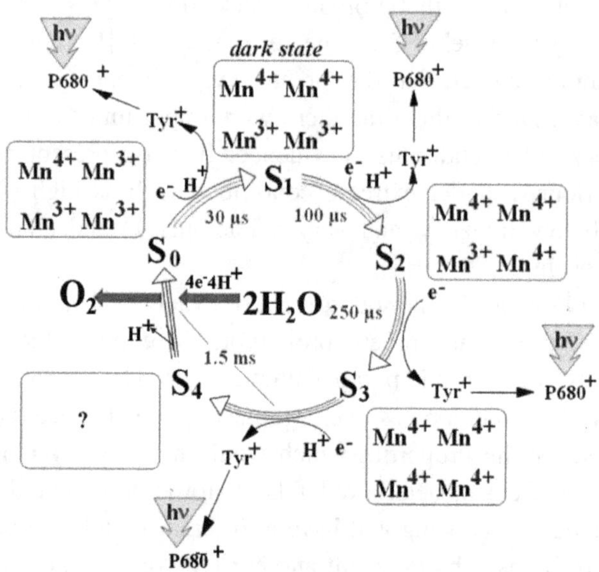

Figure 2. The S-state cycle showing how the absorption of four photons of light (hν) by P680 drives the splitting of two water molecules and formation of O_2 through a consecutive series of five intermediates (S_0, S_1, S_2, S_3 and S_4). Protons (H^+) are released during each step of the cycle except for the S_2 to S_3 transition. Electron donation from the metal-cluster to $P680^+$ is aided by the redox active tyrosine (Tyr). Each step involves a single oxidation of a Mn ion in the cluster, starting at S_0 with 3 × Mn^{3+}plus Mn^{4+} advancing to S_3 with 4 × Mn^{4+}. The exact oxidation state of S_4 is unknown but could be 3 Mn^{4+} plus Mn^{5+} or a Mn^{4+}-oxyl radical (see Figure 3). Also shown are half-times for the various steps of the cycle.

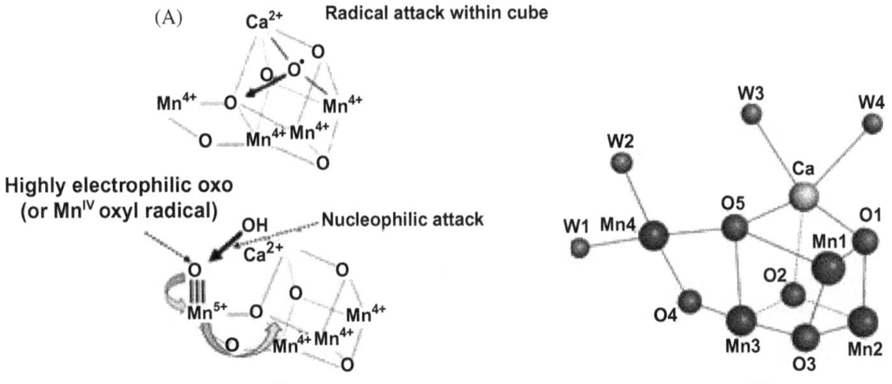

Figure 3. (A) Mechanism 1 for S_4 to S_0 transition. The formation of a terminal oxyl radical [Siegbahn, 2006, 2008, 2009] originating from a substrate water incorporated during the S_2 to S_3 transition, leads to a radical attack of an adjacent oxo-ligand formed within the Mn_3CaO_4-cubane from the other substrate water during the establishment of the cubane geometry for the S_0 state. (B) Mechanism 2 for S_4 to S_0 transition. The very high oxidation state of the Mn-cluster, particularly the Mn ion outside the Mn_3CaO_5-cubane, leads to a high electron deficient oxo (after deprotonation of water molecules during the S-state cycle). Nucleophilic attack by water or a hydroxide of the second substrate water within the coordination sphere of Ca^{2+} leads to O_2 formation [Barber *et al.*, 2004; Barber 2017a; Vinyard *et al.*, 2015]. (C) Diagrammatic representation of the Mn_4CaO_5 cluster and coordinating ligands determined by Umena *et al.* [2011]. Unfortunately the numbering of the atoms differs from the original introduced seven years earlier by Ferreira *et al.* [2004] and Barber *et al.* [2004].

linked to the cubane by an oxo-bond [Ferreira *et al.*, 2004; Umena *et al.*, 2011] (see Figure 3). This unusual structure has recently been synthesised by Zhang *et al.* [2015].

The S_4-state stores the four oxidizing equivalents needed to oxidise two water molecules to produce O_2 and in so doing reverts back to the S_0-state. As shown in Figure 2, the oxidising equivalents are stored by the Mn ions in the catalytic centre leading to four Mn^{4+} ions in the S_3-state just before the last photochemical step to S_4 [Barber, 2016a, 2016b]. The precise details of this final oxidation state are a matter of debate because O–O bond formation is very fast. As shown in Figure 2 the cycle is powered by the oxidation of RC chlorophyll (P680) generated by light driven primary charge separation in the reaction centre of PSII coupled to a redox active tyrosine (Tyr), serving as an intermediate electron/proton carrier between $P680^+$ and the Mn-cluster.

At the time of writing this chapter there was no consensus on the exact nature of the S_4-state or the precise mechanism of the formation of dioxygen. The Mn ion outside the cubane (Mn4) is adjacent to the Ca^{2+} and their positioning towards the side chains of several key amino acids, including the redox active Tyr suggested to Ferreira *et al.* [2004] and expanded by Barber *et al.* [2004] and more so by Barber [2018a], that they provide the "catalytic" surface for binding the substrate water molecules and their subsequent oxidation. Indeed a 1.9 Å structure of PSII revealed that two water molecules are bound to Mn4 and Ca^{2+} as shown in Figure 3C [Umena *et al.*, 2011]. This conclusion fitted well with earlier suggestions [Messinger *et al.*, 1995; Pecoraro *et al.*, 1998; Vrettos *et al.*, 2001; Britt *et al.*, 2004] and expanded later [Barber *et al.*, 2004; Brudvig, 2008; Vinyard *et al.*, 2015; Barber 2017a, 2017b]. It was argued that a substrate water, associated with a high valency Mn ion, is deprotonated during the S-state cycle and converted to a highly electrophilic oxo (see Figure 3B). This mechanism is dependent on Mn4 being converted to a high oxidation state (possible Mn^{5+}) during progression to the S_4-state just prior to O–O bond formation. The other three Mn ions also progress to high valency states (Mn^{4+}) (for example S_0 (Mn^{3+}, Mn^{3+}, Mn^{3+}, Mn^{4+}) to S_3 (Mn^{4+}, Mn^{4+}, Mn^{4+}, Mn^{4+}) and act as a further "oxidising battery" for the Mn^{5+}-oxo species in the S4-state . In this way the reactive oxo is electron deficient, so much so that it makes an ideal target for a nucleophilic attack by the oxygen of the second substrate water bound within the coordination sphere of the Ca^{2+} (see Figure 3B). Deprotonation of the substrate waters would be aided by nearby bases such as CP43-Arg357 and by the weak Lewis acidity of Ca^{2+}. This mechanism has been thoroughly investigated by Siegbahn [2017] using in depth calculations based on Density Function Theory (DFT) and concluded that it is energetically less favourable than a mechanism involving the formation of a radical oxo within the Mn_3CaO_4 cubane as depicted in Figure 3A. Therefore the Siegbahn mechanism [2006, 2008, 2009] concludes that the O–O bond formation during the S_4 to S_0 transition involves bridging oxo species and has some experimental support based on resonance spectroscopy [Cox *et al.*, 2014; Pérez-Navarro *et al.*, 2016] while other similar studies favour the nucleophilic mechanism [Britt *et al.*, 2004]. Recently the details of the S_4 to S_0 transition have been modified by Li and Siegbahn [2015] to allow for conformational changes in and around the metal cluster but the general concept of an oxyl–oxo mechanism for O–O bond formation involves the same two manganese atoms and the central oxo group as in the original scheme. This mechanism requires that the cubane structure to be open during the S_3 to S_4 transition to allow a substrate water molecule to enter. Recent X-ray Free Electron Laser [XFEL] measurements seem to give some support for [Suga *et al.*, 2015, 2017] and against [Young *et al.*, 2016] this mechanism. Therefore there are pros and cons for both mechanisms as discussed recently by Siegbahn [2017] and myself [Barber, 2017a, 2017b].

3. Efficiency of Aerobic Cellular Bioenergetics

The complexity of the chemical reactions which occur in a single cell is simply amazing. In the case of oxygenic organisms these reactions occur at ambient temperatures and are powered directly or indirectly by the reducing potential of high

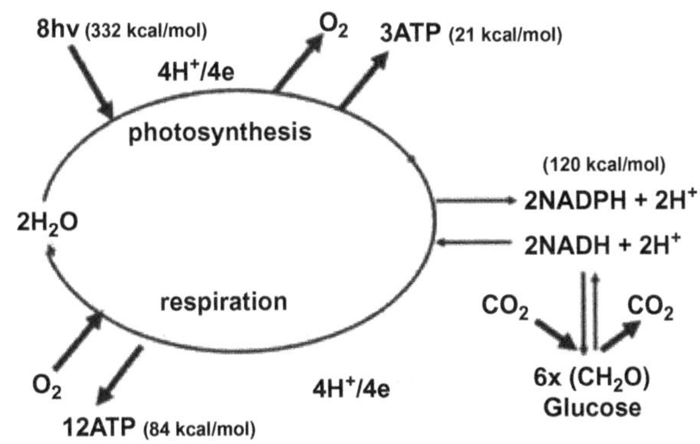

Figure 4. A diagrammatic representation of the cyclic nature of cellular bioenergetics powered by sunlight initiated by water splitting to give O_2 and "hydrogen" in the form of reduced NADP (plus a proton H^+) and how the "hydrogen" is combined with O_2 to return to water with the transfer of chemical energy to ADP to form ATP (released by its hydrolysis back to ADP). Also shown are the Gibbs free energy of the various reactants and products in calories per mole.

energy electrons derived from the water splitting reaction of PSII. The electron/proton carriers are reduced forms of nicotinamide adenosine dinucleotide (NAD) or nicotinamide adenosine dinucleotide phosphate (NADP) which have a redox potential of –320 meV. To achieve this reducing potential and avoid wasteful back reactions, PSI and PSII each require the energy of one photon of long wave visible light with both corresponding to about 1.8 eV. The downhill energy flow between PSII and PSI is captured by the conversion of ADP to ATP. Taking this into account gives an overall efficiency of about 40% for energy conversion of eight quanta of 680 nm light to give one O_2, 3ATP and 2NADPH. The oxidation of 2NADH back to water by respiration produces 18 ATP corresponding to an efficiency of about 30% while the "wasted" heat energy is vital for optimizing the temperature of warm blooded animals (Figure 4). Although these are impressive numbers for this biological H_2/H_2O cycle, in reality the overall conversion of solar energy to the very large variety of organic molecules which constitutes biomass, is much lower. Also we have assumed a single wavelength of light and in practice only a portion of visible light is available either by direct absorption or by internal conversion of shorter wavelength quanta. Therefore at very best the efficiency of photosynthesis at the level of biomass formation is only about 2% under optimal conditions and much less for normal environmental conditions, typically 0.1% [Barber, 2009].

4. The Future of *Homo sapiens*

In some ways the human race is approaching a crisis similar to that which anaerobic life had to deal with three billion years ago as the global population grows and resources become limiting. This was recognised by the Club of Rome and recorded in their publication "*The Limits to Growth*" by Meadows *et al.* [1972] which detailed a computer simulation of exponential economic and population growth with a limited supply of resources. It was concluded that if we continued to have business as usual with a continuous striving for economic growth, then by mid-21st century we would completely run out of vital materials with devastating consequences for human society. According to Google Scholar this book has been cited 18,480 (as of September 2018) and over 30,000 copies were sold in various languages. At the time it was taken very seriously by environmentalists and academics but less so by policy makers and industrialists. Nonetheless the conclusion of the Club of Rome was amplified by the oil crisis which occurred at about that time and spurred non-fossil fuel based energy research in many countries to enable security of supply and for sustainability.

Of course what has happened since then is a continuation of "business as usual" since plentiful fossil fuel energy came on stream with growth/GDP continuing to be the major driver. Nobody can argue that this approach has not brought great benefits and prosperity. In 1972 the population of our planet was just less than four billion and has almost doubled since then. Even so, the demand for extra food has been satisfied and the overall standard of living has increased for a large proportion of the global population especially with the introduction of new technologies.

The development of a country is usually measured by the Human Development Index (HDI) which has three main elements, life expectancy, educational level and GDP per capita. In recent years gender equality has been added to give the IHDI index (see [hdr.undp.org/en/content/human-development-index-hdi]).

As Figure 5 shows the level of HDI achieved is closely related to the energy consumption per capita except for highly developed countries where some use energy far in excess of that required for a high HDI. Clearly as countries strive for a higher level of development and the global population increases, the requirement for energy will rise.

At present the average annual power requirement of the world's population is about 17 terawatts (TW) equivalent to a total energy of 0.7×10^{21} joules per year. (See: BP Energy Outlook [https://www.bp.com/content/dam/bp/en/corporate/pdf/energy-economics/energy-outlook/bp-energy-outlook-2018.pdf] and IEA International Energy Agency Key world energy statistics 2017 [http://www.iea.org/publications/freepublications/publication/key-world-energy-statistics.html].

As shown in Figure 6, somewhere in the region of 85% of this power comes from fossil fuels. It has often been argued that fossil fuel resources are limited and that this is why we need to develop sustainable energy supplies by exploiting nuclear and solar energy technologies. However in the case of nuclear fusion, a

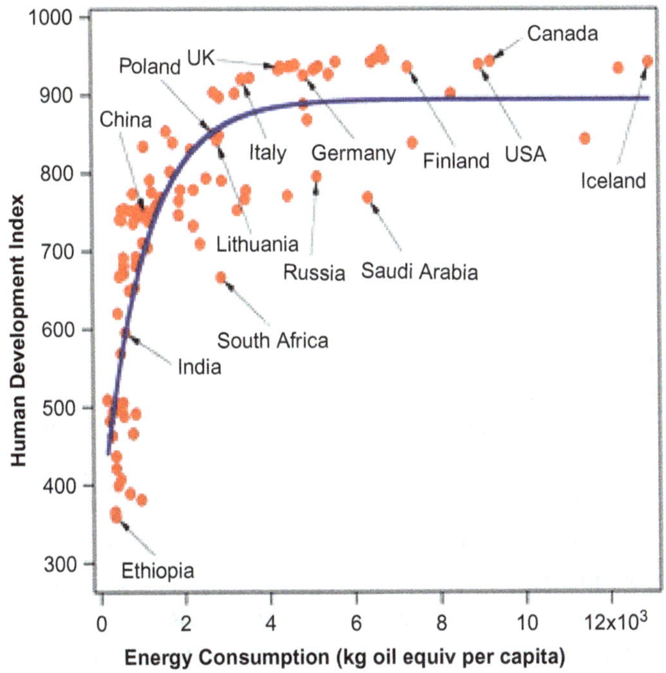

Figure 5. Relationship between the Human Development Index (measurement of the development of a particular country) and Energy Consumption per capita.

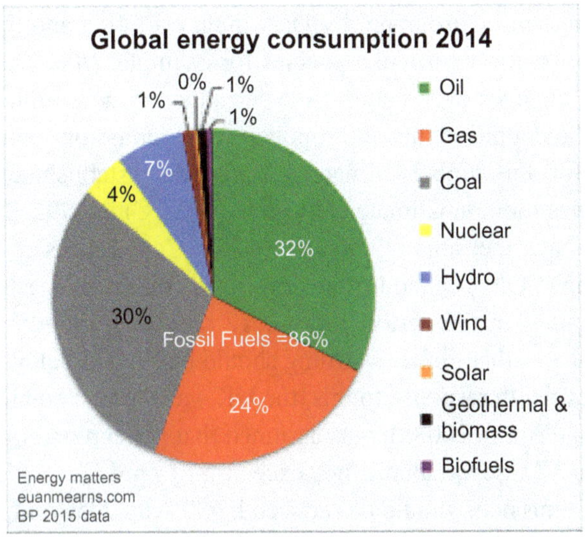

Figure 6. Global energy consumption estimated in 2014. During the past three years the relative proportions have not changed significantly with fossil fuels dominating at more than 85% (taken from BP Statistical Review 2015), [http://euanmearns.com/global-energy-trends-bp-statistical-review-2015].

technological breakthrough is required while nuclear fission is limited in the long run by the finite supply of uranium fuel. But it is important to understand that fossil fuels are derived from oxygenic photosynthesis. The overall equation of photosynthesis below (Eq.1) shows that for every molecule of O_2 produced from the splitting of two water molecules, one molecule of CO_2 is converted into organic molecules such as carbohydrate (CH_2O).

$$2H_2O + CO_2 \xrightarrow{8h\nu} O_2 + [CH_2O] + H_2O \quad (1)$$

Since today there is 21% O_2 in the atmosphere and as far as we know this is derived entirely from photosynthesis, then the amount of reduced carbon on our planet must be enormous. This means that there is essentially an almost unlimited supply of fossil fuels although presumably a great deal of the reduced carbon will not be readily obtainable. If it were, and if it was all burnt, then our planet would return to its original anaerobic state with a very high level of CO_2 in the atmosphere. This recognition of the magnitude of fossil fuel reserves is not generally appreciated and certainly not by the 1972 Club of Rome analysis of limits of growth. Of course, we now realise that it is not the availability of fossil fuels which is limiting but their continued use as an energy source with the consequential release of CO_2 and other greenhouse gases into the atmosphere resulting in the anticipated detrimental effects on global temperature and climate change. As emphasised above, life is very robust and biological organisms have adapted to large changes in their environment by Darwinian evolution and thus created new species. The option for a particular species for example, *Homo sapiens*, to adapt through genetic changes over a short time period is far more limited.

The official recognition that anthropogenic CO_2 emissions could have serious consequences for the global climate system was establishment by the UN Framework Convention on Climate Change (UNFCCC) in 1992. The members of the convention have met annually since 1997 in Conferences of Parties (COP). The first meeting (COP1) called for the developed countries to reduce their greenhouse gas emissions in the period 2008–2012 and was ratified by 50 countries. The overall goal was that global warming should be limited below 2°C relative to pre-industrial levels. In response to this, the UK was the first country to formulate a Climate Change Act in 2008. It was assumed that to keep within the 2°C target, the atmospheric CO_2 level should not exceed 550 ppm. To achieve this it was agreed that CO_2 emissions should be reduced by 80% by 2050 and zero by the end of the century. In 2015, COP21 was held in Paris and 195 countries adopted the same targets [http://unfccc.int/paris_agreement/items/9485.php]. These targets are a significant challenge. In 1972 when the Club of Rome was formed the level of CO_2 in the atmosphere was 330 ppm. Since then the level has risen to over

400 ppm and now (June 2018) is 411 ppm which is the highest for 800,000 years [https://www.co2.earth/].

If other greenhouse gases are included then the CO_2 equivalent is higher. This is an enormous change over such a short period of time and the movement is in the wrong direction. This was inevitable due to the increase in global population which is predicted to continue for several decades until levelling off at about 11 billion by 2100. So far, the efforts to offset this continuing rise in atmospheric CO_2 with the increase of non-fossil sustainable fuels (solar, wind, waves, hydro and nuclear) has not been able to keep up with the ever increasing demand for energy and the relative cheapness and availability of gas, oil and coal. This is not a good omen. If by 2050 the total global energy demand doubled from that of today, that is to 34 TW, and 80% was to come from non-fossil fuels as required by COP21, then that would amount to 27 TW. To emphasise how difficult it will be to achieve, assume all of it is generated from nuclear fission. Typically a nuclear power plant generates 1500 megawatts (MW). Therefore 18,000 nuclear power stations would be required worldwide over a 30 year period. Since this time span is 10,950 days then the conclusion is that more than one nuclear power station has to be built every day from now on. Also not included in this calculation is the decommissioning of old reactors and their replacement. This impossible goal would also rely on a vast amount of available nuclear fuel. The number of power stations would not change if nuclear fusion reactors were used since their outputs would also be in the region of 1500 megawatts and, in any case, this fusion technology does not exist at the moment. Clearly a business as usual scenario cannot be continued even if there were a breakthrough with fusion technology in the near future as often assumed by policy makers.

Fortunately we do have a very large source of fusion energy 186 million kilometers from our planet, namely the Sun. It has been calculated that the average power of the Sun arriving on the surface of earth is 120,000 TW. This means that just over one hour of sunlight provides all the power currently used globally over a year. That is planet Earth is "energy rich". Unfortunately this energy is very diffuse and therefore difficult to harvest in enormous amounts. Nevertheless solar energy is captured and stored as heat in the oceans and lakes. This gives rise to evaporation and global air currents. Thus, solar energy can be captured as hydroelectric, wind and wave power. At present this amounts to about 8% of the current global power consumption as indicated in Figure 6. It is often assumed that solar energy capture is only by photovoltaic (PV) cells, solar concentrators and water heaters, but at the moment these are very small contributors to global energy demand (just under 1%) (Figure 6). There is also biomass, the level of which is difficult to estimate. It is shown as 1% in Figure 6. However this percentage increases if the burning of wood and other organic materials for cooking and heating in undeveloped countries, is taken into account. Understandably, as these countries move out of poverty they too are turning to the use of fossil fuels. The use of biofuels is being encouraged in developed countries but this approach competes with food production and can lead to undesirable ecological consequences as a result of, for example, deforestation and loss of habitat for vulnerable plant and animal species. Nevertheless some countries are blessed with a large land mass and ideal growing conditions for ecologically managed, sustainable, biomass/biofuel production.

Over a period of a year the rate of energy storage for the whole of global photosynthesis is less than the amount used by human society. This is because the global photosynthetic efficiency is very low due to the fact that much of the Earth's surface is not, or poorly, active in photosynthesis (deserts, polar regions and deep oceans) as well as seasonal changes (winter senescence). On the other hand the efficiency can be as high as 2% when plants are grown all year round under optimal conditions, such as sugar cane in the plantations of Brazil [Barber, 2007, 2009]. The average long term global energy storage is mainly achieved by terrestrial plants where storage can be in the form of wood and other types of biomass. The photosynthetic activity of the oceans is about the same as for land plants but the storage time of energy is much less because of the rapid turnover rate of phytoplankton and algae [Falkowski and Raven, 2011]. The relatively low rate of energy storage by global photosynthesis (less than 5 TW) contrasts with the current yearly average power demand of about 17 TW. Therefore of the 40 billion tons of CO_2 released into the atmosphere per year at the present time, less than half of it is absorbed by the land biosphere/oceans (Figure 7). As a consequence there is a steady rise in atmospheric CO_2 levels.

Clearly the challenge is to develop carbon free technologies which can dramatically slow down, or prevent, the continuing release of CO_2 resulting from

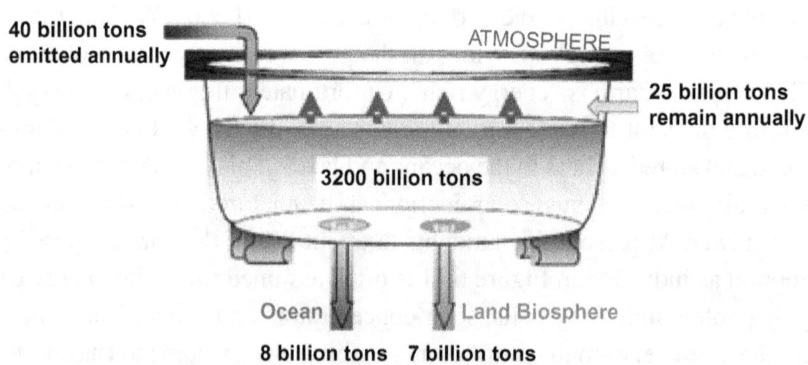

Figure 7. Diagrammatic representation of the fate of anthropogenic CO_2 emitted into the atmosphere. These approximate numbers gives an annual rise of atmospheric CO_2 of about 3 ppm which is consistent with recent measurements (*e.g.* 2016 to 2017) [https://www.co2.earth].

burning fossil fuels. In the case of solar driven technology a year-round average efficiency significantly higher than that of current plants or algae is needed to reduce land-area requirements and so be independent of food production. In this context it should be noted that the land area of Earth is 29% of which a third is desert.

It is encouraging and unprecedented for the majority of the countries that make up our planet to have come together *via* COP21 to support a global effort to reduce and ultimately eliminate man-made greenhouse gas emissions into the atmosphere. Therefore there must be an urgent global effort directed at developing scalable technology which can compete in financial terms with the relatively low cost and abundance of fossil fuels. As emphasised in this chapter, biology solved its energy problem by using sunlight to release hydrogen from water. For large scale energy storage this seems to be an ideal solution [Ismail and Bahnemann, 2014]. That is to use electric current derived from renewable sources to power water splitting using a new generation of efficient low cost electrochemical cells (Figure 8) [Acar *et al.*, 2016]. Currently, electrolysis of water to obtain "green" hydrogen on a large scale is expensive. However, low cost and non-toxic materials are being identified as alternatives to platinum for the electrochemical splitting of water [Eisenberg and Gray, 2008; Kanan and Nocera, 2008; Artero *et al.*, 2011; Wang *et al.*, 2012; Tran *et al.*, 2012, 2013, 2016; Sivula *et al.*, 2011; Reece *et al.*, 2011; Merki and Hu, 2011; Du and Eisenberg, 2012; Bassi *et al.*, 2014; Hunter *et al.*, 2016] and also see the chapter by Septina[a] *et al.* in this volume. The

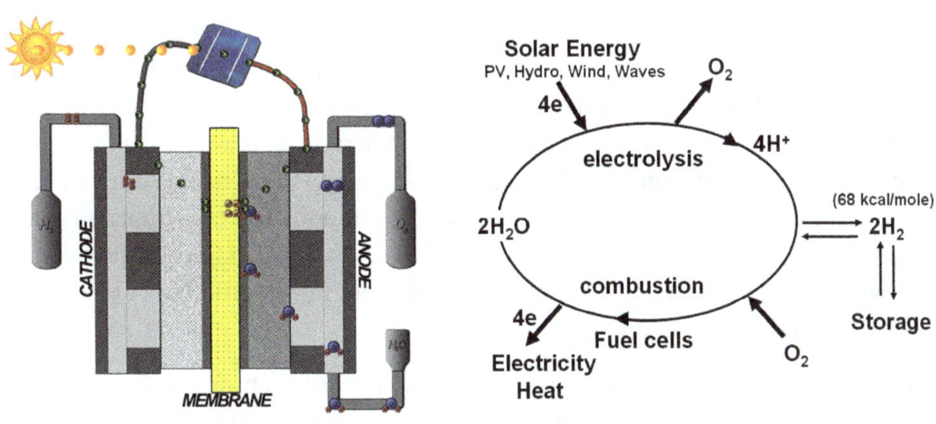

(A) (B)

Figure 8. (A) Solar driven electrolysis of water using a proton exchange membrane to separate oxygen (blue atoms) formed at the anode and hydrogen generated at the cathode *via* proton (red) reduction by electrons energised by light in a PV cell. The electricity could also come from wind-, hydro-, wave- as well as from nuclear. (B) Diagramatic representation of hydrogen fuel cycle driven by the electrolytic splitting of water to compare with Figure 4.

establishment of a new generation of electrolysers resulting from this research can therefore store electricity derived from PV-, wind-, hydro-, wave- as well as from nuclear-power. However, they must have very good turnover rates to cope with high levels of electricity supply. For a PV system coupled to an electrolyser it has been calculated with present technology that the conversion efficiency to hydrogen is about 10–11% [Blankenship *et al.*, 2011].

Hydrogen, generated by the electrolysis of water, is an excellent clean energy carrier with a high energy density in its compressed form [Acar *et al.*, 2016] compared with solid state batteries. Thus it is ideal for transport systems including automobiles, buses, motor cycles, ships and aircraft, the feasibility of which has been demonstrated by prototypes. Compressed H_2 can be used directly in an internal combustion engine or more efficiently to produce electricity *via* fuel cells which is essentially the opposite of electrolysis. Both can have efficiencies in the region of 70% to 80% [Genovese *et al.*, 2009] and have electrochemical mechanisms which resemble those of photosynthetic and respiratory electron flow (compare Figures 4 and 8A).

Hydrogen based fuel cell electrical vehicles (FCEV) are now being built commercially, like the Toyota Miria and very recently the 2017 Honda Clarity FCV and the Hyundai ix35 FCEV. South Korea plans to invest $2.33 billion US dollars over the next five years in a public-private partnership to speed up the development of the country's hydrogen fuel cell vehicle. Japan is also moving in this direction. Compared to battery electric vehicles (EV), the FCEV has a longer distance capacity and more importantly can be fully refueled in just a few minutes (as in the case of gasoline and diesel) instead of hours. Alternatively, an on-board generator powered by hydrogen, can be used to maintain the charge in a battery. Hybrid cars using this principle are now being commercially manufactured as low emission vehicles. See: [http://www.theelectrictaxi.co.uk]. Using hydrogen instead of petroleum results in no emission other than water. There is also a lot of information available about the potential of hydrogen as an energy carrier in a series of four white papers published through the UK hydrogen and fuel cell Supergen Hub [http://www.h2fcsupergen.com/our-work/whitepapers/]. Also see Brandon and Kurban [2017], Mehta [2018] and the Royal Society Policy Briefing "Options for producing low-carbon hydrogen at scale"

[https://royalsociety.org/~/media/policy/projects/hydrogen-production/energy-briefing-green-hydrogen.pdfv] http://rsta.royalsocietypublishing.org/content/375/2098/20160400.

Another very relevant briefing document was published as I was finishing this chapter by Friends of the Earth and can be accessed *via*:

https://cdn.friendsoftheearth.uk/sites/.../Future%20of%20Heating_August_2018.pdf.

Hydrogen filling stations are starting to appear in some specific areas, for example California, but there is a need to install this infrastructure worldwide as soon as possible. At the moment there are only 10 in the whole of the UK. Another problem is that the main source of hydrogen comes from steam reforming of methane derived from the combustion of coal or directly from natural gas [Kleijn and van der Voet, 2010]. Therefore the use of this "brown" hydrogen does not help tackle the problem of releasing CO_2 into the atmosphere. On the other hand, the splitting of water by electrolysis using electricity from renewable sources provides "green" hydrogen [Barber, 2018].

As stated above, the 2015 Paris Accord (COP21) adopted the same targets as the 2008 UK Climate Change Act for CO_2 emission reduction. These goals are daunting when compared with the demands and expectations of an ever increasing world population. Nevertheless the targets have been agreed by 195 countries and so far ratified by most of them. Unfortunately the US Federal government at present is less committed despite the efforts of the US Department of Energy and several States to encourage renewable energy research and development. Fortunately, US private industry is actively responding with the Tesla electric cars being a good example. In fact introduction of battery only electric cars is already well established in many countries and without doubt will become more common in the near future if the number of public charging stations can be increased significantly. Also battery technology needs to be improved to decrease charging time and increase their energy storage density. Wireless electromagnet induction should be developed to complement hard wire charging and R&D into this is underway by most of the major car makers.

The introduction of hydrogen FCEVs is lagging behind at the moment and will do so until the necessary infrastructure is established. However there are many examples of FCEV buses around the world and Germany (Alstom) is just about to operate hydrogen driven trains (Coradia iLint). In my opinion, despite the present efforts to reduce greenhouse gas emissions, it is unlikely that we will achieve the COP21 recommendations unless we immediately start to put in place the infrastructure based on two key energy carriers, "green" electricity and "green" hydrogen with the aim to shift away from fossil fuels as a primary energy source as soon as possible. This is an economic, cultural and political challenge given the uncertainties of the future and our current addiction to fossil fuels. At the moment, there is uncertainty whether policy makers will be willing to adjust priorities away from this reliance. The alternative has to be to develop carbon capture and storage at power plants. The smart way to do this will be to convert CO_2 efficiently into non-degradable long life polymers which can be used as chemically stable products, for example, as building materials including the construction of roads. Research in this area is underway [https://energy.gov/fe/articles/recycling-carbon-dioxide-makeplastics,

Banerjee *et al.*, 2016] and I recommend the reader to the very recently published commentary by Thomas [2018]. Even then this method of capturing and storing CO_2 can only be a temporary solution. I do not see the point of using hydrogen derived from water to convert CO_2 into a fuel which is the main objective of US DOE Joint Center for Artificial Photosynthesis (JCAP) [https://solarfuelshub.org/] at the moment and therefore introduce carbon back into the energy cycle when hydrogen can be used directly. Moreover we must learn not to be so wasteful with energy and develop technologies which are more efficient. It should also be remembered that coal, oil and biomass are a rich source of complex high energy molecules and provide feed stock into the chemical industries without any significant release of CO_2 and therefore must continue to serve this purpose.

Although the road ahead looks extremely demanding, we should reflect that like all living organisms we, as a species, have undergone Darwinian evolution over millions of years. We have arrived at a stage where we dominate all other forms of life on our planet and we can now dictate our future through our knowledge base and technologies even to the extent of manipulating our genes and travelling into space. Thus, we have well and truly entered the Anthropocene period [Crutzen, 2002] where the future evolution of our planet is now in the hands of one species, *Homo sapiens*. This responsibility means we must recognise that the route we are now on [more-or-less business as usual] cannot be sustained without serious consequences. We must be smart and tackle the energy/greenhouse gas problem with urgency. Fortunately this is being recognised worldwide through the UN Climate Change Conferences. The underlying message is that global action on a massive scale is now needed and just talking is not enough. If action is not taken there will inevitably be a destabilisation of global order which will involve detrimental climate change leading to mass migration, wide scale starvation and possibly serious military conflicts. Indeed if action is not taken on a large scale immediately then we may see the demise of the human race in a matter of a hundred years or so. Such a strong statement has been said many times by environmentalists and scientists such as James Lovelock and his vision of our planet as Gaia [1994]. The awareness of our responsibility to our planet and our species became more focused from the exploration of space as emphasised by the iconic first photographs of Earth taken from Apollo 8 and Voyage I (Figure 9) and the associated statements of William Anders [1968] and Carl Sagan [1997].

After seeing the image of Earth from Apollo 8 crew member William Anders commented [Anders, 1968]:

> *"We came all this way to explore the Moon, and the most important thing is that we discovered the Earth."*

(A) (B)

Figure 9. (A) Photograph taken on December 24th, 1968 was the first picture of Earth taken from the moon during Apollo 8 manned spacecraft orbiting the moon. (B) Photograph of planet Earth taken on February 14, 1990, by the Voyager 1 space probe from a record distance of about six billion kilometers (3.7 billion miles). Earth's apparent size is less than a pixel; our planet appears as a tiny blue dot against the vastness of space, among bands of sunlight reflected by the camera. Voyage 1 had completed its primary mission and was leaving the Solar System. The astronomer and author Carl Sagan, requested that NASA to turn its camera around and take one last photograph of Earth across a great expanse of space.

Similarly when Carl Sagan saw the picture taken from Voyager 1 he stated [Sagan, 1997]:

> *"Look again at that dot. That's here. That's home. That's us. On it everyone you love, everyone you know, everyone you ever heard of, every human being who ever was, lived out their lives. The aggregate of our joy and suffering, thousands of confident religions, ideologies, and economic doctrines, every hunter and forager, every hero and coward, every creator and destroyer of civilization, every king and peasant, every young couple in love, every mother and father, hopeful child, inventor and explorer, every teacher of morals, every corrupt politician, every "superstar," every "supreme leader," every saint and sinner in the history of our species lived there-on a mote of dust suspended in a sunbeam. The Earth is a very small stage in a vast cosmic arena. Think of the endless cruelties visited by the inhabitants of one corner of this pixel on the scarcely distinguishable inhabitants of some other corner, how frequent their misunderstandings, how eager they are to kill one another, how fervent their hatreds. Think of the rivers of blood spilled by all those generals and emperors so that, in glory and triumph, they could become the momentary masters of a fraction of a dot. Our posturings, our imagined self-importance, the delusion that we*

have some privileged position in the Universe, are challenged by this point of pale light. Our planet is a lonely speck in the great enveloping cosmic dark. In our obscurity, in all this vastness, there is no hint that help will come from elsewhere to save us from ourselves. The Earth is the only world known so far to harbor life. There is nowhere else, at least in the near future, to which our species could migrate. Visit, yes. Settle, not yet. Like it or not, for the moment the Earth is where we make our stand. It has been said that astronomy is a humbling and character-building experience. There is perhaps no better demonstration of the folly of human conceits than this distant image of our tiny world. To me, it underscores our responsibility to deal more kindly with one another, and to preserve and cherish the pale blue dot, the only home we've ever known."

The seriousness of what we are doing to our planet has been emphasised by many high profile people including Carl Sagan [1983], Pope Frances [2015] in his book "*Laudato Si. Care of Our Common Home*", Yuval Harari [2014] in his book "*Sapians*" and by Al Gore [2006, 2017] in making the documentary films "An Inconvenient Truth" and "Truth and Power" and his book "*An Inconvenient Truth: The Crisis of Global Warming*" published in 2006. The most recent book emphasising the seriousness of the situation is by the UK Astronomer Royal, Martin Rees [2018] entitled "*On the Future: Prospects for Humanity*". And of course the host of senior scientists specializing in environmental, atmospheric and climate sciences who contributed to the many publications of the Intergovernmental Panel on Climate Change (IPCC) have been warning us dating back to 1990. On the other hand there is another group of *bona fide* scientists and scholars who refute the conclusions of IPCC through the Nongovernmental Panel on Climate Change (NIPCC) and its 13 reports [http://climatechangereconsidered.org/].

Despite this there is a major consensus among scientists and a wide spectrum of scholars and intellectual thinkers, that if we continue on our present trajectory, the continuation of organised human society is at serious risk and at worse, the demise of our very special species and the beginning of the sixth extinction [Pimm and Brooks, 2000]. We cannot afford to take this risk.

5. Epilogue

I drafted this article during the summer of 2018. During that period several relevant things happened:

1. CNN report 23rd May 2018 entitled "Mercury rising: India records its highest temperature ever".
2. The Beijinger 6th June 2018 "Beijing breaks 50-year record as summer temperatures soar to 39°C".

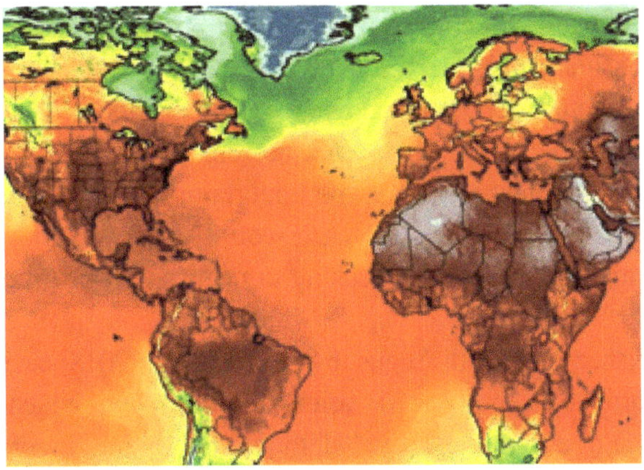

Figure 10. All-time hottest temperature records set all over the world this week. Evening Standard 6th July 2018. This map of maximum temperatures from 3rd July showed the northern hemisphere "roasting" (Climate Reanalyzer from the University of Maine).

3. In the UK and elsewhere there were severe droughts having serious consequences on agricultural output. BBC report 4th July 2018 report "Farmers warn of crisis amid drought".
4. Record high temperatures were recorded in the northern hemisphere and Figure 10 shows the situation on 6th July as reported in the London Evening Standard with the title "All-time hottest temperature records set all over the world this week".
5. The highest temperature (51.3°C) ever recorded in Africa was measured in Algeria reported on 6th July by the Washington Post.
6. July 13th The Guardian reported "Heatwave sees record high temperatures around world this week".
7. Record temperatures were recorded in Japan and reported by the BBC on 23rd July entitled "Japan heat wave: Temperatures break national records".
8. An unprecedented number of bush and forest fires occurred in many countries with 91 deaths in Greece. CBS/AP report on 29th July "Death toll rises to 91 in deadly Greece wildfires".
9. Record temperatures in Sweden reported in The Local with title "Sweden had its hottest ever July, breaking several weather records".
10. CNN report on 1st August 2018 entitled "California wildfires kill 8 people and brutal heat spells more danger".

And many more similar reports from around the world including Russia and Canada. The World Meteorological Organization (WMO) said "the rising temperatures were at odds with the global cyclical climate phenomenon known as

La Niña, which is usually associated with cooling" [https://www.theguardian.com/environment/2018/jul/13/heatwave sees].

I am not a climate scientist but these extreme events, and others which have been occurring over the past recent years, are consistent with man-made climate change although there must remain some uncertainty whether these recent events are direct evidence. My own view is that it would be unwise to dismiss this trend and its correlation with the ever increasing concentration of greenhouse gases in the atmosphere.

A paper published on the 7th August by Steffen *et al.* [2018] caught my eye while completing this article. It emphasised various tipping points which would cause an irreversible cascade of events leading to "Hothouse Earth" (see Figure 11 and legend).

As stated before, the 2°C mentioned in the legend of Figure 11 was set in the UK 2008 Climate Change Act and agreed by the international community in the 2015 Paris Accord (COP21). Unfortunately at the moment, it seems very unlikely

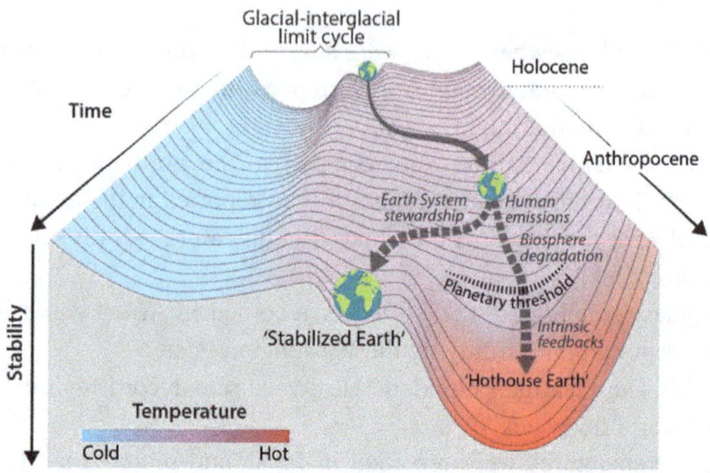

Figure 11. Stability landscape showing the pathway of the Earth System out of the Holocene and thus, out of the glacial–interglacial limit cycle to its present position in the hotter Anthropocene. Currently, the Earth System is on a Hothouse Earth pathway driven by human emissions of greenhouse gases and biosphere degradation toward a planetary threshold at 2°C beyond which the system follows an essentially irreversible pathway driven by intrinsic biogeophysical feedbacks. The other pathway leads to Stabilized Earth, a pathway of Earth System stewardship guided by human-created feedbacks to a quasistable, human-maintained basin of attraction. "Stability" (vertical axis) is defined here as the inverse of the potential energy of the system. Systems in a highly stable state (deep valley) have low potential energy, and considerable energy is required to move them out of this stable state. Systems in an unstable state (top of a hill) have high potential energy, and they require only a little additional energy to push them off the hill and down toward a valley of lower potential energy. Taken from Steffen *et al.* [2018] with permission.

that this target will be met even though it is generally now agreed that exceeding 1.5°C is considered the limit to avoid the scenario predicted by Steffen *et al.* [2018] and others.

On 11 September 2018: UN Secretary-General Antonio Guterres delivered a very passionate major policy statement on climate change consistent with my own concerns about the situation and the urgent need for global unity [https://www.un.org/sg/en/content/sg/statement/2018-09-10/secretary-generals-remarks-climate-change-delivered]. On the other hand, addressing the UN just two weeks later (25th October 2018), President Trump stated that the US "rejects the ideology of globalism and embrace the doctrine of patriotism". And, of course he had already previously pledged that the US will withdraw as a signatory to the 2015 Paris Accord (COP21) when it is legal to do so (November 2020). Whether he will change his mind so as to keep other countries on board such as Brazil, Australia and Canada will depend on his response to the latest and dire warnings announced by the IPCC on the 8th October following a five day meeting held in Incheon, South Korea [https://ipcc.ch/news_and_events/pr_181008_P48_spm.shtml].

Acknowledgements

I wish to thank the UK National Health Service and the doctors and nurses who have, and continue to help me fight metastatic colon cancer diagnosed six years ago. In particular Sarah Burton (surgeon at Frimley Park Hospital) and Sharadah Essapen (oncologist at St. Luke's Cancer Centre, Guildford). Of course I also thank my wife Lyn for her support and love.

References

Acar, C., Dincer, I. and Naterer, G.F. (2016). Review of photocatalytic water-splitting methods for sustainable hydrogen production. *Int. J. Energy Res.*, 40, 1449–1473.

Anders, W. (1968). "Lunar Module pilot on Apollo 8, the first mission to circumnavigate the Moon." nmspacemuseum.org. New Mexico Museum of Space History.

Ananyev, G.M. and Dismukes, G.C. (1996). Assembly of the tetra-Mn site of photosynthetic water oxidation by photoactivation: Mn stoichiometry and detection of a new intermediate. *Biochemistry*, 35, 4102–4109.

Appel, A.M., Bercaw, J.E., Bocarsly, A.B., Dobbek, H., DuBois, D.L., Dupuis, M., Ferry, J.G., Fujita, E., Hille, R., Kenis, P.J. and Kerfeld, C.A. (2013). Frontiers, opportunities, and challenges in biochemical and chemical catalysis of CO_2 fixation. *Chem. Rev.*, 113, 6621–6658.

Artero, V., Chavarot-Kerlidou, M. and Fontecave, M. (2011). Splitting water with cobalt. *Angew. Chem. Int. Ed.*, 50, 7238–7266.

Banerjee, A., Dick, G.R., Yoshino, T. and Kanan, M.W. (2016). Carbon dioxide utilization *via* carbonate-promoted C–H carboxylation. *Nature*, 531, 215.

Barber, J., Ferreira, K., Maghlaoui, K. and Iwata, S. (2004). Structure of the oxygen evolving center of photosystem II and its mechanistic implications. *Phys. Chem. Chem. Phys.,* 6, 4737–4742.

Barber, J. (2003). Photosystem II: The engine of life. *Biophys. Quart. Revs.*, 36, 71–89.

Barber, J. (2009). Photosynthetic energy conversion: natural and artificial. *Chem. Soc. Revs.*, 38, 185–196.

Barber, J. (2016a). Photosystem II: The water splitting enzyme of photosynthesis and the origin of oxygen in our atmosphere. *Quart. Rev. Biophys.*, 49, 1–21.

Barber, J. (2016b). Mn_4Ca cluster of photosynthetic oxygen-evolving center: Structure, function and evolution. *Biochemistry*, 55, 5901–5906.

Barber, J. (2017a). A mechanism for water splitting and oxygen production in photosynthesis. *Nature Plants*, 3, 17041–17046.

Barber, J. (2017b). Water splitting by the $Mn_4Ca^{2+}OX$ catalyst of Photosystem II: Its structure, robustness and mechanism. *Quat. Rev. Biophys.*, 50, 1–19.

Barber, J., Ferreira, K., Maghlaoui, K. and Iwata, S. (2004). Structural model of the oxygen-evolving centre of photosystem II with mechanistic implications. *Phys. Chem. Chem. Phys.*, 6, 4737–4742.

Barber, J. (2018). Hydrogen derived from water as a sustainable solar fuel: learning from biology. *Sustain. Energy Fuels,* 2, 927–935.

Bassi, P.S., Wong, L.H. and Barber, J. (2014). Iron based photoanodes for solar fuel production. *Phys. Chem. Chem. Phys.*, 16, 11834–11842.

Baymann, F., Lebrun, E., Brugna, M., Schoepp–Cothenet, B., Giudici–Orticoni, M.T. and Nitschke, W. (2003). The redox protein construction kit: pre-last universal common ancestor evolution of energy-conserving enzymes. *Philos. Trans. R. Soc. Lond. B Biol. Sci.*, 358, 267–274.

Blankenship, R.E. (2001). Molecular evidence for the evolution of photosynthesis. *Trends Plant Sci.,* 6, 4–6.

Blankenship, R.E. (2014). *Molecular Mechanisms of Photosynthesis*. John Wiley & Sons.

Blankenship, R.E., Tiede, D.M., Barber, J., Gary, W., Brudvig, G.W., Fleming, G., Ghirardi, M., Gunner, M.R., Junge, W., Kramer, D.M., Melis, A., Moore, T.A., Mosee, C.C., Nocera, D.G., Nozik, A.J., Ort, D.R., Parson, W.W., Prince, R.C. and Sayre, R.T. (2011). Comparing the efficiency of photosynthesis with photovoltaic devices: Recognizing opportunities for improvement. *Science*, 332, 805–809.

Brandon, N.P. and Kurban, Z. (2017). Clean energy and the hydrogen economy. *Philos. Trans. R. Soc. A*, 375, 20160400.

Britt, R.D., Campbell, K.A., Peloquin, J.M., Gilchrist, M.L., Aznar, C.P., Dicus, M.M., Robblee, J. and Messinger, J. (2004). Recent pulsed EPR studies of the photosystem II oxygen-evolving complex: implications as to water oxidation mechanisms. *BBA-Bioenergetics*, 1655, 158–171.

Brudvig, G.W. (2008). Water oxidation chemistry of photosystem II. *Philos. Trans. R. Soc. Lond. B,* 363, 1211–1218.

Canfield, D.E. (2005). The early history of atmospheric oxygen: homage to Robert M. Garrels. *Annu. Rev. Earth Planet. Sci.*, 33, 1–36.

Cox, N., Retegan, M., Neese, F., Pantais, D.A., Boussac, A. and Lubitz, W. (2014). Electronic structure of the oxygen-evolving complex in photosystem II prior to O–O bond formation, *Science*, 345, 804–808.

Crutzen, P.J. (2002). Geology of mankind. *Nature,* 415,415 (6867) p. 23.

Debus, R.J. (1992). The manganese and calcium-ions of photosynthetic oxygen evolution. *Biochim. Biophys. Acta,* 1102, 269–352.

Du, P. and Eisenberg, R. (2012). Catalysts made of earth-abundant elements (Co, Ni, Fe) for water splitting: recent progress and future challenges. *Energy Environ.l Sci.,* 5, 6012–6021.

Eisenberg, R. and Gray, H.B. (2008). Preface on making oxygen. *Inorg. Chem.,* 47, 1697–1699.

Falkowski, P.G. and Raven J.A. (2011). *Aquatic Photosynthesis*, 2nd Ed., Princeton University Press, Princeton, USA.

Ferreira, K.N., Iverson, T.M., Maghlaoui, K., Barber, J. and Iwata, S. (2004). Architecture of the photosynthetic oxygen-evolving center. *Science*, 303, 1831–1838.

Fischer, W.W., Hemp, J. and Johnson, J.E. (2016). Evolution of oxygenic photosynthesis. *Annu. Rev. Earth Plane. Sci.*, 44, 647–683.

Genovese, J., Harg, K., Paster, M. and Turner, J. (2009). Current state of the art hydrogen production cost estimate using water electrolysis. NREL/BK-6A1-46676. http://nrelpubs.nrel.gov/Webtop/ws/nich/www/public/SearchForm.

Goldsworthy, A. (1987). Why did Nature select green plants? *Nature,* 328, 207–208.

Gore, A. (2006). *An Inconvenient Truth: The Planetary Emergency of Global Warming and What Can We Do About It*. Rodale, New York.

Gore, A. (2017). *The Assault on Reason: Our Information Ecosystem, from the Age of Print to the Era of Trump*. Bloomsbury Publishing.

Harari, Y.N. (2014). Sapians: *A Brief History of Humankind*. Pub. Vintage, London.

Hohmann-Marriott, M.F. and Blankenship, R.E. (2011). "Evolution of photosynthesis." *Ann. Rev. Plant Physiol.,* 62, 515–548.

Hunter, B.M., Gray, H.B. and Muller, A.M. (2016). Earth-abundant heterogeneous water oxidation catalysts. *Chem. Rev.,* 1160, 14120–14136.

Ismail, A.A. and Bahnemann, D.W. (2014). Photochemical splitting of water for hydrogen production by photocatalysis: a review. *Sol. Energy Mater. Sol. Cells,* 128, 85–101.

Kanan, M.W. and Nocera, D.G. (2008). *In situ* formation of an oxygen-evolving catalyst in neutral water containing phosphate and Co^{2+}. *Science,* 321, 1072–1075.

Kleijn, R. and Van der Voet, E. (2010). Resource constraints in a hydrogen economy based on renewable energy sources: An exploration. *Renew. Sustain. Energy Rev.* 14, 2784–2795.

Kump, L.R. (2008). The rise of atmospheric oxygen. *Nature*, 451, p. 277.

Li, X. and Siegbahn, P.E. (2015). Alternative mechanisms for O_2 release and O–O bond formation in the oxygen evolving complex of photosystem II. *Phys. Chem. Chem. Phys.*, 17, 12168–12174.

Lovelock, J. (1994). *The Age of Gaia*. Norton. New York.

Lyons, T.W., Reinhard, C.T. and Planavsky, N.J. (2014). The rise of oxygen in Earth's early ocean and atmosphere. *Nature,* 506, 307.

Meadows, D.H., Meadows, D.L., Randers, J. and Behrens, W.W. (1972). *The Limits to Growth*. New York, 102, 27.

Mehta, A. (2018). Hydrogen: Still the fuel of the future? *Chem. World*, 6th Aug. 23–26.

Merki, D. and Hu, X. (2011). Recent developments of molybdenum and tungsten as hydrogen evolution catalysts. *Energy, Environ. Sci.*, 4, 3878–3888.

Messinger, J., Badger, M. and Wydrzynski, T. (1995). Detection of one slowly exchanging substrate water molecule in the S3 state of photosystem II. *Proc. Natl. Acad. Sci.USA*, 92, 3209–3213.

Mulkidjanian, A.Y., Bychkov, A.Y., Dibrova, D.V., Galperin M.Y. and Koonin, E.V. (2012). Origin of first cells at terrestrial, anoxic geothermal fields. *Proc. Natl. Acad. Sci. USA*, 109, 821–830.

Najafpour, M.M. (2011). Hollandite as a functional and structural model for the biological water oxidizing complex: manganese-calcium oxide minerals as a possible evolutionary origin for the $CaMn_4$ cluster of the biological water oxidizing complex. *Geomicrobiol. J.*, 28, 714–718.

Nelson, N. (2011). Photosystems and global effects of oxygenic photosynthesis. *BBA-Bioenergetics,* 1807, 856–863.

Olson, J.M. and Blankenship, R.E., (2005). Thinking about the evolution of photosynthesis. In *Discoveries in Photosynthesis*. 1073–1086. Springer, Dordrecht.

Olson, J.M. (2006). Photosynthesis in the Archean era. *Photosynth. Res.*, 88, 109–117.

Ouzounis, C.A., Kunin, V., Darzentas, N. and Goldovsky, L. (2006). "A minimal estimate for the gene content of the last universal common ancestor — exobiology from a terrestrial perspective." *Res. Microbiol.,* 157, 57–68.

Pecoraro, V.L., Baldwin, M.J., Caudle, M.T., Hsieh, W.Y. and Law, N.A. (1998). A proposal for water oxidation in photosystem II. *Pure Appl. Chem.*, 70, 925–929.

Pérez-Navarro, M., Neese, F., Lubitz, W., Pantazis, D.A. and Cox, N. (2016). Recent developments in biological water oxidation. *Curr. Opin. Chem. Biol.*, 31, 113–119.

Pimm, S.L. and Brooks, T.M. (2000). The sixth extinction: How large, where, and when. (*Nature and Human Society: The Quest for a Sustainable World*. National Academy Press, Washington DC, pp. 46–62.

Pope Frances (2015). *Laudato Si. On Care of Our Common Home*. Our Sunday Visitor.

Reece, S.Y., Hamel, J.A., Sung, K., Jarvi, T.D., Esswein, A.J., Pijpers, J.J. and Nocera, D.G. (2011). Wireless solar water splitting using silicon-based semiconductors and earth-abundant catalysts. *Science*, 334, 645–648.

Rees, M. (2018). *On the Future: Prospect for Humanity*. Princeton University Press.

Sagan, C. (1997). *Pale Blue Dot: A Vision of the Human Future in Space*. Random House Digital, Inc.

Sagan, C. (1983). Nuclear war and climatic catastrophe: Some policy implications. *Foreign Aff.*, 62, 257–292.

Sarchet, P. (2018). Life's true cradle. *New Sci.* June 16th, 31–34.

Sauer, K. and Yachandra, V.K. (2002). A possible evolutionary origin for the Mn4 cluster of the photosynthetic water oxidation complex from natural MnO_2 precipitates in the early ocean. *Proc. Natl. Acad. Sci. USA*, 99, 8631–8636.

Sharma, A.K., Spudich, J.L. and Doolittle, W. F. (2006). Microbial rhodopsins: functional versatility and genetic mobility. *Trends Microbiol.*, 14, 463–469.

Siegbahn, P.E.M. (2006). O–O bond formation in the S4-state of the oxygen evolving complex in *Photosystem II. Chemi.-A Eur. J.*, 12, 9217–9237.

Siegbahn, P.E.M. (2008). A structure-consistent mechanism for dioxygen formation in photosystem II. *Chemistry,* 14, 8290–8302.

Siegbahn, P.E. (2009). Structures and energetics for O_2 formation in photosystem II. *Acc. Chem. Res.*, 42, 1871–1880.

Siegbahn, P.E. (2017). Nucleophilic water attack is not a possible mechanism for O–O bond formation in photosystem II. *Proc. Naltl. Acad. Sci. USA*, 201617843.

Sivula, K., Le Formal, F. and Grätzel, M. (2011). Solar water splitting: progress using hematite (α-Fe2O3) photoelectrodes. *ChemSusChem.*, 4, 432–449.

Sojo, V., Herschy, B., Whicher, A., Camprubi, E. and Lane, N. (2016). The origin of life in alkaline hydrothermal vents. *Astrobiology*, 16, 181–197.

Steffen, W., Rockström, J., Richardson, K., Lenton, T.M., Folke, C., Liverman, D., Summerhayes, C.P., Barnosky, A.D., Cornell, S.E., Crucifix, M. and Donges, J.F. (2018). Trajectories of the earth system in the Anthropocene. *Proc. Natl. Acad. Sci. USA,* 115, 8252–8259.

Suga, M., Akita, F., Hirata, K., Ueno, G., Murakami, H., Nakajima, Y., Shimizu, T., Yamashita, K., Yamamoto, M., Ago, H. and Shen, J.R. (2015). Native structure of photosystem II at 1.95 A resolution viewed by femtosecond X-ray pulses. *Nature*, 517, 99–103.

Suga, M., Akita, F., Sugahara, M., Kubo, M., Nakajima, Y., Nakane, T., Yamashita, K., Umena, Y., Nakabayashi, M., Yamane, T. and Nakano, T. (2017). Light-induced structural changes and the site of O=O bond formation in PSII caught by XFEL. *Nature*, 543, 131–135.

Thomas, J.M. (2018). The enduring relevance and academic fascination of catalysis. *Nat. Catal.*, 1, 2.

Tran, P.D. and Barber, J. (2012). Proton reduction to hydrogen in biological and chemical systems. *Phys. Chem. Chem. Phys.*, 14, 13772–13784.

Tran, P.D., Wong, L.H., Barber, J. and Loo, J.S. (2012). Recent advances in hybrid photocatalysts for solar fuel production. *Energy Environ. Sci.*, 5, 5902–5918.

Tran, P.D., Chiam, S.Y., Boix, P.P., Ren, Y., Pramana, S.S., Fize, J., Artero, V. and Barber, J., (2013). Novel cobalt/nickel–tungsten-sulfide catalysts for electrocatalytic hydrogen generation from water. *Energy Environ. Sci.*, 6, 2452–2459.

Tran, P.D., Tran, T.V., Orio, M., Torelli, S., Truong, Q.D., Nayuki, K., Sasaki, Y., Chiam, S.Y., Yi, R., Honma, I., Barber, J. and Artero, V. (2016). Coordination polymer structure and revisited hydrogen evolution catalytic mechanism for amorphous molybdenum sulfide. *Nat. Mat.*, 15, 640–646.

Umena, Y., Kawakami, K., Shen, J.R. and Kamiya, N. (2011). Crystal structure of oxygen-evolving photosystem II at a resolution of 1.9 Å. *Nature*, 473, 55–60.

Vinyard, D.J., Khan, S. and Brudvig, G.W. (2015). Photosynthetic water oxidation: binding and activation of substrate waters for O–O bond formation. *Faraday Discuss*, 185, 37–50.

Vrettos, J.S., Limburg, J. and Brudvig, G.W. (2001). Mechanism of photosynthetic water oxidation: combining biophysical studies of photosystem II with inorganic model chemistry. *Biochim. Biophys. Acta*, 1503, 229–245.

Wang, M., Chen, L. and Sun, L. (2012). Recent progress in electrochemical hydrogen production with earth-abundant metal complexes as catalysts. *Energy Environ. Sci.*, 5, 6763–6778.

Weiss, M.C., Sousa, F.L., Mrnjavac, N., Neukirchen, S., Roettger, M., Nelson-Sathi, S. and Martin, W.F. (2016). The physiology and habitat of the last universal common ancestor. *Nature Microbiol.* 1, 16116.

Young, I.D., Ibrahim, M., Chatterjee, R., Gul, S., Fuller, F.D., Koroidov, S., Brewster, A.S., Tran, R., Alonso-Mori, R., Kroll, T. and Michels-Clark, T. (2016). Structure of photosystem II and substrate binding at room temperature. *Nature*, 540, 453–457.

Zhang, C., Chen, C., Dong, H., Shen, J.R., Dau, H. and Zhao, J. (2015). A synthetic Mn_4Ca-cluster mimicking the oxygen-evolving center of photosynthesis. *Science*, 348, 690–693.

Chapter 2

Theory of Rate Constants of Substeps in Single Molecule Experiments on F_1-ATPase

Sándor Volkán-Kacsó[*,†,‡] and Rudolph A. Marcus[†,§]

*Segerstrom Science Center, Azusa Pacific University
901 E. Alosta Ave., Azusa, CA 91702, USA

†Noyes Laboratory of Chemical Physics,
California Institute of Technology
1200 E. California Blvd., Pasadena, CA 91125, USA

‡svk@caltech.edu

§ram@caltech.edu

We review a recently proposed elastic group transfer theory and its application to free, stalling, and controlled rotation single molecule experiments on F_1-ATPase. In these experiments ATP is hydrolyzed to form ADP and inorganic phosphate, Pi. The main task is how to combine the mechanical and the chemical processes so as to form a theory, largely analytic, of the overall process and its substeps. Using the theory predictions can be made and tested within this class of experiments for the angle dependent rate constants in this motor enzyme. Data from some experiments are used to predict and compare with data from different experiments. The theory also serves as a basis for the statistical analysis of single molecule imaging and force spectroscopy trajectories, in which the enzyme kinetic processes are a convolution of processes. The theory is also used to correct for time-resolution limitations and "shot noise" often inherent in single-molecule experiments. Some new features involve (1) the unidirectionality of the rotation, (2) a postulated approximate conservation of open cleft space during the rotation of the γ-shaft in the motor, and (3) the relevance of single-molecule experiments, in which the ATPase is fixed on a microscope slide, to the physiological system.

1. Introduction

The biological function of the motor enzyme F_0F_1-ATP synthase is to regenerate, from ADP and inorganic phosphate Pi, the ATP molecules used as "energy currency" that fuel a multitude of cellular processes [Boyer, 1993]. The ATP synthase works as a nanomotor to synthesize ATP, an energetically uphill process which is achieved by a unidirectional mechanical rotation of a central γ shaft. The rotation of the shaft is prompted by a local proton transfer between offset ion channels in the F_0 ring (*e.g.*, [Junge and Nelson, 2015]), and this rotation serves to drive the chemistry for ATP synthesis in the F_1 ring. The water-soluble part of the ATP synthase, the F_1-ATPase [*e.g.*, Noji *et al.*, 1997; Yasuda *et al.*, 2001; Adachi *et al.*, 2012; Sielaff *et al.*, 2008; Spetzler *et al.*, 2009; Watanabe *et al.*, 2010] is a complex formed by the F_1 ring and the γ shaft, which functions in a reverse direction to ATP synthesis by converting the free energy of the inverse process, ATP hydrolysis, to unidirectional rotation in the opposite direction.

F_1-ATPase is robust and produces mechanical rotation when deposited on glass substrates *in vitro*. Single-molecule imaging studies have been therefore mainly focused on the study of F_1-ATPase with the broader perspective in mind that understanding the workings of this "functional" subsystem will ultimately lead to the elucidation of the detailed chemo-mechano-biology of the complete ATP synthase. Single molecule techniques complement ensemble biochemical and structural experiments [Braig *et al.*, 2000; Boyer, 1993; Senior, 2007; Weber, 2010]. An essential experiment here [Yasuda *et al.*, 2001], was to show early on the relevancy of the single molecule experiments to the ensemble experiments by showing that as a function of the ATP concentration (Michaelis-Menten plots) the rate of disappearance of the ATP in an ensemble experiment in which the F1-ATPase freely diffuses in solution are practically identical with the rate of rotation of the rotor in the single molecule experiment in which the F1-ATPase is deposited on some surface, namely that 1 ATP reacts in the former for each 120° rotation in the latter.

The rotation of the central "rotor" shaft was been monitored by dark field and fluorescent single molecule microscopy [Noji *et al.*, 1997]. In these experiments, the F_1 "stator" ring subunit is attached to a microscope slide, as in Figure 1a. Using a 200 nm diameter probe attached to the end of the shaft (where the F_0 would be), a counter-clockwise stepping rotation was observed. The stepping dwell angles were seen to be 120° apart [Adaci *et al.*, 2000], with substeps of 40° and 80° at low ATP concentration [Yasuda *et al.*, 2001].

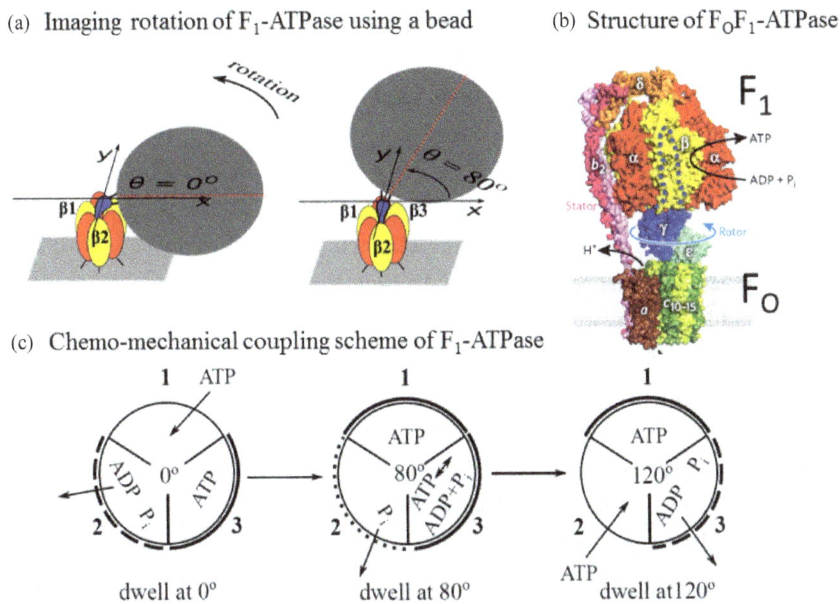

Figure 1. (a) *In vitro* construct for imaging the F_1-ATPase. The rotor is chemically bound to a microscope slide and imaging bead is attached to the rotor. A step of 80° is shown. (b) Structure of the complete membrane embedded F_1F_0-ATPase, seen upside-down relative to the F_1-ATPase in (a). (c) Scheme of coupled processes in F_1-ATPase during free rotation. The dwell angle increases in the counter clockwise direction. The species occupying the pockets of ring β subunits 1, 2 and 3.

A kinetic chemo-mechanical scheme (Figure 1c) has been formulated from a series of single-molecule experiments [Nishizaka *et al.*, 2004; Adachi *et al.*, 2007; Senior, 2007; Watanabe *et al.*, 2010]. Accordingly, ATP binding and ADP release (from a different β subunit) are concerted events (80° substep) and ATP hydrolysis and Pi release (again, from a different β subunit) occur during a 40° substep. In the scheme, when following the events in the same subunit, ATP binding occurs during the 80° step following a binding dwell set to 0° [Yasuda *et al.*, 2001]. The ATP does not undergo hydrolysis until much later in the rotation, during the 200° to 240° step, a 40° step. ADP release follows during the subsequent 80° step, from 240° to 320°. Finally, Pi release occurs during the last 40° step, from 320° to 360°. A question that naturally arises is why is there such a delay in the release of the ADP and Pi, long after an ATP has entered. One possibility is that there is an approximate conservation of total empty cleft space in the three principal clefts. Each active ATP binding site is a pocket largely in a β subunit in a cleft-like structure that can undergo a hinge-bending structural change [Braig *et al.*, 2000]. When one cleft is closing during the addition of an ATP another cleft is opening to facilitate the release of an ADP. However, that ATP cannot enter the cleft until is cleft has released an ADP. Research in the hydrolysis direction of the F_1-ATPase motor can be directly relevant to ATP synthase function by assuming reversibility in the chemo-mechanical rotation scheme [Adachi *et al.*, 2007, 2012].

The remainder of this chapter is organized as follows. In Section 2, we review a chemical-mechanical theory [Volkán-Kacsó and Marcus, 2015] that we formulated to treat several types of single molecule experiments on a biomolecular motor [Adachi *et al.*, 2007, 2012]. The theory is adapted from a theory of electron transfer and other types of transfer in solution [Marcus, 1968; Marcus and Sutin, 1985; Marcus, 1993] and contains, in addition, to treat the mechanical aspects a torsional elasticity effect [Sielaff *et al.*, 2008] in the motor in the various physical and chemical processes, such as ATP binding and ATP hydrolysis. It does so by detailed consideration of the various free energy changes including those due to the elasticity.

In Section 2.2 we discuss the relation between single molecule experiments involving ATPase deposited on a microscope slide and a freely diffusing F_1-ATPase system, and so one that has little load. The relevance of the slide-bound ATPase experiments to the freely moving ATPase is also discussed in Section 2.2.

An idea used in the theory is that binding of a nucleotide induces a closing of a binding cleft that "seals" the binding channel, and if we introduce an assumption of approximate conservation of total open cleft space in this tightly knit system causes simultaneously another cleft to open and, in this case, release its ADP. In Section 2.3 we discuss how this closing motion can result in the unidirectional rotation in the F1-ATPase.

We next consider in Section 3 single molecule experiments in which the rotor angle is both monitored as in Figure 1, and stalled and rotation-controlled by the used of magnetic tweezers. These techniques have been termed stalling and controlled rotation experiments. We have shown how controlled rotation data were predicted from the stalling data using a theory that contains the chemical and mechanical behavior [Volkan-Kacso and Marcus, 2016]. To do so, we derived an equation that is key to relating the mechanical and chemical properties of the system to the quasistatic rate constants are the various steps.

In Section 4, we review a method proposed to treat non-equilibrium controlled rotation data and relate this single-molecule data to equilibrium quantities such as rate constants.

2. Chemo-Mechanical Coupling, Directionality of Rotation and Role of Single Molecule Experiments

The theory will be described [Volkán-Kacsó and Marcus 2015, 2016, 2017a] by introducing into the thermodynamic free energy equations a mechanical elastic coupling term arising from the interaction of the subunits with the γ subunit of the F_1-ATPase, and deriving an equation for the dependence of the rate constant of a substep on the standard free energy of reaction. The theory and its application to predict controlled rotation data are summarized in this section to make the present review self-contained. The theory is described in more detail in a review [Volkán-Kacsó and Marcus, 2017b].

2.1. *Rate constants and free energy relations*

We first note that the equations of the "weak-overlap" theory of electron transfer reactions were adapted to treat "strong overlap" reactions, such as the transfer of an atom B in the atom transfer reaction $A + BC \rightarrow AB + C$ [Marcus, 1968] and used for the latter [*e.g.*, Sutin, 1966; Lewis and Hu 1984; Murdoch, 1986]. In this case the topology is such that there is in effect only one relevant potential energy surface (the lower adiabatic surface) instead of two diabatic surfaces. For such transfer reactions in solution [Marcus 1968, 1993; Marcus and Sutin 1985] the free energy barrier for the reaction, ΔG^*, is given by an equation which has been approximated in the literature by[1]:

$$\Delta G^* = \frac{\lambda}{4}\left(1 + \Delta G^0 / \lambda\right)^2, \tag{1}$$

where, for simplicity of presentation in this article, we have omitted the "work terms" that occur in transitioning from a pair of reactants (r) or products (p) in solution to form a collision complex with the F_1-ATPase. In Equation 1, ΔG^0 denotes the standard free energy reaction of the transfer reaction and λ denotes the "reorganization energy" for the reaction. The rate constant k is given by:

$$K = A\exp\left(-\Delta G^* / kT\right), \tag{2}$$

where ΔG^* is given by Equation 1 and A depends on the nature of the process, *e.g.*, whether the reaction is first or second order, and on the details of the transition state of the process. A particular example would be to use a collision theory picture [Volkán-Kacsó and Marcus, 2015] for the ATP binding to the weak binding site of an empty β subunit in which the pre-exponential term A is a collision frequency Z, and an additional factor related to an ATP binding outside of the β subunit. For the binding and release of ATP there are extensive data obtained by different experimental methods [Yasuda *et al.* 2001; Spetzler *et al.* 2009; Watanabe *et al.* 2010, 2012; Braig *et al.* 2000; Adachi *et al.* 2012]. In the ATP binding an ATP can be regarded as first forming a collision complex [Oster and Wang, 2000]

[1]The relevant equation is a bond energy-bond order equation. Its functional form has no "inverted" effect [Marcus, 1968].

with the F_1-ATPase, followed by an ATP binding as the next step (an 80° step) in the overall hydrolysis.

The simplest adaptation of Equation 1 is to use the quadratic relationship, but, because of the different topology of the surfaces for weak and strong interaction, to set $\Delta G^* = 0$ when $-\Delta G^0 \geq \lambda$, and $\Delta G^* = \lambda$ when $\Delta G^0 \geq \lambda$, between the reactants. Whereas the ET reaction rate decreases at highly negative ΔG^0 (the inverted effect) strong overlap" reaction rate reaches some limiting maximum value, a result clear from the topology of the adiabatic surface, instead of experiencing the "Marcus inverted effect." When, instead, ΔG^0 is very positive ($\geq \lambda$) we use for this limiting case $\Delta G^* = \Delta G^0$. Accordingly, in passing we note that the relation between this model and a bond energy–bond order model (BEBO) has been discussed [Marcus, 1968, and reference cited therein]. In the BEBO model the sum of the bond orders in a reaction $AB + C \rightarrow A + BC$ is assumed constant throughout the reaction step, so for an atom or group transfer a cooperative effect is seen. Here, the activation energy when B is transferred from A to C is roughly 10% of the AB bond energy for a thermoneutral reaction [Marcus, 1968, and reference cited therein]. A somewhat more detailed account of the application of the BEBO idea for a substep in the F_1-ATPase is found elsewhere [Volkán-Kacsó, and Marcus, 2017b]. The relevance to processes such as ATP binding is that when an ATP enters a channel in an empty β subunit and finally locks into place in the subunit some hydrogen bonds are broken while new hydrogen bonds are formed, perhaps approximately conserving the total number of hydrogen bonds. In this extension many H bonds are broken and many formed, with an approximate conservation of the total, thereby extending the idea if the BEBO model in which one chemical bond is broken and one is formed. For example, using the structural results of Walker [Braig *et al.*, 2000] we estimated that for the ATP binding step approximately 100 hydrogen bonds are broken in the β subunit hosting the ATP and approximately the same number are formed.

2.2. *Relevance of single-molecule experiments on the F_1-ATPase deposited on the slide to function in physiological conditions*

In single-molecule experiments the behavior of the actual F_1-ATPase system is studied by depositing the F_1-ATPase on some surface rather than diffusing it in solution. We consider two rate constants, the bimolecular ATP binding rate constant and the hydrolysis rate constant, as a benchmark that can be used to establish the relevance of single-molecule experiments.

Firstly, in imaging experiments with small (20 nm radius) beads the bimolecular ATP binding rate constant and the hydrolysis rate constants coincide with those reported in ensemble measurements [Volkan-Kacso and Marcus, 2017c]. These

Michaelis-Menten curves from single-molecule rotation experiments [Yasuda *et al.*, 2001] are reportedly identical with the corresponding data from ensemble measurements.

Secondly, catalytic dwell time distributions observed in single-molecule imaging experiments using polarization of scattered light from gold nanorods or experiments using dark field microscopy with small beads as probes yielded the rate constant for hydrolysis under rotation not limited by large viscous load. These rate constants (0.5 s^{-1} for the thermophilic bacterium PS3 and 0.12 s^{-1} for *E. coli* F_1-ATPase) have been found to be consistent in these two types of single molecule experiments as well as in the ensemble experiments [Spetzler *et al.*, 2009].

This important property illustrates that F_1-ATPase molecules attached to a microscope slide and imaged using a filament or a micro bead behave in this respect as if they were simply suspended in solution. Hence, on that basis, the kinetic and dynamic information extracted from single-molecule rotation experiments are relevant to the physiological functionality of the F_1-ATPase.

2.3. *Directionality of rotation*

An important question is the detailed mechanism of the unidirectional rotation in the F_1-ATPase. X-ray crystallography structures clearly indicated that the rotor shaft rotation is consistent with the structural changes that the stator ring subunits undergo as ATP binds, undergoes hydrolysis and products are released in the latter [Braig *et al.*, 2000]. This unidirectional (counter clockwise when viewed from where the F_0 would be) rotation was first demonstrated in single molecule imaging experiments using fluorescent filaments attached to the rotor shaft [Masaike *et al.*, 2000]. The unidirectional nature of the rotation was later shown to be "encoded" in the structure of the stator ring subunits, when the conformational kinetics of rotorless F_1-ATPase was imaged in high-speed AFM experiments [Uchihashi *et al.*, 2011].

The question then arises on the coupling between the unidirectional rotation of the shaft and the specific hinge-bending conformational changes in the subunits. A cleft that closes upon binding of an ATP molecule to the "internal" binding pocket is roughly formed at the interface between neighboring α and β subunits. A cross section in the "sagittal" plane of the barrel structure of the stator ring and rotor shaft is shown in Figure 2 for an open and a closed cleft. In these X-ray crystallography structures the binding channel straddles the two subunits and it appears to close upon the binding of an ATP molecule.

An inspection of the changes in the α and β subunits shows that the latter undergo a more substantial conformational change than the former. This asymmetry leads the side of the cleft nearest the β subunit rotating more than the side nearest the α subunit during the closure of the β subunit, contributing to the angular momentum of the rotor.

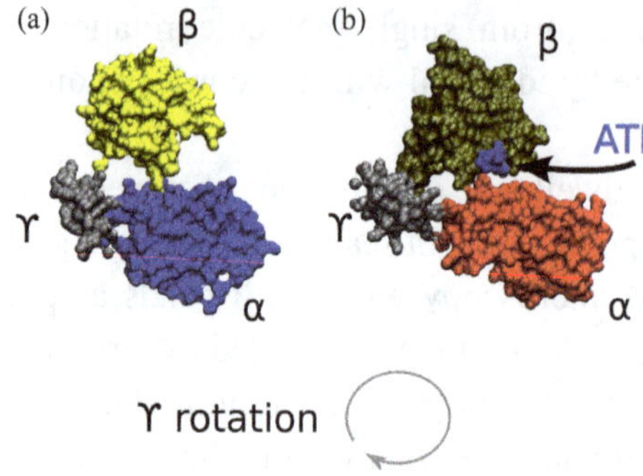

Figure 2. "Sagittal" cross section of the F_1-ATPase showing the binding cleft for two different α–β interfaces: (a) empty pocket (b) ATP-bound pocket. The same crystal structure is used for both, but a rotation of 120° has been applied to allow a direct comparison of the stator subunits and the rotor shaft. A closing of the binding channel is also apparent as would be expected to occur during the 80° step. is shown.

The small viscous load from a nanorod or nanobead probe attached to the rotor does not limit the observed rotation velocity [Spetzler *et al.*, 2009], which otherwise, with larger probes, is found to be in the overdamped limit [Oster and Wang, 2000]. If then the resulting angular momentum about the γ shaft described in the previous paragraph is compensated by an opposite angular momentum and hence opposite rotation of the shaft, a net unidirectional rotation of the γ subunit should occur during the ATP binding substep. This shaft rotation would thereby be deterministic in its directionality, dictated by the angular velocity of rotation of the side of the β subunit. The exploration of this idea of an approximate conservation of angular momentum has become feasible using the high temporal resolution imaging of rotation in nanorod single molecule experiments [Spetzler *et al.*, 2009; Martin *et al.*, 2014; Sielaff *et al.*, 2016].

3. Elastic Transfer Theory for Experiments with Magnetic Tweezers

The group transfer theory was first proposed to treat the stalling experiments [Volkán-Kacsó and Marcus, 2015] and then used to predict controlled rotation experiment data [Volkán-Kacsó and Marcus, 2016]. Later, the theory was used to treat composite events in rotation control experiments and to extract hydrolysis and synthesis data from the trajectories [Volkán-Kacsó and Marcus, 2017].

3.1. *Stalling experiments*

Magnetic tweezers, can be used to yield an arbitrary rotor angle θ. The "standard" imaging-only probe is replaced in these experiments by double beads with a

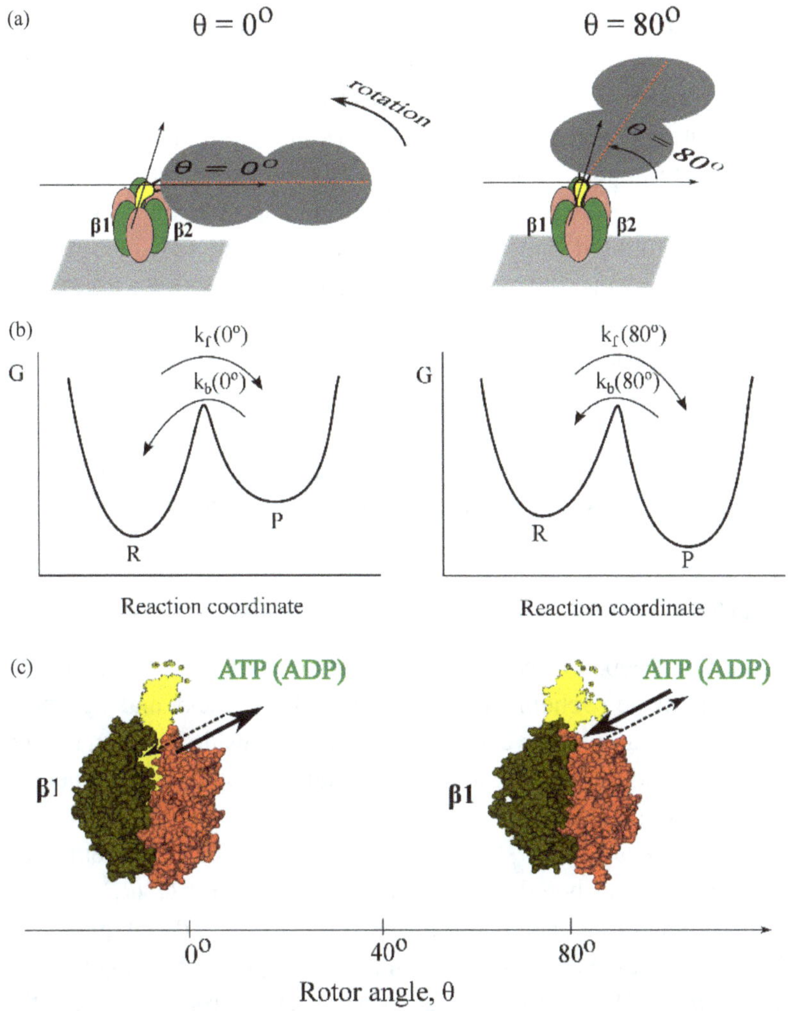

Figure 3. (a) F_1-ATPase in single molecule imaging and controlled rotation experiments at two rotor angles: 0° and 80°. A double-bead is attached to the γ rotor shaft (in yellow) and rotated against the stator ring (active subunits β1–β3). (b) Free energy profile for nucleotide binding (k_f) and release (k_b) rate constants at the two angles. (c) Open-to-close changes in the nucleotide binding β1 subunit as a function of rotor angle.

permanent magnetic dipole moment (Figure 3a) [Adachi *et al.*, 2012]. When an external magnetic field is applied the double bead attached to the rotor aligns itself to the direction of the "controlling" external field.

Using magnetic tweezers, Noji and coworkers have devised an ingenious stalling technique, which permitted the measurement of rate constants as a function of rotor angle for ATP binding and release, ATP hydrolysis and synthesis and Pi release and binding [Watanabe *et al.*, 2010]. In these stalling experiments, a freely rotating shaft was allowed to reach a dwell angle (*e.g.*, the binding dwell), then the tweezers were immediately activated to set the rotor to an arbitrary angle θ. The rotor was then stalled at θ for a given time, after which it was released and it either moved back to the original dwell or forward to the next dwell.

After repeating these experiments for a series of dwell times, Adachi and coworkers [2012] interpreted the statistics of forward and back rotations in terms of a two-state model. They assumed that (1) during the stall time, the system can undergo several forward and backward processes, and (2) the rotations upon release depended on the state of the underlying process (*e.g.*, for ATP binding either ATP-bound or empty) at the moment of release. The observed relative number of forward and back events as a function of stall time yielded the forward (k_f) and backward (k_b) rate constants of ATP binding, or other processes. The model of elastic transfer of nucleotides was proposed to explain this rate constants *versus* θ data in terms of the effect of the elastic response of the rotor structure on the free energy surface (Figure 3b) [Volkán-Kacsó and Marcus, 2015].

3.2. *Elastic transfer theory*

To calculate the θ dependence of the standard free energy of reaction $\Delta G^0(\theta)$ a thermodynamic cycle (Scheme 1) was used [Volkán-Kacsó and Marcus, 2015]. The energy balance around the cycle provides a relationship between the free energies of a ligand binding in free rotation, ΔG_0^0, and the binding free energy $\Delta G^0(\theta)$ at a constant rotor angle θ.

In Scheme 1 $G_r^0(\theta)$ and $G_p^0(\theta)$ are the free energies of the system in the reactant and product states when the rotor is at an angle θ, while $G_r^0(\theta_i)$ and G_p^0 are the reactant and product free energies in the relaxed system at dwell angles θ_i and θ_f, respectively. We assume an elastic coupling first suggested by Junge and coworkers [Panke *et al.*, 2001; Sielaff *et al.*, 2008] between the rotor angle and the stator subunit conformation [Volkán-Kacsó and Marcus, 2015] involved in each of these steps. Accordingly, the system, while in the reactant state, can be rotated from a dwell angle θ_i to an arbitrary angle θ by elastically distorting the structure, by doing a work $w_r(\theta)$ provided externally by the external magnetic field of the magnetic tweezers. To achieve a rotation from θ_i to θ, parts of the structure (a structural

$$
\begin{array}{ccc}
G_r^0(\theta) & \xrightarrow{\Delta G^0(\theta)} & G_p^0(\theta) \\
-w^r\downarrow & & \uparrow w^p \\
G_r^0(\theta_i) & \xrightarrow{\Delta G_0^0} & G_p^0(\theta_f)
\end{array}
$$

Scheme 1.

linkage formed by the rotor and the lever arm of the β subunit) undergo an elastic twisting deformation of rotation stiffness κ.

Analogously, the F_1-ATPase in the product state, can be rotated from an angle θ to the relaxed state at θ_f by by doing a work $w_p(\theta)$, a work done by the system. To achieve this rotation an elastic twisting deformation of stiffness κ must occur. The elastic work terms then are:

$$w^r = \frac{k}{2}(\theta - \theta_i)^2 \quad \text{and} \quad w^p = \frac{k}{2}(\theta - \theta_f)^2. \tag{3}$$

The energy balance in Scheme 1 requires that:

$$\Delta G^0(\theta) = \Delta G_0^0 + w^p(\theta) - w^r(\theta) = \Delta G_0^0 - \kappa(\theta_f - \theta_i)(\theta - \theta_c), \tag{4}$$

where the notation $\theta_c = (\theta_f + \theta_i)/2$ was used and ΔG_0^0 is the free energy change corresponding to the difference between the free energy minima in the final dwell state *versus* that in the initial dwell state. Using Equations 1–5 we have shown [Volkán-Kacsó and Marcus, 2017] that the theory provides a symmetric expression for the rate constant *versus* rotor angle dependence,

$$\begin{aligned} kT\ln k_f(\theta) &= kTk_f(\theta_c) + (1/2 + \Delta G_0^0/2\lambda)\Delta\theta_r - (\Delta\theta_r)^2/4\lambda, \\ kT\ln k_b(\theta) &= kTk_b(\theta_c) + (-1/2 + \Delta G_0^0/2\lambda)\Delta\theta_r - (\Delta\theta_r)^2/4\lambda, \end{aligned} \tag{5}$$

where $\Delta\theta_r = \kappa(\theta_f - \theta_i)(\theta - \theta_c)$ and $\theta_c = (\theta_i + \theta_f)/2$, as defined in Volkán-Kacsó and Marcus [2015]. The quadratic term is small in the case of the ATP binding step (less than 8% contribution). For the equilibrium constant this dependence is predicted to be exponential,

$$kT\ln K(\theta) = kT\ln K(\theta_c) - \kappa(\theta_f - \theta_i)(\theta - \theta_c), \tag{6}$$

in agreement with the experimental data [Volkán-Kacsó and Marcus, 2015]. Thus, the group transfer relates the exponential dependence of the rate constants in the stalling experiments with the harmonic response to external torques described by spring constant κ and the step size $\theta_f - \theta_i$ associated with the process (e.g., 80° for ATP binding). With data from other ensemble [Boyer, 1993; Weber and Senior, 1997] and single molecule [Yasuda *et al.*, 2001; Adachi *et al.*, 2012; Spetzler *et al.*, 2009] experiments Equations 5–6 were used to predict [Volkán-Kacsó and Marcus, 2015] for the stalling experiments on ATP binding the Bronsted slope $\alpha = \partial \ln k_f / \partial \ln K$ of 0.47, which compares with the value of 0.48 in the stalling experiment. The calculated slope of the ln $k_f(\theta)$ *versus* θ was found to be in good agreement with experiment [Volkán-Kacsó and Marcus, 2015]. A value of $\kappa = 16$ pN nm rad^{-2} was obtained from the stalling experiments. We note that on the timescale of milliseconds accessible in the stalling experiments that binding of ATP and release of ADP from a different pocket appear to occur as concerted processes. However, in other higher time resolution experiments probing timescales of microseconds, ATP binding and ADP release may be resolved as subsequent processes [Sielaff *et al.* 2016].

3.3. *Controlled rotation experiments*

In these experiments single-molecule fluorescence microscopy was used to monitor individual nucleotide (ATP and ADP) binding and release events (Figure 2c) in the F_1-ATPase [Adachi *et al.*, 2007]. To do so, the nucleotides were modified by attaching a fluorescent Cy3 moiety *via* a flexible alkane tether. When the Cy3-nucleotide was bound to the F_1-ATPase the Cy3 moiety was fluorescently active but when it was in the solution its fluorescence was quenched.

Using magnetic tweezers, Adachi *et al.* [2012] induced a slow, constant rate rotation of the γ shaft and monitored, at the same time, the individual binding and release events of the Cy3-nucleotide. Events of occupancy changing between 0 and 1 were determined and analyzed. The number of occupancy change ($0 \rightarrow 1$) and ($1 \rightarrow 0$) events yielded forward and reverse rate constants as a function of rotor angle.

There is a region of rotor angles θ roughly –45° to +45° in which the controlled rotation and stalling data overlap. An important difference between the controlled rotation and stalling experiments is that in the former a fluorescent species, Cy3-ATP and Cy3-ADP, is used instead of the regular ATP and ADP. The nucleotide concentration in the controlled rotation experiments is kept low (occupancy varies between 0 and 1), but in the stalling experiments the occupancy varied between 2 to 3.

Assuming that these differences affected the absolute rate constants, but left the angle-dependence unchanged the rate constants in the controlled rotation experiments can be predicted, without using adjustable parameters. The comparison is given in Figure 4, in the θ-range of overlap. Good agreement was found between these experiments and our calculations, seen in Figure 4, where the points (symbols) are the experimental data and the solid curves are theoretical calculations.

The comparison between theory and experiment is given in Figure 4, covering the θ-range of ~–40° to ~+40°. In the single molecule fluorescence data for monitoring the binding and release events there will be missed events when the rate of release is very fast. These missed events are artifacts that arise in single-molecule trajectories, and theory has been used as an instrument to account for such artifacts [see *e.g.*, Volkan-Kacso, 2015, and references therein]. The results, after correcting for these events, are given in Figure 4. The observations of the rates without this correction for the missed events is given by the dashed lines in that figure.

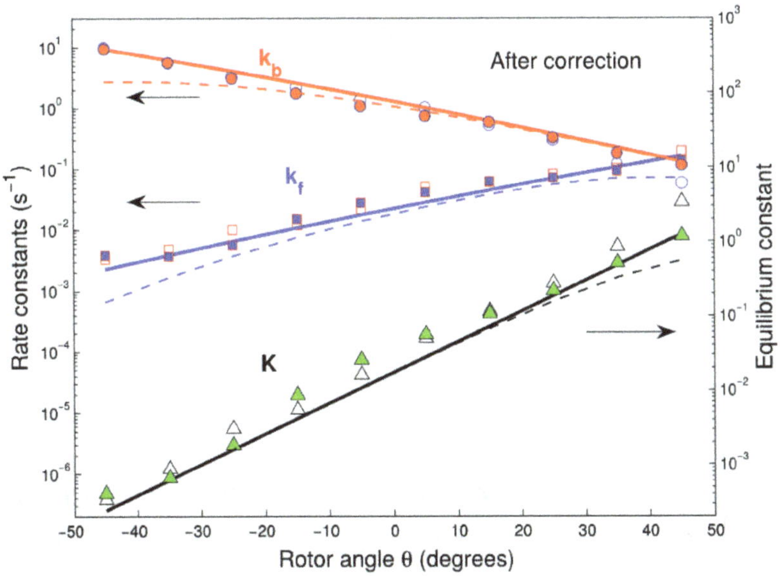

Figure 4. Corrected binding and release rate and equilibrium rate constants *versus* θ angle for Cy3-ATP in the presence (solid squares, circles and triangles) and absence of Pi (open symbols) in solution adapted from Volkán-Kacsó and Marcus [2016]. The experimental data of Adachi *et al.* [1012] corrected for missed events (and an error due to replacing the time spent in the empty state by total time of a trajectory) are compared with their theoretical counterparts (solid lines). Dashed lines show the data without corrections.

3.4. *Rate constant of a free rotation experiment and the relation between controlled and free rotation experiments*

We have recently considered the case of freely rotating F_1-ATPase, *i.e.*, when no probe is attached to the rotor shaft [Volkán-Kacsó and Marcus, 2017b]. In a free rotation experiment the system jumps from θ_ι to θ_ϕ during some fluctuation from its initial value θ_ι. We assumed that the probability of a fluctuation in θ is proportional to Boltzmann factor due to an elastic term $C(\theta)=\kappa(\theta-\theta_i)^2/2$. Using Equation 2 for the forward rate constant $\kappa_f(\theta)$ the average value of the forward rate constant in a free rotation is then given by:

$$\left\langle k_f \right\rangle = \int_{\theta_L}^{\theta_H} A e^{-[\Delta G^*(\theta)+C(\theta)]/kT} d\theta \Bigg/ \int_{\theta_L}^{\theta_H} e^{-C(\theta)/kT} d\theta. \tag{7}$$

If one neglects the quadratic term in $\Delta G^0(\theta)$ in Equation 4, valid, in the case of ATP binding, with Bronsted slope α close to 0.5, one finds that the integrand of the exponent in the numerator of Eqation 7 is peaked at an angle $\theta = \theta^* = (\theta_i + \theta_f)/2$. Using the saddle point method to evaluate the integral we find that

$$\langle k_f \rangle \approx k_f(\theta^*), \tag{8}$$

a result used in [Volkán-Kacsó and Marcus, 2015] to obtain an estimate of the ΔG_0^0 from the experimental data. Equation 8 thus establishes a relationship between free rotation and stalling experiments. The analysis of the reverse process, ATP release, can be performed in an analogous manner and yields the analog of Equation 9,

$$\langle k_b \rangle \approx k_b(\theta^*), \tag{9}$$

For the ATP binding step the maximum will be at $\theta^* = 40°$.

3.5. *Exponential rate and equilibrium constant versus rotor angle dependence of binding/release and catalysis*

As described in Section 3 stalling and controlled rotation experiments revealed exponential-like forward and back rate constants for ATP binding in the region of angular overlap in these experiments [Watanabe *et al.*, 2012; Adachi *et al.*, 2012]. Such exponential functions apply to the hydrolysis, Pi release, and various nucleotide binding rates in regions where these events occur in spontaneous rotation. The forward and back rate constants have opposite angle-dependent trends, and the equilibrium constant is strictly exponential, as predicted by theory [Volkán-Kacsó and Marcus, 2016].

The turnover in the ATP binding rate constants *versus* rotor angle plot seen in Figure 6a indicates that the closing of the binding channel continues as the angle is rotated beyond 80°. The turnover is consistent with structural data and the idea of a continuum of open-to-close states between –40° and 140°.

We suggested [Volkán-Kacsó and Marcus, 2016] that a diffusion process while the ATP is entering the channel in the subunit after the initial binding of the ATP to the outside of the F_1-ATPase. The latter then serves as the reactant state in the transfer process. The binding channel leading to the binding pocket (Figure 5c) which closes as the rotor is moved in the positive direction (Figure 5b) retards the diffusion process to the extent that it becomes the bottleneck during binding for angles beyond 50° or so. Conversely, in a standard diffusion-reaction scheme [Marcus, 1960; Noyes, 1961], the transfer reaction is the bottleneck in the θ-region from about –45° to +45° where stalling and controlled rotation experiments overlap.

As seen in Figure 5c the β subunit is the most open at –40° and the most closed presumably at +140° (0° is defined as the binding dwell angle at which the β subunit in question is empty and is about to undergo ATP binding associated with a subsequent 80° substep). The minimum at –40° does not correspond to a specific dwell angle. The structural changes of opening and closing are consistent

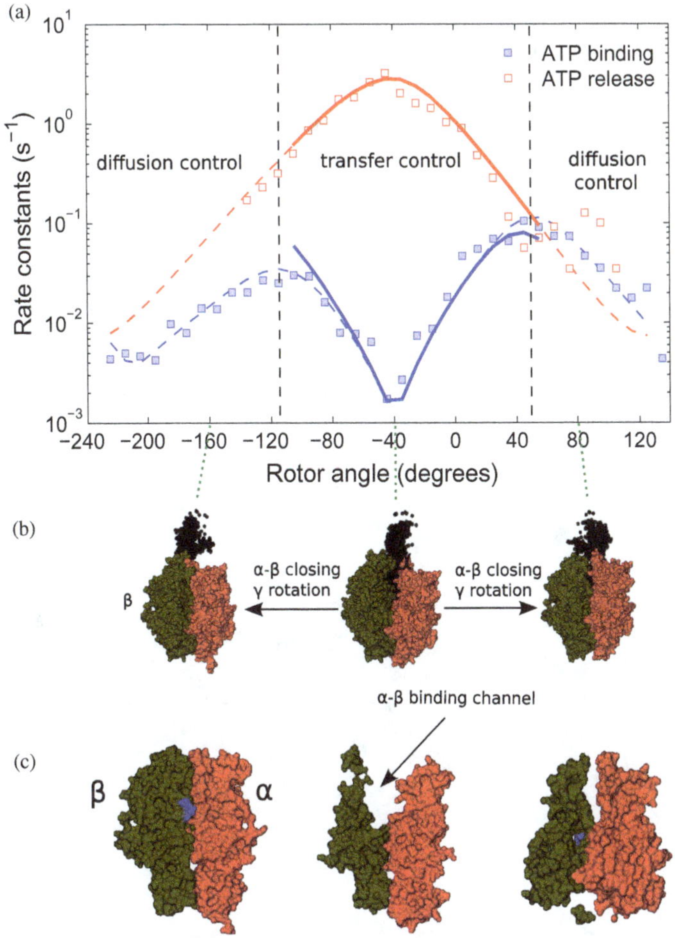

Figure 5. (a) Reported binding and release rate constants *versus* controlled rotation angle for fluorescent ATP in the presence of Pi in solution. The reported uncorrected experimental data (squares) are compared with theoretical counterparts (solid lines) by calculating missed events and also correcting for an error due to replacing the time in the empty state T_0 by the total time T. Dashed lines show a fit to the experimental data. (b) F_1-ATPase structure at three different rotor angles with β subunits in green, α subunits in red and the γ subunit in black. (c) Cutaway of the three structures revealing the binding channel at the α–β interface and its narrowing as the rotor angle is changed. The figure was reproduced from [Volkán-Kacsó and Marcus, 2017b].

with the trends of binding/release *versus* rotor angle data and with spontaneous free rotation experiments. In particular:

1. Moving the rotor to the left or right starting from –40° in both cases induces a closing of the β subunit which accelerates binding. This behavior is consistent with the spontaneous forward rotation induced by binding of ATP in free rotation experiments occurring in a θ-range to the right of –40° (from 0° to 80°).

2. Rotating the γ rotor towards –40° from the left or right in both cases induces an opening of the β subunit which accelerates release. This result, in turn, is consistent with the spontaneous forward rotation induced by the release of ADP in free rotation experiments occurring in a θ-range to the left of –40° (from of –120° to –40°).

A degree of asymmetry is also seen in the binding/release rate *versus* rotor angle data relative to the angle from –40° in both directions. Notably, in the presence of Pi in the solution, ATP binding to the right in Figure 6a occurs at much

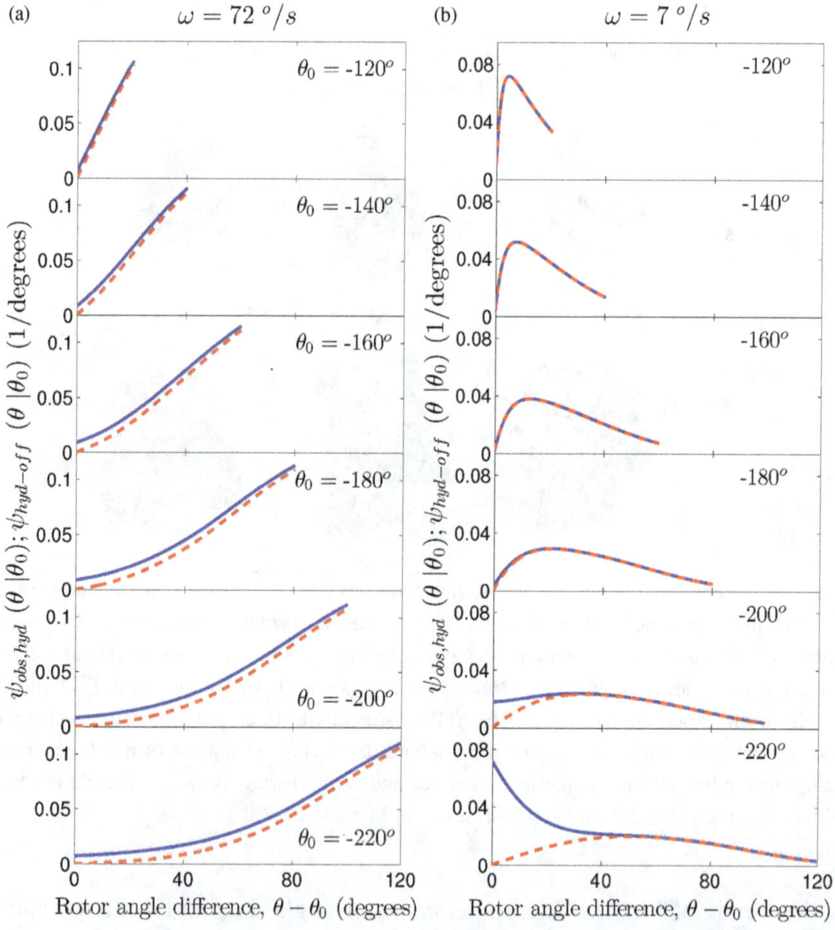

Figure 6. Lifetime distributions for binding events for rotation in the hydrolysis direction for a series of long delay events ranging from rotor angle values 140 to 240. The rotation rate in A 0.2 Hz (used by Adachi and coworkers in experiment) and in B it was 0.02 Hz. The distributions (to be observed in experiment) are shown as solid lines and the contribution from events with hidden hydrolysis are shown as dashed lines.

higher rates than ATP binding to the left. Thus, Pi binding hinders ATP binding in the synthesis direction, but does not prevent ADP binding, a behavior that is consistent with the hydrolysis scheme being the inverse of the synthesis scheme established by single-molecule experiments [Yasuda *et al.*, 2007; Watanabe *et al.*, 2010]. (In the hydrolysis direction Pi is the last to be released during the final 320°–360° substep.)

4. Theory of Long Time Binding Events in Controlled Rotation

4.1. *Lifetime distributions and extraction of angle-dependent rate constants from long time binding events*

Small intervals of angles were used in the controlled rotation experiments to detect changes in the F_1-ATPase nucleotide occupancy from fluorescence trajectories. This procedure is often termed as "binning". To extract rate constants of the individual steps in the composite events systems we first obtain the lifetime distributions involved in the steps. In the experiments by [Adachi *et al.*, 2012] the shaft is rotated at a fixed angular velocity ω, resulting in a time-dependent angle $\theta(t) = \omega(t-t_0) + \theta_0$.

During rotation in the hydrolysis direction, if a binding of ATP from solution occurs in the vicinity of the binding dwell at 0° and is followed by a long binding event of more than 120° rotation, the subsequent release event is classified as ADP release, meaning that hydrolysis occurred between binding and release. A long binding event of ATP involves a series of 3 events: ATP binding, hydrolysis and ADP release. To treat these long binding events, we use the lifetime density $\psi(\theta|\theta_0)$ defined from the survival probability p as $-dp/dt$. For the successive events, binding at θ_0, hydrolysis of ATP at θ_1, with a lifetime density ψ_{hyd}, and ADP release at θ, with a lifetime density $\psi_{off,ADP}$, the lifetime density of these combined events, $\psi_{hyd\text{-}off}$, is given as:

$$\psi_{hyd-off}(\theta\,|\theta_o) = \int_\theta^\theta \psi_{off,ADP}(\theta\,|\theta_1)\psi_{hyd}(\theta_1\,|\theta_0)d\theta_1. \tag{10}$$

Since $dp(\theta\,|\theta_0)/dt$ can be calculated from the rate equation $\omega\,(\theta|\theta_0)/d\theta = -\text{k}(\theta)$ $dp(\theta\,|\theta_0)$, hence:

$$p(\theta\,|\theta_0) = \exp\left[-(1/\omega)\int_{\theta_0}^\theta k(\theta')d\theta'\right]. \tag{11}$$

For a pair of events during rotation in the hydrolysis direction, the lifetime density for hydrolysis is given by:

$$\psi_{hyd}(\theta \mid \theta_0) = -dp_{hyd}(\theta \mid \theta_0)/d\theta$$
$$= [\mathrm{k}_{hyd}(\theta)/\omega]\exp\left[-(1/\omega)\int_{\theta_0}^{\theta} k_{hyd}(\theta')d\theta'\right]. \tag{12}$$

An expression similar to Equation 12 applies to the composite events during rotation in the synthesis direction.

Using the expressions for lifetime distributions of composite events, the rate constant of intermediate "hidden" process can be extracted from the binding and release events, monitored in the controlled rotation experiments. For the detailed steps of the procedure we refer to [Volkán-Kacsó and Marcus, 2017].

4.2. *Predictive calculation of lifetime distributions*

We consider as an example a calculation of $\psi_{hyd\text{-}off}$ in an angular window where hydrolysis and synthesis rates are known from stalling experiments. In the calculations shown in Figure 6 we also use the fluorescent ATP and ADP release rate constants in this window reported in the controlled rotation experiments. In order to extract the angle-dependent hydrolysis rate constant data a functional form is used for ADP release rate from that follow the rate constants reported in the controlled rotation experiments.

5. Conclusions and Outlook

In conclusion we have illustrated how an elastic group transfer theory can be used to treat single molecule experiments using magnetic tweezers: stalling and controlled rotation. To do so we first reviewed the relations for the rate constants of substeps such as ATP binding or ADP release provided using the theory to treat and predict the stalling and controlled rotation data in previous studies [Volkán-Kacsó and Marcus, 2015, 2016, 2017].

In the present work, we reviewed methods that can be applied to controlled rotation experiments in two ways. (1) From the lifetime distributions of the long time binding events the hydrolysis or synthesis rate constants *versus* rotor angle data can be extracted. (2) Lifetime distribution can be predicted using hydrolysis (or synthesis) rate constants from stalling experiments and these predictions can be tested against distributions estimated from the controlled rotation experiments.

In recent advanced imaging assay a high-speed, microsecond resolution, monitoring of rotation was demonstrated using gold nanorods as imaging probes, instead of the "traditional" microbeads [Spetzler *et al.*, 2009]. These methods permitted the study of the angular velocity of the rotor during the transition between dwells (timescales of μs). While in this article we focused on the stepping kinetics found in lower time-resolution experiments and the angle-dependent rate constant data extracted from the dwell times, more will undoubtedly be learned about the physiologically relevant function of the F-ATPase from developing a dynamical model for the short timescale of these "fast" transitions.

Acknowledgments

It is a real pleasure to contribute to this volume honoring a longtime friend and colleague, Bertil Andersson, in recognition of his seminal contributions to science and to the administration of science. This work was supported by the Office of the Naval Research, the Army Research Office, and the James W. Glanville Foundation.

References

Adachi, K., Oiwa, K., Nishizaka, T., Furuike, S., Noji, H., Itoh, H., Yoshida, M. and Kinosita, K., Jr. (2007). Coupling of rotation and catalysis in F1-ATPase revealed by single molecule imaging and manipulation, *Cell*, 130, 309–321.

Adachi, K., Oiwa, K., Yoshida, M., Nishizaka, T. and Kinosita, K., Jr. (2012). Controlled rotation of the F1-ATPase reveals differential and continuous binding changes for ATP synthesis, *Nat. Comm.*, 3, 1022.

Adachi, K., Yasuda, R., Noji, H., Itoh, H., Harada, Y.S., Yoshida, M. and Kinosita, K., Jr. (2000). Stepping rotation of F1-ATPase visualized through angle-resolved single-fluorophore imaging, *Proc. Natl. Acad. Sci. USA*, 97, 7243–7247.

Boyer, P.D. (1993). The binding change mechanism for ATP synthase — some probabilities and possibilities, *Biochim. Biophys. Acta*, 1140, 215–250.

Braig, K., Menz, R.I., Montgomery, M.G., Leslie, A.G. and Walker, J.E. (2000). Structure of bovine mitochondrial F1-ATPase inhibited by Mg2+ADP and aluminium fluoride, *Structure*, 8, 567–573.

Chidsey, C.E.D. (1991). Free energy and temperature dependence of electron transfer at the metal-electrolyte interface, *Science*, 251, 919–922.

Junge, W. and Nelson, N. (2015). ATP Synthase, *Annu. Rev. Biochem.*, 84, 631–657.

Lewis, E.S. and Hu, D.D. (1984). Methyl transfers. 8. The Marcus equation and transfers between aerosulfonates, *J. Am. Chem. Soc.*, 106, 3292–3296.

Marcus, R.A. (1960). Discussion comment on mixed reaction-diffusion controlled rates, *Discuss. Farad. Soc.*, 29, 129.

Marcus, R.A. (1968). Theoretical relations among rate constants, barriers, and Brønsted slopes of chemical reactions, *J. Phys. Chem.*, 72, 891–899.

Marcus, R.A. and Sutin, N. (1985). Electron transfers in chemistry and biology, *Biochim. Biophys. Acta*, 811, 265–322 (and references cited therein).

Marcus, R.A. (1993). In *Les Prix Nobel*, Frangsmyr, T., ed. Nobel Foundation Almqvist and Wiksell, Stockholm, Sweden), p. 63.

Martin, L.J., Ishmukhametov, R., Hornung, T., Ahmad, Z. and Frasch, W.D. (2014). Anatomy of F1-ATPase powered rotation, *Proc. Natl. Acad. Sci. USA*, 111, 3715–3720.

Masaike, T., Mitome, N., Noji, H., Muneyuki, E., Yasuda, R., Kinosita, K., Jr. and Yoshida, M. (2000). Rotation Of F1-ATPase and the hinge residues of the β subunit, *J. Exp. Biol.*, 203, 1–8.

Masaike, T., Koyama-Horibe, F., Oiwa, K., Yoshida, M. and Nishizaka, T. (2008). Cooperative three-step motions in catalytic subunits of F1-ATPase correlate with 80 degrees and 40 degrees substep rotations, *Nat. Struct. Mol. Biol.*, 15, 1326–1333.

Murdoch, J.R. (1983). A simple relationship between empirical theories for predicting barrier heights of electron-transfer, proton-transfer, atom-transfer, and group transfer reactions, *J. Am. Chem. Soc.*, 105, 2159–2164.

Nishizaka, T., *et al.* (2004). Chemomechanical coupling in F1-ATPase revealed by simultaneous observation of nucleotide kinetics and rotation, *Nat. Struct. Mol. Biol.*, 11, 142–148.

Noji, H., Yasuda, R., Yoshida, M. and Kinosita, K., Jr. (1997) *Nature*, 386, 299.

Noyes, R.M. (1961). Effects of diffusion rates on chemical kinetics, *Progr. React. Kinet.*, 1, 129–60.

Oster, G. and Wang, H. (2000). Reverse engineering a protein: The mechanochemistry of ATP synthase, *Biochim. Biophys. Acta.*, 1458, 482–510.

Panke, O., Cherepanov, D.A., Gumbiowski, K., Engelbrecht, S. and Junge, W. (2001). Viscoelastic dynamics of actin filaments coupled to rotary F-ATPase: Angular torque profile of the enzyme. *Biophys. J.*, 81, 1230–1233.

Sakaki, N., Shimo-Kon, R., Adachi, K, Itoh, H., Furuike, Muneyuki, E.H., Yoshida, M. and Kinosita, K., Jr. (2005). One rotary mechanism for F1-ATPase over ATP concentrations from millimolar down to nanomolar, *Biophys. J.*, 88, 2047–2056.

Senior, A.E. (2007). ATP synthase: Motoring to the finish line, *Cell*, 130, 220–221.

Sielaff, H., Rennekamp, H., Engelbrecht, S. and Junge, W. (2008). Functional halt positions of rotary F_0F_1-ATPase correlated with crystal structures, *Biophys. J.*, 95, 4979–4987.

Sielaff, H., Martin, J., Singh, D., Biukovic, G., Grueber, G. and Frasch, W.D. (2016). Power stroke angular velocity profiles of archaeal A-ATP synthase *versus* thermophilic and mesophilic F-ATP synthase molecular motors, *J. Biol. Chem.*, 291, 25351–25363.

Spetzler, D., Ishmukhametov, R., Hornung, T., Day, L.J., Martin, J. and Frasch, W.D. (2009). Single molecule measurements of F1-ATPase reveal an interdependence between the power stroke and the dwell duration, *Biochem.*, 48, 7979.

Sugawa, M., Okazaki, K.-I., Kobayashi, M., Matsui, T., Hummer, G., Masaike, T. and Nishizaka, T. (2016). F1-ATPase conformational cycle from simultaneous single-molecule FRET and rotation measurements, *Proc. Natl. Acad. Sci. USA*, 113, E2916–E2924.

Sutin, N. (1966). The kinetics of organic reactions in solution, *Annu. Rev. Phys. Chem.*, 17, 119–172.

Uchihashi, T., Iino, R., Ando, T. and Noji, H. (2011). High-speed atomic force microscopy reveals rotary catalysis of rotorless F1-ATPase H, *Science*, 333, 755–758.

Volkán-Kacsó, S. (2014). Two-state theory of binned photon statistics for a large class of waiting time distributions and its application to quantum dot blinking, *J. Chem. Phys.*, 140, 224110.

Volkán-Kacsó, S. and Marcus, R.A. (2015). Theory for rates, equilibrium constants, and Brønsted slopes in F1-ATPase single molecule imaging experiments, *Proc. Natl. Acad. Sci. USA*, 112, 14230–14235.

Volkán-Kacsó, S. and Marcus, R.A. (2016). Theory of controlled rotation experiments, predictions, tests and comparison with stalling experiments in F1-ATPase, *Proc. Natl. Acad. Sci. USA*, 113, 12029–12034.

Volkán-Kacsó, S. and Marcus, R.A. (2017). Long binding events in single molecule controlled rotation in F1-ATPase: Theory and experiment, *Proc. Natl. Acad. Sci. USA*, 114, 7272–7277.

Volkán-Kacsó, S. and Marcus, R.A. (2017b). Free, stalled, and controlled rotation single molecule experiments on F1-ATPase and their relationships, in *Photosynthesis and Bioenergetics* (J. Barber and A.V. Ruban, eds.), World Scientific Publishing Co., Singapore, pp. 35–53.

Volkán-Kacsó, S. and Marcus, R.A. (2017c). "What can be learned about the enzyme ATPase from single molecule studies of its subunit F1?" *Qart. Rev. Biophys.*, 50, e14.

Walker, J.E. (2013). The ATP synthase: The understood, the uncertain and the unknown, *Biochem. Soc. Trans.*, 41, 1–16

Watanabe, R., Iino, R. and Noji, H. (2010). Phosphate release in F1-ATPase catalytic cycle follows ADP release, *Nat. Chem. Biol.*, 6, 814–820.

Watanabe, R., Okuno, D., Sakakihara, S., Shimabukuro, K., Iino, R., Yoshida, M. and Noji, H. (2012). Mechanical modulation of catalytic power on F1-ATPase, *Nat. Chem. Biol.*, 8, 86.

Watanabe, R. and Noji, H. (2014). Timing of inorganic phosphate release modulates the catalytic activity of ATP-driven rotary motor protein, *Nat. Commun.*, 5, 3486.

Weber, J. and Senior, A.E. (1997). Catalytic mechanism of F1-ATPase. *Biochim. Biophys. Acta*, 1319, 19–58.

Weber, J. (2010). Structural biology: Toward the ATP synthase mechanism, *Nat. Chem. Biol.*, 6, 794–795.

Yasuda, R., Noji, H., Yoshida, M., Kinosita, K., Jr. and Itoh, H. (2001). Resolution of distinct rotational substeps by submillisecond kinetic analysis of F1-ATPase, *Nature*, 410, 898–904.

Zimmermann, B., Diez, M., Zarrabi, N., Gräber, P. and Börsch, M. (2005). Movements of the e-subunit during catalysis and activation in single membrane-bound H^+-ATP synthase, *EMBO*, 24, 2053–2063.

Chapter 3

Redox- and Light-Driven Hydration Dynamics in Biological Energy Transduction

Ville R. I. Kaila

Department of Chemistry, Technical University of Munich
Lichtenbergstr. 4, Garching, D-85747, Germany
ville.kaila@ch.tum.de

Biological energy transduction is catalyzed by proteins that electrochemically charge up biological membranes, powered by a chemical reaction or light-capturing process. Recent atomic structures of the respiratory enzymes, the photosynthetic machinery, as well as many light-driven ion pumps, provide together with biophysical and computational experiments a starting point for deriving molecular mechanisms of these fascinating proteins. Based on computational and experimental data, I propose here that general molecular mechanisms are employed to catalyze key charge separation processes in biological energy transduction. An electrostatic perturbation at the redox- or light-activated center is suggested to lead to coupled conformational and hydration changes that in turn modulate the barriers for the proton/ion transfer processes. The mechanistic model is discussed in context of four central energy-transducing proteins: the respiratory complex I (NADH: ubiquinone oxidoreductase), complex IV (cytochrome *c* oxidase), a light-driven sodium pump (KR2), and photosystem II.

1. Introduction

All biological systems require an input of energy that needs to be captured and stored. To this end, biological systems have evolved specialized membrane-bound protein complexes that transduce chemical and light energy into energy forms that the cell can use to thermodynamically drive endergonic processes.

The vast majority of the energy-transducing proteins operate by catalyzing a coupled transfer of protons and electrons or so-called proton-coupled electron transfer (PCET) reactions [Mitchell, 1961, 2011; Hammes-Schiffer and Stuchebrukhov, 2010; Kaila *et al.*, 2010]. The protons and electrons can be transferred in a stepwise manner (proton followed by electron or *vice versa*), or concertedly as a hydrogen atom ($H^{\bullet}$) or hydride ion (H^{-}), between the same or different donor and acceptor sites. These elementary PCET reactions are employed to create an electrochemical proton/ion gradient across a biological membrane that subsequently is used to thermodynamically drive energy-requiring processes in the cell [Mitchell, 1961; Yoshida *et al.*, 2001].

The bacterial and mitochondrial respiratory chains, together with the photosynthetic machinery of cyanobacteria, algae, and plants, form the main PCET chains that power the energy flow in nature. During the last decades, advances in X-ray crystallography and cryo-electron microscopy (cryo-EM) have elucidated molecular structures of respiratory [Baradaran *et al.*, 2013; Fiedorczuk *et al.*, 2016; Zhu *et al.*, 2016; Agip *et al.*, 2018; Iwata *et al.*, 1995, 1998; Lange *et al.*, 2001; Tsukihara *et al.*, 2003] and photosynthetic enzymes [Ferreira *et al.* 2004; Umena *et al.* 2011; Suga *et al.* 2015]. These structures, together with bacterial light-driven ion pumps [Luecke *et al.* 1999; Kato *et al.*, 2012, 2015] provide a blueprint for deriving a molecular understanding of biological energy conversion.

Atomic structures are central for deriving functional principles of biological energy transduction, but this also requires a detailed energetic and kinetic understanding of intermediate steps involved in the catalytic cycle. X-ray free-electron laser experiments (XFEL) have in recent years provided exciting data on "time-ordered" structures, particularly for light-activated proteins [Young *et al.*, 2016; Fuller *et al.*, 2017; Suga *et al.*, 2017]. Moreover, methods of computational biochemistry provide a powerful complementary approach to probe mechanistic principles of these proteins [Kaila *et al.*, 2018b]. In these approaches, the laws of classical or quantum mechanics are employed to probe the energetic and dynamic feasibility of different reaction mechanisms based on experimentally resolved static snapshots of the protein. Multi-scale molecular simulations allow to establish a connection between the structure and function that can be further tested and validated experimentally.

Based on insights obtained in recent years from a combination of structural, biophysical, and computational data, I suggest here that energy-transducing proteins employ common molecular principles to establish an electrochemical ion gradient across a membrane. This process is suggested to involve electrostatic transitions that create electric fields, which further trigger pK_a-, hydration, and conformational changes. Relaxation of the transiently stored electrostatic and strain energy is suggested to result in the charge translocation process. This chapter is ordered as follows: after reviewing general features of the proposed model, the principles are tested by quantum mechanical calculations, followed by discussion on the structure, function and mechanistic principles of complex I, cytochrome *c* oxidase, light-driven ion pumps, and photosystem II in context of the proposed model.

2. Physical Principles of Redox- and Light-Driven Charge Transport

2.1. *The electrostatic foundations*

Energy-transducing proteins operate at the interface between a low dielectric lipid membrane ($\varepsilon = 2$–4) and a high dielectric ($\varepsilon = 80$) aqueous environment. A redox- or protonation change, or a light-capturing process in the buried active site of such proteins, creates an electrostatic perturbation in the protein environment that induces an electrostatic potential (δV) and electric field ($\delta \boldsymbol{E}$) perturbation at a distance $\boldsymbol{r}$ from the site that are proportional to,

$$\delta V = k\ \delta q/\varepsilon \boldsymbol{r} \tag{1a}$$

$$\delta \boldsymbol{E} = k\ \delta q/\varepsilon \boldsymbol{r}^2 \tag{1b}$$

where δq is the charge inducing the perturbation, ε is the dielectric constant of the medium, that describes how much the charge is screened by the medium, and $\boldsymbol{r}$ is the distance between the charge and the point of interest. The constant, $k = 1/4\pi\varepsilon_0 = 332$ kcal mol^{-1} Å^{-1} or 14.4 V Å, when $\boldsymbol{r}$ is expressed in units of Å. The electric field is defined as the gradient of the electrostatic potential,

$$\boldsymbol{E}(\boldsymbol{r}) = -\nabla V(\boldsymbol{r}). \tag{1c}$$

The electric field can be considered as an *electric force* that a charged particle experiences at a position $\boldsymbol{r}$.

The protein interior often behaves as a low dielectric medium with $\varepsilon = 4$–10, which implies that a buried charge is 10–20 times more strongly screened in bulk water as compared to the protein interior. This also implies that a charge perturbation 10–15 Å away from the active site can induce electrostatic changes that are comparable to the strength of a hydrogen bond, *ca.* 3 kcal mol^{-1}, or redox/pK_a-shifts of *ca.* 140 mV/2 p*K*-units (Figure 1B). Such electrostatic shifts can trigger protonation or redox changes even at the surface of the protein. The distance dependence of the electrostatic potential and field is illustrated in Figure 2B. Although this simple picture applies to many systems, it is also important to note that the dielectric constant can also have a distance dependence, $\varepsilon(\boldsymbol{r})$ [Bonthuis *et al.*, 2011].

Charging up the protein medium has an energy cost that is comparable to the electric field strength,

$$U = 1/2\ \varepsilon\ \varepsilon_0\ |\boldsymbol{E}|^2. \tag{2}$$

In order to minimize the energy of the system, ions or charged residues will move into the electric field with a force that is proportional to the electric field,

$$\boldsymbol{F} = q\boldsymbol{E} = m\boldsymbol{a}. \tag{3}$$

This leads to an acceleration, $\boldsymbol{a}$, that is proportional to the charge (q) and mass (m) of the moving particles.

The electric field will also affect molecular dipoles ($\boldsymbol{\mu} = q\boldsymbol{d}$, where $\boldsymbol{d}$ is the distance separating the charges, $\pm q$, in the dipole), in particular water molecules that have a significant dipole moment of 1.85 D (1 Debye $\approx 3.33 \times 10^{-30}$ C m). If the electric field is uniform, it will not exert a net force on the dipole, but instead lead to a torque (τ) that orients the dipoles according to the field lines to minimize the overall electrostatic potential energy (U) of the system,

$$\tau = \boldsymbol{\mu} \times \boldsymbol{E} \tag{5a}$$

$$U = -\boldsymbol{\mu} \bullet \boldsymbol{E} \tag{5b}$$

Similarly, the electrostatic energy for placing a charge in an electric potential V is,

$$U = qV. \tag{5c}$$

A water molecule in an electric field of 1 V Å^{-1} that arises a few ångströms away from a perturbing unit charge, is thus expected based on Eq. 5b to experience an energy difference of up to *ca.* 18 kcal mol^{-1} between the parallel/antiparallel orientations of the dipole with respect the electric field. It is important to note, however, that the dipole itself also creates an electric field of its own that opposes the external field,

$$\boldsymbol{E} = k\ (-\boldsymbol{\mu}/\boldsymbol{r}^3 + 3\ (\boldsymbol{\mu} \bullet \boldsymbol{r})/\boldsymbol{r}^5). \tag{6}$$

Non-uniform electric fields that arise, *e.g.*, from a redox-active group will pull in dipoles (and charges) with a force that is proportional to the field gradient,

$$F = (\mu \bullet \nabla E). \tag{7}$$

2.2. *The electrostatic model applied to biological charge-separation*

Figure 1 illustrates these fundamental electrostatic principles applied to a generic model of an energy-transducing protein. The charge perturbation in a buried active site induces a shift in the electrostatic potential (Eq. 1a) and electric field (Eq. 1b) within the protein environment, pulling in water molecules (Eq. 7) and ions (Eq. 3) towards the perturbing charge. In order for this to take place, the surrounding helices must have some dynamic flexibility to account for the hydration changes. The water array itself provides a pre-requisite for transferring protons, which diffuse along the water molecules by a Grotthuss mechanism [Agmon *et al.*, 1995]. The ordered water chain and transferred ion oppose the electric field arising from the perturbing charge, and the subsequent charge recombination reaction closes the conduction channel. In classical redox-loops, the

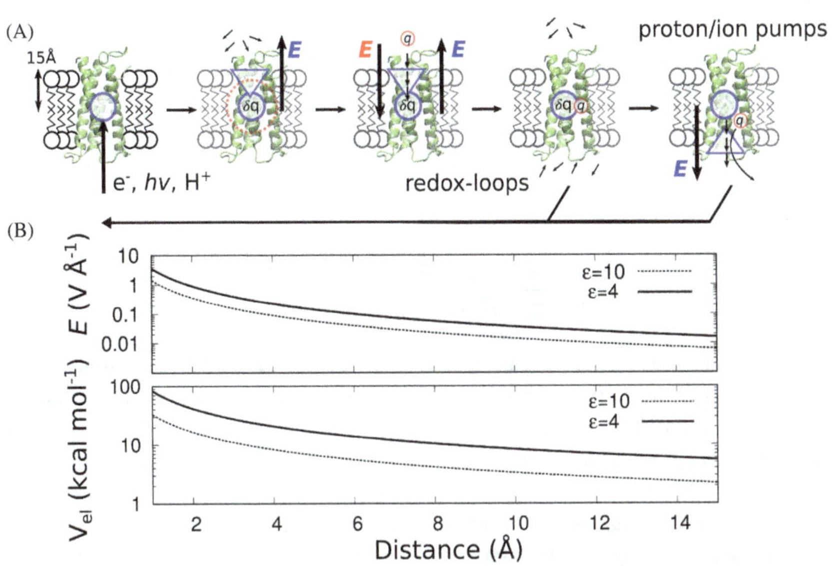

Figure 1. (A) Molecular principles for electrostatically-induced hydration transitions applied to biological charge-separation. (i) A redox-, light-absorption, or protonation change creates a charge perturbation, δq, in the buried protein interior. This gives rise to an electric field and potential perturbation (shown in panel B) on the surface of the protein. The lower panel illustrates the interaction of a unit charge with that potential (Eq. 5c).

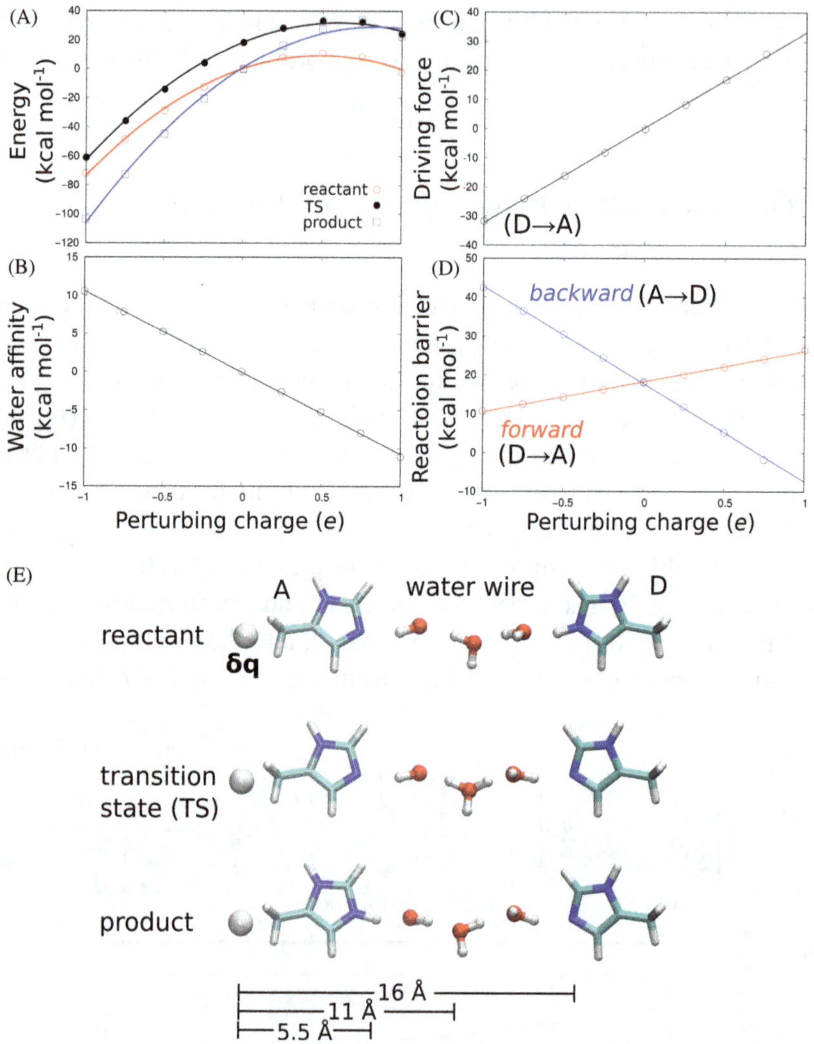

Figure 2. Quantum chemical DFT calculations (B3LYP-D3/def2-TZVP) illustrating the electrostatically-induced hydration effects on a redox-coupled proton transfer reaction. The model system, (shown in panel E), consists of a proton donor (D), a water wire, a proton acceptor (A), and a perturbing redox center (δq). The figure shows the structures of the reactant, transition state (TS), and product states. (A) Relative energies of the reactant, TS, and product states as a function of the charge perturbation. (B) Stability of water wire as a function of the charge perturbation, (C) The chemical driving force and (D) the forward and backward reaction barriers for the proton transfer as a function of the charge perturbation. The field-induced shift in geometry has been omitted in these calculations for simplicity. The figure was prepared using VMD [Humphrey *et al,* 1996]. From: Kaila *et al.*, 2018c.

proton recombines directly with the electron in the active site [Mitchell, 1961, 1991], whereas in redox- and light-driven proton pumps, the direct charge re-combination can be prevented by kinetic gates (see below). In these proton/ion pumps, relaxation of the perturbing charge couples to ejection of the ion across the membrane, by similar field-induced principles as discussed above.

2.3. *The molecular model*

The basic physical principles of the charge separation model are illustrated with quantum chemical density functional theory (DFT) calculations in Figure 2. The calculations show how a charge perturbation (δq) modulates the energetics of a proton transfer reaction between a proton donor (D) and an acceptor site (A). The donor site is placed 16 Å away from the perturbing charge (δq) and it connects *via* a water array to the acceptor site (A), located next to the redox-active group. Figure 2A shows how the redox site perturbs the energy levels of the reactant state, the transition state, and the product state with a parabolic dependence on the perturbing charge. The dependence of the chemical driving force (Figure 2C) and the reaction barrier (Figure 2D) on the perturbing charge are shown in Figures 2C and 2D, respectively. Interestingly, upon reduction of the redox center ($\delta q = 0 \rightarrow -1$) the forward reaction barrier (D→A) decreases from *ca.* 15 kcal mol^{-1} to 10 kcal mol^{-1}, which changes the reaction rate from milliseconds to microseconds according to transition state theory. The calculations also predict that the backward reaction barrier becomes chemically inaccessible due to the large product state stabilization (Figure 2C). The stability of the water array, which provides a pre-requisite for the proton transfer reaction, increases linearly with the perturbing charge, with a stabilization up to 10 kcal mol^{-1} relative to the oxidized state. The calculations thus suggest that the redox reaction leads to an electrostatic effect that (*a*) favors formation of a water wire connecting the proton donor with the proton acceptor, (*b*) lowers the reaction barrier for the forward proton transfer reaction, (*c*) increases the reaction barrier for the backward (leakage) reaction, and (*d*) increases the thermodynamic driving force for the reaction. These simple physical principles are discussed below in context of four complex energy-transducing proteins.

3. Long-Range Proton-Coupled Electron Transfer in Respiratory Complex I

3.1. *Structure and function of complex I*

Complex I or NADH:ubiquinone oxidoreductase is a redox-driven proton pump that reduces ubiquinone (Q) to ubiquinol (QH_2), and employs the free energy to

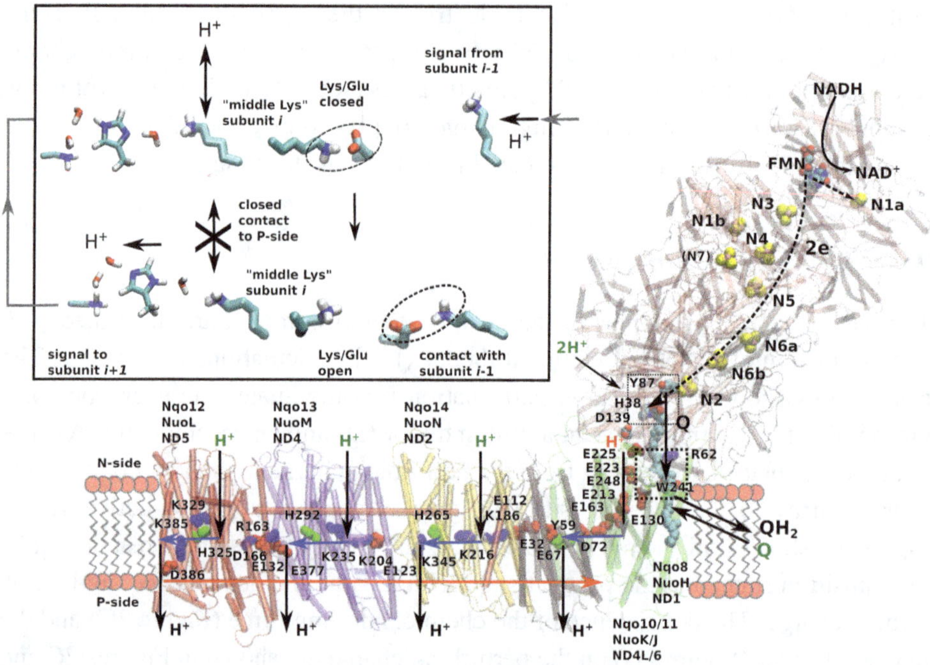

Figure 3. Structure and function of complex I. Electron transfer from NADH along a wire of FeS centres leads to reduction of quinone to quinol in the hydrophilic domain of complex I. The QH_2 moves to a second binding site (thicker dotted box) within the membrane domain that could activate the proton pumping module in Nqo8/10/11, Nqo14, Nqo13, Nqo12 by forward (blue arrow) and backward (red arrow) signal propagation. Protons and Q taken up for the next reaction cycle are marked in green. *Inset*: opening of the Lys/Glu ion-pair triggers proton transfer from the "middle Lys" to the terminal edge of each antiporter-like subunit (see main text).

translocate four protons across the mitochondrial inner membrane or bacterial cytoplasmic membrane (Figure 3) [Wikström, 1984; Brandt, 2006; Hirst, 2013; Sazanov, 2015; Kaila, 2018]. The electron transfer reaction takes place in the hydrophilic arm of the enzyme along a 100 Å chain of iron-sulfur (FeS) centers that bridges the NADH site with Q [Ohnishi, 1998; Sazanov and Hinchliffe, 2006; Verkhovskaya *et al.*, 2008]. The proton pumping module is located in the 200 Å wide membrane domain, comprising the antiporter-like subunits NuoL (Nqo12/ND5), NuoM (Nqo13/ND4), NuoN (Nqo14/ND2), and NuoH (Nqo8/ND1) (*E. coli/T. thermophilus/*bovine nomenclature).

The electron and proton transfer processes are tightly coupled, and complex I can also operate in reverse by oxidizing QH_2 to Q using a pH-gradient across the membrane, an operation mode that is important, *e.g.*, under anoxic and hypoxic conditions [Pryde *et al.*, 2011]. Interestingly, perturbation of the proton pumping elements at the far end of complex I by point mutations, reduces the quinone oxidoreductase activity

[Torres-Bacete *et al.*, 2007, 2009; Euro *et al.*, 2008, Nakamaru-Ogiso *et al.*, 2010; Michel *et al.*, 2011]. Although these observations are consistent with the principle of *microscopic reversibility*, it nevertheless remains puzzling how a tight coupling across such large distances is achieved at a molecular level.

As discussed above, electrostatic interactions are long-range due to their 1/distance decay, but if the coupling sites are separated by >15 Å, the pK_a-shifts are expected to be comparable to thermal fluctuations (Figure 1). It is therefore unlikely that direct electrostatic interactions between the Q reduction site and the terminal proton pumping elements could *alone* establish a tight coupling across the complete 200 Å protein framework. Electrostatic interactions, however, induce strong effects in their local environment (Figure 1), and a stepwise signal propagation could provide a basis for the long-range coupling effects, taken that the energy does not dissipate during the propagation process. Nevertheless, even with 5–10% dissipation across each coupling element distributed, *e.g.*, 10 Å apart from each other, the pumping signal could still be transmitted up to *ca.* 200 Å, which is comparable to the distance separating the Q-site from the terminal antiporter-like subunit, Nqo12. Mechanistic principles based on *electrostatic couplings* are often referred to as *indirect coupling* mechanisms, as these effects are transmitted through space rather than by direct structural elements.

Conformational changes can also transmit long-range effects, *e.g.*, by a structural shift on a helix that results in affinity changes on the other end of the helix. Such *direct coupling* models have also been suggested to provide a functional basis for proton pumping in complex I [Efremov and Sazanov, 2011; Baradaran *et al.*, 2013]. Complex I has a 100 Å long amphipathic lateral helix (the "*HL-helix*") that runs parallel to the membrane plane. It was suggested that this helix could function as a piston that pushes the proton pumping elements [Efremov and Sazanov, 2011; Baradaran *et al.* 2013]. However, no structural transitions have so far been experimentally observed in this helix, and site-directed mutagenesis experiments indicate that residues can be inserted without affecting the pumping activity, until a certain length threshold, after which complex I does not properly assemble [Belevich *et al.*, 2011; Zhu and Vik, 2015]. If the lateral helix is removed in molecular dynamics (MD) simulations, the antiporter like subunits partially dissociate, and complex I adopts a twisted shape. These findings suggest that the lateral helix might serve as a molecular clamp that secures a tight electrostatic coupling within the membrane domain (see below).

The mammalian complex I, nevertheless, undergoes subtle conformational changes that involve twisting and bending motions around the hydrophilic and membrane domains [Zhu *et al.*, 2016; Fiedorczuk *et al.*, 2016; Di Luca and Kaila, 2018]. It currently remains unclear if these motions are functionally important for the proton pumping process, as some complex I isoforms are suggested to be rather rigid [Parey *et al.*, 2018]. However, these twisting and bending modes have been linked to the *active* and *deactive* states in the mammalian enzyme that strongly modulate the activity [Vinogradov, 1998; Maklashina *et al.*, 2003].

3.2. *Coupled electrostatic, hydration, and conformational changes trigger proton pumping*

The long-range PCET reaction in complex I is initiated by transfer of two electrons and one proton between nicotinamide adenine dinucleotide (NADH) and a protein-bound flavin mononucleotide (FMN) cofactor, producing a $NAD^+/FMNH^-$ species [Saura and Kaila, 2018] at a high rate of up to 15,000 s^{-1} [Birrell *et al.*, 2009, 2013]. The two electrons enter the FeS chain, and are stepwise transferred to the terminal iron sulfur center N2 (Figure 3), which has a 120–180 mV-higher potential than the other FeS centers at *ca.* –320 mV [Verkhovskaya *et al.*, 2008, 2013], with each FeS center functioning as a one-electron acceptor [Ohnishi, 1998]. The N2 cluster transfers electrons to the quinone, which is bound *ca.* 20–30 Å above the membrane plane [Hunte *et al.*, 2010] in a narrow cavity, *ca.* 10 Å away from the terminal N2 center [Baradaran *et al.*, 2013]. The quinone headgroup forms contacts with the conserved Tyr-87 and His-38 residues of Nqo4 (*Thermus thermophilus* numbering) [Tocilescu *et al.*, 2010; Baradaran *et al.*, 2013] that function as proton donors in the Q-reduction process to quinol (QH_2 or QH^-, see Sharma *et al.*, 2015; Gamiz-Hernandez *et al.*, 2017]. Protonation of the reduced quinone, weakens the ion-pair between His-38 and Asp-139, and leads to its dissociation in molecular simulations (Sharma *et al.*, 2015; Warnau *et al.*, 2018). This charge re-distribution in the Q site (δq) effectively propagates to the membrane domain by a chain of conserved charged residues (Glu/Asp; Arg) within Nqo8, and triggers conformational changes in a region of conserved glutamates, the "E-quartet" region [Sharma *et al.*, 2015]. These conformational changes attract water molecules and protons from the N-side of the membrane by the electrostatic principles discussed in Section 2. Conformational and hydration changes in the Nqo8 and Nqo10 subunits are supported by both MD simulations [Sharma *et al.*, 2015; Haapanen and Sharma, 2017, Di Luca *et al.*, 2017; Saura and Kaila 2018b] as well as in different experimental structures of complex I [Baradaran *et al.*, 2013; Zickermann *et al.*, 2015; Agip *et al.*, 2018; Parey *et al.*, 2018].

Thermodynamically, Q is at a low potential in its binding site near Tyr-87/His-38, possibly close to –300 mV for both the Q/SQ and SQ/QH_2 redox-pairs [Verkhovskaya and Wikström, 2014; Gamiz-Hernandez, *et al.*, 2017], suggesting that no energy is transduced upon reduction of Q by electron transfer from NADH

(–320 mV). However, the redox potential of Q in the membrane is +90 mV [Brandt, 2006], which implies that the major energy transduction step could take place upon motion of the ubiquinol towards the membrane [Wikström *et al.*, 2015; Kaila *et al.*, 2018; Warnau *et al.*, 2018]. Molecular simulations suggest [Warnau *et al.*, 2018] that ubiquinol is destabilized in the active site, which pushes it towards the membrane domain on μs-ms timescale. We recently identified a second Q-binding site within Nqo8, in a region surrounded by many aromatic and charged residues [Warnau *et al.*, 2018]. When ubiquinol reaches this site, it could induce local conformational changes, which pushes the proton onward along Nqo8 towards the Nqo14 interface. Notably, conformational changes have recently been observed in cryo-EM structures [Agip *et al.*, 2018] and biophysical experiments [Cabrera-Orefice *et al.*, 2018] in nearby regions. Taken together, these findings suggest that motion of the quinone to the second binding site could trigger the proton pumping process.

The three antiporter-like subunits, Nqo14, Nqo13, and Nqo12, each contain repeating structural elements with a conserved Lys/Glu ion-pair, a middle Lys residue, which bridges *via* a histidine residue to a titratable residue (Lys in Nqo14/12, Asp in Nqo13) (Figure 3). The middle Lys and the terminal charged residue are located on broken helix elements (TM7 and TM12) and their protonation states have been suggested to regulate the hydration level and protonic connectivity to the N- and P-sides of the membrane [Di Luca *et al.*, 2017, 2018b], following the electrostatic principles outlined in section 2. The broken helix elements are dynamically flexible and allow them to undergo subtle conformational changes required for hydration transitions.

Protonation of the terminal residues in Nqo10 could trigger dissociation of the initial Lys/Glu ion-pair in Nqo14 that electrostatically destabilizes the protonated middle lysine residue (Lys-216_{14}) [Kaila, 2018]. This in turn lowers the barrier for the proton transfer reaction (Di Luca *et al.*, 2018b) and pushes the proton onwards to the terminal edge of Nqo14. Deprotonation of the middle lysine weakens the electric field directed towards the N-side, destabilizing the water connectivity to the N-side. This dehydration effect could provide a kinetic gate that traps the released proton from leaking to the wrong side of the membrane. Protonation of the terminal residue further opens the subsequent ion-pair in Nqo13 that propagates by similar principles onwards to the terminal edge of Nqo12.

This "*forward pulse*" could charge up the terminal edges of Nqo10, Nqo14, Nqo13, and Nqo12, with four protons. To this end, MD simulations [Kaila *et al.*, 2014, Di Luca *et al.*, 2017] suggest that the terminal edge of Nqo12 is close to the P-side of the membrane with well-hydrated connectivity for the proton to exit across the membrane. Based on the principle of *microscopic reversibility*, release of the proton stored at the terminal edge of Nqo12 triggers protonation of the middle Lys and closing of the Lys/Glu ion pair in Nqo12. The proton uptake and ion-pair closing, destabilize the proton stored at the terminal edge of Nqo13 that is ejected to the P-side, followed by re-protonation of the middle lysine and closing of the Lys/Glu ion pair within Nqo13. This charge relaxation could subsequently release the protons at the Nqo14 and Nqo10 subunit by similar processes, completing the "*reverse pulse*" [Kaila, 2018]. To complete the pumping cycle, the formed quinol must further exit complex I and exchange with an oxidized species, either directly or indirectly [*cf.* Wikström *et al.*, 2015]. Moreover, the protons lost in the initial charge propagation cascade must be taken up from the N-side of the membrane for complex I to enter the next catalytic cycle. Although exact mechanistic details of these processes still remain unclear, the electrostatic-coupled hydration model satisfies key features of the long-range proton pumping in complex I.

4. Redox-Coupled Proton-Pumping in Cytochrome *c* Oxidase

4.1. *Structure and function of redox-driven proton pumping in CcO*

Cytochrome *c* oxidase (C*c*O) is a redox-driven proton pump that reduces oxygen (O_2) to water (H_2O) and employs the free energy to pump protons across the membrane [Wikström, 1977; Kaila *et al.*, 2010, Wikström *et al.*, 2015]. The reaction cycle of C*c*O is initiated by binding of molecular oxygen to the binuclear heme a_3/Cu_B center (BNC) in the R → A step (Figure 4), leading to splitting of the dioxygen bond by transfer of four electrons and one proton from the heme a_3 ($Fe^{II} \rightarrow Fe^{IV}$), Cu_B, ($Cu^{I} \rightarrow Cu^{II}$), and a redox-active tyrosine (Tyr-244, YOH → YO$^{\bullet}$, *Bos taurus* numbering), which is covalently bound to the active site. The oxygen splitting reaction produces two deprotonated water molecules ($O^{2-} + OH^{-}$) in the A → P_M step. In the remaining reaction cycle, the electrons donated in this initial reaction are stepwise transferred back to the active site by the soluble cytochrome *c via* the Cu_A center and a low-spin heme *a*. Each electron transfer reaction between the electron-queuing heme *a* and the BNC couples to transfer of a *pumped proton* across the membrane, and a *chemical proton* that re-protonates the BNC. The overall PCET-reactions are expected to be similar in each reaction step ($P_M \rightarrow F$, $F \rightarrow O_H$, $O_H \rightarrow E_H$, and $E_H \rightarrow R$, Figure 4B), with the exception of a switch in the proton uptake channel (see below). Experiments suggest [Belevich *et al.*, 2007] that before the pumped proton is ejected across the membrane, it is transiently stored at a proton loading site (PLS) near the propionic groups of heme a_3. C*c*O employs the D- and K-channels, named after the conserved Asp-61 and Lys-316 residues, for proton uptake [Lee *et al.*, 2012] (Figure 4). All pumped protons originate from the

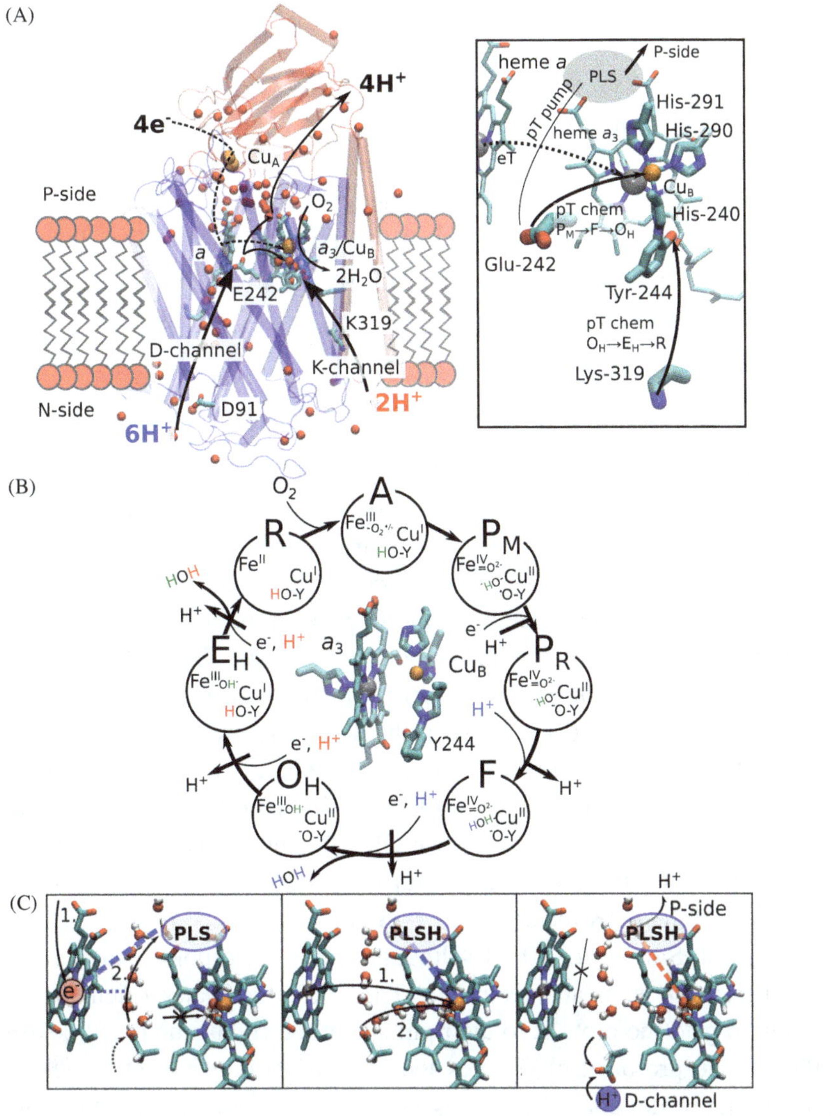

Figure 4. (A) Structure and function of C*c*O. *Inset*: proton transfer reaction along pumping and chemical pathways. All pumped protons are transferred *via* the D-channel, chemical protons are taken up from the K-channel in the $O_H \rightarrow E_H$ and $E_H \rightarrow R$ steps. (B) The catalytic cycle of C*c*O. Chemical protons taken up from the D- and K-channels are marked in blue and red, respectively. (C) PCET-reaction in the active site of C*c*O. *Left*: 1. Reduction of the heme *a* polarizes the proton loading site (PLS), and lowers the proton transfer barrier from Glu-242 to the PLS. The water contacts between Glu-242 and the BNC are rather weak due to the directed electric field originating from heme *a*. Glu-242 is rapidly re-protonated from the D-channel. *Middle*: Protonation of the PLS triggers (1) electron transfer to BNC, producing an electric field that (2) attracts *both* water molecules and the proton to the BNC. *Right*: The affinity of the PLS proton is lowered due to BNC reduction and protonation. The PLS proton is ejected across the membrane favoured by the oxidized heme *a*, rapid re-protonation, and/or flipping of Glu-242.

D-channel that ends at Glu-242 [Iwata *et al.*, 1995], which shuttles protons both to the PLS and the BNC. Interestingly, C*c*O switches proton uptake channels after transfer of the two initial electrons to the BNC. While the pumped protons continue to be transferred along the D-channel, the chemical protons are taken up in the $O_H \rightarrow E_H$, and $E_H \rightarrow R$ steps from the K-channel, with Lys-319 as an intermediate proton donor [Vygodina *et al.*, 1998] (Figure 4A, *inset*). While many aspects of the pumping cycle are well understood, it still remains unclear how the protons are *gated* to the right place at the right time, and what prevents the protons from *leaking* backwards during pumping process.

4.2. *Electrostatically-driven proton transfer dynamics*

Reduction of the electron-queuing heme *a* neutralizes the electric charge within the heme macrocycle, $[Fe^{III}\ Porph]^{+1} \rightarrow [Fe^{II}\ Porph]^{0}$, that lowers the proton transfer barrier between Glu-242 and the proton loading site (PLS) (Figure 4C) [Blomberg and Siegbahn, 2012; Goyal *et al.*, 2015; Supekar *et al.*, 2016; Liang *et al.*, 2017]. The exact location of the PLS still remains unclear, but it must be electrostatically strongly coupled to both heme *a* and the BNC. To this end, the D- and A-propionates of heme a_3, a nearby histidine residue, as well as a structurally conserved water cluster between the propionic groups, have been suggested as possible candidates [Popovic *et al.*, 2005; Kaila *et al.*, 2011; Lu and Gunner, 2014; Supekar and Kaila, 2016]. Molecular simulations suggest that reduction of heme *a* favors water contacts between Glu-242 and the D-propionate of heme a_3, whereas the contacts between the oxidized BNC and Glu-242 are weaker due to electric field effects [Kaila *et al.*, 2010; Sharma *et al.*, 2015; Supekar *et al.*, 2016; 2018, *cf.* also Son *et al.*, 2017]. Although the water dipoles experience a strong orientational force in an electric field (Eq. 5), the water orientation *alone* is unlikely to gate the protons to the right site at the right time [*cf.* Wikström *et al.*, 2003], as the proton itself creates locally even stronger electric fields (Eq. 6). However, the water contact provides a central pre-requisite for the proton transfer between Glu-242 and the BNC/PLS, which are located about 10–15 Å from the glutamate [Wikström *et al.*, 2003; Kaila *et al.*, 2010; Sharma *et al.*, 2015; Liang *et al.*, 2016; *cf.* also Pisliakov *et al.*, 2008]. Breaking and forming the water contacts between the "*chemical*" and "*pump*" sites could provide an important element of a molecular gate that transfers the protons to the right site at the right time. Notably, the proton dissociation barrier from Glu-242 is also expected to depend on the hydration state of the non-polar cavity [Kaila *et al.*, 2010, Liang *et al.*, 2016].

Protonation of the PLS by Glu-242 increases the redox potential of the BNC that leads to an electron transfer from heme *a* to the BNC on nanosecond timescales

[Jasaitis *et al.*, 2005]. The electron transfer is favored by a low reorganization energy [Kaila *et al.*, 2010b; Tipmanee and Blumberger, 2012] and a nearly isoenergtic driving force [Marcus and Sutin, 1985], whereas the heme groups are almost at *van-der-Waals* contact, making their electronic coupling strong.

Reduction of the BNC produces an electric field that stabilizes the water contact between Glu-242 and Cu_B (Figure 4C) [Kaila *et al.*, 2010; Sharma *et al.*, 2015; Supekar *et al.*, 2016; 2018], consistent with the proposed electrostatic model. Son *et al.* [2017] recently suggested that protonation of the D-propionate also increases the hydration state of the non-polar cavity above Glu-242. The electric field originating from the reduced BNC stabilizes the hydrogen-bonded water contacts between Glu-242 and Cu_B, which also lowers the proton transfer barrier between the two sites. The recombination of the electron and the proton weakens the electric field, that in turn, lowers the affinity of the proton stored at the PLS. At this reaction step, it is central that the PLS proton does not recombine with the transiently deprotonated Glu-242. This may be achieved by the net positive charge of the oxidized heme *a* that increases the barrier for the backward proton transfer reaction [Supekar *et al.*, 2017]. Moreover, the deprotonated Glu-242 has a tendency to flip towards the D-channel [Kaila *et al.*, 2007, but *cf.* Yang and Cui, 2011; Woelke *et al.*, 2013; Samudino *et al.*, 2016], a conformation that is also observed in all available X-ray structures of C*c*O [Iwata *et al.*, 1996; Tsukihara *et al.*, 2003]. The anionic Glu-242 is expected to collect both water molecules and protons from the well hydrated D-channel, favoring its rapid re-protonation. Molecular simulations also suggest that the opening of the ion-pair between the D-propionate of heme a_3 and Arg-438 is sensitive to the redox state of heme *a* [Wikström *et al.*, 2005; Kaila *et al.*, 2011], which could provide an additional gating element that prevents charge recombination.

When the BNC in reduced in the P and F states, Glu-242 forms stable hydrogen-bonded contacts with the oxygenous ligand of Cu_B [Kaila *et al.*, 2010; Sharma *et al.*, 2015; Supekar *et al.*, 2017, 2018]. However, after formation of a weakly bound aquo-ligand on Cu_B in the O_H state (Figure 4), the contact between the glutamate and the Cu_B becomes weaker [Supekar and Kaila, 2018]. The H_2O-ligand is likely to dissociate away from Cu_B in the O_H state [Sharma and Wikström, 2013], further weakening the hydrogen-bonded contact between Glu-242 and Cu_B [Supekar and Kaila, 2018]. DFT calculations show that reduction of the Cu_B in the $O_H \rightarrow E_H$ transition produces an electric field on Tyr-244 due to the special electronic equilibrium between $Cu^{II/I}$ and $YO^{-}/^{\bullet}$ [Kaila *et al.*, 2009; Sharma and Wikström, 2013; Blomberg and Siegbahn, 2015; Supekar and Kaila, 2018], an effect that leads to a side-chain flip of Lys-319 towards the BNC [Lepp *et al.*, 2008]. This electric field increases the hydration state of the K-channel, and lowers the proton transfer barrier to the BNC [Supekar and Kaila, 2018], suggesting that the electrostatically-induced hydration effect could also be responsible for activation of the K-channel.

Although the molecular architecture of C*c*O is very different from complex I, it is interesting to note that the fundamental mechanistic similarities that arise from the electrostatic perturbation-model, could drive hydration and protonation dynamics in C*c*O.

5. Light-Driven Ion Pumping in Microbial Rhodopsins

A large class of bacterial light-driven ion-pumps transforms light energy into an electrochemical ion gradient, powered by a retinal chromophore that is embedded within a trans-membrane (TM) helical environment [Ernst *et al.*, 2014]. Since detailed structural elucidation of the proton-pumping bacteriorhodopsin [Luecke *et al.*, 1999], other members of the rhodopsin family that function as outward proton pumps or inward chloride pumps have been identified [Kato *et al.*, 2012, 2015]. The recently discovered Na^+-pumping rhodopsin (KR2) [Kato *et al.*, 2015; Gushchin *et al.*, 2015] is particularly interesting from a bioenergetics perspective, as the positively charged sodium ion is transferred towards the positively charged retinal Schiff base. In the following, I will discuss some electrostatically-driven conformational and hydration changes in context of KR2 [Suomivuori *et al.*, 2017], but similar principles could also govern ion translocation in other light-gated ion-pumps [Lórenz-Fonfría *et al.*, 2015].

In all rhodopsin proteins, photoexcitation of the retinal transfers positive charge from the Schiff base towards the β-ionone unit of the chromophore [Zhou *et al.*, 2014 and references therein]. In KR2, this results in a photo-isomerization of the *all*-trans retinal to the 13-*cis* form that further triggers proton transfer between the Schiff base and a nearby conserved carboxylate, Asp-116 [Kato *et al.*, 2015]. These electrostatic perturbations lead to conformational changes in the TM helices, that expose an electric field on the surface of the protein [Suomivuori *et al.*, 2017], consistent with the electrostatic model. The process is facilitated by conformational changes in a conserved buried leucine residue, Leu-120, that allows water molecules and the Na^+ ion to move towards the buried protein interior.

Site-directed mutagenesis experiments [Inoue *et al.*, 2015] and molecular simulations [Suomivuori *et al.*, 2017] suggest that KR2 employs three ion-binding sites in the pumping process (Figure 5): *Site* 1 is located close to the intracellular side and contains several serine residues, *site* 2 is located at the center of the membrane plane, close to Asp-116 and the Schiff base, and *site* 3 is located next to Asp-251 on the other side of the retinal before the exit to the extracellular side of the membrane. The electrostatic attraction triggered by the photoexcitation and

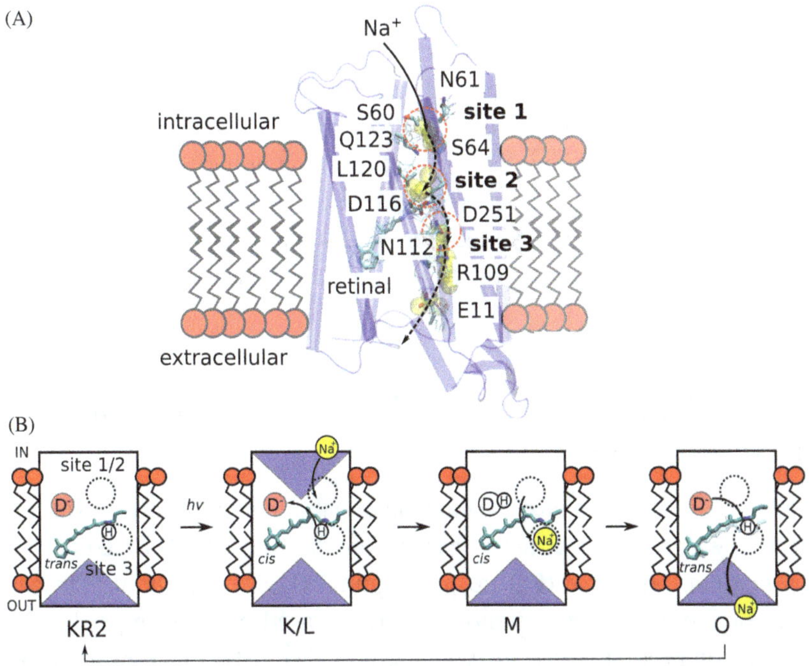

Figure 5. Structure and function of the sodium-pumping rhodopsin, KR2. (A) Putative sodium binding *sites* 1–3 and key residues involved in the ion transport. Average Na^+ positions during MD simulations are marked in yellow spheres. (B) Putative mechanism: The photoexcitation of the dark state (KR2) leads to retinal isomerization, proton transfer from the retinal to D116 (D^-), and sodium uptake to the primary binding site(s) by electrostatically-induced hydration and conformational changes. Protonation of D116 destabilizes the sodium ion, which is pushed to *site* 3. Retinal re-protonation and re-isomerization ejects the sodium across the membrane. Water contacts between intra- and extracellular sides are marked with blue triangles. Adapted from [Suomivuori *et al.*, 2017].

subsequent conformational changes, allows the Na^+ ion to move from the intracellular side of the membrane to *site* 1 and onwards to *site* 2. The strain energy stored in the initial K/L intermediate (Figure 5), destabilizes the protonated Schiff base, which in turn triggers proton transfer from the retinal to Asp-116. Neutralization of Asp-116 and the retinal, switches off the electric field, which in turn dehydrates the contact to the intracellular side in MD simulations, and closes the conduction channel. Similar to the deprotonation of the "middle Lys" in complex I, this could provide a gating element that prevents the ion from leaking. The charge re-distribution destabilizes the Na^+ at *site* 2 and pushes the ion towards *site* 3 on the other extracellular side of the retinal. This process is likely to take place in the M intermediate that has a strongly blue-shifted retinal due to its deprotonation [Inoue *et al.*, 2015; Bogachev *et al.*, 2016]. Upon formation of the O intermediate, the proton is transferred from Asp-116 back to the Schiff-base, which re-isomerizes to the *all*-trans form. This destabilizes the Na^+ ion at *site* 3 and leads to release of the ion to the extracellular side by opening of a conserved Arg-109/Glu-11 ion-pair.

The electrostatic perturbations and subsequent hydration effects in KR2 have thus also interesting resemblance to the proton pumping steps in complex I and cytochrome *c* oxidase. Although only the sodium pumping KR2 is discussed here, recent FTIR experiments [Lórenz-Fonfría *et al.*, 2015] suggest that ion transport in the related channelrhodopsin and possibly other rhodopsins is also driven by hydration changes, supporting the general mechanistic features in light-driven ion-transport.

6. Redox-Coupled Substrate Water Deprotonation in Photosystem II

Photosystem II is the only enzyme capable of catalyzing light-driven splitting and oxidation of water. The electrons released from the chlorophyll-based photochemistry reduce plastoquinone (PQ) to plastoquinol (PQH_2), providing the basis for synthesis of biomass, whereas dioxygen is released as a "side-product" in the water oxidation process by the enzyme's Mn_4O_5Ca cluster that powers respiratory chains in all aerobic life forms [Cardona *et al.* 2012].

The light-absorption process is initiated at the dimeric P680 chlorophyll center that decays into a charge-separated $P_{D1}/P_{D2}{}^{+/-}$ radical pair. The electron is transferred along the D1 subunit of the protein along chlorophyll and pheophytin pigments (Chl_{D1}, $Pheo_{D1}$) *via* plastoquinone, PQ_A to PQ_B, which is stepwise reduced to plastoquinol (PQH_2) (Figure 6). The highly oxidizing $P680^{+/\bullet}$ is re-reduced *via* a redox active tyrosine, Tyr-161 (Y_Z), which in turn, oxidizes the Mn_4O_5Ca center in the four distinct S states ($S_0 \rightarrow S_4$), each one triggered by a light flash. The dioxygen release takes place in the $S_4 \rightarrow S_0$ transition [Cardona *et al.*, 2012; Cox *et al.*, 2013, Dau *et al.*, 2012], most likely by a radical coupling mechanism [Siegbahn, 2006; 2013; Ugur *et al.*, 2016], although acid-base mechanisms have also been suggested [Messinger *et al.*, 1995; *cf.* also Vinyard *et al.*, 2015; Barber, 2017 and *refs*. therein]. However, the exact chemical identity of the substrate waters and the proton acceptor sites involved still remain unclear. The recently discovered chlorophyll *f*-containing photosystems, provide new exciting PSII-isoforms that enable us to understand the eneregtic and kinetic boundaries of oxygenic photosynthesis [Nürnberg *et al.*, 2018].

Recent experiments [Nilsson *et al.*, 2014] suggest that the substrate water molecules are already bound in the so-called S_2 state, which is obtained by a light flash from the dark stable S_1 state. The S_2 state has both open and closed cubane isoforms, with the O5 oxo-ligand connected to Mn1 or Mn4, favoring the low-spin (S = 1/2) and high-spin (S = 5/2) states, respectively [Boussac *et al.*, 1996; Pantazis *et al.*, 2012; Bovi *et al.*, 2013] (Figure 6). The S_2 state comprises three

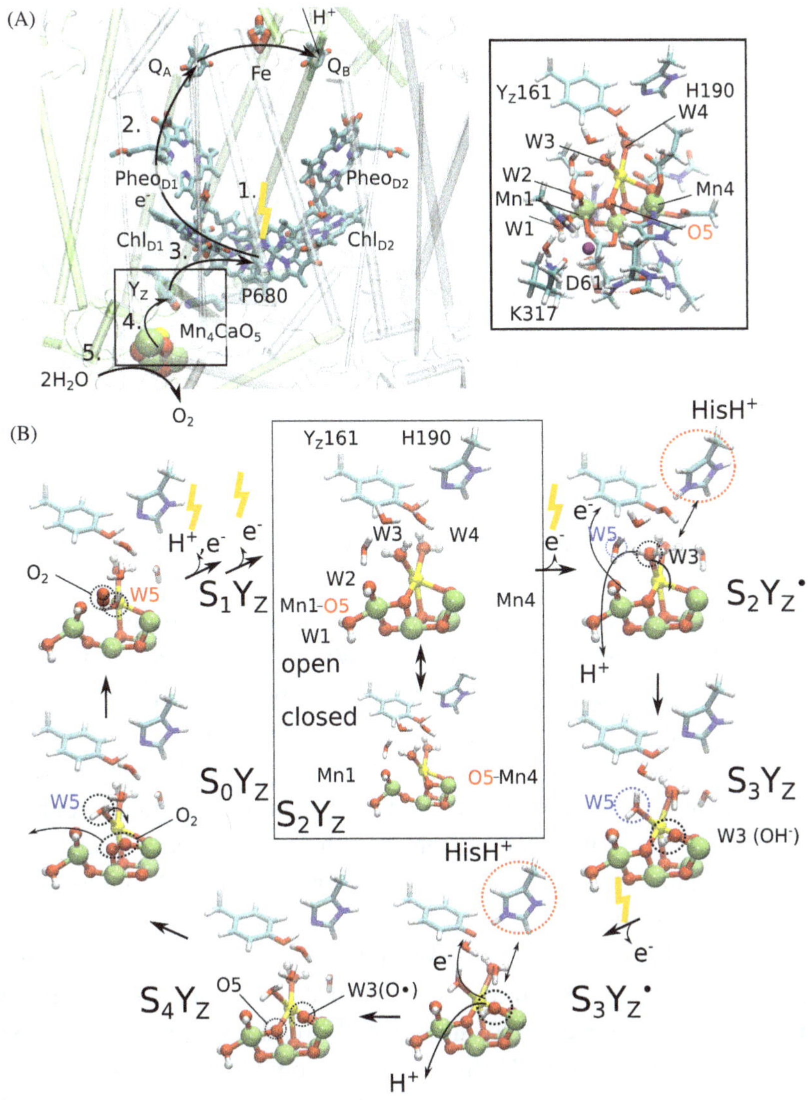

Figure 6. (A) Structure and function of PSII. Photoexcitation of P680 leads to 1. charge separation and 2. electron transfer towards Q_B. 3. Y_z re-reduces $P680^{+\bullet}$, and 4. the Mn_4CaO_5 re-reduces Y_z. 5. After four light flashes, O_2 is produced. (B) Putative mechanism of water splitting and oxidation in PSII. Formation $Y_Z^{\bullet}/HisH^+$ produces a perturbing charge/electric field (red dotted circle) that deprotonates W3, which moves between Mn4 and O5. This motion is accompanied by binding of W5 to Ca. Further deprotonation of W3 produces an oxyl radical that forms the O–O bond. W5 replaces the O_2 molecule, forming a new O5-oxo ligand upon deprotonation.

Mn^{IV} centers and one Mn^{III} on Mn1 (open form) or Mn4 (closed form). Moreover, the Mn4-bound water molecule W2 is likely to be in a hydroxide state, whereas the other metal-bound water molecules (W1, W3, W4) are in their H_2O form [Cox *et al.*, 2013].

Tyr-161 functions as oxidant for Mn_4O_5Ca and an electron donor for $P680^{+/\bullet}$. Oxidation of Tyr-161 produces a neutral tyrosyl radical ($Y_Z\bullet$) by proton transfer from Tyr-161 to His-190 (Figure 6b). The positive charge on the imidazole ring induces an electric field that upshifts the pK_a of cluster-bound water molecules. To this end, DFT calculations [Ugur *et al.*, 2016] suggest that the calcium-bound W3 undergoes a strong pK_a shift upon $Y_Z\bullet$ formation that could trigger its deprotonation *via* a water array connecting to Asp-61 (Figure 6). The deprotonation reaction destabilizes W3, which upon formation of the S_3 state moves between Mn1 and O5, when all metal centers take the Mn^{IV} form. Recent EPR experiments [Cox *et al.*, 2014] support a symmetrical ligand structure on all Mn centers in the S_3 state. The PCET-coupled motion is also supported by two independent time-resolved X-ray structures of photosystem II [Young *et al.*, 2016; Suga *et al.*, 2017] in which electron density accumulates at the site predicted by the DFT calculations [Ugur *et al.*, 2016, *cf.* also Shoij *et al.*, 2015, Siegbahn, 2018]. The W3 involvement has recently also been supported by FTIR experiments [Kim *et al.*, 2017, 2018].

The motion of the W3 towards the Mn1 leads to a void coordination site at Ca, that is filled by a nearby water molecule (W5) according to our DFT models (Figure 6) [Ugur *et al.*, 2016], and FTIR experiments [Kim *et al.*, 2017]. An additional flash regenerates the $Y_Z^\bullet$ state, which similarly creates an electric field that polarizes the surrounding water molecules. The Mn1-bound W3 further donates its proton to Asp-61, which could be facilitated by rotation of ligands connected to Mn4 [Askerka *et al.*, 2015]. Deprotonation of the W3-hydroxide leads to an oxyl radical that can attack the O5 group with a low barrier of *ca.* 5 kcal mol^{-1}, forming a dioxygen bond [Ugur *et al.*, 2016]. A similar coupling mechanism was earlier suggested by Siegbahn [2006, 2013]. When the O_2 diffuses out from the Mn_4O_5Ca center, the calcium-bound W5 moves into the space between Mn1 and Mn4, forming a new O5 ligand upon deprotonation, which could result in a O5-hydroxide in the S_1 state [Suga *et al.*, 2015]. Diffusion of W5, couples to re-binding of new W3 that forms the substrate water molecule in the subsequent reaction cycle. These findings suggest that the calcium ion could serve as gate, controlling substrate water access to the active site of PSII. More generally, the electrostatic effects, triggered by the $Y_Z^\bullet/HisH^+$ formation could control the hydration and protonation dynamics in PSII, with a remarkable resemblance to the molecular principles governing other energy transducing proteins.

7. Conclusions

I have discussed here the electrostatic basis of redox- and light-driven proteins that are responsible for primary energy transduction in nature. Although the overall structure and function of these proteins are different, it is suggested that they employ similar electrostatically-induced hydration transitions to drive ion transport across the membrane. In the proposed model, a redox change or photoexcitation in the active site, triggers an electrostatic perturbation that results in hydration- and conformational changes in the protein surroundings, which in turn provide a conduction pathway for the charge separation process. The charge state of the active site is suggested to regulate the barrier for the proton/ion transfer process. Non-productive charge re-combination reactions are avoided by molecular gates that kinetically modulate the barriers for forward and backward reactions by coupled electrostatic and hydration changes. The general features of the model are illustrated by quantum mechanical density functional theory calculations. Although the thermodynamic and kinetic effects are over-estimated in these simplified models, the calculations suggest that water contact between proton donor and acceptor sites are significantly stabilized due to charge perturbation effects, which also lower the proton transfer barrier from milliseconds to microseconds. I have discussed four central energy-transducing proteins in terms of these simple principles: the *respiratory complex I* that reversibly employs quinone reduction to pump protons up to 200 Å from the active site; *cytochrome c oxidase*, which sorts both chemical and pumped protons based on redox-changes within its active site region; the *Na^+-pumping rhodopsin* KR2 that employs a light-capturing process to create an alternate access state for the sodium ion across the membrane; and *photosystem II*, where the chlorophyll-based photochemistry is employed to stepwise oxidize and deprotonate water molecules. All systems are suggested to operate using similar hydration and conformational transitions that are triggered by electrostatic perturbations.

Acknowledgments

I wish to thank Prof. Bertil Andersson for stimulating discussion at NTU in March 2018, and my research team for many insightful discussions during preparation of this manuscript. Our research is supported by the European Research Council (ERC) under the European Union's Horizon 2020 research and innovation program/grant agreement 715311. I also acknowledge the German Research Foundation, the Jane and Aatos Erkko Foundation, and the DAAD program for supporting our work. The European Supercomputing Centers, SuperMUC at the Leibniz-Rechenzentrum (LRZ) (projects: pr48de and pr27xu) and the CSC — The IT center for science, are also acknowledged for supporting our research by computational resources.

References and Citations

Agip, A.A., Blaza, J.N., Bridges, H.R., Viscomi, C., Rawson, S., Muench, S.P. and Hirst, J. (2018). Cryo-EM structures of complex I from mouse heart mitochondria in two biochemically defined states, *Nat. Struct. Mol. Biol.*, 25, 548–556.

Agmon, N. (1995). The Grotthuss mechanism, *Chem. Phys. Lett.*, 244, 456–462.

Askerka M., Vinyard, D.J., Brudvig, G.W. and Batista, V.S. (2015). NH_3 binding to the S2 state of the O_2-evolving complex of photosystem II: analogue to H_2O binding during the S2 → S3 transition, *Biochemistry*, 54, 5783–5610.

Baradaran, R., Berrisford, J.M., Minhas, G.S. and Sazanov, L.A. (2013). Crystal structure of the entire respiratory complex I, *Nature,* 494, 443–448.

Barber, J. (2017). A mechanism for water splitting and oxygen production in photosynthesis, *Nat. Plants*, 3, 17041.

Belevich, I., Bloch, D.A., Belevich, N., Wikström, M. and Verkhovsky, M.I. (2007). Exploring the proton pump mechanism of cytochrome *c* oxidase in real time. *Proc. Natl. Acad. Sci. USA*, 104, 2685–2690.

Belevich, G., Knuuti, J., Verkhovsky, M.I., Wikström, M. and Verkhovskaya, M. (2011). Probing the mechanistic role of the long α-helix in subunit L of respiratory complex I from *Escherichia coli* by site-directed mutagenesis, *Mol. Microbiol.*, 82, 1086–1095.

Birrell, J.A. and Hirst, J. (2013). Investigation of NADH binding, hydride transfer, and NAD^+ dissociation during NADH oxidation by mitochondrial complex I using modified nicotinamide nucleotides, *Biochemistry*, 52, 4048–4055.

Birrell, J.A., Yakovlev, G. and Hirst, J. (2009). Reactions of the flavin mononucleotide in complex I: a combined mechanism describes NADH oxidation coupled to the reduction of $APAD^+$, ferricyanide, or molecular oxygen, *Biochemistry*, 48, 12005–12013.

Bogachev, A.V., Bertsova, Y.V., Verkhovskaya, M.L., Mamedov, M.D. and Skulachev, V.P. (2016). Real-time kinetics of electrogenic Na(+) transport by rhodopsin from the marine flavobacterium *Dokdonia sp.* PRO95, *Sci. Rep.* 6, 21397.

Bovi, D., Narzi, D. and Guidoni, L. (2013). The S2 state of the oxygen-evolving complex of photosystem II explored by QM/MM dynamics: spin surfaces and metastable states suggest a reaction path towards the S3 state, *Angew. Chem. Int. Ed.*, 52, 11744–11749.

Boussac, A., Girerd, J.-J. and Rutherford, A.W. (1996). Conversion of the spin state of the manganese complex in photosystem II induced by near-infrared light, *Biochemistry*, 35, 6984–6989.

Boussac, A., Ugur, I., Marion, A., Sugiura, M., Kaila, V.R.I. and Rutherford, A.W. (2018). The low spin — high spin equilibrium in the S2-state of the water oxidizing enzyme, *Biochim. Biophys. Acta — Bioenerg.*, 1859342–1859356.

Blomberg, M.R.A. and Siegbahn, P.E.M. (2012). The mechanism for proton pumping in cytochrome *c* oxidase from an electrostatic and quantum chemical perspective, *Biochim. Biophys. Acta — Bioenerg.*, 1817, 495–505.

Blomberg, M.R.A. and Siegbahn, P.E.M. (2015). Protonation of the binuclear active site in cytochrome *c* oxidase decreases the reduction potential of Cu_B, *Biochim. Biophys. Acta — Bioenerg.*,1847, 1173–1180.

Brandt, U. (2006). Energy converting NADH: ubiquinone oxidoreductase (complex I), *Annu. Rev. Biochem.*, 75, 69–92.

Bonthuis, J.D, Gekle, S. and Netz, R. (2011). Dielectric profile of interfacial water and its effect on double-layer capacitance, *Phys. Rev. Lett.*, 107, 166102.

Cardona, T., Sedoud, A., Cox, N. and Rutherford, A.W. (2012). Charge separation in photosystem II: a comparative and evolutionary overview, *Biochim. Biophys. Acta — Bioenerg.*, 1817, 26–43.

Cabrera-Orefice, A., Yoga, E.G., Wirth, C., Siegmund, K., Zwicker, K., Guerrero-Castillo, S., Zickermann, V., Hunte, C. and Brandt, U. (2018). Locking loop movement in the ubiquinone pocket of complex I disengages the proton pumps, *Nat. Commun.*, 9, 4500.

Cox, N., Pantazis, D.A., Neese, F. and Lubitz, W. (2013). Biological water oxidation, *Acc. Chem. Res.*, 46, 1588–1596.

Cox, N., Retegan, M., Neese, F., Pantazis, D.A., Boussac, A. and Lubitz, W. (2014). Photosynthesis. Electronic structure of the oxygen-evolving complex in photosystem II prior to O–O bond formation, *Science*, 345, 804–808.

Dau, H., Zaharieva, I. and Haumann, M. (2012). Recent developments in research on water oxidation by photosystem II, *Curr. Opin. Chem. Biol.*, 16, 3–10.

Di Luca, A., Gamiz-Hernandez, A.P. and Kaila, V.R.I. (2017). Symmetry-related proton transfer pathways in respiratory complex I, *Proc. Natl. Acad. Sci. USA*, 114, E6314–E6321.

Di Luca, A., Mühlbauer, M.E., Saura, P. and Kaila, V.R.I. (2018). How inter-subunit contacts in the membrane domain of complex I affect proton transfer energetics, *Biochim. Biophys. Acta — Bioenerg.*, 1859, 734–741.

Di Luca, A. and Kaila, V.R.I. (2018) Global collective motions in the mammalian and bacterial respiratory complex I, *Biochim. Biophys. Acta — Bioenerg.*, 1859, 326–332.

Efremov, R.G. and Sazanov, L.A. (2011). Structure of the membrane domain of the respiratory complex I, *Nature,* 465, 414–420.

Euro, L., Belevich, G., Verkhovsky, M.I., Wikström, M. and Verkhovskaya, M. (2008) Conserved lysine residues of the membrane subunit NuoM are involved in energy conversion by the proton-pumping NADH: ubiquinone oxidoreductase (Complex I), *Biochim. Biophys. Acta — Bioenerg.*, 1777, 1166–1172.

Ernst, O.P., *et al.* (2014). Microbial and animal rhodopsins: structures, functions, and molecular mechanisms, *Chem. Rev.*, 114, 126–163.

Ferreira, K.N, Iverson, T.M., Maghlaoui, K., Barber, J. and Iwata, S. (2004). Architecture of the photosynthetic oxygen-evolving center, *Science*, 303, 1831–1838.

Fiedorczuk, K., Letts, J.A., Degliesposti, G., Kaszuba, K., Skehel, M. and Sazanov, L.A. (2016). Atomic structure of the entire mammalian mitochondrial complex I, *Nature*, 538, 406–410.

Fuller, F.D. *et al.* (2017). Drop-on-demand sample delivery for studying biocatalysts in action at X-ray free-electron lasers, *Nat. Meth.*, 14, 443–449

Gamiz-Hernandez, A.P., Jussupow, A., Johansson, M.P. and Kaila, V.R.I. (2017). Terminal electron-proton transfer dynamics in the quinone reduction of respiratory complex I, *J. Am. Chem. Soc.*, 139, 16282–16288.

Goyal, P., Yang, S. and Cui, Q. (2015). Microscopic basis for kinetic gating in cytochrome *c* oxidase: insights from QM/MM analysis, *Chem. Sci.*, 6, 826–841.

Gushchin, I., Shevchenko, V., Polovinkin, V., Kovalev, K., Alekseev, A., Round, E., Borshchevskiy, V., Balandin, T., Popov, A., Gensch, T., Fahlke, C., Bamann, C., Willbold, D., Büldt, G., Bamberg, E. and Gordeliy, V. (2015). Crystal structure of a light-driven sodium pump, *Nat. Struct. Mol. Biol.*, 22, 390–395.

Haapanen, O. and Sharma, V. (2017). Role of water and protein dynamics in proton pumping by respiratory complex I, *Sci. Rep.*, 7, 1–12.

Hammes-Schiffer, S. and Stuchebrukhov, A.A. (2010). Theory of coupled electron and proton transfer reactions, *Chem. Rev.*, 110, 6939–6960.

Hirst, J. (2013). Mitochondrial Complex I, *Annu. Rev. Biochem.*, 82, 551–575.

Humphrey, W., Dalke, A. and Schulten, K. (1996). VMD: visual molecular dynamics, *J. Mol. Graph.*, 14, 33–38.

Hunte, C., Zickermann, V. and Brandt, U. (2010). Functional modules and structural basis of conformational coupling in mitochondrial complex I, *Science,* 329, 448–451.

Inoue, K., Ono, H., Abe-Yoshizumi, R., Yoshizawa, S., Ito, H., Kogure, K. and Kandori, H. (2013). A light-driven sodium ion pump in marine bacteria, *Nat. Commun.*, 4, 1678.

Iwata, S., Lee, J.W., Okada, K., Lee, J.K., Iwata, M., Rasmussen, B., Link, T.A., Ramaswamy, S. and Jap, B.K. (1998). Complete structure of the 11-subunit bovine mitochondrial cytochrome *bc1* complex, *Science*, 281, 64–71.

Iwata, S., Ostermeier, C., Ludwig, B. and Michel, H. (1995). Structure at 2.8 A resolution of cytochrome *c* oxidase from *Paracoccus denitrificans*, *Nature*, 376, 660–669.

Jasaitis, A., Rappaport, F., Pilet, E., Liebl, U. and Vos, M.H. (2005). Activationless electron transfer through the hydrophobic core of cytochrome *c* oxidase, *Proc. Natl. Acad. Sci. USA*, 102, 10882–10886.

Kato, H.E., Zhang, F., Yizhar, O., Ramakrishnan, C., Nishizawa, T., Hirata, K., Ito, J., Aita, Y., Tsukazaki, T., Hayashi, S., Hegemann, P., Maturana, A.D., Ishitani, R., Deisseroth, K. and Nureki, O. (2012). Crystal structure of the channelrhodopsin light-gated cation channel, *Nature*, 482, 369–374.

Kato, H.E., *et al.* (2015). Structural basis for Na(+) transport mechanism by a light-driven Na(+) pump, *Nature,* 521, 48–53.

Kaila, V.R.I., Verkhovsky, M.I., Hummer, G. and Wikström, M. (2008). Glutamic acid 242 is a valve in the proton pump of cytochrome *c* oxidase, *Proc. Natl. Acad. Sci. USA*, 105, 6255–6259.

Kaila, V.R.I., Johansson, M.P., Sundholm, D., Laakkonen, L. and Wikström, M. (2009). The chemistry of the CuB site in cytochrome *c* oxidase and the importance of its unique His-Tyr bond, *Biochim. Biophys. Acta — Bioenerg.*, 1787, 221–233.

Kaila, V.R.I., Verkhovsky, M.I. and Wikström, M. (2010). Proton-coupled electron transfer in cytochrome oxidase, *Chem. Rev.*, 110, 7062–7081.

Kaila, V.R.I., Johansson, M.P., Sundholm, D. and Wikström, M. (2010b). Interheme electron tunneling in cytochrome *c* oxidase, *Proc. Natl. Acad. Sci. USA,* 107, 21470–21475.

Kaila, V.R.I., Sharma, V. and Wikström, M. (2011). The identity of the transient proton loading site of the proton-pumping mechanism of cytochrome *c* oxidase, *Biochim. Biophys. Acta — Bioenerg.*, 1807, 80–84.

Kaila, V.R.I., Wikström, M. and Hummer, G. (2014). Electrostatics, hydration and proton transfer dynamics in the membrane domain of respiratory complex I, *Proc. Natl. Acad. Sci. USA*, 111, 6988–6993.

Kaila, V.R.I. (2018). Long-range proton-coupled electron transfer in biological energy conversion: towards mechanistic understanding of respiratory complex I, *J. R. Soc. Interface*, 15, 20170916.

Kaila, V.R.I. (2018b). Multi-scale molecular simulations on respiratory complex I. In Mechanisms of Primary Energy Transduction in Biology (Ed. M Wikström), London, UK: the Royal Society of Chemistry, pp. 81–103.

Kaila, V.R.I. *et al.* (2018c). In preparation.

Kim, C.J., Bao, H., Burnap, R.L. and Debus, R.J. (2018). Impact of D1-V185 on the water molecules that facilitate O_2 formation by the catalytic Mn_4CaO_5 cluster in photosystem II, *Biochemistry*, 57, 4299–4311.

Kim, C.J. and Debus, R.J. (2017). Evidence from FTIR difference spectroscopy that a substrate H_2O molecule for O_2 formation in photosystem II is provided by the Ca ion of the catalytic Mn_4CaO_5 cluster, *Biochemistry*, 56, 2558–2570.

Lange, C., Nett, J.H., Trumpower, B.L. and Hunte, C. (2001). Specific roles of protein-phospholipid interactions in the yeast cytochrome *bc1* complex structure, *EMBO J.*, 20, 6591–6600.

Lee, H.J., Reimann, J., Huang, Y. and Ädelroth, P. (2012). Functional proton transfer pathways in the heme-copper oxidase superfamily. *Biochim. Biophys. Acta — Bioenerg.*, 1817, 537–544.

Lepp, H., Svahn, E., Faxén, K. and Brzezinski, P. (2008). Charge transfer in the K proton pathway linked to electron transfer to the catalytic site in cytochrome *c* oxidase, *Biochemistry*, 47, 4929–4935

Lórenz-Fonfría, V.A., Bamann, C., Resler, T., Schlesinger, R., Bamberg, E. and Heberle J. (2015). Temporal evolution of helix hydration in a light-gated ion channel correlates with ion conductance, *Proc. Natl. Acad. Sci. USA,* 112E5796–E5804.

Lu, J. and Gunner, M.R. (2014). Characterizing the proton loading site in cytochrome *c* oxidase, *Proc. Natl. Acad. Sci. USA*, 111, 12414–12419.

Luecke, H., Schobert, B., Richter, H.T., Cartailler, J.P. and Lanyi, J.K. (1999). Structural changes in bacteriorhodopsin during ion transport at 2 angstrom resolution, *Science*, 286, 255–261.

Liang, R., Swanson, J.M., Peng, Y., Wikström, M. and Voth, G.A. (2016). Multiscale simulations reveal key features of the proton-pumping mechanism in cytochrome *c* oxidase, *Proc. Natl. Acad. Sci. USA*, 113, 7420–7425.

Maklashina, E., Kotlyar, A.B. and Cecchini, G. (2003). Active/de-active transition of respiratory complex I in bacteria, fungi and animals, *Biochim. Biophys. Acta — Bioenerg.*, 1606, 95–103.

Marcus, R.A. and Sutin, N. (1985). Electron transfers in chemistry and biology, *BBA Rev. Bioenerg.*, 811, 265–322.

Messinger, J., Badger, M. and Wydrzynski, T. (1995). Detection of one slowly exchanging substrate water molecule in the S3 state of photosystem II. *Proc. Natl. Acad. Sci. USA*, 92, 3209–3213.

Michel, J., DeLeon-Rangel, J., Zhu, S., Van Ree, K. and Vik, S.B. (2011). Mutagenesis of the L, M, and N subunits of complex I from *Escherichia coli* indicates a common role in function, *PLoS One*, 6(2), e17420.

Mitchell, P. (2011). Chemiosmotic coupling in oxidative and photosynthetic phosphorylation, *Biochim. Biophys. Acta — Bioenerg.*, 1807, 1507–1538.

Mitchell, P. (1961). Coupling of phosphorylation to electron and hydrogen transfer by a chemi-osmotic type of mechanism, *Nature,* 191, 144–148.

Nakamaru-Ogiso, E., Kao, M.C., Chen, H., Sinha, S.C., Yagi, T. and Ohnishi, T. (2010). The membrane subunit NuoL (ND5) is involved in the indirect proton pumping mechanism of Escherichia coli Complex I, *J. Biol. Chem.*, 285, 39070–39078.

Nilsson, H., Krupnik, T., Kargul, J. and Messinger, J. (2014). Substrate water exchange in photosystem II core complexes of the extremophilic red alga *Cyanidioschyzon merolae*, *Biochim. Biophys. Acta — Bioenerg.*, 1837, 1257–1262.

Nürnberg, D.J., Morton, J., Santabarbara, S., Telfer, A., Joliot, P., Antonaru, L.A., Ruban, A.V., Cardona, T., Krausz, E., Boussac, A., Fantuzzi, A. and Rutherford, A.W. (2018). Photochemistry beyond the red limit in chlorophyll f-containing photosystems, *Science*, 360, 1210–1213.

Ohnishi, T. (1998). Iron-sulfur clusters/semiquinones in complex I, *Biochim. Biophys. Acta — Bioenerg.*, 1364, 186–206.

Parey, K., Brandt, U., Xie, H., Mills, D.J., Siegmund, K., Vonck, J., Kühlbrandt, W. and Zickermann, V. (2018). Cryo-EM structure of respiratory complex I at work, *Elife*, 7, pii: e39213.

Pantazis, D.A., Ames, W., Cox, N., Lubitz, W. and Neese, F. (2012). Two interconvertible structures that explain the spectroscopic properties of the oxygen-evolving complex of photosystem II in the S2 state, *Angew. Chem. Int. Ed.*, 51, 9935–9940.

Pisliakov, A.V., Sharma, P.K., Chu, Z.T., Haranczyk, M. and Warshel, A. (2008). Electrostatic basis for the unidirectionality of the primary proton transfer in cytochrome *c* oxidase, *Proc. Natl. Acad. Sci. USA*, 105, 7726–7731.

Popović, D.M., Quenneville, J. and Stuchebrukhov, A.A. (2005). DFT/electrostatic calculations of pK(a) values in cytochrome *c* oxidase, *J. Phys. Chem. B*, 109, 3616–3626.

Pryde, K.R. and Hirst, J. (2011). Superoxide is produced by the reduced flavin in mitochondrial complex I: a single, unified mechanism that applies during both forward and reverse electron transfer, *J. Biol. Chem.*, 286, 18056–18065.

Samudio, B.M., Couch, V. and Stuchebrukhov, A.A. (2016). Monte Carlo simulations of glu-242 in cytochrome *c* oxidase, *J. Phys. Chem. B*, 120, 2095–2105.

Saura, P. and Kaila, V.R.I. (2018). Energetics and dynamics of NADH-driven proton-coupled electron transfer in respiratory complex I, *(in review).*

Saura, P. and Kaila, V.R.I. (2018b). Molecular dynamics and structural models of the cyanobacterial NDH-1 complex, *Biochim. Biophys. Acta — Bioenerg., (in press).*

Sazanov, L.A. (2015). A giant molecular proton pump: structure and mechanism of respiratory complex I, *Nat. Rev. Mol. Cell. Biol.*, 16, 375–388.

Sazanov, L.A. and Hinchliffe, P. (2006). Structure of the hydrophilic domain of respiratory complex I from *Thermus thermophilus*, *Science,* 311, 1430–1536.

Sharma, V., Karlin, K.D. and Wikström, M. (2013). Computational study of the activated O_H state in the catalytic mechanism of cytochrome *c* oxidase, *Proc. Natl. Acad. Sci. USA*, 110, 16844–16849.

Sharma, V., Belevich, G., Gamiz-Hernandez, A.P., Róg, T., Vattulainen, I., Verkhovskaya, M.L., Wikström, M., Hummer, G. and Kaila, V.R.I. (2015). Redox-induced activation of the proton pump in the respiratory complex I, *Proc. Natl. Acad. Sci. USA*, 112, 11571–11576.

Sharma, V., Enkavi, G., Vattulainen, I., Róg, T. and Wikström, M. (2015b). Proton-coupled electron transfer and the role of water molecules in proton pumping by cytochrome *c* oxidase, *Proc. Natl. Acad. Sci. USA*, 112, 2040–2045.

Siegbahn, P.E.M. (2006). O–O bond formation in the S4 state of the oxygen evolving complex in photosystem II, *Chem. Eur. J.*, 12, 9217–9237.

Siegbahn, P.E.M. (2013). Water oxidation mechanism in photosystem II, including oxidations, proton release pathways, O–O bond formation and O_2 release, *Biochim. Biophys. Acta — Bioenerg.*, 1827, 1003–1019.

Siegbahn, P.E.M. (2018). The S2 to S3 transition for water oxidation in PSII (photosystem II), revisited, *Phys. Chem. Chem. Phys.*, 20, 22926–22931.

Shoji, M., Isobe, H. and Yamaguchi, K. (2015). QM/MM study of the S2 to S3 transition reaction in the oxygen-evolving complex of photosystem II, *Chem. Phys. Lett.*, 636, 172–179.

Son, C.Y., Yethiraj, A. and Cui, Q. (2017). Cavity hydration dynamics in cytochrome *c* oxidase and functional implications, *Proc. Natl. Acad. Sci. USA*, 114, E8830–E8836.

Suga, M., Akita, F., Hirata, K., Ueno, G., Murakami, H., Nakajima, Y., Shimizu, T., Yamashita, K., Yamamoto, M., Ago, H. and Shen, J.-R. (2015). Native structure of photosystem II at 1.95 Å resolution viewed by femtosecond X-ray pulses, *Nature*, 517, 99–103.

Suga, M., *et al.* (2017). Light-induced structural changes and the site of O=O bond formation in PSII caught by XFEL, *Nature*, 543, 131–135.

Supekar, S., Gamiz-Hernandez, A.P. and Kaila, V.R.I. (2016). A protonated water cluster as a transient proton-loading site in cytochrome *c* oxidase, *Angew. Chem. Int. Ed*,. 55, 11940–11944.

Supekar, S. and Kaila, V.R.I. (2018). Dewetting transitions coupled to K-channel activation in cytochrome *c* oxidase, *Chem. Sci.*, 9, 6703–6710.

Suomivuori, C.M., Gamiz-Hernandez, A.P., Sundholm, D. and Kaila, V.R.I. (2017). Energetics and dynamics of a light-driven sodium-pumping rhodopsin, *Proc. Natl. Acad. Sci. USA*, 114, 7043–7048.

Tocilescu, M.A., Fendel, U., Zwicker, K., Dröse, S., Kerscher, S. and Brandt, U. (2010). The role of a conserved tyrosine in the 49-kDa subunit of complex I for ubiquinone binding and reduction, *Biochim. Biophys. Acta — Bioenerg.*, 1797, 625–632.

Torres-Bacete, J., Nakamaru-Ogiso, E., Matsuno-Yagi, A. and Yagi, T. (2007). Characterization of the NuoM (ND4) subunit in *Escherichia coli* NDH-1: conserved charged residues essential for energy-coupled activities, *J. Biol. Chem.*, 282, 36914–36922.

Torres-Bacete, J., Sinha, P.K., Castro-Guerrero, N., Matsuno-Yagi, A. and Yagi, T. (2009). Features of subunit NuoM (ND4) subunit in *Escherichia coli* NDH-1: topology and implication of conserved Glu144 for coupling site 1, *J. Biol. Chem.*, 284, 33062–33069.

Tipmanee, V. and Blumberger, J. (2012). Kinetics of the terminal electron transfer step in cytochrome *c* oxidase, *J. Phys. Chem. B*, 116, 1876–1883.

Tsukihara, T., Shimokata, K., Katayama, Y., Shimada, H., Muramoto, K., Aoyama, H., Mochizuki, M., Shinzawa-Itoh, K., Yamashita, E., Yao, M., Ishimura, Y. and Yoshikawa, S. (2003). The low-spin heme of cytochrome *c* oxidase as the driving element of the proton-pumping process, *Proc. Natl. Acad. Sci. USA*, 100, 15304–15309.

Ugur, I., Rutherford, A.W. and Kaila, V.R.I. (2016). Redox-coupled substrate water reorganization in the active site of Photosystem II-The role of calcium in substrate water delivery, *Biochim. Biophys. Acta — Bioenerg.*, 1857, 740–748.

Umena, Y., Kawakami, K., Shen, J.-R. and Kamiya, N. (2011). Crystal structure of oxygen-evolving photosystem II at a resolution of 1.9 Å, *Nature*, 473, 55–60.

Verkhovskaya, M. and Bloch, D.A. (2013). Energy-converting respiratory complex I: on the way to the molecular mechanism of the proton-pump, *Int. J. Biochem. Cell. Biol.*, 45, 491–511.

Verkhovskaya, M.L., Belevich, N.P., Euro, L., Wikström, M. and Verkhovsky, M.I. (2008). Real-time electron transfer in respiratory complex I, *Proc. Natl. Acad. Sci. USA,* 105, 3763–3767.

Verkhovskaya, M. and Wikström, M. (2014). Oxidoreduction properties of bound ubiquinone in complex I from *Escherichia coli*, *Biochim. Biophys. Acta — Bioenerg.*, 1837, 246–250.

Vinogradov, A.D. (1998). Catalytic properties of the mitochondrial NADH-ubiquinone oxidoreductase (complex I) and the pseudo-reversible active/inactive enzyme transition, *Biochem. Biophys. Acta — Bioenerg.*, 1364, 169–185.

Vinyard, D.J., Khan, S. and Brudvig, G.W. (2015). Photosynthetic water oxidation: binding and activation of substrate waters for O–O bond formation, *Faraday Discuss.*, 185, 37–50.

Vygodina, T.V., Pecoraro, C., Mitchell, D., Gennis, R. and Konstantinov, A.A. (1998). Mechanism of inhibition of electron transfer by amino acid replacement K362M in a proton channel of Rhodobacter sphaeroides cytochrome *c* oxidase, *Biochemistry*, 37, 3053–3061.

Warnau, J., Sharma, V., Gamiz-Hernandez, A.P., Di Luca, A., Haapanen, O., Vattulainen, I., Wikström, M., Hummer, G. and Kaila, V.R.I. (2018). Redox-coupled quinone dynamics in the respiratory complex I, *Proc. Natl. Acad. Sci. USA,* 115, E8413–E8420.

Wikström, M.K.F. (1977). Proton pump coupled to cytochrome *c* oxidase in mitochondria, *Nature*, 266, 271–273.

Wikström, M. (1984). Two protons are pumped from the mitochondrial matrix per electron transferred between NADH and ubiquinone, *FEBS Lett.*, 169, 300–304.

Wikström, M., Verkhovsky, M.I. and Hummer, G. (2003). Water-gated mechanism of proton translocation by cytochrome c oxidase, *Biochim. Biophys. Acta — Bioenerg.*, 1604, 61–65.

Wikström, M., Ribacka, C., Molin, M., Laakkonen, L., Verkhovsky, M. and Puustinen, A. (2005). Gating of proton and water transfer in the respiratory enzyme cytochrome *c* oxidase, *Proc. Natl. Acad. Sci. USA,* 102, 10478–10481.

Wikström, M., Sharma, V., Kaila, V.R.I., Hosler, J.P. and Hummer, G. (2015). New perspectives on the proton pumping in cellular respiration, *Chem. Rev.*, 115, 2196–2221.

Woelke, A.L., Galstyan, G., Galstyan, A., Meyer, T., Heberle, J. and Knapp, E.W. (2013). Exploring the possible role of Glu286 in C*c*O by electrostatic energy computations combined with molecular dynamics, *J. Phys. Chem. B*, 117, 12432–12441.

Yoshida, M., Muneyoki, E. and Hisabori, T. (2001). ATP synthase — a marvellous rotary engine of the cell, *Nat. Rev. Mol. Cell. Biol.*, 2, 669–677.

Zhou, X., Sundholm, D., Wesołowski, T.A. and Kaila, V.R.I. (2014). Spectral tuning of rhodopsin and visual cone pigments, *J. Am. Chem. Soc.*, 136, 2723–2726.

Zhu, S. and Vik, S.B. (2015). Constraining the lateral helix or respiratory complex I by cross-linking does not impair enzyme activity or proton translocation, *J. Biol. Chem.*, 290, 20761–20773.

Zhu, J., Vinothkumar, K.R. and Hirst, J. (2016). Structure of mammalian respiratory complex I, *Nature*, 536, 354–358.

Yang, S. and Cui Q. (2011). Glu-286 rotation and water wire reorientation are unlikely the gating elements for proton pumping in cytochrome *C* oxidase, *Biophys. J.*, 101, 61–69.

Young, I.D. *et al.* (2016). Structure of photosystem II and substrate binding at room temperature, *Nature*, 540, 453–457.

Zickermann, V., Wirth, C., Nasiri, H., Siegmund, K., Schwalbe, H., Hunte, C. and Brandt, U. (2015). Mechanistic insight from the crystal structure of mitochondrial complex I, *Science*, 347, 44–49.

Chapter 4

Reflections on Redox States in Enzymes

Per E. M. Siegbahn

Department of Organic Chemistry, Arrhenius Laboratory,
Stockholm University, SE-106 91, Stockholm, Sweden
ps@organ.su.se

The mechanisms for two of the most important enzymes in nature have recently been characterized by DFT model calculations. Photosystem II oxidizes water to O_2, while nitrogenase reduces N_2 to ammonia. In the present chapter, comparisons are made between these mechanisms with an emphasis on redox states. Both enzymes change their redox states in a few initial steps to reach the last stable state before activation. That state has Mn(IV) on all atoms in PSII, while the corresponding state for nitrogenase has Fe(II) on all iron atoms. After this, PSII uses an additional oxidation to reach a very unstable oxyl radical state, while nitrogenase releases H_2 to reach a state with two Fe(I), also very unstable and reactive.

1. Introduction

The concept of redox states is of fundamental value when discussing mechanisms for redox active enzymes. In the present paper, two extreme cases of redox reactions are discussed and compared. In Photosystem II (PSII), one of the most stable molecules in nature, H_2O, is oxidized to O_2. In nitrogenase on the other hand, a likewise very stable molecule, N_2, is instead reduced to NH_3. In PSII, the Mn_4Ca complex used as catalyst is oxidized as far as possible using the strongest oxidant known in nature, with a redox potential of +1.25 V. An Mn-complex at the highest oxidation level, which is still stable, for this type of weak ligand field type of Mn-cluster, should then have all Mn-atoms in the Mn(IV) state. Such metal complexes are well known as stable systems in the laboratory. It should be added here that manganese complexes with even higher oxidation states, such as Mn(V) and even Mn(VII) are known, but only for complexes with strong ligand fields. In nitrogenase, the FeMo complex used as catalyst is reduced as far as possible using the strongest reductant known in nature, with a redox potential of –1.6 V. An Fe-complex at the lowest redox level, which is still stable, should then have all iron atoms in the Fe(II) state.

Again, such complexes are well known in nature and in the laboratory. Already with these simple facts, it is possible to make far-reaching conclusions concerning the mechanisms of these enzymes.

The extreme redox potentials used for PSII and nitrogenase are obviously one important factor for being able to perform these demanding reactions. However, there are also strong demands on the catalysts. To reach the above mentioned redox levels is not enough. These redox states are still stable and not reactive enough. The ability to reach even more extreme redox levels, is the key for catalysis. How this is achieved will be discussed below, in some detail.

2. Photosystem II

In PS II, water is oxidized to oxygen in four steps, each one requiring one flash of light. This light flash excites the chlorophyll P_{680} in the reaction center to an ionized state, which has a redox potential of 1.25 V. The oxidant is used to oxidize the oxygen evolving complex (OEC) where O_2 will form in the final fourth step. The states are termed S_0 to S_4. The OEC is shown in Figure 1 in its S_3 state, the most oxidized state characterized experimentally before O_2 formation. It consists of a cuboidal structure of three manganese and one calcium linked by oxo bridges, with a dangling fourth manganese linked to the cube [Ferreira, 2004; Umena, 2011; Siegbahn, 2008].

Since S_3 has a stable structure and only one of the four oxidations remains, the conclusion can be drawn that the S_3 state should have all manganese in its Mn(IV) state, the highest possible oxidation state which is still stable. This conclusion has been proven correct by both theory and spectroscopy. Since S_3 is preceded by three oxidations, this means that the starting S_0 state should have three Mn(III) and one Mn(IV), also shown to be the correct state. The Mn(IV) state is on Mn2. The S_0 structure is shown in Figure 2.

When the S_0 and S_3 structures are compared, there are some notable differences. For every oxidation, there should normally be also a deprotonation to keep the charge constant. In the present case, there is one deprotonation of the hydroxide between Mn3 and Mn4. This occurs in the S_0 to S_1 transition. To be able to make additional deprotonations a water must be added, and this occurs in the S_2 to S_3 transition. This water becomes bound as an hydroxide to Mnl, which is

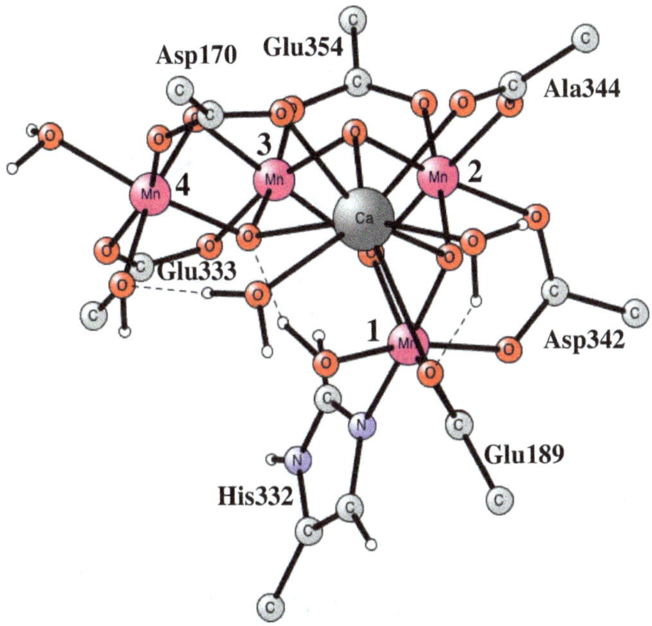

Figure 1. The core structure for the oxygen evolving complex in PSII in the S_3 state.

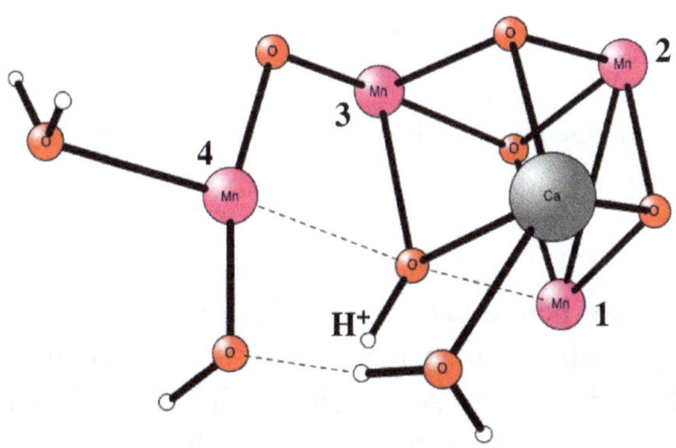

Figure 2. The core structure for the oxygen evolving complex in PSII in the S_0 state. The redox state is 3Mn(III)1Mn(IV).

only 5-coordinated in S_0 to S_2. This means that from S_0 to S_3 there are only two, rather than the expected three, deprotonations. Instead, the additional deprotonation occurs is the step directly after S_3.

The S_3 state has four Mn(IV) atoms, which is still expected to be quite unreactive and not suitable for water oxidation. The critical transition is therefore the one from S_3 to S_4. Rather than oxidizing manganese further, the hydroxide ligand on Mnl is deprotonated and oxidized to a radical oxyl state, see Figure 3. It is

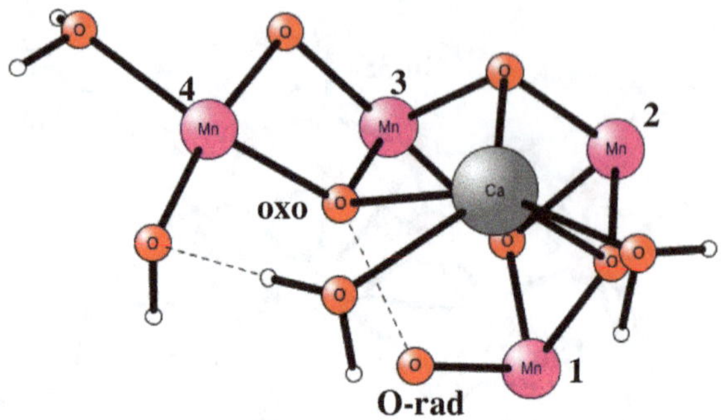

Figure 3. The core structure for the oxygen evolving complex in PSII in the S_4 state. The redox state is 4Mn(IV)0(rad).

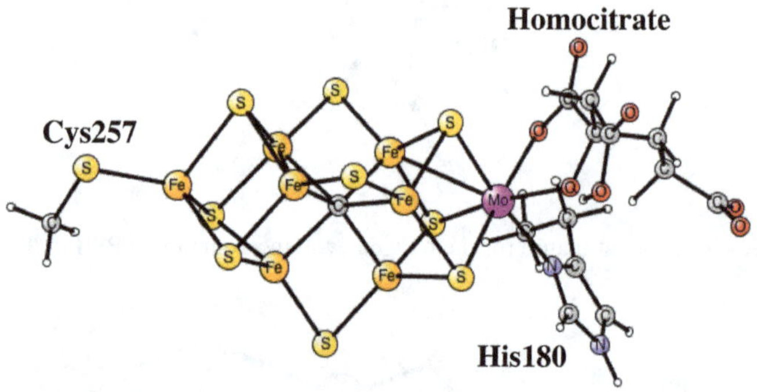

Figure 4. The core structure for the MoFe-cofactor of nitrogenase in its ground state. The redox state is Mo(III)3Fe(II)4Fe(III).

important to note that this transition is quite endergonic. This means that S_4 is not formed until precisely before and concertedly with O–O bond formation. The presence of a long-lived, extremely reactive, S_4 state would otherwise be likely to lead to other undesired reactions. A very similar situation is found for nitrogenase, see below. Another view on this is that the very high redox potential of +1.25 V is still not high enough to reach the required oxygen radical state exergonically. Once the oxygen radical has been formed endergonically, it will react with an oxo group to form O_2 with a quite small barrier.

3. Nitrogenase

The core structure of the nitrogenase FeMo-cofactor is shown in Figure 4. It has seven irons and one molybdenum connected by sulfide bridges [Kim, 1992; Spatzal, 2011]. Two amino acids and one unusual cofactor, homocitrate, are bound to the FeMo-cofactor. The most striking feature of the structure is that there is a central carbide bound to the irons [Spatzal, 2011; Lancaster, 2011]. The redox state of the cofactor has Mo(III), three Fe(II) and four Fe(III). This is a quite high redox state, in particular, considering that it should catalyze the reduction of one of the most stable molecules in nature, N_2.

The metal cofactor is reduced by a nearby P-cluster. To obtain the very low redox potential required, a very intricate mechanism is used. The binding between two proteins and the consumption of two ATP are needed for giving the P-cluster the lowest redox potential in nature of –1.6 V.

The experimentally suggested mechanism, starts with the ground state in Figure 4 termed E_0, and reduces it four times to reach the active E_4 state. Two hydrides have been shown by EPR to be present in E_4 [Hoffman, 2013]. Since four reductions require four protonations, the remaining protons are suggested to be bound to the bridging sulfides. For each protonation of a sulfide, the redox state decreases by one unit. However, when an added proton becomes bound as a hydride, the cofactor is oxidized. The surprising conclusion made is therefore that the FeMo-cofactor in its most reduced state has the same oxidation state as the ground state, with three Fe(III). In analogy with PSII, where the last observable state has only Mn(IV) atoms, the expected last observable state of the MoFe-cofactor should have only Fe(II).

The first set of model calculations were performed using the experimentally suggested mechanism. However, no stable hydride positions were found and N_2 was very far from being activated. These results were not surprising since the FeMo-cofactor has three Fe(III). Instead, the calculations suggest that the catalytic cycle should be preceded by four initial activation reductions, only made once before turnover, starting with the ground state in Figure 4 [Siegbahn, 2016, 2018]. These reductions should, as usual, be accompanied by four protonations. The by far most stable positions for the protons were very surprisingly found to be on the carbide, with three protons on the carbide and one on a sulfide. After these reductions, the FeMo-cofactor has the expected redox state with all Fe(II) atoms. This state represents the activated E_0 state.

Going into the catalytic cycling with the above activated E_0 state, four reductions are necessary for reaching the final state E_4, which is shown in Figure 5. This E_4 state is similar to the E_4 state suggested by experiments, except for the protonated carbide, as CH_3, and one more protonated sulfide. Most importantly, there are two hydrides. This E_4 state has Fe(II) on all irons.

Again, in parallel to the situation in PSII, the last observable state in nitrogenase is not enough for activating N_2. In PSII, another oxidation is necessary, leading to the oxygen radical S_4 state, which is very reactive. In contrast in nitrogenase, another reduction is not necessary for reaching a reactive state. Instead, as suggested by EPR experiments, the two hydrides are removed as H_2 [Hoffman, 2013]. This means that two electrons are left behind on the cofactor, which then obtains a redox state with two Fe(I), the rest with Fe(II). This active state is, as the S_4 state

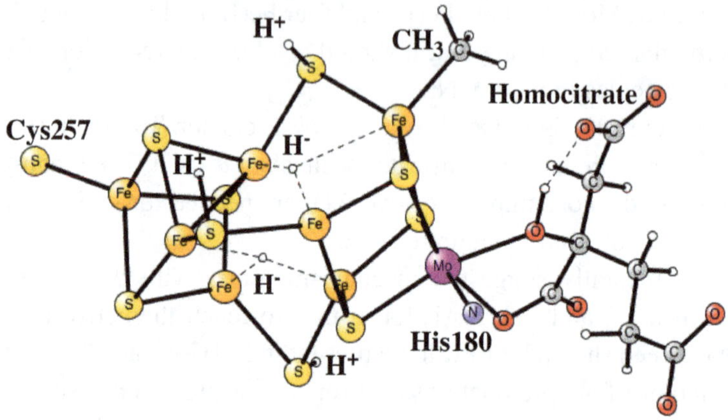

Figure 5. The core structure for the MoFe-cofactor of nitrogenase in the E_4 state, still not fully activated. The redox state is Mo(III)7Fe(II).

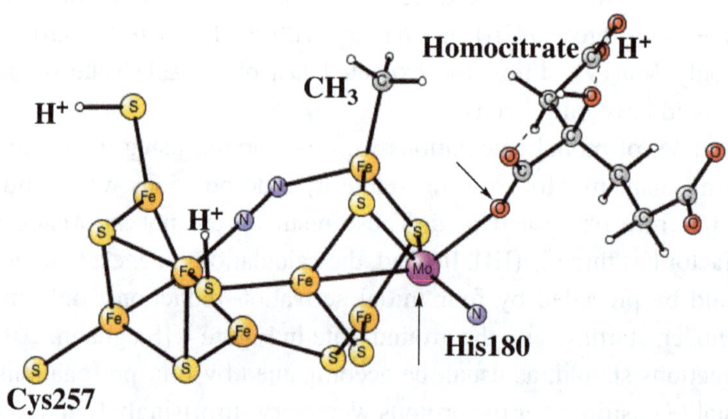

Figure 6. The core structure for the MoFe-cofactor of nitrogenase in the E_4-N_2 state, fully activated. The redox state is Mo(III)5Fe(II)2Fe(I). The arrows indicate the change of the coordination of the homocitrate.

in PSII, very unstable and can easily fall down to an unreactive state with all Fe(II), but it is able to reduce and activate N_2 before that happens. There is one additional feature which is necessary for the binding of N_2, and this is a rotation of the large homocitrate cofactor, leading to an open site on Mo. Going from the six-coordinated to the five-coordinated molybdenum leads to an additional reduction of the metal cofactor. The final bound state with N_2 is shown in Figure 6.

4. Discussion

There are interesting parallels between PSII, where water is oxidized, and nitrogenase, where N_2 is reduced. Both require extreme redox potentials to be active and cycles through four states. In both cases the last observable state is not active. In PSII this occurs for the S_3 state, one oxidation before O–O bond formation. In nitrogenase it occurs for the E_4 state. For PSII, the final oxidation creates un unstable oxygen radical which is very reactive. For nitrogenase, the active state is reached by a loss of two hydrides as H_2, which also leads to an unstable very reactive state, here with two Fe(I) atoms.

Since the final, very reactive, states in both cases are very unstable, they can not be directly observed experimentally. However, there is no problem to study these states by model calculations, which therefore fill an important role. In PSII very much is known about the states preceding the last one, allowing ample possibilities to investigate if the model calculations are accurate for this system. Starting these calculations with the low-resolution structure initially available [Ferreira, 2004], showed that modifications of the structure were necessary. Most importantly an additional oxo, not seen in the X-ray structure because of the low resolution, needed to be introduced into the structure. Also, an hydroxide becomes bound in the S_3 state and is therefore not present in the X-ray structure for the ground state either. After these modifications, a mechanism for O–O bond formation was found, that stands all experimental tests so far [Siegbahn, 2006]. It was found that the O–O bond is formed between an oxo-group and an oxyl radical.

For nitrogenase, essentially all experimental information concerns the ground state. There is one exception, and this is an EPR study of the E_4 state. It was demonstrated that there should be two bridging hydrides in this structure, which should leave almost thermoneutrally as H_2 just before N_2 becomes bound and activated [Hoffman, 2013]. This is a very strong requirement for any mechanism suggested. To fulfill this demand was a major reason the experimentally suggested mechanism had to be rejected and four initial activation steps had to be introduced. Otherwise, the mechanism did not fit the calculated energetics at all. The four initial activation steps led to the important requirement for protonating the central carbide, which had not been suggested before.

Acknowledgments

This work was supported by grants from the Swedish research council.

References and Citations

Ferreira, K.N., Iverson, T.M., Maghlaoui, K., Barber, J., Iwata, S. (2004). Architecture of the photosynthetic oxygen evolving center, *Science,* 303, 1831–1838.

Umena, Y., Kawakami, K., Shen, J.-R., Kamiya, N. (2011). Crystal structure of oxygen-evolving photosystem II at a resolution of 1.9 A, *Nature,* 473, 55–60.

Siegbahn, P.E.M. (2008). A structure consistent mechanism for dioxygen formation in photosystem II, *Chem. Eur. J.,* 27, 8290–8302.

Kim, J., Rees, D.C. (1992). Structural models for the metal centers in the nitrogenase molybdenum-iron protein, *Science,* 257, 1677–1682.

Spatzal, T., Aksoyoglu, M., Zhang, L., Andrade, S.L.A., Schleicher, E., Weber, S., Rees, D.C., Einsle, O. (2011). Evidence for interstitial carbon in nitrogenase FeMo cofactor. *Science,* 334, 940.

Lancaster, K.M., Roemelt, M., Ettenhuber, P., Hu, Y., Ribbe, M.W., Neese, F., Bergmann, U., DeBeer, S. (2011). X-ray emission spectroscopy evidences a central carbon in the nitrogenase iron-molybdenum cofactor. *Science,* 334, 974–977.

Hoffman, B.M., Lukoyanov, D., Dean, D.R., Seefeldt, L.C. (2013). Nitrogenase: A draft mechanism. *Ace. Chem. Res.,* 46, 587–595.

Siegbahn, P.E.M. (2016). Model calculations suggest that the central carbon in the FeMo-cofactor of nitrogenase becomes protonated in the process of nitrogen fixation. *J. Am. Chem. Soc,* 138, 10485–10495.

Siegbahn, P.E.M. (2018). A major structural change of the homocitrate ligand of probable importance for the nitrogenase mechanism, *Inorg. Chem.,* 57, 1090–1095.

Siegbahn, P.E.M. (2006). Structures and energetics for O_2 formation in photosystem II, *Chem. Eur. J.,* 12, 9217–9227.

Chapter 5

The Role of Singlet Oxygen in Photoinhibition of Photosystem II

Imre Vass

Institute of Plant Biology, Biological Research Centre,
Hungarian Academy of Sciences,
Temesvári krt. 62, 6726 Szeged, Hungary
vass.imre@brc.mta.hu

Singlet oxygen ($^1O_2$) is an important reactive oxygen species, formed in photosynthetic systems mostly *via* interaction of the triplet state of chlorophylls and other pigments. A specific site for $^1O_2$ production is P_{680}, the primary electron donor of Photosystem II, whose triplet state [$^3P_{680}$] is continuously formed during light energy conversion. $^1O_2$ is a highly reactive molecule, which can damage organic molecules (pigments, lipids, proteins, *etc.*) in the vicinity of its production site, and has been implicated in the important process of photoinhibition, *i.e.* the light induced decrease of Photosystem II activity. In this review, data concerning the involvement of $^1O_2$ in damaging the function of the Photosystem II complex, and also in the inhibition of its repair are discussed. Available literature data show that in the absence of Photosystem II repair the rate of $^1O_2$ production is positively correlated with the rate of photodamage under a wide range of experimental conditions. These data point to a cause-effect relationship between $^1O_2$ production and Photosystem II photodamage. $^1O_2$ has also been shown to inhibit a key step in repair cycle of Photosystem II, at the *de novo* synthesis of the D1 reaction center protein. While there is a consensus on the mechanism of $^1O_2$ induced inhibition of Photosystem II repair, being at the translational elongation step of *psbA* mRNA encoding the D1 protein, the mechanism of photodamage is still debated with arguments that Photosystem II damage is caused by direct photoinactivation of the catalytic Mn cluster of water oxidation. These unresolved questions are discussed in detail with the aim to provide unifying explanation for the apparently contradicting data. On the basis of compelling evidence for the involvement of $^1O_2$ in photodamage of PSII it can be safely concluded that photoinactivation of the Mn cluster cannot be the sole mechanism of photodamage of PSII. $^1O_2$ might play a role in the inactivation of the Mn cluster, or $^1O_2$-mediated and direct photoinactivation mechanisms might proceed in parallel in a wavelength dependent fashion.

1. Introduction

Although light is the primary driving force of photosynthetic energy conversion light can also damage the energy converting molecular machinery. Light induced decrease of photosynthetic activity, commonly referred to as photoinhibition, is a well-known phenomenon, which has been studied in detail since the 1960-ies. During more than half a century considerable understanding of this highly important phenomenon has accumulated, as summarized in several detailed reviews [Andersson *et al.*, 1992; Barber and Andersson, 1992; Aro *et al.*, 1993; Adir *et al.*, 2003; Tyystjarvi, 2008; Ohad *et al.*, 2011; Vass, 2011; Murata *et al.*, 2012; Vass, 2012; Tyystjarvi, 2013; Zavafer *et al.*, 2015b]. However, important questions are still under debate regarding the actual mechanistic principles and details.

One of the important targets of light induced photosynthetic activity loss is the Photosystem II (PSII) complex. PSII is a large pigment-protein assembly embedded in the thylakoid membrane, which performs the light induced oxidation of water to molecular oxygen and protons, and delivers the electrons liberated in the process to the plastoquinone electron acceptors located in the lipid phase of the membrane [Shen, 2015]. The PSII complex, consists of more than 20 protein subunits, about 40 chlorophyll (Chl) molecules, protein-bound plastoquinone (PQ) as well as Mn, Ca, Cl and Fe, which act as redox cofactors of light-driven electron transfer [Ferreira *et al.*, 2004; Loll *et al.*, 2005]. The reaction center of PSII is made up of the heterodimer of the D1 and D2 proteins, which bind or contain the redox cofactors involved in light-induced electron transport.

During light-driven PSII electron transfer, the primary event is charge separation between the excited reaction center Chl assembly [P_{680}*] and pheophytin (Phe) molecule producing the primary charge separated state [$P_{680}^+Phe^-$], which is followed by rapid charge stabilization processes. On the acceptor side of PSII, the electron from reduced Phe is transferred to the first, Q_A, and second, Q_B, plastoquinone (PQ) electron acceptor and then to the mobile pool of PQ in the lipid phase of the thylakoid membrane. On the donor side, P_{680}^+ is reduced by Tyr-Z, a redox active tyrosine of the D1 protein, which then extracts an electron from the water-oxidizing complex. Water oxidation is catalyzed by a cluster of four Mn ions that undergo light-induced changes in their oxidation states, called S-states. The complex cycles through five S-states denoted as S_0,...,S_4 and oxygen is released during the $S_3 \rightarrow S_4 \rightarrow S_0$ transition, in which S_4 is a short-lived intermediate (see [Kern *et al.*, 2018]).

PSII undergoes competing processes of photodamage and its continuous repair in intact systems [Aro *et al.*, 1993]. Photoinhibition takes place when the rate of photodamage exceeds the rate of repair leading to net loss of PSII activity. The balance of the damage and repair processes can be upset either by the increased rate of photodamage, or by the decreased rate of repair.

For the mechanism of photodamage two main ideas have been put forward. According to one idea light induces redox and structural modifications at the acceptor side of PSII, which directly inhibits electron transfer towards the PQ pool and/or leads to the formation of reactive oxygen species (ROS) which damage the PSII complex (for reviews see [Vass and Cser, 2009; Vass, 2012]). The other idea is that the absorbed photons damage directly the Mn cluster of the water oxidizing complex (WOC), leading to the loss of electron transport from the WOC towards the PSII reaction center and the PQ pool [Hakala *et al.*, 2005; Ohnishi *et al.*, 2005].

The mechanisms by which intact photosynthetic organisms can repair light induced damage of PSII are studied in great detail and summarized in excellent reviews [Barber and Andersson, 1992; Aro *et al.*, 1993; Adir *et al.*, 2003; Murata *et al.*, 2007; Takahashi and Murata, 2008; Vass and Aro, 2008; Nixon *et al.*, 2010; Ohad *et al.*, 2011; Komenda *et al.*, 2012; Murata *et al.*, 2012; Tyystjarvi 2013; Järvi *et al.*, 2015; Li *et al.*, 2018]. Here the question of PSII repair will be covered only from the point of view of the involvement of ROS.

One of the important ROS forms produced in PSII and inferred in the process of photoinhibition at both the damage and repair side of the process, is singlet oxygen. Singlet oxygen ($^1\Delta gO_2$, which will be abbreviated as $^1O_2$) is produced *via* interaction of ground state molecular oxygen ($^3\Sigma gO_2$) with long lived excited triplet states of pigments, which are mostly Chls and their precursors in photosynthetic systems [Krieger-Liszkay, 2005; Krieger-Liszkay, *et al.*, 2008]. $^1O_2$ has a short halftime of about 3–3.5 μs in aqueous environment (Baier *et al.*, 2005; Skovsen *et al.*, 2005), which increases to *ca.* 10–14 μs in lipidic environment, such as cell membranes (Baier *et al.*, 2005). These lifetime values correspond to *ca.* 200 to 400 nm diffusion distance of $^1O_2$, respectively (Baier *et al.*, 2005; Skovsen *et al.*, 2005), which means that $^1O_2$ can reach significant distances from its production site in cellular environment. In the following parts of the chapter current knowledge concerning the role of $^1O_2$ in photoinhibition will be overviewed.

2. Contrasting Mechanisms of PSII Photodamage

There are two main ideas regarding the mechanisms of light induced inactivation of PSII activity. One proposed mechanism is that light induces redox and structural modifications at the acceptor side of PSII, which directly inhibit electron transfer towards the PQ pool and/or lead to the formation of $^1O_2$, which damages the PSII complex [van Mieghem *et al.*, 1989; Vass *et al.*, 1992; Koivuniemi *et al.*, 1993; Telfer *et al.*, 1994; Vass and Cser, 2009; Rehman *et al.*, 2013]. The other idea assumes that light damages directly the catalytic Mn cluster of the WOC, which blocks electron transport at the donor side of PSII leading to the loss of electron flow from water to the PQ pool [Hakala *et al.*, 2005; Ohnishi *et al.*, 2005]. Direct light induced damage of the Mn cluster is well established in the UV-B and UV-A spectral range [Jones and Kok, 1966; Vass *et al.*, 1996; Vass *et al.*, 1999; Vass *et al.*, 2002] and may occur to some extent in the blue region as well. It cannot be excluded that some light induced Mn damage may occur also in the visible or red part of the spectrum as well [Hakala *et al.*, 2005; Ohnishi *et al.*, 2005], however, it is highly unlikely that the direct hit of the Mn cluster would be the sole mechanism of PSII photodamage in the visible range as discussed below.

Murata and coworkers have proposed the so called two-step mechanism of PSII of photodamage [Ohnishi *et al.*, 2005]: The first step being the light induced damage of the Mn cluster. This is followed by the inactivation of the PSII reaction center as the second step due to the stabilization of highly reactive radical intermediates of PSII electron transfer, P_{680}^+ and Y_D^+ in the absence of electron flow from the WOC. The second step of this mechanism, was called earlier as donor side induced photoinactivation [Aro *et al.*, 1993], and was studied in great detail in isolated PSII membrane systems [Tamura *et al.*, 1989; Blubaugh and Cheniae, 1990; Blubaugh *et al.*, 1991] well before the two-step mechanism was suggested. The actual question is if the Mn cluster can indeed be damaged specifically by visible light in intact systems? The literature appears to be quite controversial in this respect. Early data indicated that release of the catalytic Mn ions from the WOC occurred in parallel with the degradation of the D1 protein, which provides most of their binding ligands. Both of these processes were much slower that the loss of the S_2 state formation in the WOC and of the O_2 evolving activity [Virgin *et al.*, 1988]. These data show that PSII activity loss is not induced by direct photoinactivation of the Mn cluster. On the other hand, the main argument of proponents of the direct Mn photodamage was the resemblance of the action spectrum of PSII photodamage and the absorption spectrum of the catalytic Mn ions in the WOC, as inferred from the absorption of organic Mn compounds (Hakala *et al.*, 2005), Mn containing enzymes [Hakala *et al.*, 2006] or Mn(III/IV) model compounds [Ohnishi *et al.*, 2005]. However, recently this idea has been challenged by showing that the water oxidizing activity was damaged practically to the same extent by blue, green and red light [Zavafer *et al.*, 2015a], which should have very different extent of damage if their effect was related to absorption by Mn. Therefore, in the latest version of the two-step mechanism of PSII photodamage the actual mode of action of visible light on the Mn cluster remained unspecified [Zavafer *et al.*, 2015a]. Detailed comparative analysis of the available photoinhibition action spectra also showed that photoinactivation of PSII could not be satisfactorily rationalized by light absorption in the catalytic Mn cluster alone [Zavafer *et al.*, 2015b]. In the following sections the experimental results, which are directly related to $^1O_2$ mediated damage of the PSII function will be overviewed.

3. Production and Detection of $^1O_2$ in Photosynthetic Systems

The idea that $^1O_2$ may be produced in PSII can be traced back to the detection of $^3P_{680}$ and its interaction with O_2 in isolated D1/D2 reaction center complexes, which are capable only of the primary charge separation reaction [Durrant *et al.*, 1990]. $^3P_{680}$ was also detected in PSII particles under highly reducing conditions leading to double reduction of Q_A, which blocks forward electron transfer [van Mieghem *et al.*, 1989]. Subsequently it was demonstrated in PSII membrane particles, which are capable of complete light induced electron transport from water to the PQ pool, that under anaerobic conditions the light induced loss of electron transport activity was accompanied by the formation of $^3P_{680}$ [Vass *et al.*, 1992]. $^3P_{680}$ yields a characteristic spin polarized EPR signal, resulting from the charge recombination of the primary radical pair $P_{680}^+Phe^-$ [Rutherford *et al.*, 1981]. Importantly, the spin polarized EPR signal was eliminated by O_2 in the photoinhibited samples [Vass *et al.*, 1992], demonstrating the interaction of $^3P_{680}$ and triplet ground state O_2, which is known to result in the formation of $^1O_2$.

The idea of light induced $^1O_2$ formation in PSII containing preparations was later confirmed by various approaches, such as the detection of its specific luminescence emission at 1270 nm in highly purified PSII core complexes [Macpherson *et al.*, 1993; Telfer *et al.*, 1999]. Since the 1270 nm luminescence could not be detected in larger photosynthetic complexes [Li *et al.*, 2012] other approaches were needed to study $^1O_2$ in PSII membranes and isolated thylakoids. One of the successful methods was the application of EPR spin traps, such as TEMP or TEMPD, which are converted into paramagnetic, EPR detectable species after capturing $^1O_2$. By applying this approach ample evidence was provided for light induced $^1O_2$ production in isolated thylakoids and PSII membranes under conditions of the so called acceptor side photoinhibition, where the donor side of PSII was initially functional [Hideg *et al.*, 1994a, 1994b; Hideg and Vass, 1995]. $^1O_2$ production in isolated PSII core complexes was also shown by NMR detection of the photooxygenated product of a benzoic acid sodium salt [Mishra *et al.*, 1996]. With the development of detection techniques measurement of the direct 1270 nm luminesce signal from $^1O_2$ has also became possible in PSII membrane particles [Tomo *et al.*, 2012].

Spin trap compounds, which change the yield of their fluorescence emission have also proven very useful for $^1O_2$ detection. One of these compounds is DanePy, developed by Éva Hideg and coworkers, responds to interaction with $^1O_2$ by fluorescence yield quenching [Kálai *et al.*, 1998]. DanePy was the first sensor, which made possible the detection of $^1O_2$ production in intact leaves, under photoinhibitory conditions [Hideg *et al.*, 1998]. A highly successful application of DanePy was the work, which led to the identification of $^1O_2$ formation in the so called *flu* mutant of *Arabidopsis* [op den Camp, *et al.*, 2003; Wagner *et al.*, 2004].

The other useful fluorescent compound is the commercially available singlet oxygen sensor green (SOSG), which increases its green fluorescence yield after capturing $^1O_2$ [Gollmer *et al.*, 2011]. SOSG was successfully applied to confirm $^1O_2$ production in intact plants [Flors *et al.*, 2006; Driever *et al.*, 2009] and in the cell-free medium of the *Symbiodinium* dinoflagellate alga [Rehman *et al.*, 2016b]. Unfortunately, the penetration of SOSG through the cell walls of unicellular cyanobacteria and the *Symbiodinium* microalgae is limited [Rehman *et al.*, 2013, 2016b]. Despite the array of $^1O_2$ sensors applicable *in vitro*, in isolated thylakoid membrane preparations, and intact leaves detection of $^1O_2$ in microalgal cells posed a significant challenge due to the inability of the commonly used sensors to penetrate through the cell wall. This problem is especially difficult in case of unicellular cyanobacteria, such as the commonly used model organism *Synechocystis* 6803, whose mucoid sheath outside the cell wall restricts the penetration of SOSG. Although some progress has been achieved recently by using SOSG in some microalgae, especially in *Chlamydononas* [Prasad *et al.*, 2016, 2018], *Synechocystis* cells require additional treatment, such as heat exposure, to enhance the uptake of the sensor [Prasad *et al.*, 2018]. However, even under such conditions the distribution of SOSG is still quite heterogeneous among the cells [Prasad *et al.*, 2018], and therefore SOSG is not suitable for quantitative detection of $^1O_2$ formation in cyanobacteria, and other microalgae with thick cell wall, such as *Symbiodinium*.

The challenge of reliable and quantitative detection of $^1O_2$ in microalgae could be finally overcome by the application of the simple His-mediated chemical trapping method. In this case interaction of $^1O_2$ with a suitable chemical trap leads to oxidation reactions resulting in O_2 uptake [Verlhac *et al.*, 1984; Méndez-Hurtado *et al.*, 2012], which can be measured with a common Clark-type electrode [Telfer *et al.*, 1994]. The first successful application of this method in intact systems was in the cyanobacterium *Synechocystis* 6803 [Rehman *et al.*, 2013; Bersanini *et al.*, 2014; Hakkila *et al.*, 2014; Sedoud *et al.*, 2014; Kusama *et al.*, 2015; Vajravel *et al.*, 2016; Tibiletti *et al.*, 2018] and later in the dinoflagellate alga *Symbiodinium* [Rehman *et al.*, 2016b]. The method can also be applied in purple bacterial systems [Hajdu *et al.*, 2018].

It is important to note that $^1O_2$ can be produced not only by the reaction center Chls of the PSII complex, but also by the Chls bound to the light harvesting complex, which can lead to $^1O_2$-mediated photodamage of LHCII [Zolla and Rinalducci, 2002; Rinalducci *et al.*, 2004]. Even though the majority of Chls are found in the antenna they are efficiently protected by nearby carotenoids against $^1O_2$ formation. Carotenoids not only quench the already formed $^1O_2$ but can also dissipate the excitation energy of nearby Chls, thereby preventing the ^{3}Chl formation, which is the precursor of $^1O_2$. This double protecting function of carotenoids is very efficient in light harvesting complexes. However, in the PSII reaction center the protective carotenoid molecules cannot be placed close enough to the P_{680} pigment assembly without decreasing the efficiency of primary charge separation. As a consequence the $^1O_2$ producing potential of PSII is much higher than that of the LHCII complex, as was demonstrated by comparative measurements [Hideg and Vass, 1995].

$^1O_2$ can also be produced by intermediates of Chl biosynthesis such as protoporphyrin or protochlorophyllide, whose accumulation can lead to programmed cell death in higher plants as was shown for the *flu* mutant of *Arabidopsis* [op den Camp *et al.*, 2003; Wagner *et al.*, 2004]. The *aba1 Arabidopsis* mutant, which is deficient in abscisic acid (ABA), and defective in zeaxanthin epoxidase responsible for the transformation of zeaxanthin into antheraxanthin and violaxanthin, shows high level of $^1O_2$ production and shares a number of coregulated genes with the *flu* mutant [Sanchez-Corrionero *et al.*, 2017]. High light exposure induced chloroplast aggregation followed by chloroplast rupture and eventual cell death in the *aba1* mutant, which was however not associated with direct $^1O_2$ cytotoxicity, but rather with a delayed stress response related to $^1O_2$-mediated signaling [Sanchez-Corrionero *et al.*, 2017].

Involvement of $^1O_2$ in the photodamage of phycobilisomes, which serve as light harvesting antenna in cyanobacteria has also been demonstrated [Rinalducci *et al.*, 2008]. $^1O_2$ production has also been suggested to occur in FeS center containing electron transport complexes, such as PSI and the cytb_6f complex, with the idea that these exogenous $^1O_2$ sources could be responsible for PSII damage [Jung and Kim, 1990; Kim, 1993; Chung and Jung, 1995]. The possibility that PSI could be a significant source of $^1O_2$ was excluded on the basis of a study, which showed that $^1O_2$ production by isolated PSI complexes was negligible in comparison to PSII and free Chl [Hideg and Vass, 1995]. This finding was later confirmed by the authors of the original articles about the possible role of PSI as $^1O_2$ source [Suh *et al.*, 2000]. Also, *Synechocystis* 6803 mutants, which lack the PSI complex are highly sensitive to light and produce more $^1O_2$ on equal Chl basis than the WT strain [Vass *et al.*, unpublished data].

The idea that the cytb$_6$f complex could be a $^1O_2$ source was supported by using isolated cytb$_6$f preparations, whose presence induced PSII photoinactivation. However, this effect was pronounced only in the UV-A and blue spectral region [Suh *et al.*, 2000], and therefore it is unlikely to play a significant role under physiological conditions.

4. Involvement of $^1O_2$ in Flashing Light Induced Photodamage of PSII

Besides the ambiguity of the conclusions drawn from the resemblance of the action spectra of photodamage and different Mn compounds (see above) there is a wide range of experimental results, which cannot be explained by idea of direct photoinactivation of the Mn cluster. Direct hit of the Mn cluster should result in a photon dose dependent effect independent of the application of continuous light or light pulses for the illumination. The photon dose dependency has indeed been confirmed under continuous illumination of intact leaves [Tyystjärvi and Aro 1996] and in *Synechocystis* 6803 cells [Nagy *et al.*, 1995; Nishiyama *et al.*, 2004]. However, Ohad and coworkers have shown an important deviation from the photon dose dependency by demonstrating that the extent of PSII photodamage is strongly enhanced if light is provided in the form of short duration flashes, instead of continuous exposure [Keren *et al.*, 1995,1997]. Moreover, the spacing of the flashes had also an important effect, being two orders of magnitude more damaging when placed tens of seconds apart instead of shorter intervals, or continuous illumination [Keren *et al.*, 1997]. These findings related the cause of flashing light induced photodamage to $^1O_2$ production arising from the recombination of the $S_2Q_B^-$ and $S_3Q_B^-$ charge pairs. The importance of flash spacing was confirmed by independent experiments in which thylakoid membranes were exposed to 80 flash packages, containing 1, 2, 3, or 4 light pulses fired at 1 Hz frequency, separated by 100 s dark intervals. Importantly, the packages which contained 1 or 2 flashes induced 3 times more PSII damage than the packages which contained 4 flashes [Szilárd *et al.*, 2005], which is just the opposite what could be expected from a dose-dependent Mn damage. Since each flash packages were fired after 100 s dark intervals the Mn cluster was always in the same S-state distribution (*ca.* 25 % S_0 and 75% S_1) before the flashes hit the thylakoids. Therefore, the 3-fold difference in the extent of PSII damage can only be explained by the difference of S-state distribution after the flashes, being mostly in S_2 and S_3 after the 1 flash containing packages and in S_1 after the 4 flash containing packages. The S_2 and S_3 states can easily recombine with the reduced Q_A of Q_B quinone electron acceptors, while the S_1 state cannot, therefore these data are in full agreement with the findings of the Ohad group and support the idea that PSII damage is related to the presence of charge separated states, which can produce $^1O_2$ during their recombination [Szilárd *et al.*, 2005]. $^1O_2$ measurements using DanePy in the same study confirmed that both the loss of PSII activity and the extent of $^1O_2$ production were enhanced with the increase of flash spacing that left more time for the $^1O_2$ producing charge recombination [Szilárd *et al.*, 2005]. The above data demonstrate that flashing light induced photodamage of PSII is related to O_2 producing charge recombinations instead of direct photoincativation of the Mn cluster.

5. Involvement of $^1O_2$ in Continuous Illumination Induced Photodamage of PSII

When the role of $^1O_2$ in inducing or mediating photodamage of PSII under continuous light is considered it is important to establish if there exists a general correlation between PSII activity loss and the rate of $^1O_2$ production, or not? By using the His-mediated chemical trapping approach for measuring the $^1O_2$ production rate several examples have been provided, which show that the extent of PSII photodamage is positively correlated with the rate of $^1O_2$ production in intact cyanobacterial cells. In the following section these examples will be described.

5.1. *Role of Phe/Phe⁻ redox potential in $^1O_2$ dependent photodamage of PSII*

The Gln130 residue of the D1 protein is a ligand to Phe and its change to Glu or Leu shifts the redox potential of Phe/Phe^- by +33 and –74 mV with respect to the WT, respectively [Merry *et al.*, 1998]. These changes have been demonstrated to affect the efficiency of charge recombination in PSII [Rappaport *et al.*, 2002; Cser and Vass, 2007], and also the light sensitivity of PSII in *Synechocystis* PCC 6803 cells [Rehman *et al.*, 2013]. In the D1-Gln130Glu mutant a non-radiative charge recombination pathway is enhanced, which competes with the pathway that leads to $^3P_{680}$ production and therefore decreases the rate of $^1O_2$ production, and the opposite effect occurs in the D1-Gln130Leu mutant [Rappaport *et al.*, 2002; Cser and Vass, 2007]. A more recent study has demonstrated that the modified charge recombination pathways lead to decreased $^1O_2$ production and photoprotection in the D1-Gln130Glu mutant, and an increased $^1O_2$ production concomitant with enhanced PSII activity loss in the presence of lincomycin in the D1-Gln130Leu mutant [Rehman *et al.*, 2013]. These data demonstrate that a single amino acid change in the PSII reaction center can modulate both $^1O_2$ production efficiency and photodamage in a positively correlated way [Vass *et al.*, 2007; Vass and Cser, 2009; Rehman *et al.*, 2013]. Since the D1–130 residue is located in H-bonding distance to Phe, far from the Mn cluster it is highly unlikely that it would affect the efficiency of the putative light induced damage of the catalytic Mn ions of the WOC. Therefore, the observed changes in the extent of photodamage cannot be assigned to a Mn-absorption mediated inactivation of WOC. On the other hand, the positive correlation between the extent of photodamage and $^1O_2$ production points to a cause-effect relationship between $^1O_2$ production and PSII photodamage [Vass 2011, 2012; Rehman *et al.*, 2013].

It is very important to note that several cyanobacteria contain a small *psbA* multigene family that encodes different D1 protein forms, which are expressed differentially under low and high light conditions [Bustos *et al.*, 1990]. The D1 protein forms, which are enhanced under high light always contain Glu at the D1–130 position, while the low light forms contain Gln. The presence of the high light D1 form provides partial photoprotection in the presence of lincomycin in *Thermosynechococcus elongatus* [Kós *et al.*, 2008; Sander *et al.*, 2010], *Synechococcus elongatus* 7942 [Campbell *et al.*, 1995, 1998] and also in *Synechocystis* 6803 containing the high light D1 form of *Synechococcus elongatus* 7942 [Tichy *et al.*, 2003]. From these data it appears that cyanobacteria actively use the modulation of the nonradiative pathway in the PSII reaction center that decreases $^1O_2$ production in order to provide photoprotection under high light conditions [Vass and Cser 2009]. The photoprotective effect can be partly ascribed to the presence of D1- Glu130 as demonstrated by the site directed *Synechocystis* 6803 mutants described above. It is also of note however, that the low- and high-light D1 forms differ not only at the 130th position (Gln *vs.* Glu, respectively), but at several places of the N-terminal, which is involved in the FtsH protease mediated degradation step of the D1 protein in the PSII repair cycle [Silva *et al.*, 2003; Komenda *et al.*, 2007]. Therefore, the exchange of low- and high-light D1 forms serves both the decrease of photodamage and the increased efficiency of PSII repair.

5.2. *Role of Q_A/Q_A^- redox potential in $^1O_2$ dependent photodamage of PSII*

Besides the effect of Phe/Pe$^-$ redox potential on $^1O_2$ formation and related photodamage similar effects are induced by the modification of the Q_A/Q_A^- redox potential. The exchange of Ala to Ser in the D2–249 position, within the Q_A

binding pocket, via site directed mutagenesis in *Thermosynechococcus elongatus* shifts the redox potential of Q_A by *ca.* –60 mV, and leads to enhanced PSII photodamage and $^1O_2$ production [Fufezan *et al.*, 2007]. The effect of herbicides, which modify the redox potential of Q_A/Q_A^- via binding to the Q_B site in PSII confirm that the yield of $^1O_2$ production is regulated by the Em[Q_A/Q_A^-]. In the presence of the phenolic herbicide bromoxynil, which lowers Em[Q_A/Q_A^-] by *ca.* 45 mV [Krieger-Liszkay and Rutherford, 1998], about twice as much $^1O_2$ was produced [Fufezan *et al.*, 2002] compared to that in the presence of DCMU, which raises the Em[Q_A/Q_A^-] by *ca.* 50 mV [Krieger-Liszkay and Rutherford, 1998]. This difference in $^1O_2$ production is positively correlated with the extent of photodamage, which is enhanced in the presence of bromoxynil and retarded by DCMU [Jansen *et al.*, 1993].

5.3. *The role of carotenoid mediated $^1O_2$ quenching and alternative electron transport pathways in photoprotection of PSII*

The orange carotenoid protein (OCP) is responsible for photoprotection by non-photochemical quenching in most cyanobacteria, including the model species *Synechocystis* 6803 [Wilson *et al.*, 2006; Kirilovsky, 2007; Kerfeld and Kirilovsky, 2013; Kirilovsky and Kerfeld, 2013]. By using a combination of OCP deficient and OCP overexpressing mutants together with EPR detection and His-mediated chemical trapping of $^1O_2$ it has been shown that OCP is not only the mediator of NPQ, but at the same time a very efficient physical quencher of $^1O_2$ [Sedoud *et al.*, 2014]. It has also been demonstrated that this $^1O_2$ scavenging ability of OCP protects *Synechocystis* 6803 against photodamage even in red light when the NPQ activity of OCP, which can be induced only by blue-green light, is not active [Sedoud *et al.*, 2014]. The OCP protein, which interacts with phycobilisomes on the stromal side of PSII [Harris *et al.*, 2016], has no direct interaction with the Mn cluster, which is located at the lumenal side of PSII. In addition, the protein matrix of OCP absorbs in the UV, while the bound echinenone carotenoid molecule absorbs in the blue range. Therefore, it can also be excluded that OCP could protect the Mn cluster against absorbing red photons. As a consequence, the most plausible explanation for the photoprotective effect of OCP in the red region that it is correlated with its $^1O_2$ scavenging ability [Sedoud *et al.*, 2014].

OCP-mediated $^1O_2$ quenching has been confirmed in *Synechocystis* 6803, however, its protective effect against photoinhibition was assigned to the effect of $^1O_2$ on PSII repair instead of protection against photodamage [Kusama *et al.*, 2015]. However, in these experiments chloramphenicol was used as protein synthesis inhibitor, which has been shown to produce superoxide in both PSI [Okada *et al.*, 1991] and PSII [Rehman *et al.*, 2016b]. Chloramphenicol induced superoxide production enhances PSII photodamage and therefore unsuitable for photoinhibition studies [Rehman *et al.*, 2016b]. In a more recent work the effect of OCP induced scavenging of $^1O_2$ production on PSII repair was revisited by using lincomycin as protein synthesis inhibitor [Takahashi *et al.*, 2018], which is free from the superoxide producing artifact, with the conclusion that OCP has no effect on photodamage. However, the presented data do not agree with this conclusion since is a clear protective effect is shown in the OCP overexpressing strain in the presence of lincomycin [Takahashi *et al.*, 2018]. Therefore, the argument that the OCP induced decrease of $^1O_2$ production would affect photoinhibition only at the PSII repair side remains a hypothesis, which contradicts the data shown in the paper [Takahashi *et al.*, 2018], and also the earlier reported photoprotective effect of OCP [Sedoud *et al.*, 2014].

A further example for the photoprotective role of $^1O_2$ scavenging by a carotenoid component was demonstrated in a *Synechocystis 6803* mutant that lacks the carotene isomerize enzyme (CrtH), which results in a largely decreased β-carotene content. Light sensitivity of this mutant was significantly enhanced relative to its WT strain, concomitant with a ca. twofold increase in the $^1O_2$ production rate [Vajravel *et al.*, 2016].

A group 2 σ factor mutant of *Synechocystis* 6803 (ΔsigCDE) is more resistant to photodamage in the presence of lincomycin than the WT due to the enhanced expression of photoprotective carotenoids (OCP) and of the Flv4, sll0218, Flv2 proteins [Hakkila *et al.*, 2014]. Carotenoids eliminate $^1O_2$ directly as discussed above, while the flavodiiron proteins alleviate the excitation pressure by channeling up to 30% of PSII-originated electrons to O_2 without the production of harmful ROS [Zhang *et al.*, 2012]. The flv4-2/OE mutant is more resistant to photodamage of PSII in the presence of lincomycin and exhibits a more oxidized state of the PQ pool and reduced production of $^1O_2$ compared with control strains [Bersanini *et al.*, 2014].

6. Involvement of $^1O_2$ in the Inhibition of PSII Repair

The net loss of photosynthetic activity in intact systems is determined by the balance of photodamage of PSII and its protein synthesis dependent repair. The repair cycle of PSII is a rather complex process, which has the following main steps: (i) Degradation of the photodamaged D1 (and D2) reaction proteins of PSII and partial disassembly of the photodamaged PSII complex. (ii) Production of *psbA* (and *psbD*) transcripts. (iii) Translation of the *psbA* (and *psbD*) mRNA to produce new D1 (and D2) protein copies followed by their insertion into PSII. (iv) Binding of Mn ions to form the catalytic Mn cluster of water oxidation followed by photoactivation of the complex to restore fully functional PSII. These processes are summarized in excellent reviews [Barber and Andersson, 1992; Aro *et al.*, 1993; Adir *et al.*, 2003; Murata *et al.*, 2007; Takahashi and Murata, 2008; Vass and Aro, 2008; Nixon *et al.*, 2010; Ohad *et al.*, 2011; Komenda *et al.*, 2012; Murata *et al.*, 2012; Tyystjarvi, 2013; Järvi *et al.*, 2015].

$^1O_2$ and other forms of ROS have been shown to be involved in the repair of photodamaged PSII. In an important paper it was found [Nishiyama *et al.*, 2001] that exposure of *Synechocystis* 6803 cells to H_2O_2 and methyl viologen (MV), which produces superoxide at the acceptor side of PSI, enhances significantly the loss of PSII activity under light treatment at 1500 μmol photons $m^{-2}s^{-1}$ intensity. However, the presence of MV or H_2O_2 had apparently no effect on the light induced PSII activity loss if the protein synthesis inhibitor chloramphenicol was also present [Nishiyama *et al.*, 2001]. It was also shown that a mutant, which lacked the H_2O_2 scavenging enzymes catalase peroxidase and thioredoxin peroxidase enzymes, encoded by the *katG* and *tpx* genes, respectively showed enhanced PSII activity loss in the absence, but not in the presence of chloramphenicol [Nishiyama *et al.*, 2001]. Addition of H_2O_2 in the recovery phase of the experiments strongly inhibited the restoration of PSII activity, in parallel with the decrease of the incorporation of newly synthesized D1 copies into the thylakoid membrane. Interestingly H_2O_2 had only small effect on *psbA* transcript production, but decreased significantly the relative level of *psbA* mRNAs in membrane associated polysomes. From these data it was concluded that the H_2O_2 inhibitory effect occurs mainly at the initiation and elongation steps in the translation of *psbA* mRNAs [Nishiyama *et al.*, 2001]. In a later study [Nishiyama *et al.*, 2004] similar experiments were performed in the presence of externally added rose bengal as $^1O_2$ sensitizer. It was found that the $^1O_2$ produced by the added sensitizer inhibited PSII repair practically in the same way as was shown earlier for H_2O_2, leading to the conclusion that the primary target of $^1O_2$ might be the elongation step of *psbA* mRNA translation [Nishiyama *et al.*, 2004, 2005, 2006]. In these publications it was even suggested that the only effect of $^1O_2$ in photoinhibition is the inhibition of the repair cycle without any effect on photodamage. This discrepancy will be discussed further below.

7. Unresolved Questions

As discussed above, literature data reveal a serious disagreement regarding the involvement of $^1O_2$ in photodamage of PSII. Part of the publications conclude that $^1O_2$ does not induce photodamage of PSII at all, and the $^1O_2$ induced inhibition of PSII activity in intact systems arises solely from the inhibition of PSII repair (see

[Nishiyama *et al.*, 2006]). Other data, however, demonstrate that the extent of photodamage, which is measured in the presence of lincomycin or in isolated photosynthetic membranes, is correlated with the rate of $^1O_2$ production (see [Rehman *et al.*, 2013]).

Reconciliation of this disturbing disagreement requires scrutinization of the respective experimental conditions. One obvious observation is that the works, which concluded the lack of $^1O_2$ on photodamage used very high light intensity (1500 μmol photons $m^{-2}s^{-1}$) in combination with chloramphenicol as protein synthesis inhibitor. While in the other studies, where the effect of $^1O_2$ on photodamage was demonstrated, lower light intensity (500 μmol photons $m^{-2}s^{-1}$) and lincomycin as protein synthesis inhibitor was used (see [Rehman *et al.*, 2013]). Chloramphenicol has been shown to act as an electron acceptor both in PSI [Okada *et al.*, 1991] and PSII [Rehman *et al.*, 2016b]. This effect on the one hand interferes with the O_2 evolution measurements, which are used to monitor PSII activity. On the other hand, the electrons from chloramphenicol are transferred to O_2 and produce superoxide [Okada *et al.*, 1991; Rehman *et al.*, 2016a], which enhances the rate of PSII photodamage [Rehman *et al.*, 2016a], as also shown for superoxide generated by PSI specific illumination in thylakoids [Tjus *et al.*, 2001], or by methylviologen in *Chlamydomonas* [Krieger-Liszkay *et al.*, 2011]. These adverse effects of chloramphenicol together with the high light intensity lead to a very steep loss of PSII activity, which makes it difficult to observe smaller changes. An example for the problem of drawing conclusions from rapidly changing photodamage curves can be found in a recent paper [Takahashi *et al.*, 2018] in which the presented data (Figure 3) show an at least 10% protection by the overexpression of the OCP protein in *Synechocystis* 6803, yet the conclusion from these data was that OCP does not protect against photodamage.

A further problem arises from the accessibility of $^1O_2$ produced from externally added photosensitizers to reach protein embedded targets, such as the Mn cluster. Despite its high reactivity $^1O_2$ has a long diffusion distance in cellular systems. In animal cells $^1O_2$ generated in the cytoplasm was capable of moving across cell membranes [Snyder *et al.*, 2005]. While, in *Chlamydomonas* cells high light induced $^1O_2$ was shown to leave the thylakoid membrane and involved in nuclear signaling [Fischer *et al.*, 2007]. Therefore, externally added $^1O_2$ sensitizers can be applied to study the role of $^1O_2$ in photoinhibition. A commonly used photosensitizer is rose bengal, which reaches the chloroplasts in tobacco leaves, and accelerates PSII damage concomitant with the oxidative modification of the D1 protein [Hideg *et al.*, 2007; Kovács *et al.*, 2014]. Illumination of isolated thylakoids in the presence of rose bengal has also been shown to induce the modification of the PSII acceptor side and D1 protein degradation [Chung and Jung, 1995]. However, in these studies rose bengal was used at 100 μM – 2 mM concentrations, which is much higher than the 2–10 μM used in the cyanobacterial studies in which no $^1O_2$ was detected on PSII photodamage [Nishiyama *et al.*, 2004]. Therefore, it is quite possible that despite the relatively large diffusion distance of $^1O_2$ in cellular environment externally generated $^1O_2$ in the cyanobacterial systems could not reach efficiently the Mn cluster, which is buried under a protective protein shield on the lumenal side of the thylakoid membrane. This situation is in a stark contrast with the internal generation of $^1O_2$ by $^3Chl_{D1}$ in the PSII reaction center, which is located only 2.0–2.5 nm from the Mn cluster. On the other hand, the $^1O_2$ generated by rose bengal in the cytosol of *Synechocystis* cells can easily interact with the components of the protein synthesis machinery and can inhibit the PSII repair cycle at the translational elongation step explaining the effect of rose bengal generated $^1O_2$ on PSII repair [Nishiyama *et al.*, 2004]. Further experiments with careful control of the accessibility of external $^1O_2$ photosensitizers to different cell compartments will be needed to clarify this important point.

Accessibility can also cause problems in the interpretation of data regarding the effect of $^1O_2$ scavengers on photodamage. α-tocopherol and plastoquinone are well known endogenous $^1O_2$ scavengers in photosynthetic systems, which are irreversibly oxidized upon interaction with this reactive molecule [Kruk and Trebst, 2008; Nowicka and Kruk, 2012]. A rapid turnover of both plastoquinone and α-tocopherol was observed under high light stress in *C. reinhardtii* and had been correlated to the conditions of the D1 protein turnover [Krieger-Liszkay and Trebst, 2006; Kruk and Trebst, 2008]. The role of α-tocopherol in the protection of PSII from photodamage has been confirmed by [Havaux *et al.*, 2000; Golan *et al.*, 2006]. However, other data showed that in *Synechocystis* 6803 and *Arabidopsis* mutants, which lack α-tocopherol only the efficiency of PII repair is suppressed, but the rate of photodamage was not enhanced [Hakala-Yatkin *et al.*, 2011; Inoue *et al.*, 2011).

This discrepancy could be caused by the inability of α-tocopherol to access the heart of the PSII RC where $^3P_{680}$ is acting as $^1O_2$ photosensitizer. However, α-tocopherol can scavenge $^1O_2$ inside thylakoid membrane, which decreases the $^1O_2$ level in the cytosol where it can interact with the protein synthesis machinery. This idea is supported by the finding that Trolox, a water-soluble analog of α-tocopherol protects against light induced damage of the surface exposed regions of the D1–D2 reaction center heterodimer in isolated PSII reaction center complexes, but not the Chls embedded in the protein complex (Arellano *et al.*, 2011). These data suggest that $^1O_2$ scavengers which are unable to access the membrane embedded regions of the PSII complex cannot provide protection against $^1O_2$ mediated damage to the reaction centre Chls or the Mn cluster. Such scavengers can protect only against $^1O_2$ which is diffused out from the inner part of the PSII RC, or produced outside of this region, e.g. from light harvesting

Chls, or Chl biosynthesis intermediates. A nice example for such an effect is the partial protection against light induced Chl bleaching by externally added histidine, which eliminates $^1O_2$ *via* chemical trapping, in the dinoflagellate alga *Symbiodinium* [Rehman *et al.*, 2016b].

The presence of $^1O_2$ outside the thylakoid membranes is demonstrated by the ability of externally added histidine to induce O_2 uptake in intact microalgal cells [Rehman *et al.*, 2013,2016b]. The modulation of the rate of $^1O_2$ production by changes in the redox state of membrane embedded electron transport components when measured *via* histidine oxidation (see above) also shows that most of the $^1O_2$, which is captured by histidine, originates from the PSII reaction center. This idea is supported by the application of external magnetic field, which lowers the yield of $^3P_{680}$ and therefore should suppress $^1O_2$, decreases protein synthesis dependent loss of PSII activity [Hakala-Yatkin *et al.*, 2011]. Although no $^1O_2$ was measured in this study the data indicate that $^1O_2$, which is produced in the PSII reaction center can inhibit PSII repair outside the thylakoid membranes.

The controversial part of this study is that the magnetic field had apparently no effect on PSII photodamage in the presence of lincomycin. The situation is further complicated by the effect of OCP, which is attached to the phycobilisomes on the stromal side of the thylakoid membrane, but can still protect against PSII photodamage [Sedoud *et al.*, 2014]. This effect could be explained by assuming that $^1O_2$ present outside the thylakoids can damage the PSII complex, which could be alleviated by external scavengers. This idea would be in line with hypotheses that $^1O_2$ sources located outside of PSII, *e.g.* in the cytb$_6$f complex [Suh *et al.*, 2000] can contribute to PSII damage.

8. The Mechanism of $^1O_2$ Mediated Photoinhibition of PSII

As described in detail in earlier reviews [Vass, 2011, 2012] the initial event of PSII photodamage under strong illumination is the reduction of the PQ pool in the absence of suitable electron sink capacity (Calvin-Benson cycle, flavodiiron proteins, *etc.*). The highly reduced state of the PQ pool slows down forward electron transport at the acceptor side of PSII, and can even block it if Q_A becomes doubly reduced [van Mieghem *et al.*, 1989; Vass *et al.*, 1992]. The modification of PSII acceptor side facilities direct conversion of $^1P_{680}$* to $^3P_{680}$* and also the backflow of electrons leading to the recombination of the $P_{680}^+Phe^-$ primary charge separated state, which also yields $^3P_{680}$* with high probability. $^3P_{680}$* in turn interacts with O_2 and converts it to highly reactive $^1O_2$. The triplet state, which is produced upon charge recombination from $^3[P_{680}^+Phe^-]$, has been shown to be located on the so called Chl_{D1} component of the P_{680} Chl assembly [van Mieghem *et al.*, 1991; Noguchi, 2002], which is in the middle of the PSII complex.

An important observation is the linear light intensity dependence of the rate of photodamage even above the intensities, which saturate photosynthesis [Tyystjärvi and Aro, 1996]. An often used argument is that only the direct photoinactivation of the Mn cluster can explain this observation, but $^1O_2$ production which is facilitated by changes in the redox state of electron transport components, cannot. As described in detail by [Rehman *et al.*, 2013] the rate of $^1O_2$ production is linearly dependent on light intensity even above the saturation of photosynthesis. At high light intensities, when photosynthesis is saturated, forward electron transport from Phe^- to Q_A, Q_B and the PQ pool is blocked, and PSII is in the closed state. Under these conditions primary charge separation still takes place, but the $^1[P_{680}{}^+Phe^-]$ state is stabilized and there is an increased probability for spin conversion to the $^3[P_{680}{}^+Phe^-]$ state, whose recombination yields $^3P_{680}$, the sensitizer of $^1O_2$. The rate of primary charge separation is light intensity dependent. Therefore, the steady state level of $^3P_{680}$ linearly increases with increasing light intensity in a very wide range, i.e. until all P_{680} is converted to the triplet state, which explains the linear light intensity dependence of $^1O_2$ production [Rehman *et al.*, 2013].

The lifetime of $^1O_2$ ranges from 3–3.5 μs in aqueous environment [Baier *et al.*, 2005; Skovsen *et al.*, 2005], which increases to *ca.* 10–14 μs in lipidic environment, such as cell membranes [Baier *et al.*, 2005]. These lifetime values correspond to *ca.* 200 to 400 nm diffusion distance of $^1O_2$, respectively [Baier *et al.*, 2005; Skovsen *et al.*, 2005], which make all important electron transport components accessible for $^1O_2$ produced from $^3Chl_{D1}$. $^1O_2$ is known to interact with many organic molecules, including pigments, lipids and amino acids.

The $^1O_2$ induced damage of amino acid residues is very important, since this effect can be involved in the oxidative modification of the D1 (and D2) reaction center protein(s) [Lupínková and Komenda, 2004] leading to chemical cleavage of the polypeptide change [Okada *et al.*, 1996], or marking it for proteolytic degradation. Even more important is that $^1O_2$ sensitive aromatic amino acid residues, such as His are involved in the ligation of the catalytic Mn cluster [Barber *et al.*, 2004; Ferreira *et al.*, 2004; Suga *et al.*, 2015], which is located only *ca.* 20–25 Å (2–2.5 nm) away from $^3Chl_{D1}$, the primary source of $^1O_2$ in PSII. Although direct experimental demonstration of $^1O_2$ mediated damage to the catalytic Mn cluster is missing synthesis of available indirect evidence (see [Zavafer *et al.*, 2017]) supports the idea that $^1O_2$ may be responsible for light induced damage of the Mn cluster instead of its direct hit by photons in the visible spectral range.

The data discussed in this section provide ample evidence in various experimental systems for the idea that enhanced $^1O_2$ production is accompanied by enhanced photodamage of PSII in the absence of protein synthesis dependent repair. Even more importantly, elimination of $^1O_2$ by various protective systems (quenching, scavenging, competing charge recombination pathways, and elimination of excitation pressure) protect against photodamage, which supports the idea that $^1O_2$ actually causes the functional and structural damage of PSII. Although these findings do not rule out that direct light induced damage of the Mn cluster might also be involved in PSII photodamage, they make it absolutely clear that direct photon hit of the Mn cluster cannot be the sole mechanism of photodamage, or even the specific first step in the so called two-step mechanism suggested by [Ohnishi *et al.*, 2005].

A plausible reconciliation of these seemingly contradicting data and ideas could be the involvement of $^1O_2$ in the inactivation of the Mn cluster as was proposed recently [Zavafer *et al.*, 2017]. However, further experimental work will be needed to clarify this important point.

As regards the involvement of $^1O_2$ in the repair cycle the situation is more clear than in the case of photodamage, There is a consensus in the literature that $^1O_2$ inhibits PSII repair at the translational elongation step of *psbA* mRNA during *de novo* synthesis of photodamaged D1 [Nishiyama *et al.*, 2006]. It also of note, however, that this type of inhibition of PSII repair is not specific for $^1O_2$ since other ROS (H_2O_2, $O_2^{-\bullet}$ $OH^{-\bullet}$) have been suggested to exert the same effect on D1 protein synthesis [Nishiyama *et al.*, 2001, 2006].

9. Conclusions

$^1O_2$ undoubtedly plays an important role in photoinhibition. Since the main source of $^1O_2$ in PSII is $^3P_{680}$ ($^3Chl_{D1}$), which is located in the heart of the reaction center $^1O_2$ can easily reach vital targets of the light energy converting machinery. These targets could include the catalytic Mn cluster, which is ligated partly by amino acid residues, e.g. histidine, which are efficiently attacked by $^1O_2$. Such an effect could provide an explanation for the discrepancy between observations, which show initial damage of the Mn cluster and the correlation of PSII inactivation with $^1O_2$ production. $^1O_2$ can also diffuse out from the PSII reaction center and inhibit the repair cycle of PSII at the *de novo* synthesis step of the D1 protein.

It is also of note that despite the compelling evidence concerning the involvement of $^1O_2$ in photodamage the contribution of other, $^1O_2$-independent, processes cannot be excluded. There is a possibility that $^1O_2$-mediated and direct light induced damaging processes proceed in parallel in a wavelength dependent manner. However, clarification of this important question will require further, carefully designed experiments.

Acknowledgments

Work on this chapter was supported by a grant from the Hungarian National Research, Development and Innovation Office (NKFIH K-116016).

References and Citations

Adir, N., Zer, H., Shochat, S. and Ohad, I. (2003). Photoinhibition — a historical perspective. *Photosynth. Res.*, 76, 343–370.

Andersson, B., Salter, H., Virgin, I., Vass, I. and Styring, S. (1992). Photodamage to Photosystem II — primary and secondary events. *J. Photochem. Photobiol.*, 15B,15–31.

Arellano, J.B., Li, H., González-Pérez, S., Gutiérrez, J., Melo, T.B., Vacha, F., Naqvi, K.R. (2011). Trolox, a water-soluble analogue of α-tocopherol, photoprotects the surface-exposed regions of the photosystem II reaction center *in vitro*. Is this physiologically relevant? *Biochemistry*, 50, 8291–8301.

Aro, E.-M., Virgin, I. and Andersson, B. (1993). Photoinhibition of Photosystem II. Inactivation, protein damage and turnover. *Biochim. Biophys. Acta*, 1143, 113–134.

Baier, J., Maier, M., Engl, R., Landthaler, M. and Bäumler, W. (2005). Time-resolved investigations of singlet oxygen luminescence in water, in phosphatidylcholine, and in aqueous suspensions of phosphatidylcholine or HT29 cells. *J. Phys. Chem.*, 109, 3041–3046.

Barber, J. and Andersson, B. (1992). Too much of a good thing: light can be bad for photosynthesis. *Trends Biochem. Sci.*, 17, 61–66.

Barber, J., Ferreira, K., Maghlaoui, K. and Iwata, S. (2004). Structural model of the oxygen-evolving centre of Photosystem II with mechanistic implications. *Phys. Chem. Chem. Phys.*, 6, 4737–4742.

Bersanini, L., Battchikova, N., Jokel, M., Rehman, A.U., Vass, I., Allahverdiyeva, Y. and Aro, E.-M. (2014). Flavodiiron protein Flv2/Flv4-related photoprotective mechanism dissipates excitation pressure of PSII in cooperation with phycobilisomes in cyanobacteria. *Plant Physiol.*, 164 (2), 805–818.

Blubaugh, D.J., Atamian, M., Babcock, G.T., Golbeck, J.H. and Cheniae, G.M. (1991). Photoinhibition of hydroxylamine-extracted Photosystem II membranes: Identification of the sites of photodamage. *Biochemistry*, 30, 7586–7597.

Blubaugh, D.J. and Cheniae, G.M. (1990). Kinetics of photoinhibition in hydroxylamine-extracted Photosystem II membranes: Relevance to photoactivation and sites of electron donation. *Biochemistry*, 29, 5109–5118.

Bustos, S.A., Schaefer, M.R. and Golden, S.S. (1990). Different and rapid responses of four cyanobacterial *psbA* transcripts to changes in light intensity. *J. Bacteriol*, 172, 1998–2004.

Campbell, D., Erikson, M.-J., Öquist, G., Gustafsson, P. and Clarke, A.K. (1998). The cyanobacterium *Synechochoccus* resists UV-B by exchanging Photosystem II reaction-center D1 proteins. *Proc. Natl. Acad. Sci. USA*, 95, 364–369.

Campbell, D., Zhou, G.Q., Gustafsson, P., Oquist, G. and Clarke, A.K. (1995). Electron transport regulates exchange of two forms of photosystem II D1 protein in the cyanobacterium *Synechococcus*. *EMBO J.*, 14, 5457–5466.

Chung, S.K. and Jung, J. (1995). Inactivation of the acceptor side and degradation of the D1 protein of Photosystem II by singlet oxygen photogenerated from the outside. *Photochem. Photobiol.*, 61, 383–389.

Cser, K. and Vass, I. (2007). Radiative and non-radiative charge recombination pathways in Photosystem II studied by thermoluminescence and chlorophyll fluorescence in the cyanobacterium *Synechocystis* 6803. *Biochim. Biophys. Acta*, 1767, 233–243.

Driever, S.M., Fryer, M.J., Mullineaux, P.M. and Baker, N.R. (2009). Imaging of reactive oxygen species *in vivo*. In: *Plant Signal Transduction. Methods in Molecular Biology*, Pfannschmidt T (ed.). Humana Press, a part of Springer Sience + Business Media, LLC 2009, pp. 109–116.

Durrant, J.R., Giorgi, L.B., Barber, J., Klug, D.R. and Porter, G. (1990). Characterization of triplet states in isolated Photosystem II reaction centres: oxygen quenching as a mechanism for photodamage. *Biochim. Biophys. Acta*, 1017, 167–175.

Ferreira, K.N., Iverson, T.M., Maghlaoui, K., Barber, J. and Iwata, S. (2004). Architecture of the photosynthetic oxygen-evolving center. *Science*, 303, 1831–1838.

Fischer, B.B., Krieger-Liszkay, A., Hideg, É., Šnyrychová, I., Wiesendanger, M. and Eggen, R.I.L. (2007). Role of singlet oxygen in chloroplast to nucleus retrograde signaling in *Chlamydomonas reinhardtii*. *FEBS Lett.*, 581, 5555–5560.

Flors, C., Fryer, M.J., Waring, J., Reeder, B., Bechtold, U., Mullineaux, P.M., Nonell, S., Wilson, M.T. and Baker, N.R. (2006). Imaging the production of singlet oxygen *in vivo* using a new fluorescent sensor, Singlet Oxygen Sensor Greenr®. *J. Exp. Bot.*, 57, 1725–1734.

Fufezan, C., Rutherford, A.W. and Krieger-Liszkay, A. (2002). Singlet oxygen production in herbicide-treated Photosystem II. *FEBS Lett.*, 532, 407–410.

Golan, T., Müller-Moulé, P. and Niyogi, K.K. (2006). Photoprotection mutants of *Arabidopsis thaliana* acclimate to high light by increasing photosynthesis and specific antioxidants. *Plant Cell Environ.*, 29, 879–887.

Gollmer, A., Arnbjerg, J., Blaikie, F.H., Pedersen, B.W., Breitenbach, T., Daasbjerg, K., Glasius, M. and Ogilby, P.R. (2011). Singlet oxygen sensor green (R): Photochemical behavior in solution and in a mammalian cell. *Photochem. Photobiol.*, 87(3), 671–679. doi:10.1111/j.1751-1097.2011.00900.x.

Hajdu, K., Rehman, A.U., Vass, I. and Nagy, L. (2018). Detection of singlet oxygen formation inside photoactive biohybrid composite material. *Materials*, 11 (28; doi:10.3390/ma11010028).

Hakala-Yatkin, M., Sarvikas, P., Paturi, P., Mantysaari, M., Mattila, H., Tyystjarvi, T., Nedbal, L. and Tyystjarvi, E. (2011). Magnetic field protects plants against high light by slowing down production of singlet oxygen. *Physiol. Plant*, 142(1), 26–34. doi:10.1111/j.1399-3054.2011.01453.x.

Hakala, M., Rantamaki, S., Puputti, E.M., Tyystjarvi, T. and Tyystjarvi, E. (2006). Photoinhibition of manganese enzymes: insights into the mechanism of photosystem II photoinhibition. *J. Exp. Bot.*, 57(8), 1809–1816.

Hakala, M., Tuominen, I., Keranen, M., Tyystjarvi, T. and Tyystjarvi, E. (2005). Evidence for the role of the oxygen-evolving manganese complex in photoinhibition of Photosystem II. *Biochim. Biophys. Acta*, 1706, 68–80.

Hakkila, K., Antal, T., Rehman, A.U., Kurkela, J., Wada, H., Vass, I., Tyystjärvi, E. and Tyystjärvi, T. (2014). Oxidative stress and photoinhibition can be separated in the cyanobacterium *Synechocystis* sp. PCC 6803. *Biochim. Biophys. Acta*, 1837, 217–225.

Harris, D., Tal, O., Jallet, D., Wilson, A., Kirilovsky, D. and Adir, N. (2016). Orange carotenoid protein burrows into the phycobilisome to provide photoprotection. *Proc. Natl. Acad. Sci. USA*, 113(12), E1655-E1662.

Havaux, M., Bonfils, J.-P., Lutz, C. and Niyogi, K.K. (2000). Photodamage of the photosynthetic apparatus and its dependence on the leaf developmental stage in the *npq*1 *Arabidopsis* mutant deficient in the xanthophyll cycle enzyme violaxanthin de-epoxidase. *Plant Physiol.*, 124, 273–284.

Hideg, É., Kálai, T., Hideg, K. and Vass, I. (1998). Photoinhibition of photosynthesis *in vivo* results in singlet oxygen production. Detection *via* nitroxide-induced fluorescence quenching in broad bean leaves. *Biochemistry*, 37 (33), 11405–11411.

Hideg, É., Kós, P. B. and Vass, I. (2007). Photosystem II damage induced by chemically generated singlet oxygen in tobacco leaves. *Physiol. Plant*, 131, 33–40.

Hideg, É., Spetea, C. and Vass, I. (1994a). Singlet oxygen and free radical production during acceptor- and donor-side-induced photoinhibition. Studies with spin trapping EPR spectroscopy. *Biochim. Biophys. Acta*, 1186, 143–152.

Hideg, É., Spetea, C. and Vass, I. (1994b). Singlet oxygen production in thylakoid membranes during photoinhibition as detected by EPR spectroscopy. *Photosynth. Res.*, 39, 191–199.

Hideg, É. and Vass, I. (1995). Singlet oxygen is not produced in Photosystem I under photoinhibitory conditions. *Photochem. Photobiol.*, 62, 949–952.

Inoue, S., Ejima, K., Iwai, E., Hayashi, H., Appel, J., Tyystjärvi, E., Murata, N. and Nishiyama, Y. (2011). Protection by α-tocopherol of the repair of photosystem II during photoinhibition in *Synechocystis* sp. PCC 6803. *Biochim. Biophys. Acta*, 1807, 236–241.

Jansen, M.A.K., Depka, B., Trebst, A. and Edelman, M. (1993). Engagement of specific sites in the plastoquinone niche regulates degradation of the D1 protein in Photosystem II. *J. Biol. Chem.*, 268, 21246–21252.

Järvi, S., Suorsa, M. andAro, E.-M. (2015). Photosystem II repair in plant chloroplasts — Regulation, assisting proteins and shared components with photosystem II biogenesis. *Biochim. Biophys. Acta*, 1847(9), 900–909.

Jones, L.W. and Kok, B. (1966). Photoinhibition of chloroplast reactions. I. Kinetics and action spectra. *Plant Physiol.*, 41, 1037–1043.

Jung, J. and Kim, H.S. (1990). The chromophores as endogeneous sensitizers involved in the photogeneration of singlet oxygen in spinach thylakoids. *Photochem. Photobiol.*, 52, 1003–1009.

Kálai, T., Hideg, É., Vass, I. and Hideg, K. (1998). Double (flruorescent and spin) sensors for detection of reactive oxygen species in the thylakoid membrane. *Free Rad. Biol. Med.*, 24, 649–652.

Keren, N., Berg, A., van Kan, P.J.M., Levanon, H. and Ohad, I. (1997). Mechanism of Photosystem II photoinactivation and D1 protein degradation at low light: The role of back electron flow. *Proc. Natl. Acad. Sci. USA*, 94, 1579–1584.

Keren, N., Gong, H. and Ohad, I. (1995). Oscillations of reaction center II-D1 protein degradation *in vivo* induced by repetitive light flashes. *J. Biol. Chem.*, 270, 806–814.

Kerfeld, C.A. and Kirilovsky, D. (2013). Structural, mechanistic and genomic insights into OCP-mediated photoprotection. In: *Genomics of Cyanobacteria,* Chauvat F, Cassier-Chauvat C (eds.). 1st edn. Academic Press, Elsevier Ltd., pp. 1–26.

Kern, J., Chatterjee, R., Young, I.D., Fuller, F.D., Lassalle, L., Ibrahim, M., Gul, S., Fransson, T., Brewster, A.S., Alonso-Mori, R., Hussein, R., Zhang, M., Douthit, L., de Lichtenberg, C., Cheah, M.H., Shevela, D., Wersig, J., Seuffert, I., Sokaras, D., Pastor, E., Weninger, C., Kroll, T., Sierra, R.G., Aller, P., Butryn, A., Orville, A.M., Liang, M.N., Batyuk, A., Koglin, J. E., Carbajo, S., Boutet, S., Moriarty, N.W., Holton, J.M., Dobbek, H., Adams, P.D., Bergmann, U., Sauter, N.K., Zouni, A., Messinger, J., Yano, J. and Yachandra, V.K. (2018). Structures of the intermediates of Kok's photosynthetic water oxidation clock. *Nature*, 563(7731), 421.

Kim, C.S.H., G.H. Kim, J.M. andJung, J. (1993). *In situ* susceptibilities of Photosystems I and II to photosensitized inactivation *via* singlet oxygen mechanism. *Photochem. Photobiol.*, 57, 1069–1074.

Kirilovsky, D. (2007). Photoprotection in cyanobacteria: the orange carotenoid protein (OCP)-related non-photochemical-quenching mechanism. *Photosynth. Res.*, 93, 7–16.

Kirilovsky, D. and Kerfeld, C.A. (2013). The orange carotenoid protein: A blue-green light photoactive protein. *Photochem. Photobiol. Sci.*, 12, 1135–1143.

Koivuniemi, A., Swiezewska, E., Aro, E.M., Strying, S. and Andersson, B. (1993). Reduced content of the quinone acceptor Q_A in Photosystem II complexes isolated from thylakoid membranes after prolonged photoinhibition under anaerobic conditions. *FEBS Lett.*, 327, 343–346.

Komenda, J., Sobotka, R. and Nixon, P.J. (2012). Assembling and maintaining the Photosystem II complex in chloroplasts and cyanobacteria. *Curr. Opin. Plant Biol.*, 15, 245–251.

Komenda, J., Tichy, M., Prasil, O., Knoppová, J., Kuviková, S., de Vries, R. and Nixon, P.J. (2007). The exposed N-terminal tail of the D1 subunit is required for rapid D1 degradation during Photosystem II repair in *Synechocystis* sp PCC 6803. *Plant Cell*, 19, 2839–2854.

Kós, P.B., Deák, Z., Cheregi, O. and Vass, I. (2008). Differential regulation of *psbA* and *psbD* gene expression, and the role of the different D1 protein copies in the cyanobacterium *Thermosynechococcus elongatus* BP-1. *Biochim. Biophys. Acta*, 1777, 74–83.

Kovács, L., Kálai, T., Tandori, J., Kós, P.B. and Hideg, É. (2014). Assessing the applicability of singlet oxygen photosensitizers in leaf studies. *Photochem. Photobiol.*, 90, 129–136.

Krieger-Liszkay, A. (2005). Singlet oxygen production in photosynthesis. *J. Exp. Bot.*, 56, 337–346.

Krieger-Liszkay, A., Fufezan, C. and Trebst, A. (2008). Singlet oxygen production in photosystem II and related protection mechanism. *Photosynth. Res.*, 98, 551–564.

Krieger-Liszkay, A., Kós, P.B. and Hideg, É. (2011). Superoxide anion radicals generated by methylviologen in photosystem I damage photosystem II. *Physiol. Plant*, 142, 17–25.

Krieger-Liszkay, A. and Rutherford, A.W. (1998). Influence of herbicide binding on the redox potential of the quinone acceptor in Photosystem II.: Relevance to photodamage and phytotoxicity. *Biochemistry*, 37(50), 17339–17344.

Krieger-Liszkay, A. and Trebst, A. (2006). Tocopherol is the scavenger of singlet oxygen produced by the triplet states of chlorophyll in the PSII reaction centre. *J. Exp. Bot.* 57(8), 1677–1684.

Kruk, J. and Trebst, A. (2008). Plastoquinol as a singlet oxygen scavenger in photosystem II. *Biochim. Biophys. Acta*, 1777, 154–162.

Kusama, Y., Inoue, S., Jimbo, H., Takaichi, S., Sonoike, K., Hihara, Y. and Nishiyama, Y. (2015). Zeaxanthin and echinenone protect the repair of Photosystem II from inhibition by singlet oxygen in *Synechocystis* sp. PCC 6803. *Plant Cell Physiol.*, 1–11.

Lei Li, Eva-Mari Aro. and Harvey Millar, A. (2018). Mechanisms of photodamage and protein turnover in photoinhibition. *Trends Plant Sci.*, 23(8), 667–676.

Li, H., Melo, T.B., Arellano, J.B. and Naqvi, K.R. (2012). Temporal profile of the singlet oxygen emission endogenously produced by photosystem II reaction centre in an aqueous buffer. *Photosynth. Res.*, 112, 75–79.

Loll, B., Kern, J., Saenger, W., Zouni, A. and Biesiadka, J. (2005). Towards complete cofactor arrangement in the 3.0′ resolution structure of Photosystem II. *Nature*, 438, 1040–1044.

Lupínková, L. and Komenda, J. (2004). Oxidative modifications of the Photosystem II D1 protein by reactive oxygen species: From isolated protein to cyanobacterial cells. *Photochem. Photobiol.*, 79, 152–162.

Macpherson, A.N., Telfer, A., Barber, J. and Truscott, T.G. (1993). Direct detection of singlet oxygen from isolated Photosystem II reaction centres. *Biochim. Biophys. Acta*, 1–9.

Méndez-Hurtado, J., López, R., Suárez, D. and Menéndez, M.I. (2012). Theoretical study of the oxidation of histidine by singlet oxygen. *Chem. Eur. J.*, 18, 8437–8447.

Merry, S.A.P., Nixon, P.J., Barter, L.M.C., Schilstra, M., Porter, G. and Barber, J. (1998). Modulation of quantum yield of primary radical pair formation in Photosystem II by site-directed mutagenesis affecting radical cations and anions. *Biochemistry*, 37, 17439–17447.

Mishra, R.K., Mishra, N.P., Kambourakis, S., Orfanopoulos, M. and Ghanotakis, D. F. (1996). Generation and trapping of singlet oxygen during strong illumination of a Photosystem II core complex. *Plant Sci.*, 115, 151–155.

Murata, N., Allakhverdiev, S.I. and Nishiyama, Y. (2012). The mechanism of photoinhibiton *in vivo*: Re-evaluation of the roles of catalase, α-tocopherol, non-photochemical quenching, and electron transport. *Biochim. Biophys. Acta*, 1817, 1127–1133.

Murata, N., Takahashi, S., Nishiyama, Y. and Allakhverdiev, S.I. (2007). Photoinhibition of photosystem II under environmental stress. *Biochim. Biophys. Acta*, 1767, 414–421.

Nagy, L., Bálint, E., Barber, J., Ringler, A., Cook, K.M. and Maróti, P. (1995). Photoinhibition and law of reciprocity in photosynthetic reaction of *Synechocystis* sp. PCC 6803*. *J. Plant Physiol.*, 145, 410–415.

Nishiyama, Y., Allakhverdiev, S.I. and Murata, N. (2005). Inhibition of the repair of Photosystem II by oxidative stress in cyanobacteria. *Photosyn. Res.*, 84(1–3), 1–7.

Nishiyama, Y., Allakhverdiev, S.I. and Murata, N. (2006). A new paradigm for the action of reactive oxygen species in the photoinhibition of Photosystem II. *Biochim. Biophys. Acta*, 1757, 742–749.

Nishiyama, Y., Allakhverdiev, S.I., Yamamoto, H. and Hayashi, H. (2004). Singlet oxygen inhibits the repair of Photosystem II by suppressing the translation elongation of the D1 protein in *Synechocystis* sp. PCC 6803. *Biochemistry*, 43, 11321–11330.

Nishiyama, Y., Yamamoto, H., Allakhverdiev, S.I., Inaba, H., Yokota, A. and Murata, N. (2001). Oxidative stress inhibits the repair of photodamage to the photosynthetic machinery. *EMBO J*, 20, 5587–5594.

Nixon, P.J., Michoux, F., Yu, J., Boehm, M. and Komenda, J. (2010). Recent advances in understanding the assembly and repair of photosystem II. *Ann. Bot.*, 106, 1–16.

Noguchi, T. (2002) Dual role of triplet localization on the accessory chlorophyll in the photosystem II reaction center: Photoprotection and photodamage of the D1 protein. *Plant Cell Physiol.*, 43, 1112–1116.

Nowicka, B. and Kruk, J. (2012). Plastoquinol is more active than α-tocopherol in singlet oxygen scavenging during high light stress of *Chlamydomonas reinhardtii*. *Biochim. Biophys. Acta*, 1817, 389–394.

Ohad, I., Berg, A., Berkowicz, S.M., Kaplan, A. and Keren, N. (2011). Photoinactivation of photosystem II: is there more than one way to skin a cat? *Physiol. Plant*, 142, 79–86.

Ohnishi, N., Allakhverdiev, S.I., Takahashi, S., Higashi, S., Watanabe, M., Nishiyama, Y. and Murata, N. (2005). Two-step mechanism of photodamage to Photosystem II: Step 1 occurs at the oxygen-evolving complex and step 2 occurs at the photochemical reaction center. *Biochemistry*, 44, 8494–8499.

Okada, K., Ikeuchi, M., Yamamoto, N., Ono, T.A. and Miyao, M. (1996). Selective and specific cleavage of the D1 and D2 proteins of Photosystem II by exposure to singlet oxygen: factors responsible for the susceptibility to cleavage of the proteins. *Biochim. Biophys. Acta*, 1274, 73–79.

Okada, K., Satoh, K. and Katoh, S. (1991). Chloramphenicol is an inhibitor of photosynthesis. *FEBS Lett.*, 295, 155–158.

op den Camp, R.G.L., Przybyla, D., Ochsenbein, C., Laloi, C., Kim, C., Danon, A., Wagner, D., Hideg, É., Gőbel, C., Feussner, I., Nater, M. and Apel, K. (2003). Rapid induction of distinct stress responses after the release of singlet oxygen in *Arabidopsis*. *Plant Cell*, 15, 2320–2332.

Prasad, A., Ferretti, U., Sedlarova, M. and Pospisil, P. (2016). Singlet oxygen production in *Chlamydomonas reinhardtii* under heat stress. *Sci Rep.*, 6, 20094.

Prasad, A., Sedlarova, M. and Pospisil, P. (2018). Singlet oxygen imaging using fluorescent probe Singlet Oxygen Sensor Green in photosynthetic organisms. *Sci Rep.*, 8(1),13685.

Rappaport, F., Guergova-Kuras, M., Nixon, P.J., Diner, B.A. and Lavergne, J. (2002). Kinetics and pathways of charge recombination in Photosystem II. *Biochemistry*, 41, 8518–8527.

Rehman, A. U., Cser, K., Sass, L. and Vass, I. (2013). Characterization of singlet oxygen production and its involvement in photodamage of Photosystem II in the cyanobacterium *Synechocystis* PCC 6803 by histidine-mediated chemical trapping. *Biochim. Biophys. Acta*, 1827, 689–698.

Rehman, A.U., Kodru, S. and Vass, I. (2016a). Chloramphenicol mediates superoxide production in Photosystem II and enhances Its photodamage in isolated membrane particles. *Front. Plant Sci.*, 7, 479.

Rehman, A.U., Szabó, M., Deák, Z., Sass, L., Larkum, A., Ralph, P. and Vass, I. (2016b). *Symbiodinium* sp. cells produce light-induced intra- and extracellular singlet oxygen, which mediates photodamage of the photosynthetic apparatus and has the potential to interact with the animal host in coral symbiosis. *New Phytol.*, 212, 472–484.

Rinalducci, S., Pedersen, J.Z. and Zolla, L. (2004). Formation of radicals from singlet oxygen produced during photoinhibition of isolated light-harvesting proteins of Photosystem II. *Biochim. Biophys. Acta*, 1608, 63–73.

Rinalducci, S., Pedersen, J.Z. and Zolla, L. (2008). Generation of reactive oxygen species upon strong visible light irradiation of isolated phycobilisomes from *Synechocystis* PCC 6803. *Biochim. Biophys. Acta*, 1777, 417–424.

Rutherford, A.W., Paterson, D.R. and Mullet, J.E. (1981). A light-induced spin-polarized triplet detected by EPR in Photosystem II reaction centers. *Biochim. Biophys. Acta*, 635, 205–214.

Sanchez-Corrionero, A., Sanchez-Vicente, I., Gonzalez-Perez, S., Corrales, A., Krieger-Liszkay, A., Lorenzo, O. and Arellano, J.B. (2017). Singlet oxygen triggers chloroplast rupture and cell death in the zeaxanthin epoxidase defective mutant aba1 of *Arabidopsis thaliana* under high light stress. *J. Plant Physiol.*, 216, 188–196.

Sander, J., Nowaczyk, M., Buchta, J., Dau, H., Vass, I., Deak, Z., Dorogi, M., Iwai, M., Rögner, M. (2010). Functional characterization and quantification of the alternative PsbA copies in *Thermosynechococcus elongatus* and their role in photoprotection. *J. Biol. Chem.*, 285, 29851–29856.

Sedoud, A., Lopez-Igual, R., Rehman, A.U., Wilson, A., Perreau, F., Boulay, C., Vass, I., Krieger-Liszkay, A. and Kirilovsky, D. (2014). The cyanobacterial photoactive orange carotenoid protein is an excellent singlet oxygen quencher. *Plant Cell*, 26(4), 1781–1791.

Shen, J. R. (2015). The structure of Photosystem II and the mechanism of water oxidation in photosynthesis. *Annu. Rev. Plant Biol.*, 66, 23–48.

Silva, P., Thompson, E., Bailey, S., Kruse, O., Mullineaux, C.W., Robinson, C., Mann, N.H. and Nixon, P.J. (2003). FtsH is involved in the early stages of repair of Photosystem II in *Synechocystis* sp. PCC 6803. *Plant Cell*, 15, 2152–2164.

Skovsen, E., Snyder, J.W., Lambert, J.D.C. and Ogilby, P.R. (2005). Lifetime and diffusion of singlet oxygen in a cell. *J. Phys. Chem. B*, 109, 8570–8573.

Snyder, J.W., Skovsen, E., Lambert, J.D.C. and Ogilby, P.R. (2005). Subcellular, time-resolved studies of singlet oxygen in single cells. *J. Am. Chem. Soc.*, 127, 14558–14559.

Suga, M., Akita, F., Hirata, K., Ueno, G., Murakami, H., Nakajima, Y., Shimizu, T., Yamashita, K., Yamamoto, M., Ago, H. and Shen, J.-R. (2015). Native structure of photosystem II at 1.95 Â resolution viewed by femtosecond X-ray pulses. *Nature*, 517, 99–103.

Suh, H.-J., Kim, C.S. and Jung, J. (2000). Cytochrome b_6/f complex as an indigenous photodynamic generator of singlet oxygen in thylakoid membranes. *Photochem. Photobiol.*, 71, 103–109.

Szilárd, A., Sass, L., Hideg, É. and Vass, I. (2005). Photoinactivation of Photosystem II by flashing light. *Photosynth. Res.*, 84, 15–20.

Takahashi, H., Kusama, Y., Li, X., Takaichi, S. and Nishiyama, Y. (2018). Overexpression of orange carotenoid protein protects the repair of Photosystem II under strong light in *Synechocystis* sp. PCC 6803. *Plant Cell Physiol.* doi:10.1093/pcp/pcy218.

Takahashi, S. and Murata, N. (2008). How do environmental stresses accelerate photoinhibition? *Trends Plant Sci.*, 13, 178–182.

Tamura, N., Inoue, Y. and Cheniae, G.M. (1989). Photoactivation of the water-oxidizing complex in Photosystem II membranes depleted of Mn, Ca and extrinsic proteins. II. Studies on the function of Ca^{2+}. *Biochim. Biophys. Acta*, 976, 173–181.

Telfer, A., Bishop, S.M., Phillips, D. and Barber, J. (1994). Isolated photosynthetic reaction center of Photosystem II as a sensitizer for the formation of singlet oxygen. Detection and quantum yield determiantion using a chemical trapping technique. *J. Biol. Chem.*, 269, 13244–13253.

Telfer, A., Oldham, T.C., Phillips, D. and Barber, J. (1999). Singlet oxygen formation detected by near-infrared emission from isolated photosystem II reaction centres: direct correlation between P680 triplet decay and luminescence rise kinetics and its consequences for photoinhibition. *J. Photochem. Photobiol. B*, 48, 89–96.

Tibiletti, T., Rehman, A.U., Vass, I. and Funk, C. (2018). The stress-induced SCP/HLIP family of small lightharvesting-like proteins (ScpABCDE) protects Photosystem II from photoinhibitory damages in the cyanobacterium *Synechocystis* sp. PCC 6803. *Photosynth. Res.*, 135, 103–112.

Tichy, M., Lupínková, L., Sicora, C., Vass, I., Kuvikova, S., Prasil, O., Komenda, J. (2003). *Synechocystis* 6803 mutants expressing distinct forms of the Photosystem II D1 protein from *Synechococcus* 7942: relationship between the *psbA* coding region and sensitivity to visible and UV-B radiation. *Biochim. Biophys. Acta*, 1605, 55–66.

Tjus, S.E., Scheller, H.V., Andersson, B. and Moller, B.J. (2001). Active oxygen produced during selective excitation of Photosystem I is damaging not only to Photosystem I, but also to Photosystem II. *Plant Physiol.*, 125, 2007–2015.

Tomo, T., Kusakabe, H., Nagao, R., Ito, H., Tanaka, A., Akimoto, S., Mimuro, M. and Okazaki, S. (2012). Luminescence of singlet oxygen in photosystem II complexes isolated from cyanobacterium *Synechocystis* sp. PCC6803 containing monovinyl or divinyl chlorophyll *a*. *Biochim. Biophys. Acta*, 1817, 1299–1305.

Tyystjarvi, E. (2008). Photoinhibition of Photosystem II and photodamage of the oxygen evolving manganese cluster. *Coord. Chem. Rev.*, 252(3–4), 361–376.

Tyystjarvi, E. (2013). Photoinhibition of Photosystem II. In: *International Review of Cell and Molecular Biology*, Jeon K (ed). 1st edn. Academic Press, Elsevier Inc., pp. 243–303.

Tyystjärvi, E. and Aro, E.M. (1996). The rate constant of photoinhibition, measured in lincomycin- treated leaves, is directly proportional to light intensity. *Proc. Natl. Acad. Sci. USA*, 93, 2213–2218.

Vajravel, S., Kovács, L., Kis, M., Rehman, A.U., Vass, I., Gombos, Z. and Tóth, T. N. (2016). b-carotene influences the phycobilisome antenna of cyanobacterium *Synechocystis* sp. PCC 6803. *Photosynth. Res.*, 130, 403–415.

van Mieghem, F.J.E., Nitschke, W., Mathis, P. and Rutherford, A.W. (1989). The influence of the quinone-iron electron-acceptor complex on the reaction center photochemistry of Photosystem-II. *Biochim. Biophys. Acta*, 977(2), 207–214.

van Mieghem, F.J.E., Satoh, K. and Rutherford, A.W. (1991). A chlorophyll tilted 30-degree relative to the membrane in the Photosystem-II reaction center. *Biochim. Biophys. Acta*, 1058(3), 379–385.

Vass, I. (2011). Role of charge recombination processes in photodamage and photoprotection of the photosystem II complex. *Physiol. Plant.*, 142, 6–16.

Vass, I. (2012). Molecular mechanisms of photodamage in the Photosystem II complex. *Biochim. Biophys. Acta*, 1817, 209–217.

Vass, I. and Aro, E.-M. (2008). Photoinhibition of Photosystem II electron transport. In: *Primary Processes of Photosynthesis: Basic Principles and Apparatus*, Renger G (ed.). Comprehensice Series in Photochemical and Photobiological Sciences. Roy. Soc. Chem., Cambridge, pp. 393–411.

Vass, I. and Cser, K. (2009). Janus-faced charge recombinations in photosystem II photoinhibition. *Trends Plant Sci.*, 14, 200–205.

Vass, I. and Cser, K., Cheregi, O. (2007). Molecular mechanisms of light stress of photosynthesis. *Ann. NY Acad. Sci.*, 1113, 114–122.

Vass, I., Máté, Z., Sass, L., Viczián, A., Nagy, F., Kirilovsky, D. and Etienne, A.-L. (1999). UV-B induced damage of Photosystem II and its repair *via de novo* synthesis of the D1 and D2 reaction center subunits. In: *Photosynthesis: Mechanisms and Effects*, Garab G (ed.). Kluwer, pp. 2109–2114.

Vass, I., Sass, L., Spetea, C., Bakou, A., Ghanotakis, D. and Petrouleas, V. (1996). UV-B induced inhibition of Photosystem II electron transport studied by EPR and chlorophyll fluorescence. Impairment of donor and acceptor side components. *Biochemistry*, 35, 8964–8973.

Vass, I., Styring, S., Hundal, T., Koivuniemi, A., Aro, E.-M. and Andersson, B. (1992). Reversible and irreversible intermediates during photoinhibition of Photosystem II: Stable reduced Q_A species promote chlorophyll triplet formation. *Proc. Natl. Acad. Sci. USA*, 89, 1408–1412.

Vass, I., Turcsányi, E., Touloupakis, E., Ghanotakis, D. and Petroluleas, V. (2002). The mechanism of UV-A radiation-induced inhibition of Photosystem II electron transport studied by EPR and chlorophyll fluorescence. *Biochemistry*, 41, 10200–10208.

Verlhac, J.B., Gaudemer, A. and Kraljic, I. (1984). Water-soluble porphyrins and metalloporphyrins as photosensitizers in aerated aqueous solutions. I. Detection and determination of quantum yield of formation of singlet oxygen. *Nouv. J. Chim.*, 8, 401–406.

Virgin, I., Styring, S. and Andersson, B. (1988). Photosystem II disorganization and manganese release after photoinhibition of isolated spinach thylakoid membranes. *FEBS Lett.*, 233, 408–412.

Wagner, D., Przybyla, D., op den Camp, R., Kim, C., Landgraf, F., Lee, K.P., Würsch, M., Laloi, C., Nater, M., Hideg, E. and Apel, K. (2004). The genetic basis of singlet oxygen-induced stress responses of *Arabidopsis thaliana*. *Science*, 306, 1183–1185.

Wilson, A., Ajlani, G., Verbavatz, J.-M., Vass, I., Kerfeld, C.A. and Kirilovsky, D. (2006). A soluble carotenoid protein involved in phycobilisome-related energy dissipation in cyanobacteria. *Plant Cell*, 18, 992–107.

Zavafer, A., Cheah, M.H., Hillier, W., Chow, W.S. and Takahashi, S. (2015a). Photodamage to the oxygen evolving complex of photosystem II by visible light. *Sci. Rep.*, 5, 8.

Zavafer, A., Chow, W.S. and Cheah, M.H. (2015b). The action spectrum of Photosystem II photoinactivation in visible light. *J. Photochem. Photobiol. B*, 152, 247–260.

Zavafer, A., Koinuma, W., Chow, W.S., Cheah, M.H. and Mino, H. (2017). Mechanism of photodamage of the oxygen evolving Mn cluster of Photosystem II by excessive light energy. *Sci. Rep.*, 7(1), 7604.

Zhang, P., Eisenhut, M., Brandt, A.-M., Carmel, D., Silén, H.M., Vass, I., Allahverdiyeva, Y., Salminen, T.A. and Aro, E.-M. (2012). Operon *flv4-flv2* provides cyanobacterial Photosystem II with flexibility of electron transfer. *Plant Cell*, 24, 1952–1971.

Zolla, L. and Rinalducci, S. (2002). Involvement of active oxygen species in degradation of light-harvesting proteins under light stresses. *Biochemistry*, 41, 14391–14402.

Chapter 6

Modular Assembly of ATP Synthase

John E. Walker, Jiuya He and Joe Carroll

Medical Research Council Mitochondrial Biology Unit,
University of Cambridge, Cambridge Biomedical Campus,
Hills Road, Cambridge CB2 0XY, UK
walker@mrc-mbu.cam.ac.uk

The ATP synthases in human mitochondria are assembled as separate modules representing the F_1-catalytic domain, the stator and a ring of eight c-subunits, an essential part of the enzyme's rotor. Then an intermediate template structure containing the F_1-domain the c_8-ring and the stator is built up from these modules. Into this structure the membrane subunits ATP6 and ATP8 are inserted, ATP6 forming the transmembrane proton pathway with the adjacent c_8-ring that is required for generating its rotation from the proton motive force. The association of ATP6 and ATP8 with this intermediate is stabilized by the addition of the 6.8 kDa proteolipid. The resulting complex can make ATP and form dimers at the apices of the cristae by interactions in the membrane domain between the two 6.8 kDa proteolipids, and the two ATP6 subunits. In a subsequent step subunit DAPIT (diabetes associated protein in insulin sensitive tissue) is added linking the dimers in long rows along the cristal edges. The ATP synthase in *Saccharomyces cerevisiae* is assembled by a similar, but non-identical, pathway. This modular assembly may reflect a modular evolution of ATP synthases, where the proto-typical complex evolved by the association of pre-existing modules incapable individually of catalyzing ATP synthesis.

1. Introduction

The subunit compositions and the arrangement of subunits in ATP synthases differ between the simpler eubacterial and chloroplast enzymes, and the more complex assemblies found in the mitochondria of eukaryotes (Figure 1).

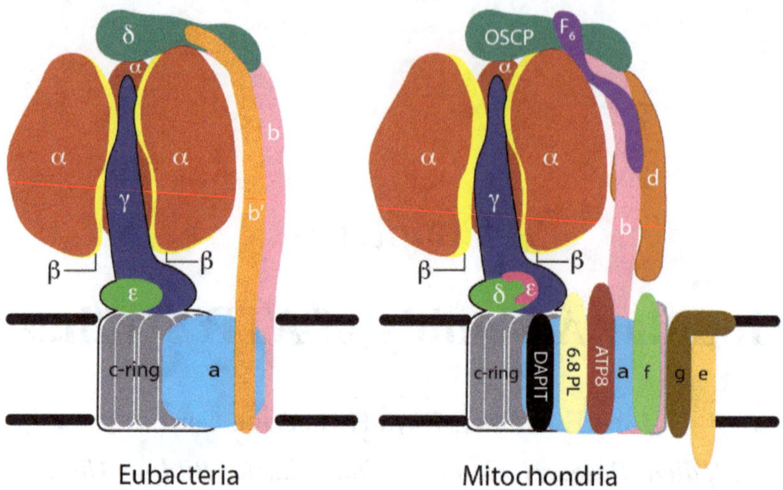

Figure 1. Subunit compositions and the arrangement of subunits in ATP synthases from eubacteria and chloroplasts (left) and human mitochondria (right). In each panel, in order to expose the γ-subunit in the central stalk, an α- and a β-subunit have been removed from the front of the upper F_1-catalytic domain which contains three α-subunits and three β-subunits and single copies of γ-, δ- and ε-subunits. Some eubacteria have identical b and b′-subunits, whereas others and chloroplasts have non-identical, but related b- and b′-subunits. In the mitochondrial complex, the supernumerary subunits e, f, g, DAPIT, 6.8 kDa proteolipid and ATP8 are components of the membrane domain.

The eubacterial and chloroplast complexes being the simplest and smallest ATP synthases, are assemblies of eight or nine kinds of subunits [Foster and Fillingame, 1979, 1982; Walker *et al.*, 1982, 1984; Falk *et al.*, 1985; Cozens and Walker, 1987; Van Walraven *et al.*, 1993]. Those with eight kinds of subunits have two identical b-subunits; those with nine subunits, represented by enzymes from photosynthetic bacteria and chloroplasts, have two non-identical, but related, subunits, b and b′. The number of c-subunits in the membrane domains of bacterial and chloroplast ATP synthases varies from 10–17, depending on the species [Hermolin and Fillingame, 1989; Jiang *et al.*, 2001; Meier *et al.*, 2005; Pogoryelow *et al.*, 2005; Vollmar *et al.*, 2009; Matthies *et al.*, 2009; Preiss *et al.*, 2015; Sobti *et al.*, 2016]. Their total molecular masses are in the range of 500–550 kDa. In contrast, the bovine enzyme, as an example of a mitochondrial ATP synthase, has 29 subunits of 18 types (including the regulatory protein, IF_1) with a combined molecular mass of about 650 kDa [Walker *et al.*, 1982, 1985, 1991; Fearnley and Walker, 1986; Collinson *et al.*, 1994; Chen *et al.*, 2007; Meyer *et al.*, 2007]. The core of the mitochondrial enzymes consists of a set of orthologs of the eight or nine subunits found in bacterial and chloroplast enzymes, with one exception. In the mitochondrial complex, the two bacterial b-subunits (or the single b and b′-subunits) are replaced by the mitochondrial b-subunit plus subunits d and F_6, which, although they have related functions to bacterial b- and b′-subunits in the peripheral stalk, are unrelated in sequence [Walker *et al.*, 1987, 1991; Collinson *et al.*, 1994]. The membrane domains of ATP synthases from vertebrates contain a c_8-ring, a feature that probably persists throughout metazoans [Walpole *et al.*, 2015]. In contrast, the ATP synthases from *Saccharomyces cerevisiae* and other characterized fungal enzymes have c_{10}-rings [Stock *et al.*, 1999; Hahn *et al.*, 2016; Vinothkumar *et al.*, 2016]. In addition, to these core subunits, depending on the species, the mitochondrial enzymes have up to six additional supernumerary subunits associated with their membrane domains known as subunits e, f, g, DAPIT (diabetes associated protein in insulin sensitive tissue), 6.8 kDa proteolipid (6.8PL) and ATP8 (see Figure 1) [Fearnley and Walker, 1986; Walker *et al.*, 1991; Collinson *et al.*, 1994; Chen *et al.*, 2007; Meyer *et al.*, 2007; Liu *et al.*, 2015].The supernumerary subunits have no known direct role in the synthesis or hydrolysis of ATP, although they are required for the structural integrity required of a fully functioning ATP synthase.

Irrespective of their source, the constituent subunits of all ATP synthase complexes that have been characterized biochemically and structurally are organized as shown in Figure 1. They have two major domains representing the membrane extrinsic and the membrane intrinsic sectors of the enzyme, and the two domains are linked together by central and peripheral stalks. The membrane extrinsic sector, known as the F_1-domain, is the catalytic part where ATP is formed from ADP and inorganic phosphate. The membrane intrinsic sector of ATP synthases, sometimes called F_o, contains a rotary motor driven by the transmembrane proton force, Δp, generated by respiration. These membrane extrinsic and intrinsic domains are connected by a central stalk composed of single copies of the γ-, δ- and ε-subunits [Stock *et al.*, 1999; Abrahams *et al.*, 1994; Watt *et al.*, 2010] and a peripheral stalk [Collinson *et al.*, 1994; Vinothkumar *et al.*, 2016; Karrasch *et al.*,1999; Dickson *et al.*, 2006; Rees *et al.*, 2009; Zhou *et al.*, 2015; Morales-Rios *et al.*, 2015]. The central stalk, in close association with the ring of c-subunits in the membrane domain, constitutes the enzyme's rotor, and *via* its rotary action it transmits energy from the membrane domain to the catalytic domain to drive the formation of ATP from ADP and phosphate. The rest of the enzyme constitutes the enzyme's stator against which the rotor turns. The peripheral stalk, composed of single copies of subunits b, d, F_6 and OSCP (oligomycin sensitivity conferral protein), is a key component of the stator. It is an elongated and largely α-helical structure linking the topmost extremity of the $\alpha_3\beta_3$-component of the F_1-domain to the a-subunit (ATPase-6 or ATP6 in mitochondria and subunit IV in chloroplasts) in the membrane domain [Sobti *et al.*, 2016; Hahn *et al.*, 2016; Vinothkumar *et al.*, 2016; Dickson *et al.*, 2006; Rees *et al.*, 2009; Zhou *et al.*, 2015; Morales-Rios *et al.*, 2015]. At the top of the F_1-domain of mitochondrial enzymes, the N-terminal domain of the OSCP binds to the N-terminal α-helical regions of the three α-subunits [Vinothkumar *et al.*, 2016; Rees *et al.*, 2009; Carbajo *et al.*, 2005, 2007]. In bacterial and chloroplast ATP synthases, the δ-subunit is bound in a similar way [Sobti *et al.*, 2016; Morales-Rios *et al.*, 2015]. In the membrane domain of the peripheral stalk, the N-terminal regions of the b-subunit (or subunits) bind to the a-subunit, and hold it against the rotating c-ring in the active enzyme [Sobti *et al.*, 2016; Hahn *et al.*, 2016; Vinothkumar *et al.*, 2016; Zhou *et al.*, 2015; Morales-Rios *et al.*, 2015]. The two extremities of the peripheral stalk are linked by a largely α-helical and probably mostly rigid section. The integrity of the peripheral stalk is essential for keeping Δp coupled to ATP synthesis, and, in enzymes where the action of the enzyme can be reversed, for maintaining the coupling of ATP hydrolysis to the generation of Δp. In mitochondrial enzymes, ATP8 is also part of the peripheral stalk [Hahn *et al.*, 2016; Vinothkumar *et al.*, 2016; Fearnley and Walker, 1986; Collinson *et al.*, 1994; Zhou *et al.*, 2016; Lee *et al.*, 2015]. It is anchored in the membrane by an N-terminal α-helix, and the rest of the subunit extends into the membrane extrinsic region of the peripheral stalk. ATP8 is essential for a fully functioning mitochondrial enzyme, probably by providing a brace to augment the role of the b-subunit of maintaining contact between the c-ring and the a-subunit [Vinothkumar *et al.*, 2016].

Another important feature of mitochondrial ATP synthases is that they are dimerized via specific interactions in their membrane domains between the ATP6 subunits and subunit j (yeast) or 6.8PL (human) subunits in each monomer [Guo *et al.*, 2017; He *et al.*, 2018], and the dimers themselves associate *via* front to back interactions to form long rows of dimers along the edges of cristae [Dudkina *et al.*, 2006; Davies *et al.*, 2011]. Removal of the capacity of the ATP synthase complexes to dimerize prevents the formation of the cristae [Paumard *et al.*, 2002].

2. Assembly of Mitochondrial ATP Synthases

The way in which mitochondrial ATP synthases are put together requires that proteins emanating from two genomes, the nuclear genome and the mitochondrial genome, are assembled together in the inner membranes of the organelle. The ATP synthases from yeast and human mitochondria have been studied, and the names of the genes and encoded subunits are summarized in Table 1.

In *S. cerevisiae*, three of the proteins, subunits Atp6p (ATP6), Atp8p (ATP8 or Aap1) and Atp9p (subunit c) are encoded in the mitochondrial genome (mt-DNA), and the remainder are nuclear gene products. In mammals, exemplified by humans, ATP6 and ATP8 are both encoded in mt-DNA, but subunit c is produced from three different nuclear genes, ATP5G1, ATP5G2 and ATP5G3 [Dyer and Walker, 1993; Yan *et al.*, 1994], and the other subunits are encoded in single

Table 1. Equivalent genes and orthologous subunits in human and yeast ATP synthases.

Human		*S. cerevisiae*	
Gene	**Subunit**	**Gene**	**Subunit**
ATP5A1	alpha	ATP1	alpha
ATP5B	beta	ATP2	beta
ATP5C1	gamma	ATP3	gamma
ATP5D	delta	ATP16	delta
ATP5E	epsilon	ATP15	epsilon
ATP5F1	F_o subunit b	ATP4	4 or b
ATP5G1	F_o subunit C1[a], or c	ATP9 or OLI1	9 or c
ATP5G2	F_o subunit C2[a], or c	—	—
ATP5G3	F_o subunit C3[a], or c	—	—
ATP5H	d	ATP7	d
ATP5I	e	ATP21	e or TIM11
ATP5J	coupling factor 6 or F_6	ATP14	h
ATP5J2	f	ATP17	f
ATP5L	g	ATP20	g
ATP5O	OSCP	ATP5	5 or OSCP
MT-ATP6	a or ATP6	ATP6	a or ATP6
MT-ATP8	protein 8, ATP8 or A6L	ATP8	8 or Aap1
C14orf2	6.8 kDa mitochondrial proteolipid or 6.8PL	ATP18	i or j
USMG5	Diabetes-associated protein in insulin-sensitive tissues or DAPIT	ATP19	k

[a]C1, C2 and C3 are identical subunits encoded in three different genes.

nuclear genes. The subunit c precursor proteins encoded in the three human genes differ in their mitochondrial import sequences, but removal of the import sequences during entry into the organelle generates an identical mature c-protein (see Figure 2).

The pathway of assembly of the yeast ATP synthase, studied extensively by Tzagoloff and colleagues (Figure 3) [Rak *et al.*, 2011], involves the formation of three modules representing the F_1-domain, the c_{10}-ring and a stator domain consisting of the peripheral stalk (subunits Atp4p, Atp7p, Atp14p and the OSCP) together with Atp6p and Atp8p (Figure 3). With the assistance of assembly factors Atp11p and Atp12p [Ackerman *et al.*, 2002], and under specific conditions FMC1 [Lefebvre-Legendre *et al.*, 2001], the F_1-domain is assembled from three copies of α- and β-subunits and single copies of γ-, δ- and ε-subunits. Assembly factor

```
ATP5G1 MQTAGALFISPALIRCCTRGLIRPVSASFLNSPVNSSKQPSY------SNFPLQVARREF 54
ATP5G2 MFACSKFVSTPSLVKSTSQLLSRPLSAVVLKRPEILTDESLSSL-AVSCPLTSLVSSRSF 59
ATP5G3 MFACAKLACTPSLIRAGSRVAYRPISASVLSRPEASRTGEGSTVFNGAQNGVSQLIQREF 60
       * :.. :  :*:*::. ::   **:** .*. *                   :  *.*

ATP5G1 QTSVVSRDIDTAAKFIGAGAATVGVAGSGAGIGTVFGSLIIGYARNPSLKQQLFSYAILG 114
ATP5G2 QTSAISRDIDTAAKFIGAGAATVGVAGSGAGIGTVFGSLIIGYARNPSLKQQLFSYAILG 119
ATP5G3 QTSAISRDIDTAAKFIGAGAATVGVAGSGAGIGTVFGSLIIGYARNPSLKQQLFSYAILG 120
       ***.:*******************************************************

ATP5G1 FALSEAMGLFCLMVAFLILFAM 136
ATP5G2 FALSEAMGLFCLMVAFLILFAM 141
ATP5G3 FALSEAMGLFCLMVAFLILFAM 142
       **********************
```

Figure 2. The sequences of the three precursors of the c-subunit of human ATP synthase, produced from ATP5G1, ATP5G2 and ATP5G3. The mitochondrial import sequences, surrounded by grey, are removed during entry into the organelle to generate an identical mature c-subunit from all three genes. The red lines indicate the trans-membrane α-helices. Asterisks denote identities.

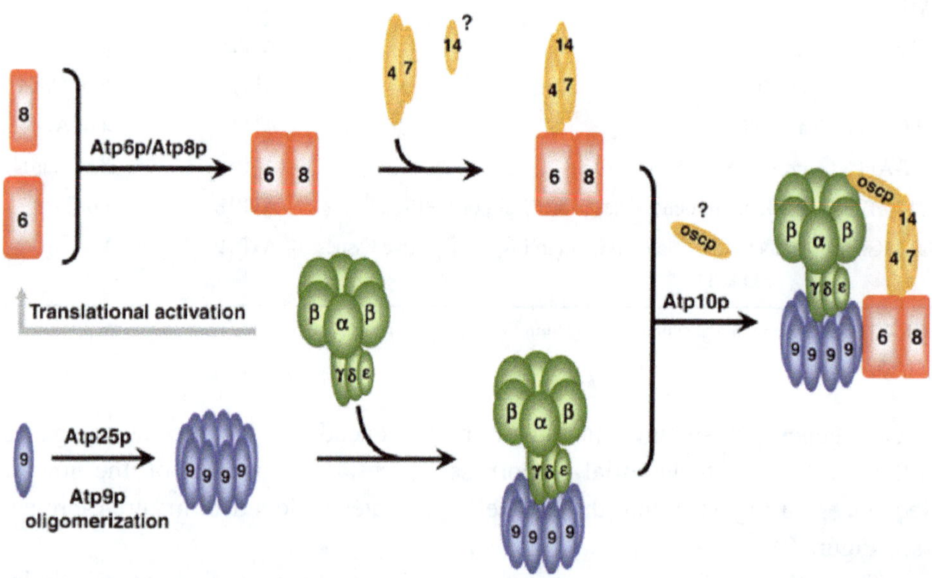

Figure 3. Pathway of assembly of ATP synthase from yeast mitochondria. Atp6p, Atp8p and Atp9 are encoded in mitochondrial genes. Atp25p is an assembly factor that assists in the formation of the ring made of 10 copies of Atp9p. Two other assembly factors, Atp11p and Atp12p assist in the assembly of the F_1-domain. Atp10p helps in the addition of the sub-complex of peripheral stalk and Atp6p and Atp8p subunits to the F_1-c_{10} sub-complex. Taken from Rak *et al.* [2011].

Atp25p, helps to put together the c_{10}-ring from 10 copies of Atp9p [Zeng *et al.*, 2008], which was proposed to then associate with the F_1-domain to form an F_1-c_{10} sub-complex. In a final step requiring the participation of Atp10p [Ackerman *et al.*, 1990], the stator module (the peripheral stalk plus Atp6p and Atp8p) associates with the F_1-c_{10} subcomplex to complete the monomeric complex. Atp23p is also required for assembly, and to cleave Atp6p into a mature form by proteolysis [Zeng *et al.*, 2007; Osman *et al.*, 2007], a step not required for human ATP6. A revised assembly pathway was proposed recently where the yeast specific inner membrane assembly complex (INAC), made from Ina17 and Ina22, is required for assembly of a stable peripheral stalk [Lytovchenko *et al.*, 2014] and formation of a complex containing F_1-peripheral stalk-Atp6/Atp8-Atp23-Atp10, to which the c_{10}-ring is added to produce the holoenzyme [Naumenko *et al.*, 2017].

Recent studies have shown that the assembly of the human ATP synthase also proceeds *via* the assembly of modules representing sub-domains of the complete complex, but the pathway is necessarily different to the pathway in *S. cerevisiae* because of the human c-subunit being a nuclear gene product. The assembly of the c-ring, made in this case, and probably in all metazoans, of eight c-subunits, has not yet been studied in detail, but TMEM70, a mitochondrial membrane protein may participate in helping to put the c-ring together [Vrbacky *et al.*, 2016], and human mutations in TMEM70 are associated with ATP synthase deficiency, and manifest in a range of symptoms including lactic acidosis, 3-methylglutaconic aciduria and cardiomyopathy [Čížková *et al.*, 2008; Torraco *et al.*, 2012]. Likewise, as yet, there is no definitive description of the assembly of the human F_1-domain. Human ATPAF1, ATPAF2 and C7orf55 are orthologs of yeast ATP11p, Atp12p and FMC1, and it is reasonable to assume that they play similar roles to the yeast orthologs in the assembly of the human F_1-domain [Wang *et al.*, 2001; Suzuki *et al.*, 2016; Li *et al.*, 2017]. Human mutations in the ε-subunit, a small 50 amino acid subunit that binds to the human δ-subunit at the foot of the central stalk, also affect assembly of the F_1-domain and lead to the accumulation of c-rings [Mayr *et al.*, 2011; Havlícková *et al.*, 2010].

By removing individual nuclear encoded membrane subunits, by disrupting their corresponding genes with CRISPR-Cas9, and characterization of the vestigial (residual) complexes, and the vestigial ATPase-complex in ρ^o-cells which lack mt-DNA and are incapable of making ATP6 and ATP8, a detailed description of the assembly of the membrane domain of the human ATP synthase has emerged (Figure 4) [He *et al.*, 2017a, 2017b, 2018]. In the absence of either of the peripheral stalk subunits OSCP or b, the cells can still assemble a sub-complex made of

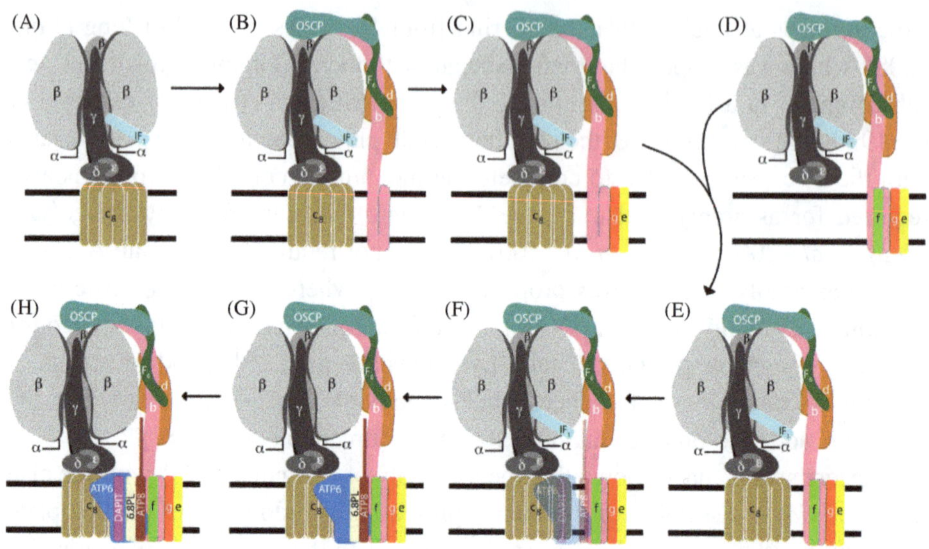

Figure 4. Pathway of assembly of the membrane domain of human ATP synthase. Parts A–D and F and G, intermediate vestigial complexes, and subunit compositions characterized from mitochondria of clonal cells where the genes for individual subunits have been disrupted by CRISPR-Cas9: A, HAP1-Δb or -ΔOSCP; B, HAP1-Δe or -Δg; C, HAP1-Δf; D, HAP1-Δc; F, HAP1-Δ6.8PL; G, HAP1-ΔDAPIT. E, vestigial complex from the mitochondria of ρ^0 cells. The horizontal black lines denote the boundaries of the inner mitochondrial membrane. Panel H, the complete monomeric ATP synthase.

the F_1-domain and the c_8-ring (Figure 4A). In a subsequent step, this complex then binds a peripheral stalk sub-complex containing subunits OSCP, b, d and F_6 (Figure 4B), but how this peripheral stalk sub-complex is assembled remains to be uncovered, although an incompletely defined sub-complex containing subunit b has been observed. In the next steps, subunits e and g associate with this intermediate in an unknown order (Figure 4C), and then in a subsequent step, subunit f binds to this complex (Figure 4E). The intermediate in Figure 4E also forms via an alternative route involving the addition of the c_8-ring to the vestigial complex in Figure 4D, which is the complex that accumulates in ρ^o-cells. The existence of the intermediate in Figure 4D provides strong evidence that the three modules F_1, c_8-ring and stator can come together by two pathways, one described in Figures 4A–4C and 4E, where first the F_1-domain and c_8-rings first associate to form F_1-c_8 (Figure 4A), and then the peripheral stalk is added (Figure 4B) and elaborated by the subsequent addition of subunits e and g, and then subunit f (Figures 4C and 4E), or alternatively, as suggested by the intermediate in Figure 4D, that the peripheral stalk and associated subunits e, f and g can be added, possibly in several steps, and that then the intermediate in Figure 4E forms by addition of the c_8-ring. This intermediate provides the template to allow the two mitochondrially encoded membrane subunits, ATP6 and ATP8 to be incorporated. However, their stable association with the complex, and the formation of a functional transmembrane proton pathway and an enzyme complex that can couple Δp to ATP synthesis, requires the further addition of subunit 6.8PL (Figure 4F). At this point the monomeric functional complex is complete, and the two subunits that form the interface between monomeric complexes, ATP6 and 6.8PL have been incorporated, and the complex has the capacity to form dimers. The final subunit to be incorporated is DAPIT, which probably provides an important link between dimers in the formation of the long rows of dimers on the cristae edges. The yeast ortholog of DAPIT is subunit k, and presumably it fulfills a similar role in yeast mitochondria. A striking property of DAPIT is that it is turned over much more rapidly than any of the other subunits of the complex, which turn over together at a slower rate [He *et al.*, 2018]. This rapid turnover of DAPIT may be associated with the re-modelling of mitochondrial membranes that occurs during the fission and fusion of mitochondria in live cells.

These recent studies of the assembly of the human ATP synthase have brought to light one additional feature of the assembly process, and that is the role of the ATPase inhibitory protein, IF_1, which can inhibit ATP hydrolysis, but not ATP synthesis [Pullman *et al.*, 1963]. In the assembly pathway in Figure 4, IF_1 was found associated with all of the intermediate sub-complexes in Figures 4A–4F. The complexes in Figure 4A–4E, lack a transmembrane pathway for protons, and therefore are incapable of making ATP, and, if unregulated, would act in reverse and hydrolyze ATP, and the complex in Figure 4F is at best partially coupled to Δp. Thus, by binding to these intermediates, IF_1 plays the important role of preventing wastage of ATP that would otherwise occur during the assembly of human ATP synthase. Although a similar role has not been remarked on for yeast IF_1, it is likely to occur.

3. Modular Evolution of ATP Synthase

Exactly how such a complex enzyme as ATP synthase evolved, probably more than 2.3 billion years ago is unknowable. However, it seems likely that the modular assembly of the mitochondrial enzyme described above may be paralleled and even related to a process whereby the ATP synthase complex evolved as separate structural modules, with functions other than ATP synthesis, which at some point in time became joined together to make a machine capable of making the vast amounts of ATP required each day for sustaining biological life on Earth. Evidence that this may be so, first came to light when the sequence of the *unc* operon encoding the eight subunits of the ATP synthase in *Escherichia coli* was first determined [Walker *et al.*, 1984; Gay and Walker, 1981a, 1981b; Saraste *et al.*, 1981].

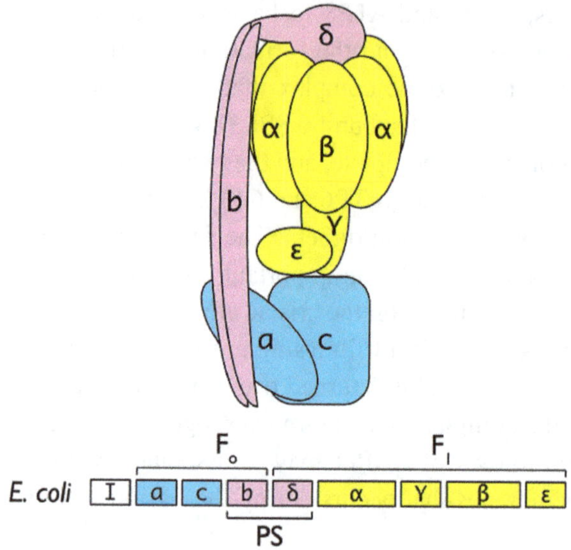

Figure 5. The organization of subunits in the ATP synthase from *E. coli* (above) into F_1-catalytic, (yellow), F_o-membrane proton channel (blue) and peripheral stalk (PS; purple) is reflected in the organization of the corresponding genes in the *unc* operon (below); *unc*I encodes a membrane protein that is not a subunit of the enzyme, but it may be involved in the assembly of the c-ring. The genes are transcribed left to right in a single-messenger RNA from a promoter to the left of *unc*I.

It was noticed that the organization of genes in the operon was related to where the gene products were probably to be found in the intact ATP synthase complex (at the time the structure was unknown, and the peripheral stalk had not been discovered) (Figure 5). Thus, there is a promoter proximal cluster of two genes for subunit a (bacterial ortholog of ATP6) and subunit c, which form the proton channel, followed by a second cluster of two genes encoding subunits b and δ, both components of the peripheral stalk, and then a cluster of four genes encoding subunits α, γ, β and ε, the constituents of the F_1-catalytic domain. Thus, it appeared that these gene orders are molecular fossils that indicate that the three sectors of the ATP synthase initially evolved independently to perform some other unknown functions, before coming together to form a machine capable of synthesizing the large quantities of ATP that are required to sustain biological life. Further evidence that this may be so is provided by the finding that the $\alpha_3\beta_3$-component of the F_1-module is found in other biological contexts carrying out functions other than the synthesis of ATP, for example, in hexameric DNA helicases [Singleton *et al.*, 2000]. As yet, there is no known example of the peripheral stalk and proton channel modules being used in other biological contexts, although, for the lack of structural evidence, it cannot be excluded yet that an evolutionarily module related to the proton channel module is present in the rotary motor that drives the turning of the flagellae in motile bacteria, which is a second rotary motor driven by Δp [Minamino *et al.*, 2008].

The assembly of ATP synthases in bacteria has been studied in *E. coli.* Pioneering work was carried out by Gibson and colleagues in the 1970s and 1980s [Gibson, 1982], and later by Deckers-Herberstreit [Deckers-Heberstreit, 2013]. One interesting corollary, however, is that the sequencing of the *E. coli unc* operon revealed the presence of an unsuspected ninth gene, *uncI* [Gay and Walker, 1981b]. It is promoter proximal and encodes a membrane protein that is not a component of the ATP synthase complex itself. More recent work indicates that it is an assembly factor required for putting the bacterial c-ring together [Suzuki *et al.*, 2007]. Whether a related protein is required for assembling c-rings in mitochondria remains to be seen.

Acknowledgements

This work was supported by the Medical Research Council (U.K.) *via* grants MC_U105663150 and MR/M009858/1 to John E. Walker.

References

Abrahams, J.P., Leslie, A.G., Lutter, R. and Walker, J.E. (1994). Structure at 2.8 Å resolution of F_1-ATPase from bovine heart mitochondria. *Nature*, 370, 621–628.

Ackerman, S.H. and Tzagoloff, A. (1990). *ATP10*, a yeast nuclear gene required for the assembly of the mitochondrial F_1-F_o complex. *J. Biol. Chem.*, 265, 9952–9959.

Ackerman, S.H. (2002). Atp11p and Atp12p are chaperones for F_1-ATPase biogenesis in mitochondria. *Biochim. Biophys. Acta*, 1555, 101–105.

Carbajo, R.J., Kellas, F.A., Runswick, M.J., Montgomery, M.G., Walker, J.E. and Neuhaus, D. (2005). Structure of the F_1-binding domain of the stator of bovine F_1F_o-ATPase and how it binds an alpha-subunit. *J. Mol. Biol.*, 351, 824–838.

Carbajo, R.J., Kellas, F.A., Yang, J.C., Runswick, M.J., Montgomery, M.G., Walker, J.E. and Neuhaus, D. (2007). How the N-terminal domain of the OSCP subunit of bovine F_1F_o-ATP synthase interacts with the N-terminal region of an alpha subunit. *J. Mol. Biol.*, 368, 310–318.

Chen, R., Runswick, M.J., Carroll, J., Fearnley, I.M. and J. E. Walker, J.E. (2007). Association of two proteolipids of unknown function with ATP synthase from bovine heart mitochondria. *FEBS Lett.*, 581, 3145–3148.

Čížková, A., Stránecký, V., Mayr, J.A.,Tesarová, M., Havlícková, V., Paul, J., Ivánek, R., Kuss, A.W., Hansíková, H., Kaplanová, V., Vrbacký, M., Hartmannová, H., Nosková, L., Honzík, T., Drahota, Z., Magner, M., Hejzlarová, K., Sperl, W., Zeman, J., Houstek, J. and Kmoch, S. (2008). TMEM70 mutations cause isolated ATP synthase deficiency and neonatal mitochondrial encephalocardiomyopathy. *Nat. Genet.*, 40, 1288–1290.

Collinson, I.R., van Raaij, M.J., Runswick, M.J., Fearnley, I.M., Skehel, J.M., Orriss, G.L., Miroux, B. and Walker, J.E. (1994). ATP synthase from bovine heart mitochondria. *In vitro* assembly of a stalk complex in the presence of F_1-ATPase and in its absence. *J. Mol. Biol.*, 242, 408–421.

Cozens, A.L. and Walker, J.E. (1987). The organization and sequence of the genes for ATP synthase subunits in the cyanobacterium *Synechococcus* 6301. Support for an endosymbiotic origin of chloroplasts. *J. Mol. Biol.* 194, 359–383.

Davies, K.M., Strauss, M., Daum, B., Kief, J.H., Osiewacz, H.D., Rycovska, A., Zickermann, V. and Kühlbrandt, W. (2011). Macromolecular organization of ATP synthase and complex I in whole mitochondria. *Proc. Natl. Acad. Sci. USA*, 108, 14121–14126.

Deckers-Heberstreit (2013). Assembly of the *Escherichia coli* FoF1 ATP synthase involves distinct subcomplex formation. *Biochem. Soc. Trans.*, 41, 1289–1293.

Dickson, V.K., Silvester, J.A., Fearnley, I.M., Leslie, A.G. and Walker, J.E. (2006). On the structure of the stator of the mitochondrial ATP synthase. *EMBO J.*, 25, 2911–2918.

Dudkina, N.V., Sunderhaus, S., Braun, H.P. and Boekema, E.J. (2006). Characterization of dimeric ATP synthase and cristae membrane ultrastructure from *Saccharomyces* and *Polytomella* mitochondria., *FEBS Lett.*, 580, 3427–3432.

Dyer, M.R. and Walker, J.E. (1993). Sequences of members of the human gene family for the c subunit of mitochondrial ATP synthase. *Biochem. J.*, 293, 51–64.

Falk, G., Hampe, A. and Walker, J.E. (1985). Nucleotide sequence of the *Rhodospirillum rubrum* atp operon., *Biochem J* 228, 391–407.

Foster, D.L. and Fillingame, R.H. (1979). Energy-transducing H^+-ATPase of *Escherichia coli*. Purification, reconstitution, and subunit composition. *J. Biol. Chem.* 254, 8230–8236.

Foster, D.L. and Fillingame, R.H. (1982). Stoichiometry of subunits in the H^+-ATPase complex of *Escherichia coli*. *J. Biol. Chem.*, 257, 2009–2015.

Fearnley, I.M. and Walker, J.E. (1986). Two overlapping genes in bovine mitochondrial DNA encode membrane components of ATP synthase. *EMBO J.*, 5, 2003–2008.

Gay, N.J. and Walker, J.E. (1981a). The atp operon: nucleotide sequence of the region encoding the alpha-subunit of *Escherichia coli* ATP-synthase. *Nucleic Acids Res.*, 9, 2187–2194.

Gay, N.J. and Walker, J.E. (1981b). The atp operon: nucleotide sequence of the promoter and the genes for the membrane proteins, and the delta subunit of *Escherichia coli* ATP-synthase. *Nucleic Acids Res.*, 9, 3919–3926.

Gibson, F. (1982). The Leeuwenhoek Lecture, 1981. The biochemical and genetic approach to the study of bioenergetics with the use of *Escherichia coli*: progress and prospects. *Proc. R. Soc. Lond. B Biol. Sci.*, 215, 1–18.

Guo, H., Bueler, S.A. and Rubinstein, J.L. (2017). Atomic model for the dimeric F_o region of mitochondrial ATP synthase. *Science*, 358, 936–940.

Hahn, A., Parey, K., Bublitz, M., Mills, D.J., Zickermann, V., Vonck, J.,Kühlbrandt, W. and Meier, T. (2016). Structure of a complete ATP synthase dimer reveals the molecular basis of inner mitochondrial membrane morphology. *Mol. Cell*, 63, 445–456.

Havlícková, V., Kaplanová, V., Nůsková, H., Drahota, Z. and Houstek, J. (2010). Knockdown of F_1 epsilon subunit decreases mitochondrial content of ATP synthase and leads to accumulation of subunit c. *Biochim. Biophys. Acta*, 1797, 1124–1129.

He, J., Ford, H.C., Carroll, J., Ding, S., Fearnley, I.M. and Walker, J.E. (2017a). Persistence of the mitochondrial permeability transition in the absence of subunit c of human ATP synthase. *Proc. Natl. Acad. Sci. USA*, 114, 3409–3414.

He, J., Carroll, J., Ding, S., Fearnley, I.M. and Walker, J.E. (2017b). Permeability transition in human mitochondria persists in the absence of peripheral stalk subunits of ATP synthase. *Proc. Natl. Acad. Sci. USA*, 114, 9086–9091.

He, J., Ford, H.C., Carroll, J., Douglas, C., Gonzales, E., Ding, S., Fearnley, I.M. and Walker, J.E. (2018). Assembly of the membrane domain of ATP synthase in human mitochondria., *Proc. Natl. Acad. Sci. USA*, 115, 2988–2993.

Hermolin, J. and Fillingame, R.H. (1989). H^+-ATPase activity of *Escherichia coli* F_1F_o is blocked after reaction of dicyclohexylcarbodiimide with a single proteolipid (subunit c) of the F_o complex. *J. Biol. Chem.*, 264, 3896–3903.

Jiang, W., Hermolin, J. and Fillingame, R.H. (2001). The preferred stoichiometry of c subunits in the rotary motor sector of *Escherichia coli* ATP synthase is 10. *Proc. Natl. Acad. Sci. USA*, 98, 4966–4971.

Karrasch, S. and Walker, J.E. (1999). Novel features in the structure of bovine ATP synthase. *J. Mol. Biol.*, 290, 379–384.

Lee, J., Ding, S., Walpole, T.B., Holding, A.N., Montgomery, M.G., Fearnley, I.M. and Walker, J.E. (2015). Organization of subunits in the membrane domain of the bovine F-ATPase revealed by covalent cross-linking. *J. Biol. Chem.*, 290, 13308–13320.

Lefebvre-Legendre, L., Vaillier, J., Benabdelhak, H., Velours, J., Slonimski, P.P. and di Rago, J.P. (2001). Identification of a nuclear gene (*FMC1*) required for the assembly/stability of yeast mitochondrial F_1-ATPase in heat stress conditions. *J. Biol. Chem.*, 276, 6789–6796.

Li, Y., Jourdain, A.A., Calvo, S.E., Liu, J.S. and Mootha, V.K. (2017). CLIC, a tool for expanding biological pathways based on co-expression across thousands of datasets. *PLoS Comput. Biol.*, 13, e1005653.

Liu, S., Charlesworth, T.J., Bason, J.V., Montgomery, M.G., Harbour, M.E., Fearnley, I.M. and Walker, J.E. (2015). The purification and characterization of ATP synthase complexes from the mitochondria of four fungal species. *Biochem. J.*, 468, 167–175.

Lytovchenko, O., Naumenko, N., Oeljeklaus, S., Schmidt, B., K. von der Malsburg, K., Deckers, M., Warscheid, B., van der Laan, M. and Rehling, P. (2014). The INA complex facilitates assembly of the peripheral stalk of the mitochondrial F_1F_o-ATP synthase. *EMBO J*, 33, 1624–1638.

Matthies, D., Preiss, L., Klyszejko, A.L., Muller, D.J., Cook, G.M., Vonck, J. and Meier, T. (2009). The c_{13} ring from a thermoalkaliphilic ATP synthase reveals an extended diameter due to a special structural region., *J. Mol. Biol.*, 388, 611–618.

Mayr, J.A., Havlícková, V., Zimmermann, F., Magler, I., Kaplanová, V., Jesina, P., Pecinová, A., Nusková, H., Koch, J., Sperl, W. and Houstek, J. (2010). Mitochondrial

ATP synthase deficiency due to a mutation in the *ATP5E* gene for the F_1 ε subunit. *Hum. Mol. Genet.*, 19, 3430–3439.

Meier, T., Polzer, P., Diederichs, K., Welte, W. and Dimroth, P. (2005). Structure of the rotor ring of F-Type Na^+-ATPase from *Ilyobacter tartaricus*. *Science*, 308, 659–662.

Meyer, B., Wittig, I., Trifilieff, E., Karas, M. and Schägger, H. (2007). Identification of two proteins associated with mammalian ATP synthase. *Mol. Cell. Proteomics*, 6, 1690–1699.

Minamino, T., Imada, K., and Namba, K. (2008). Molecular motors of the bacterial flagella., *Curr. Opin. Struct. Biol.*, 18, 693–701.

Morales-Rios, E., Montgomery, M.G., Leslie, A.G. and Walker, J.E. (2015). Structure of ATP synthase from *Paracoccus denitrificans* determined by X-ray crystallography at 4.0 Å resolution. *Proc. Natl. Acad. Sci. USA*, 112, 13231–13236.

Naumenko, N., Morgenstern, M., Rucktäschel, R., Warscheid, B. and Rehling, P. (2017). INA complex liaises the F_1F_o-ATP synthase membrane motor modules. *Nat. Commun.*, 8, 1237.

Osman, C., Wilmes, C., Tatsuta, T. and Langer, T. (2007). Prohibitins interact genetically with Atp23, a novel processing peptidase and chaperone for the F_1F_o-ATP synthase. *Mol. Biol. Cell*, 18, 627–635.

Paumard, P., Vaillier, J., Coulary, B., Schaeffer, J., Soubannier, V., Mueller, D.M., Brèthes, D., di Rago, J.P. and Velours, J. (2002). The ATP synthase is involved in generating mitochondrial cristae morphology. *EMBO J*, 21, 221–230.

Pogoryelov, D., Yu, J., Meier, T., Vonck, J., Dimroth, P. and Muller, D.J. (2005). The c_{15} ring of the *Spirulina platensis* F-ATP synthase: F_1/F_o symmetry mismatch is not obligatory. *EMBO Rep.*, 6, 1040–1044.

Preiss, L., Langer, J.D., Yildiz, Ö., Eckhardt-Strelau, L., Guillemont, J.E., Koul, A. and Meier, T. (2015). Structure of the mycobacterial ATP synthase Fo rotor ring in complex with the anti-TB drug bedaquiline. *Sci. Adv.*, 1, e1500106.

Pullman, M.E. and Monroy, G.C. (1963). A naturally occurring inhibitor of mitochondrial adenosine triphosphatase. *J. Biol. Chem.*, 238, 3762–3769.

Rak, M., Gokova, S. and Tzagoloff, A. (2011). Modular assembly of yeast mitochondrial ATP synthase. *EMBO J.*, 30, 920–930.

Rees, D.M., Leslie, A.G. and Walker, J.E. (2009). The structure of the membrane extrinsic region of bovine ATP synthase. *Proc. Natl. Acad. Sci. USA*, 106, 21597–21601.

Saraste, M., Gay, N.J., Eberle, A., Runswick, M.J. and Walker, J.E. (1981). The atp operon: nucleotide sequence of the genes for the gamma, beta, and epsilon subunits of *Escherichia coli* ATP synthase. *Nucleic Acids Res.*, 9, 5287–5296.

Singleton, M.R., Sawaya, M.R., Ellenberger, T. and D.B. Wigley. (2000). Crystal structure of T7 gene 4 ring helicase indicates a mechanism for sequential hydrolysis of nucleotides. *Cell*, 101, 589–600.

Sobti, M., Smits, C., Wong, A.S., Ishmukhametov, R., Stock, D., Sandin, S. and Stewart, A.G. (2016). Cryo-EM structures of the autoinhibited *E. coli* ATP synthase in three rotational states. *Elife*, 5.

Stock, D., Leslie, A.G. and Walker, J.E. (1999). Molecular architecture of the rotary motor in ATP synthase. *Science*, 286, 1700–1705.

Suzuki, T., Ozaki, Y., Sone, N., Feniouk, B.A. and Yoshida, M. (2007). The product of *uncI* gene in F_1F_o-ATP synthase operon plays a chaperone-like role to assist c-ring assembly. *Proc. Natl. Acad. Sci. USA*, 104, 20776–20781.

Suzuki, T., Iida, N., Suzuki, J., Watanabe, Y., Endo, T., Hisabori, T. and Yoshida, M. (2016). Expression of mammalian mitochondrial F_1-ATPase in *Escherichia coli* depends on two chaperone factors, AF1 and AF2. *FEBS Open Bio.*, 6, 1267–1272.

Torraco, A., Verrigni, D., Rizza, T., Meschini, M.C., Vazquez-Memije, M.E., Martinelli, D., Bianchi, M., Piemonte, F., Dionisi-Vici, C., Santorelli, F.M., Bertini, E. and Carrozzo, R. (2012). TMEM70: A mutational hot spot in nuclear ATP synthase deficiency with a pivotal role in complex V biogenesis. *Neurogenetics*, 13, 375–386.

Van Walraven, H.S., Lutter, R. and Walker, J.E. (1993). Organization and sequences of genes for the subunits of ATP synthase in the thermophilic cyanobacterium *Synechococcus* 6716. *Biochem. J.*, 294, 239–251.

Vinothkumar, K.R., Montgomery, M.G., Liu, S. and Walker, J.E. (2016). Structure of the mitochondrial ATP synthase from *Pichia angusta* determined by electron cryomicroscopy. *Proc. Natl. Acad. Sci. USA*, 113, 12709–12714.

Vollmar, M., Schlieper, D., Winn, M., Büchner, C. and Groth, G. (2009). Structure of the c_{14} rotor ring of the proton translocating chloroplast ATP synthase. *J. Biol. Chem.*, 284, 18228–18235.

Vrbacký, M., Kovalčíková, J., Chawengsaksophak, K., Beck, I.M., Mráček, T., Nůsková, H., Sedmera, D., Papoušek, F., Kolář, F., Sobol, M., Hozák, P., Sedlacek, S. and Houštěk, J. (2016). Knockout of TMEM70 alters biogenesis of ATP synthase and leads to embryonal lethality in mice. *Hum. Mol. Genet.*, 25, 4674–4685.

Walker, J.E., Runswick, M.J. and Saraste, M. (1982). Subunit equivalence in *Escherichia coli* and bovine heart mitochondrial F_1F_o ATPases. *FEBS Lett.*, 146, 393–396.

Walker, J.E., Saraste, M. and Gay, N.J. (1984). The unc operon. Nucleotide sequence, regulation and structure of ATP-synthase. *Biochim. Biophys. Acta*, 768, 164–200.

Walker, J.E., Fearnley, I.M., Gay, N.J., Gibson, B.W., Northrop, F.D., Powell, S.J., Runswick, M.J., Saraste, M. and Tybulewicz, V.L. (1985). Primary structure and subunit stoichiometry of F_1-ATPase from bovine mitochondria. *J. Mol. Biol.*, 184, 677–701.

Walker, J.E., Runswick, M.J. and Poulter, L. (1987). ATP synthase from bovine mitochondria. The characterization and sequence analysis of two membrane-associated sub-units and of the corresponding cDNAs. *J. Mol. Biol.*, 197, 89–100.

Walker, J.E., Lutter, R., Dupuis, A. and Runswick, M.J. (1991). Identification of the subunits of F_1F_o-ATPase from bovine heart mitochondria. *Biochemistry*, 30, 5369–5378.

Walpole, T.B., Palmer, D.N., Jiang, H., Ding, S., Fearnley, I.M. and Walker, J.E. (2015). Conservation of complete trimethylation of lysine-43 in the rotor ring of c-subunits of metazoan adenosine triphosphate (ATP) synthases. *Mol. Cell. Proteomics*, 14, 828–840.

Wang, Z.G., White, P.S., and Ackerman, S.H. (2001). Atp11p and Atp12p are assembly factors for the F_1-ATPase in human mitochondria. *J. Biol. Chem.*, 276, 30773–30778.

Watt, I.N., Montgomery, M.G., Runswick, M.J., Leslie, A.G. and Walker, J.E. (2010). Bioenergetic cost of making an adenosine triphosphate molecule in animal mitochondria. *Proc. Natl. Acad. Sci. USA*, 107, 16823–16827.

Yan, W.L., Lerner, T.J., Haines, J.L. and Gusella, J.F. (1994). Sequence analysis and mapping of a novel human mitochondrial ATP synthase subunit 9 cDNA (ATP5G3). *Genomics*, 24, 375–377.

Zeng, X., Neupert, W. and Tzagoloff, A. (2007). The metalloprotease encoded by *ATP23* has a dual function in processing and assembly of subunit 6 of mitochondrial ATPase. *Mol. Biol. Cell*, 18, 617–626.

Zeng, X., Barros, M.H., Shulman, T. and Tzagoloff, A. (2008). *ATP25*, a new nuclear gene of *Saccharomyces cerevisiae* required for expression and assembly of the Atp9p subunit of mitochondrial ATPase. *Mol. Biol. Cell*, 19, 1366–1377.

Zhou, A., Rohou, A., Schep, D.G., Bason, J.V., Montgomery, M.G., Walker, J.E., Grigorieff, N. and Rubinstein, J.L. (2015). Structure and conformational states of the bovine mitochondrial ATP synthase by cryo-EM. *Elife*, 4, e10180.

Chapter 7

Cytochrome *c* Oxidase — Remaining Questions About the Catalytic Mechanism

Mårten Wikström* and Vivek Sharma*,†

**Institute of Biotechnology, University of Helsinki, Helsinki, Finland*
marten.wikstrom@helsinki.fi
†Department of Physics, University of Helsinki, Helsinki, Finland
vivek.sharma@helsinki.fi

Cytochrome *c* oxidase harnesses dioxygen (O_2) as the terminal electron acceptor in the respiratory chains of mitochondria and many aerobic bacteria, and was appropriately named The Respiratory Enzyme (Das Atmungsferment) by Otto Warburg. Its catalytic activity completes the oxygen cycle in biology, which starts by the water-splitting reaction of photosynthesis, amply covered elsewhere in this volume. The mechanism of catalysis of cytochrome *c* oxidase is largely known today, but some uncertainty remains with respect to the ferric/cupric intermediate of the binuclear heme iron/copper centre, which is to be discussed here.

1. Introduction

Spectroscopic data gathered by several laboratories over the last half century has revealed the key aspects of the catalytic mechanism of cytochrome *c* oxidase and identified the corresponding structures of intermediates of the active binuclear centre (BNC), which consists of a heme group, a histidine-ligated copper ion nearby, and a tyrosine residue covalently bonded to one of the copper histidines [Wikström *et al.*, 2018]. The chemical nature of this important tyrosine was first

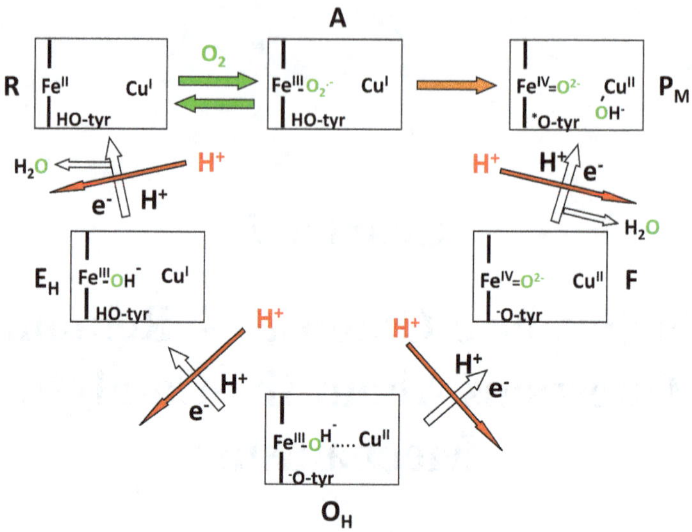

Figure 1. Catalytic cycle of cytochrome *c* oxidase. The rectangular boxes encompass the binuclear centre (BNC), including heme a_3, Cu_B and the tyrosine covalently linked to one of the three histidine ligands of Cu_B (histidines not shown). The name of the catalytic intermediate is given next to the box. Red arrows with protons (H^+) denote proton-pumping events; black H^+ denotes uptake of a "substrate proton". tyr–O* denotes a neutral tyrosine radical. The alternative structure of the activated state $\mathbf{O_H}$ (Fe[III]--OH^- Cu[I] TyrO*) and relaxation of $\mathbf{O_H}$ to the resting state **O** are not depicted, but later discussed extensively in the text.

reported by Yoshikawa *et al.* on the basis of one of their early X-ray structures of cytochrome *c* oxidase from bovine heart mitochondria [Tsukihara *et al.*, 1996].

Figure 1 depicts the catalytic cycle schematically. The reduced intermediate **R** reacts with O_2 to form the oxygen-bound intermediate **A** (about 10 μs at room temperature and 1 mM O_2) in which charge transfer from heme iron to the ligand makes this a ferric-superoxide structure. In the absence of a donor electron in the adjacent heme *a*, the next step is an exergonic splitting of the O–O bond (*ca.* 200 μs) using two electrons from the heme iron, one from the copper and the fourth from the tyrosine in the active site, which also donates a proton to form the neutral tyrosyl–histidine ligand of Cu[II]. This yields the highly oxidised so-called $\mathbf{P_M}$ state. In contrast, if the donor heme *a* is reduced it delivers the electron to the BNC instead of the tyrosine, which only donates the proton (*ca.* 25 μs). The result is then state $\mathbf{P_R}$ with ferryl heme, Cu[II]–OH^- and tyrosinate anion [Morgan *et al.*, 2001].

It is noteworthy that O_2 binding is highly reversible with a K_D as high as *ca.* 0.3 mM [Chance *et al.*, 1975; Verkhovsky *et al.*, 1994]. By contrast, the conversion of intermediate **A** to $\mathbf{P_M}$ is exergonic with a free energy change of *ca.* –5 kcal/mol [Blomberg, 2016]. The four subsequent reaction steps all involve net uptake of an electron and a proton to the BNC, and are each linked to pumping of an additional proton from the negatively charged *N*-side of the membrane to the positive *P*-side (red arrows) [Verkhovsky *et al.*, 1999]. In the first of these steps, the **P** intermediate is converted to the **F** state (100–150 μs), and then the latter transforms into state $\mathbf{O_H}$ (*ca.* 1.5–2 ms).

2. The Enigmatic O States

Whilst the BNC structures of intermediates **R**, **A**, $\mathbf{P_M}$, $\mathbf{P_R}$, and **F** are well-established (see above and Figure 1), the structures of the ferric/cupric **O** states ($\mathbf{O_H}$ in Figure 1) still presents problems. Proton translocation (red arrows, Figure 1) has been shown to occur in all four reaction steps associated with transfer of an electron to the BNC [Verkhovsky *et al.*, 1999], but three observations seemed to be in contrast to such a conclusion. First, the injection of an electron into cytochrome *c* oxidase, as isolated, where the BNC is in the resting ferric/cupric state, resulted in fast electron transfer (eT) to heme *a*, but very sluggish further eT to the BNC, and no proton translocation [Verkhovsky *et al.*, 1999; Bloch *et al.*, 2004]. Secondly, classical anaerobic equilibrium redox titrations of cytochrome oxidase, either in mitochondria or of the isolated mitochondrial or bacterial enzyme, showed midpoint redox potentials (E_m) of heme a_3 (Fe[III]/Fe[II]) and Cu_B (Cu[II]/Cu[I]), *i.e.* the two transitions on the left hand side of Figure 1, of less than 400 mV [Gorbikova *et al.*, 2006]. Since the donor, cytochrome *c*, has an E_m of *ca.* 250 mV, the approximate driving force for electron transfer between the donor and the BNC would be only some 150 mV, which is far too low to drive protonic separation of *two* charges across a typical protonmotive force across the membrane of some 200 mV (one of those charges is due to the proton pump; the other from the combination of eT from cytochrome *c* to the BNC and uptake of one proton per electron by the latter to complete oxygen reduction chemistry).

Thirdly, the sum of free energy changes in the catalytic cycle must match the free energy change of the overall reaction, *i.e.* O_2 reduction to water by cytochrome *c*. The approximate E_m values are known from experiments for the two first redox couples, $\mathbf{P_M}$/**F** and **F**/**O** to be of the order of 800 mV each [Wikström *et al.*, 2018], the free energy change of the conversion of **A** to $\mathbf{P_M}$ is *ca.* –5 kcal/mol (~220 meV) (see above), and the free energy change of O_2 binding (**R**→**A**) is close to zero (see above). This means that the average E_m value of the remaining two redox couples (**O**/**E** and **E**/**R**) must be of the order of 690 mV, *i.e.* much higher than measured in redox titrations (<400 mV).

Based on these three findings, we proposed that the ferric/cupric state of the BNC in the resting enzyme (we called it state **O**) is not normally involved in the catalytic cycle, which would instead work with an activated form called $\mathbf{O_H}$.

The subscript H stemmed from the discovery by Rousseau *et al.* [Han *et al.*, 1990] by resonance Raman spectroscopy of a high-spin ferric hydroxy intermediate that followed the **F** state in the catalytic sequence. When electron donation to the enzyme ceases, we proposed that the $\mathbf{O_H}$ state spontaneously relaxes into the **O** state with lower energy and with a considerably lowered E_m. This notion was supported by the finding that electron injection at a sufficiently early stage after oxidation of fully reduced enzyme (~100 ms), indeed led both to fast transfer of the electron to the BNC, to proton pumping, and finally to reduction of Cu_B alone, indicating a much higher E_m than that observed in redox titrations [Bloch *et al.*, 2004; Belevich *et al.*, 2007].

One confusion in the field stems from the fact that the ferric/cupric "as isolated" enzyme (*i.e.* what we here call the **O** form) can in itself exist in different forms, *pulsed versus resting*, and *slow versus fast* [Moody *et al.*, 1991]. Here, whilst the structural difference between these forms remains unknown, the fast form has been shown to best correspond to native oxidised enzyme in mitochondria, and the method of its purification is well-established.

In an attempt to test the postulated structural and functional difference between the **O** and $\mathbf{O_H}$ states, Jancura *et al.* [2006] studied the rate of heme a_3 reduction of "fast" bovine cytochrome *c* oxidase by hexaamineruthenium plus dithionite (reduction of Cu_A and heme *a* is very fast and occurred within the time of mixing in the flow apparatus). The $\mathbf{O_H}$ state was produced by mixing reduced enzyme with O_2 in the presence of the two reductants, whereas reduction of state **O** was assessed by mixing enzyme anaerobically with the reductants. No significant difference in heme a_3 reduction rate was observed in the two conditions. It is noteworthy, however, that the observed rates of heme a_3 reduction were 80–90 s^{-1} in the $\mathbf{O_H}$ conditions, and nearly the same for the state **O** conditions, which is only some 15% of the maximum turnover of bovine enzyme (see *e.g.* [Riegler *et al.*, 2005]). The corresponding rates of heme a_3 and Cu_B reduction deduced from freeze-quench EPR data [Jancura *et al.*, 2006] were only *ca.* 5 s^{-1}. The suspicion thus arises that the employed reductants may have limited the rate and therefore did not reveal the $\mathbf{O_H}$ intermediate.

In a more recent account by Vilhjálmsdóttir *et al.* [2018] the complex of cytochrome *c* with cytochrome *c* oxidase was tested for the occurrence of state $\mathbf{O_H}$. In this system the fully reduced oxidase was supplemented with an additional electron in the form of the bound cytochrome *c*. It was thus expected that when the enzyme reacted with O_2 it would immediately have an additional electron available after forming the activated $\mathbf{O_H}$ intermediate, and thus lead to a situation in which Cu_B has a high redox potential and would readily accept that electron. However, the optical spectroscopic experiments revealed no preference for reduction of Cu_B. In contrast, the "extra electron" was mainly distributed between

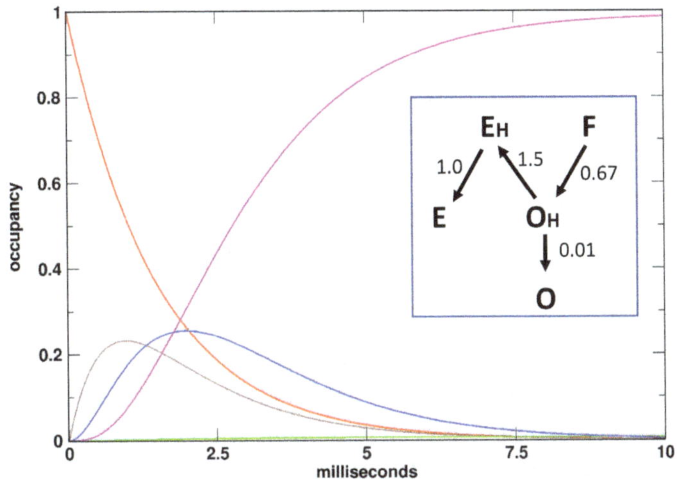

Figure 2. Hypothetical fate of the $\mathbf{O_H}$ intermediate in the experiment by Vilhjálmsdóttir *et al.* [2018]. The insert in the marked square shows a possible reaction scheme of what may happen after primary relaxation of the **F** state into $\mathbf{O_H}$. The numbers denote approximate reaction velocities in ms^{-1}. All reactions are assumed to be irreversible for simplicity. The graph shows a simulation of this system with the **F** state (red trace) fully occupied initially. The other curves show the time dependence of the **E** state (purple), the $\mathbf{E_H}$ state (blue), the $\mathbf{O_H}$ state (brown), and the **O** state (green).

cytochrome *c*, Cu_A and heme *a*, with preference for the latter, but with very little contribution from the BNC, which remained highly oxidised.

This experimental approach is interesting, and we can only suggest one reason for why it may not have revealed the nature of the $\mathbf{O_H}$ state.

Figure 2 (inset) shows a scheme of how the **F** intermediate (see Fig. 1) turns into $\mathbf{O_H}$, and then how $\mathbf{O_H}$ might initially be converted to an activated (and hypothetical) $\mathbf{E_H}$ state, which relaxes into an **E** state, where redox potentials correspond to those found in anaerobic equilibrium redox titrations. As shown in Figure 2 (main panel), under the kinetic conditions of the scheme (insert) the $\mathbf{E_H}$ state with a high redox potential for Cu_B would not be sufficiently occupied to be observed in the experiments by Vilhjálmsdóttir *et al.* [2018].

3. The Structures of the $\mathbf{O_H}$ and O States

Earlier, based on density functional theory (DFT) calculations [Sharma *et al.*, 2013], we suggested a structure of the activated $\mathbf{O_H}$ state of the BNC (see Figure 1) in which the Cu_B center possesses a high redox potential necessary to catalyze redox-coupled proton pumping in the $\mathbf{O_H} \rightarrow \mathbf{E_H}$ transition of the catalytic cycle (see [Verkhovsky *et al.*, 1999; Bloch, *et al.*, 2004]). In a considerable fraction of this structure, Cu_B attains a cuprous (Cu[I]) electronic character together with an unpaired electron on the cross-linked Tyr, formally written as Fe[III]–OH^- Cu[I] TyrO*. In this state it was observed that the distance between Cu and Fe is reduced to *ca.* 4Å, and that the hydroxy ligand of Fe is also a weak ligand of Cu, forming a µ-hydroxo bridge between the metals, also observed in a number of compounds synthesized to mimic the oxidase active site [Obias *et al.*, 1998; Hematian *et al.*, 2015]. Later on, Blomberg and Siegbahn [2015], reached the same structure of the active site in the $\mathbf{O_H}$ state, and remarkably similar results were also obtained by computational studies of the structurally similar active site of the ba_3 oxidase from *Thermus thermophilus* [Du and Noodleman, 2015]. In contrast to the overall good agreement on the structure of the active site in the $\mathbf{O_H}$ state, the identity of the relaxed **O** state remains less certain. In our earlier work, we suggested that the structure of the **O** state is (Fe[III]–OH_2 $^-$HO–Cu[II] $TyrO^-$) with a proton being shared between the OH groups of the two metals. A similar structure of the **O** state was initially proposed by Blomberg and Siegbahn based on DFT calculations [Blomberg and Siegbahn, 2014]. However, in their later studies [Blomberg and Siegbahn, 2015] it was suggested that state **O** may have the two-hydroxy structure Fe[III]–OH^- $^-$HO–Cu[II] TyrOH, which despite being higher in energy than state $\mathbf{O_H}$, has lower redox potential of the active site, a characteristic in agreement with experimental data. Two additional structures of **O** states were further considered, $\mathbf{O_B}$ and $\mathbf{O_P}^+$, the latter with an extra proton in the BNC [Blomberg and Siegbahn, 2015]. Because no magnetic coupling was found for the two-hydroxy structure, and its higher energy, it was concluded that the $\mathbf{O_P}^+$ structure (Fe[III]–OH^-...Cu[II] TyrOH) is the best candidate for the relaxed **O** state with a calculated exchange coupling (J) of ~114 cm^{-1} [Blomberg and Siegbahn, 2015].

By using the available coordinates of the two-hydroxy **O** and $\mathbf{O_H}$ states [Blomberg and Siegbahn, 2015], we performed single-point DFT calculations using the B3LYP functional [Becke, 1993; Lee *et al.*, 1988] and the def2-TZVP basis sets [Weigend and Ahlrichs, 2005] (dielectric constant of 4 with dispersion corrections [Grimme *et al.*, 2010]), and constructed a state similar to our earlier proposal of the **O** state structure, *viz.* (Fe[III]–OH_2 $^-$HO–Cu[II] $TyrO^-$). We again (*cf.* [Sharma *et al.*, 2013]) found the latter shared-proton-structure to be *ca.* 16 kcal/mol lower in energy than the two-hydroxy **O** state, which is at the same energy level as the $\mathbf{O_H}$ state (high spin) in agreement with our earlier estimates [Sharma *et al.*, 2013], as well as with those of Blomberg and Siegbahn [2015]. Moreover, we further tested our models [Sharma *et al.*, 2013] by removing an extra water molecule that hydrogen bonded the oxygenous ligands of the metals to tyrosine, and then found the $\mathbf{O_H}$ state to be *ca.* 10 kcal/mol higher in energy than the **O** state (Fe[III]–OH_2 $^-$HO–Cu[II] $TyrO^-$), which has no unpaired electron on tyrosine in contrast to the $\mathbf{O_H}$ state. Overall, the data supports the idea that (Fe[III]–OH_2 $^-$HO–Cu[II] TyrO$^-$) could be the relaxed **O** state with a low redox potential of Cu_B due to no unpaired spin on tyrosine and due to the copper hydroxyl ligand. When we calculate the magnetic coupling constant (J) of this state using the Yamaguchi formulation [Soda *et al.*, 2000], we find it to be *ca.* 7 cm^{-1}. This weak exchange coupling is actually in excellent agreement with the most recent experimental observations [Oganesyan *et al.*, 1998; Hunter *et al.*, 2000; Cheesman *et al.*, 2004], which suggest, in contrast to previous belief, that the magnetic coupling in the fully oxidised relaxed state of the heme-copper oxidases is very weak, of the order of ~1 cm^{-1}. Interestingly, this is also in line with the extent of magnetic coupling of *ca.* 4–6 cm^{-1} between iron and copper in the $\mathbf{P_R}$ intermediate [Morgan *et al.*, 2001].

4. Peroxide in the Active Site?

X-ray structures of cytochrome *c* oxidase from different sources have been interpreted to have a peroxide ligand bridging ferric Fe and cupric Cu in the BNC (see [Andersson *et al.*, 2017] for citations) in the state that we have named state **O** here. Yoshikawa *et al.* [1998], who first made this suggestion, have made it very clear that the resting ferric/cupric state in which a peroxide is postulated is not a state of the enzyme active in catalysis. Indeed, there is no proposal available about how a ferric/cupric peroxide state **O** would fit any reasonable catalytic mechanism. Yoshikawa *et al.* [Mochizuki *et al.*, 1999] reported that titration of the bovine enzyme revealed uptake of six and not four reducing equivalents on full reduction of enzyme in state **O**, in support of a peroxide ligand. Earlier work by others (references in [Mochizuki *et al.*, 1999]) had reported uptake of four reducing equivalents, but Yoshikawa *et al.* ascribed those findings to contamination of the earlier non-crystalline enzyme preparations with iron. However, that would hardly explain the classical data of Vanneste [1966], which was based on pyridine hemochrome determination. In addition, Mitchell *et al.* [1992] found that by reducing bovine heart oxidase with carbon monoxide, or treating enzyme in state **O** with hydrogen peroxide, two or three protons were taken up on forming the **P** and **F** states, respectively, from state **O** (see Figure 1), which is incompatible with an initial ferric/cupric peroxide structure. Bränden and co-workers [Andersson *et al.*, 2017] recently reported the structure of cytochrome ba_3 from *Thermus thermophilus*, using serial femtosecond crystallography with lipid cubic phase microcrystals at room temperature. This work, which showed a single water or hydroxide between the metals in the BNC, also provides an ample summary of the X-ray data of the binuclear site of heme-copper oxidases collected from different species and by different techniques.

5. Discussion

The $\mathbf{O}_P^+$ structure (Fe[III]–OH⁻...Cu[II] TyrOH) suggested by Blomberg and Siegbahn [Blomberg and Siegbahn, 2015] as the best alternative for the relaxed **O** state is *ca.* 6.5 kcal/mol lower in energy compared to the $\mathbf{O}_H$ state. However, the extra proton in the BNC is a source of problems. First, FTIR experiments [Gorbikova *et al.*, 2008] have shown that the pK_a of the tyrosine is ~6.6 in the **O** state, indicating that only a small fraction of the tyrosine may be protonated at the neutral or slightly alkaline pH of most experiments. Secondly, the extra proton would increase the net charge of the BNC to +2. Actually, we should include the heme a_3 propionates in such an assessment, in which case the total charge of the BNC is zero in all intermediates of the catalytic cycle known so far. In the $\mathbf{O}_H$ state, for example, the ferric (+3), porphyrinate (–2) and hydroxide (–1) yields a charge of zero, and the cupric copper (+2) plus tyrosinate (–1), gives a total charge of +1. However, whilst the charge of the heme a_3 D-propionate is neutralised by a conserved arginine residue, the negative charge of the A-propionate is not, so that the overall charge is zero. To invoke an additional charge in the BNC (as in the $\mathbf{O}_P^+$ structure) would be expected to be very costly in such a low dielectric environment (see *e.g.* [Kaila *et al.*, 2010]). Thirdly, the source of the postulated extra proton is problematic. Just a little less than two protons are known to be taken up from the external medium on oxidation of the fully reduced enzyme [Mitchell *et al.*, 1992; Mitchell and Rich, 1994], and uptake of one proton each must be attributed to the **P**→**F** and **F**→$\mathbf{O}_H$ steps (Figure 1) (the slight downward deviation from two protons can be attributed to the weak proton–electron linkages of the redox transitions of the two hemes and of Cu_A).

The relaxed **O** state structure that we are suggesting (Fe[III] –OH_2 ⁻HO–Cu[II] TyrO⁻) is at least 10 kcal/mol lower in energy than the $\mathbf{O}_H$ state, exhibits the deprotonated tyrosinate in accordance with FTIR data (see above), and has a redox midpoint potential considerably lower than the $\mathbf{O}_H$ state. It is presumably formed from the "activated" $\mathbf{O}_H$ state after diffusion of a water molecule from the apolar cavity next to the BNC into the active site [Sharma *et al.*, 2013].

Acknowledgements

VS acknowledges computing time support from the Center for Scientific Computing, Finland, and financial support from the Academy of Finland, the Sigrid Jusélius Foundation and the University of Helsinki. MW is indebted to the Magnus Ehrnrooth Foundation and Societas Scientiarum Fennica for financial support, and to the Nanyang Technolgical University for their hospitality.

References

Andersson, R., Safari, C., Dods, R., Nango, E., Tanaka, R., Yamashita, A., Nakane, T., Tono, K., Joti, Y., Båth, P., Dunevall, E., Bosman, R., Nureki, O., Iwata, S., Neutze, R. and Brändén, G. (2017). Serial femtosecond crystallography structure of cytochrome *c* oxidase at room temperature, *Sci. Rep.*, 7, 4518.

Becke, A.D. (1993). Densityfunctional thermochemistry. III. The role of exact exchange, *J. Chem. Phys.*, 98, 5648–5652.

Belevich, I., Bloch, D.A., Belevich, N., Wikström, M. and Verkhovsky, M.I. (2007). Exploring the proton pump mechanism of cytochrome *c* oxidase in real time, *Proc. Natl. Acad. Sci. USA*, 104, 2685–2690.

Bloch, D., Belevich, I., Jasaitis, A., Ribacka, C., Puustinen, A., Verkhovsky, M.I. and Wikström, M. (2004). The catalytic cycle of cytochrome *c* oxidase is not the sum of its two halves, *Proc. Natl. Acad. Sci. USA*, 101, 529–533.

Blomberg, M.R. (2016). Mechanism of oxygen reduction in cytochrome *c* oxidase and the role of the active site tyrosine, *Biochemistry*, 55, 489–500.

Blomberg, M.R. and Siegbahn, P.E. (2014). Proton pumping in cytochrome *c* oxidase: Energetic requirements and the role of two proton channels, *BBA-Bioenergetics*, 1837, 1165–1177.

Blomberg, M.R. and Siegbahn, P.E. (2015). How cytochrome *c* oxidase can pump four protons per oxygen molecule at high electrochemical gradient, *BBA-Bioenergetics*, 1847, 364–376.

Blomberg, M.R. and Siegbahn, P.E. (2015). Protonation of the binuclear active site in cytochrome *c* oxidase decreases the reduction potential of CuB, *BBA-Bioenergetics*, 1847, 1173–1180.

Chance, B., Saronio, C. and Leigh, J. (1975). Functional intermediates in the reaction of membrane-bound cytochrome oxidase with oxygen, *J. Biol. Chem.*, 250, 9226–9237.

Cheesman, M.R., Oganesyan, V.S., Watmough, N.J., Butler, C.S. and Thomson, A.J. (2004). The nature of the exchange coupling between high-spin Fe(III) heme *o3* and CuB(II) in *Escherichia coli* quinol oxidase, cytochrome *bo3*: MCD and EPR studies, *J. Am. Chem. Soc.*, 126, 4157–4166.

Du, W.-G.H. and Noodleman, L. (2015). Broken symmetry DFT calculations/analysis for oxidized and reduced dinuclear center in cytochrome *c* oxidase: Relating structures, protonation states, energies, and Mössbauer properties in ba3*Thermus thermophilus*, *Inorg.Chem.*, 54, 7272.

Gorbikova, E.A., Vuorilehto, K., Wikström, M. and Verkhovsky, M.I. (2006). Redox titration of all electron carriers of cytochrome *c* oxidase by Fourier transform infrared spectroscopy, *Biochemistry*, 45, 5641–5649.

Gorbikova, E.A., Wikström, M. and Verkhovsky, M.I. (2008). The protonation state of the cross-linked tyrosine during the catalytic cycle of cytochrome *c* oxidase, *J. Biol. Chem.i*, 283, 34907–34912.

Grimme, S., Antony, J., Ehrlich, S. and Krieg, H. (2010). A consistent and accurate ab initio parametrization of density functional dispersion correction (DFT-D) for the 94 elements H-Pu, *J. Chem. Phys.*, 132, 154104.

Han, S., Ching, Y.-C. and Rousseau, D.L. (1990). Ferryl and hydroxy intermediates in the reaction of oxygen with reduced cytochrome *c* oxidase, *Nature*, 348, 89–90.

Hematian, S., Garcia-Bosch, I. and Karlin, K.D. (2015). Synthetic heme/copper assemblies: toward an understanding of cytochrome *c* oxidase interactions with dioxygen and nitrogen oxides, *Acc. Chem. Res.*, 48, 2462.

Hunter, D.J.B., Oganesyan, V.S., Salerno, J.C., Butler, C.S., Ingledew, W.J. and Thomson, A.J. (2000). Angular dependences of perpendicular and parallel mode electron paramagnetic resonance of oxidized beef heart cytochrome *c* oxidase, *Biophys. J.*, 78, 439–450.

Jancura, D., Berka, V., Antalik, M., Bagelova, J., Gennis, R.B., Palmer, G. and Fabian, M. (2006). Spectral and kinetic equivalence of oxidized cytochrome *c* oxidase as isolated and "activated" by reoxidation, *J. Biol. Chem.*, 281, 30319–30325.

Kaila, V.R., Verkhovsky, M.I. and Wikström, M. (2010). Proton-coupled electron transfer in cytochrome oxidase, *Chem. Rev.*, 110, 7062–7081.

Lee, C., Yang, W. and Parr, R.G. (1988). Development of the Colle-Salvetti correlation-energy formula into a functional of the electron density, *Phys.Rev. B*, 37, 785.

Mitchell, R. and Rich, P.R. (1994). Proton uptake by cytochrome *c* oxidase on reduction and on ligand binding, *BBA-Bioenergetics*, 1186, 19–26.

Mitchell, R., Mitchell, P. and Rich, P.R. (1992). Protonation states of the catalytic intermediates of cytochrome *c* oxidase, *BBA-Bioenergetics*, 1101, 188–191.

Mochizuki, M., Aoyama, H., Shinzawa-Itoh, K., Usui, T., Tsukihara, T. and Yoshikawa, S. (1999). Quantitative reevaluation of the redox active sites of crystalline bovine heart cytochrome *c* oxidase, *J. Biol. Chem.*, 274, 33403–33411.

Moody, A.J., Cooper, C.E. and Rich, P.R. (1991). Characterisation of 'fast'and 'slow'forms of bovine heart cytochrome-*c* oxidase, *BBA-Bioenergetics*, 1059, 189–207.

Morgan, J.E., Verkhovsky, M.I., Palmer, G. and Wikström, M. (2001). Role of the PR intermediate in the reaction of cytochrome *c* oxidase with O_2, *Biochemistry*, 40, 6882–6892.

Obias, H.V., van Strijdonck, G.P., Lee, D.-H., Ralle, M., Blackburn, N.J. and Karlin, K.D. (1998). Heterobinucleating ligand-induced structural and chemical variations in [(L) FeIII–O–CuII]$^+$ μ-oxo complexes, *J. Am. Chem. Soc.*, 120, 9696–9697.

Oganesyan, V.S., Butler, C.S., Watmough, N.J., Greenwood, C., Thomson, A.J. and Cheesman, M.R. (1998). Nature of the coupling between the high-spin Fe(III) heme and CuB(II) in the active site of terminal oxidases: Dual-mode EPR spectra of fluoride cytochrome *bo3*, *J. Am. Chem. Soc.*, 120, 4232–4233.

Riegler, D., Shroyer, L., Pokalsky, C., Zaslavsky, D., Gennis, R. and Prochaska, L.J. (2005). Characterization of steady-state activities of cytochrome *c* oxidase at alkaline pH: mimicking the effect of K-channel mutations in the bovine enzyme, *BBA-Bioenergetics*, 1706, 126–133.

Sharma, V., Karlin, K.D. and Wikström, M. (2013). Computational study of the activated OH state in the catalytic mechanism of cytochrome *c* oxidase, *Proc. Natl. Acad. Sci. USA*, 110, 16844–16849.

Soda, T., Kitagawa, Y., Onishi, T., Takano, Y., Shigeta, Y., Nagao, H., Yoshioka, Y. and Yamaguchi, K. (2000). *Ab initio* computations of effective exchange integrals for H–H, H–He–H and Mn_2O_2 complex: comparison of broken-symmetry approaches, *Chem. Phys. Lett.*, 319, 223–230.

Tsukihara, T., Aoyama, H., Yamashita, E., Tomizaki, T., Yamaguchi, H., Shinzawa-Itoh, K., Nakashima, R., Yaono, R. and Yoshikawa, S. (1996). The whole structure of the 13-subunit oxidized cytochrome *c* oxidase at 2.8 Å, *Science*, 272, 1136.

Vanneste, W.H. (1966). The stoichiometry and absorption spectra of components *a* and a_3 in cytochrome *c* oxidase, *Biochemistry*, 5, 838–848.

Verkhovsky, M.I., Jasaitis, A., Verkhovskaya, M.L., Morgan, J.E. and Wikström, M. (1999). Proton translocation by cytochrome *c* oxidase, *Nature*, 400, 480–483.

Verkhovsky, M.I., Morgan, J.E. and Wikstroem, M. (1994). Oxygen binding and activation: Early steps in the reaction of oxygen with cytochrome *c* oxidase, *Biochemistry*, 33, 3079–3086.

Vilhjálmsdóttir, J., Gennis, R.B. and Brzezinski, P. (2018). The electron distribution in the "activated" state of cytochrome *c* oxidase, *Sci. Rep.*, 8, 7502.

Weigend, F. and Ahlrichs, R. (2005). Balanced basis sets of split valence, triple zeta valence and quadruple zeta valence quality for H to Rn: Design and assessment of accuracy, *Phys. Chem. Chem. Phys.*, 7, 3297–3305.

Wikström, M., Krab, K. and Sharma, V. (2018). Oxygen activation and energy conservation by cytochrome *c* oxidase, *Chem. Rev.,* 118, 2469–2490.

Yoshikawa, S., Shinzawa-Itoh, K., Nakashima, R., Yaono, R., Yamashita, E., Inoue, N., Yao, M., Fei, M.J., Libeu, C. P. and Mizushima, T. (1998). Redox-coupled crystal structural changes in bovine heart cytochrome *c* oxidase, *Science*, 280, 1723–1729.

Chapter 8

Cytochrome *c* Oxidase: Oxygen Consumption, Energy Conservation and Control

Peter R. Rich

Glynn Laboratory of Bioenergetics, Institute of Structural and Molecular Biology, University College London, Gower Street, London WC1E 6BT, UK

prr@ucl.ac.uk

Human cytochrome *c* oxidase activity consumes almost all of the oxygen that we breathe, and the human population accounts for consumption of perhaps 0.1% of the net global biological oxygen generation. It is a member of a superfamily of haem-copper oxidases, falling within the A1 subgroup with a range of bacterial homologues. All A1 members have three common core subunits with a similar catalytic binuclear centre in subunit I that catalyses oxygen reduction. Energy released is conserved by coupled proton translocation which contributes to a transmembrane protonmotive force that can be used to generate ATP. A plausible mechanistic model of the protein substructures that enable this couple proton transfer has emerged from a wide range of mechanistic studies, supported in particular by mutant bacterial A1-type cytochrome *c* oxidase properties and by more limited studies of yeast mutant mitochondrial cytochrome *c* oxidase. However, the atomic structure of bovine mitochondrial cytochrome *c* oxidase revealed an additional possible proton pathway. In combination with functional data and identification of structurally-mobile elements, this has led to an alternative suggested mechanism of proton translocation in mammalian forms of the enzyme. Key features and differences of these two major models and emerging mechanisms of activity control are reviewed.

1. Introduction: Oxygen Generation and Human Oxygen Consumption

The mass of the Earth's atmosphere is estimated at around 5.15×10^{18} kg, of which molecular oxygen is 20.95% by volume and 23.14% by mass [Lide, 1995]. Hence, total oxygen mass is roughly 1.19×10^{18} kg and turnover times of between 2000 and 9000 years have been reported [Duursma and Boisson, 1994; Walker, 1980]. Taking recent estimates of around 4000 years, maintenance of this *status quo* requires the generation of around 0.3×10^{15} kg.year^{-1} of molecular oxygen. The vast majority of this is thought to be generated biologically by oxygenic photosynthesis through the water-splitting action of the Mn_4Ca oxygen evolving enzyme that is associated with photosystem II [Barber, 2017]. The world human population in 2018 has been estimated at 7.6×10^9 and a typical human male consumes around 380 litres (0.54 kg) of oxygen per day [Rich, 2003]. This would equate to a total human oxygen consumption of around 1.5×10^{12} kg.year^{-1}, or 0.5% of the annual net oxygen production. Given the planetary age distribution towards younger ages, this is perhaps more realistically 0.1–0.2% of net oxygen production. This may be compared to a recent alternative estimate of 3.09×10^{12} kg.year^{-1} [Huang *et al.*, 2018], or ~1% of global production. The remainder of the oxygen consumption is consumed by other life forms, by natural geoprocesses and, increasingly, by modern industrial and fuel uses [Huang *et al.*, 2018].

Biological oxygen consumption occurs through a wide range of different types of enzyme oxidoreductases that can result in superoxide, hydrogen peroxide or water as the product. The standard potential E'_o for O_2/H_20 at pH 7 is +815 mV *versus* SHE [Wood, 1974] for liquid water at 25°C and pH 7 and an O_2 fugacity of 1 atmosphere. Atmospheric oxygen concentration is 21% and can be much lower in some environments, meaning that the operative E_h will be slightly lowered. In addition ambient pH will raise or lower the operative E_h by $59 \times (7 - \text{ambient pH})$ mV. However, provided the reductant source has a sufficiently low operative E_h, the overall reductant/oxygen oxidoreduction reaction can have a large negative free energy change that can be used to power other processes. A range of membrane-located terminal respiratory oxidases have evolved to utilise oxygen as a terminal electron sink to sustain continuous currents along diverse respiratory chains, reducing oxygen to the safest product water and conserving part of the energy released by oxygen reduction as a charge and proton gradient in the chemiosmotic transmembrane electrochemical proton gradient (the protonmotive force, PMF). In mammals and many other eukaryotes, this oxygen-reducing function is catalysed solely by cytochrome *c* oxidase (C*c*O), a large multiprotein complex that spans the inner mitochondrial membrane. Electrons from reduced cytochrome *c* are transferred to a buried catalytic site, the binuclear centre (BNC), formed from a haem A (haem a_3) and a copper atom (Cu_B). It uses four electrons from cytochrome *c* in the intermembrane space (IMS), together with four protons from the mitochondrial matrix, to reduce molecular oxygen to two waters. As a result of the vectorial nature of the reaction, part of the large free energy of the oxygen reduction reaction can be conserved in the transmembrane PMF, as envisaged in the original chemiosmotic proposals of Mitchell [Mitchell, 1966]. However, each full catalytic cycle also causes an additional four protons to be translocated from the matrix to the IMS [Wikström, 1977], hence increasing the fraction of free energy that can be stored. In typical conditions in mammalian cells, the free energy available from the reduction of oxygen with electrons from ferrous cytochrome *c*, the electron donor to C*c*O, is roughly 50 kcal.mol^{-1} [Rich, 2017a]. Human cell phosphate potentials can be as high as –70 kJ.mol^{-1} [Wackerhage *et al.*, 1998]. Given an $H^+/ATP_{cytoplasm}$ ratio of 3.7, dictated by the ATP synthase c ring stoichiometry of eight [Watt *et al.*, 2010] and the 1H^+/ATP cost of ATP/(ADP + Pi) exchange across the inner mitochondrial membrane, this would equate to a standing PMF of around 190 mV and, therefore, an efficiency by C*c*O of available free energy storage into the PMF of around 70%. This means that, even with such high cellular phosphate potentials and associated PMFs, the overall reaction is still highly irreversible. In environments such as exercising muscle, where the phosphate potential and the PMF may well drop considerably, the efficiency would be much lower, though the throughput rate and power output (free energy gain × rate) might increase [Rich, 1982].

2. Protein Structure and Electron Transfer

Mitochondrial C*c*Os are members of a large haem-copper oxidase (HCO) superfamily [García-Horsman *et al.*, 1994; Sousa *et al.*, 2012; Hemp and Gennis, 2008], falling within an A1 branch that includes closely-related bacterial C*c*Os and quinol oxidases. Atomic structures have been reported for bovine mitochondrial C*c*O [Tsukihara *et al.*, 1996], several A1-type bacterial C*c*Os [Iwata *et al.*, 1995; Svensson-Ek *et al.*, 2002] and an A1-type ubiquinol oxidase (cytochrome *bo* [Abramson *et al.*, 2000]). All have the same common core of three protein subunits: I, II and III (Figure 1).

Subunit I is the largest of the three core subunits. It has 12 membrane-spanning α-helices that are roughly arranged as three membrane-spanning four-helical semicircular arcs, each surrounding pore-like volumes [Iwata *et al.*, 1995]. One of these houses a *bis*-histidine coordinated low spin A-type haem (haem *a*) and its hydroxyethylfarnesyl chain. A second houses the catalytic binuclear centre (BNC) which is formed from a histidine-ligated high spin A-type haem (haem a_3) and a copper ion ligated by three histidines, one of which is covalently linked *via* its N_τ

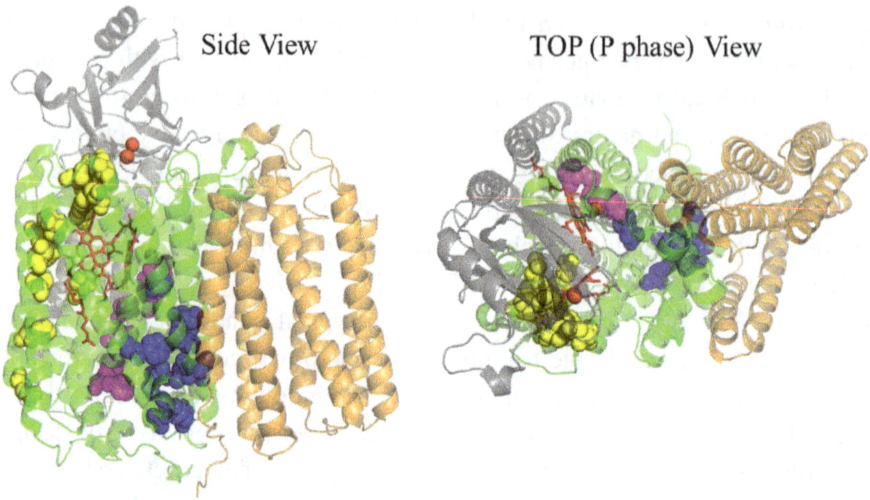

Figure 1. Overall core structure of oxidised bovine C*c*O. Only core subunits I (green), II (grey) and III (brown) are shown. Redox active metals are shown as red lines (haems) or red spheres (Cu_A and Cu_B). Amino acid residues assigned to the proposed proton channels are shown as spheres in blue (D channel), pink (K channel) and yellow (H channel). The figure was drawn with coordinates from PDB 5B1A [Yano *et al.*, 2016].

nitrogen to the *meta* ring carbon of a catalytic tyrosine [Tsukihara *et al.*, 1996]. The lower part of the third "pore" houses the proton-conducting D channel (see below) linking the N phase to residue E242 (bovine sequence numbering is used throughout unless otherwise stated). It spans roughly half of the membrane dielectric barrier but is blocked by aromatic residues at its top. The haem a_3 and Cu_B metals are separated by around 5 Å and, in their reduced states, have free coordination sites that create the dioxygen binding site. The ensuing multistep catalytic cycle of oxygen reduction and water formation with four electrons and four protons is relatively well understood [Wikström *et al.*, 2018; Rich, 2017b]. However, several outstanding issues remain concerning the chemical structures of several of the intermediates. In particular, the measurable thermodynamic properties of the oxidised, "O", and one-electron-reduced, "E", intermediates appear to have potentials that do not provide sufficient free energy for the associated coupled proton transfers with which they are linked; instead, it is suggested that they must have unstable, as yet undetected, high potential (O_H and E_H) configurations that exist only for short periods during turnover [Blomberg and Siegbahn, 2015a; Blomberg and Siegbahn, 2015b; Sharma *et al.*, 2013].

Subunit II has a large hydrophilic domain that interfaces with the P phase surface of subunit I above the haem groups. It houses a dinuclear Cu_A cofactor close to this interface (Figure 2) that acts as the initial acceptor of electrons from soluble cytochrome *c*. The structure of a horse heart cytochrome *c*/bovine C*c*O complex [Shimada *et al*., 2017b] confirms the primarily electrostatic basis for complex [Ferguson-Miller *et al*., 1978; Zhen *et al*., 1999]. The binding interface is provided almost entirely by subunit II and, surprisingly, has a large number of water molecules bound between the proteins. It is unclear whether these waters are of functional significance [Shimada *et al*., 2017b], or whether the structure may represent some kind of pre/post-binding intermediate stabilised by the low pH required for crystallisation and/or the fact that both cytochrome *c* and C*c*O were in their oxidised states. The electrons from cytochrome *c* are transferred from Cu_A into the BNC indirectly and strictly *via* haem *a*, despite similar short distances of Cu_A to the macrocycle of haem *a* (~16Å) and haem a_3 (~19Å). The physical explanation for this is still not fully settled (reviewed in [Rich, 2017b]). The edges of the haem macrocycles are separated by only ~7Å, enabling extremely fast inter-haem electron transfer [Moser *et al*., 1992] as has been confirmed experimentally [Pilet *et al*., 2004]. As discussed below, this indirect electron transfer sequence to the BNC *via* haem *a* is central to the coupled proton translocation mechanism.

Core subunit III appears to be unique to the A branch of the HCO superfamily and is possibly related to the presence of the D channel which is also unique to the A branch. It has no redox cofactors and is comprised of seven hydrophobic transmembrane helices with a distinct cleft between helices I/II and III–VII. The cleft has been suggested to provide a selective lipid-lined channel through which substrate oxygen diffuses from the membrane interior to the buried subunit I surface where a possible pathway leads to the BNC binding site [Riistama *et al*., 1996; Shinzawa-Itoh *et al*., 2007; Sharma *et al*., 2015]. Perhaps significantly, the interface of subunit III with subunit I is mostly with the arc of helices in subunit I whose lower region houses the D channel (Figure 1). Part of subunit III is close to its entrance and may aid proton collection at this site [Hosler, 2004; Alnajjar *et al*., 2014].

3. Coupled Proton Translocation

3.1. *Structural aspects*

The interior of subunit I must have pathways for intra-protein proton transfers, both into the BNC for oxygen reduction and across the entire structure for proton translocation. It is thought that such pathways are generally enabled by arrays of hydrophilic amino acids and water molecules. These can, at least transiently, form a continuous H-bonded "proton wire" between proton source and sink [Nagle and Tristram-Nagle, 1983; DeCoursey, 2003], facilitating net proton transfer by a

Grotthuss-type of mechanism [Grotthuss, 2006] involving H-bond rearrangements followed by functional group rotations. Several such arrays are evident in the crystal structures of both A1-type bacterial [Iwata *et al.*, 1995; Fetter *et al.*, 1995] and bovine mitochondrial [Yoshikawa *et al.*, 1998] C*c*O structures (Figure 1). The K channel, named from a central lysine (K319), is located in the "pore" containing the BNC and connects the N aqueous phase, possibly with an opening between subunits I and II [Hiser *et al.*, 2013; Liu *et al.*, 2017], to the catalytic tyrosine of the BNC. A second array, the D channel, is located in the third 'pore' and leads from an aspartic acid (D91) at its N phase entrance to residue E242 that is located roughly ~10 Å from both haem macrocycle edges.

A third structure, the H channel, was firstly described in bovine C*c*O [Yoshikawa and Shimada, 2015], though is also evident to some extent in bacterial A-type C*c*Os. It is associated primarily with residues on the helices that form the arc housing haem *a*. Its 'lower' domain spans from the N phase at D407 towards the ethylfarnesyl hydroxyl group and the D-ring formyl substituent edge of haem *a* where there are a number of hydrophilic amino acids and water molecules. Unusually, proton transfer is proposed to occur by hydronium ion movement through a series of hydratable cavities, at least one of which is opened only transiently at the "R" catalytic intermediate (*i.e.* both BNC metals reduced before oxygen substrate binds to form the "A" intermediate) through structural changes around residue S382 [Yano *et al.*, 2016; Shimada *et al.*, 2017a; Luo *et al.*, 2018; Ishigami *et al.*, 2017]. The "upper" region of the H channel consists of an extensive network of hydrophilic amino acids and waters that could connect the region alongside haem *a* to the P phase. This has a conformationally-flexible span that includes residues D50 and D51 that are close to the interface with subunit II (Figure 2). The region around the haem *a* edge was initially proposed to transfer protons directly to this domain through the peptide link between Y440 and S441, which could act as a unidirectional proton gate. More recently [Yano *et al.*, 2016], it has been suggested that protons are firstly transferred from the vicinity of haem *a* into a storage region above the haems — the proton trap (see below) — before being sequentially released into the P phase though the upper H channel domain.

3.2. *General mechanism*

The specific roles of the K, D and H channels in conducting protons, particularly the translocated protons, are presently disputed, with the possibility mooted that proton transfer route into the trap may differ in mammalian, compared to bacterial, A1 C*c*Os. Studies of mutant forms of bacterial A1-type C*c*Os with a wide range of time-resolved methods that measure spectra, electrometric changes and pH changes have led to a feasible molecular model involving only the D and K

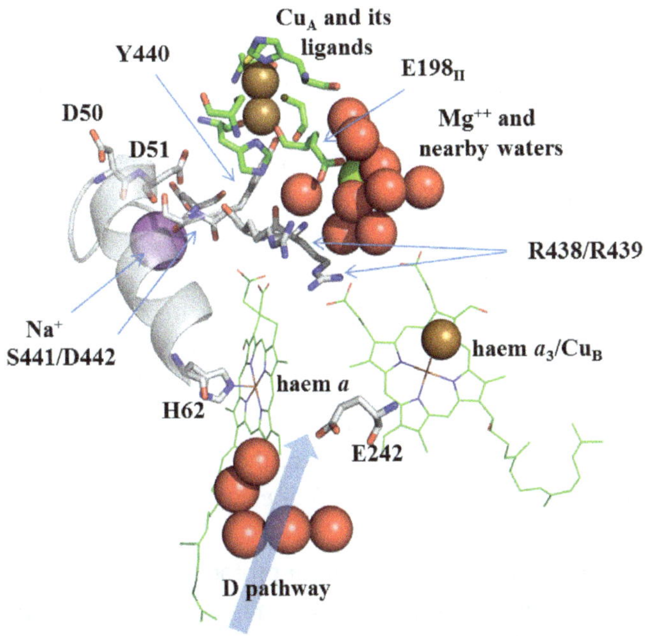

Figure 2. Structure of the region above the haems of oxidised bovine C*c*O. The figure shows a side view of haem *a* and the BNC. A short helical span connects a mobile region containing D50 and D51 with one of the histidine ligands to haem *a* (H62). A second short span consists of an arginine pair (R438/R439) that interact with haem propionates, Y440 which extends close to the Cu$_A$ centre in subunit II, and S441/D442 that ligate the bound Na$^+$ ion (D442 ligates *via* a water which is not shown). Also shown is Cu$_A$ and its immediate ligands (carbons coloured in green) and the bound Mg^{++} ion with nearby water oxygens (red spheres) that may provide the proton trap region. Residue E198 of subunit II links to both to Cu$_A$ and the Mg^{++}. Residue E242 at the top of the D channel, together with waters immediately below it, illustrates the gap that must be spanned to provide a proton pathway into the proton trap. The figure was drawn with coordinates from PDB 5B1A [Yano *et al.*, 2016].

channels, with no clear role for the H channel [Lee *et al.*, 2000; Wikström *et al.*, 2015]. More limited studies of effects of mutations in these three channels of yeast mitochondrial C*c*O support similar conclusions [Maréchal *et al.*, 2014; Dodia, 2014]. The K channel provides the pathway for transfer of one or two of the “substrate” protons required for oxygen reduction to water from the N phase into the BNC in the $O_H \rightarrow E_H \rightarrow R$ steps. It may also act as a “dielectric channel” to lower the energetics of transient uncompensated electron transfer into the BNC at several stages of the catalytic cycle [Jünemann *et al.*, 1997; Rich and Maréchal, 2013; Rich, 2017b] by reversible movement of charge within the K channel, perhaps involving the protonated K319 residue [Lepp *et al.*, 2008]. The D channel is thought to provide part of the pathway from the N phase into the BNC for the remaining substrate protons in the $P \rightarrow F \rightarrow O_H$ steps, together with the path from the N phase towards the proton trap *via* E242 for all four translocated protons [Zaslavsky and Gennis, 2000]. Strong evidence for the latter role comes from mutations in the D channel that prevent proton translocation without inhibiting the oxygen chemistry [Han *et al.*, 2006; Pfitzner *et al.*, 2000]. In contrast, deductions from the atomic structure of bovine C*c*O, together with more limited direct functional measurements, have indicated a central role of the H pathway in proton translocation, with the D and/or K channels providing possible routes only for substrate protons [Yoshikawa and Shimada, 2015].

Despite these differences, all proposed mechanisms of proton translocation in the A-type C*c*Os have several features in common. The first is that a full catalytic cycle results in four coupled proton translocations [Wikström, 1977], with one proton appearing in the P phase after each of the four one-electron reduction steps ($P \rightarrow F \rightarrow O_H \rightarrow E_H \rightarrow R$) required to regenerate the R state after the dioxygen O-O bond has been broken by reduction with four electrons from the BNC in the A→P step.

A second common feature is that all four proton translocations occur by essentially the same mechanism, regardless of the stage of the oxygen reduction cycle, with protons transferred through the same input and output pathways to/from the proton trap. Though still lacking direct experimental evidence, it is generally thought that this trap is in a region above the haem a_3 propionates, possibly involving residues/waters around the bound Mg^{++} ion [Schmidt *et al.*, 2003; Supekar *et al.*, 2016; Yano *et al.*, 2016; Seibold *et al.*, 2005] (Figure 2).

Thirdly, and an essential feature of any coupling model, is that there must be a means whereby proton transfers into and out of the trap region are "gated" so that input is always from the N phase and output is always into the P phase, even when there is a protonmotive force which is counteracting such directionality. These protonation events must also be separated from the protonations required for oxygen reduction chemistry within the BNC. Failure of any of these would result in uncoupling of proton translocation from the oxygen reduction chemistry.

In the most widely held model of the bacterial/yeast C*c*O coupling mechanism through the D channel, a key physical feature is electrostatic charge compensation [Rich, 1995]. Specifically, uncompensated electron transfer from haem *a* into the BNC is energetically unfavourable because the BNC is in a low dielectric environment. Electron transfer into the haem *a* and from there into the BNC is associated with pK changes that cause the proton on E242 at the D channel top to move towards the proton trap. In available crystal structures, a considerable gap has to be spanned above E242 (Figure 2). The most reasonable proposal for how this is achieved is by transient formation of a water wire that is too unstable to be observable in crystal structures [Wikström *et al.*, 2003]. The positive charge of the excess proton in the trap site electrostatically stabilises the negative charge of the electron in the BNC, but its physical location prevents its acting as a substrate proton for the oxygen reduction chemistry within the BNC [Rich, 1995]. Once PCET into the BNC has occurred, E242 becomes reprotonated from the N phase *via* the D channel. Arrival of the electron within the BNC creates an unstable oxygen intermediate with a high pK site that requires protonation. The proton is thought to be delivered from the N phase *via* the K channel in the $O_H \rightarrow E_H \rightarrow R$ steps but arrives *via* the D channel, E242 and a second short transient proton wire in the $P \rightarrow F \rightarrow O_H$ steps. In order to prevent uncoupling, a gating mechanism must exist to prevent its protonation by backflow of the trap proton into the BNC *via* E242. Plausible gating mechanisms involving E242/water wire conformations and the charge state of haem *a* have been proposed [Kaila *et al.*, 2009; Kim *et al.*, 2009; Blomberg and Siegbahn, 2012; Goyal *et al.*, 2015]. Once the reduced BNC has become protonated, the electrostatic stabilisation between BNC and trap charges is lost. The pK of the trap decreases and the proton dissociates into the P phase. This exit pathway, which has not been well-defined (but see below), must also be gated to prevent trap protonation from the P phase in association with electron transfer into the BNC.

In this model, one full proton translocation across the membrane occurs by this same mechanism with each of the four steps of electron transfer from haem *a* into the BNC ($O_H \rightarrow E_H$, $E_H \rightarrow R$, $P \rightarrow F$ and $F \rightarrow O$). A further key feature is the assumption that the four stable proton uptakes that are known to be associated with the formation of each of the stable oxygen intermediates (E, R, P and F) [Mitchell *et al.*, 1992; Mitchell and Rich, 1994] are substrate protons that bind within the BNC, rather than being associated with loading of the proton trap. This inventory of protons bound by the stable BNC intermediates has contributed to a scheme of chemically-reasonable structures [Wikström, 2012; Wikström *et al.*, 2018; Rich, 2017b] that are supported by a considerable body of spectroscopic data, though the details of the catalytically active but unstable O_H and E_H states remain open [Blomberg, 2015; Sharma and Wikström, 2016] as do aspects of the haem *a*/BNC PCET sequence [Faxén *et al.*, 2005], the gating between E242 and the proton trap or BNC [Wikström *et al.*, 2003; Wikström *et al.*, 2005; Blomberg and Siegbahn, 2012] and the identification of the trap site and its gated P phase exit route.

3.3. *Proton translocation in mammalian CcOs*

B- and C-type C*c*Os of the HCO superfamily do not possess a functional D channel and yet are coupled to proton translocation, though at a lower stoichiometry or, possibly, are less tightly coupled [Sousa *et al.*, 2012; Wikström *et al.*, 2015].

They have a channel spatially related to the K channel, though whether they possess a second functional channel of some type is not clear. These findings have led to the notion that, although the oxygen reduction chemistry is likely to be very similar in all HCOs, the mechanism of coupling to proton translocation may be more varied and achievable by different types of structures.

The possibility of an alternative coupling mechanism in mammalian C*c*Os was prompted by the recognition of the H channel in bovine mitochondrial C*c*O subunit I. As described above, the lower part of this structure consists of a series of potentially hydratable cavities leading from the N phase to the formyl and hydroxyethylfarnesyl substituents on the D- and A-pyrrole ring of the low spin haem [Tsukihara *et al.*, 1996]. Comparisons of oxidised, reduced and ligated structures [Yano *et al.*, 2016; Shimada *et al.*, 2017a; Kubo *et al.*, 2013; Shimada *et al.*, 2017a; Ishigami *et al.*, 2017; Shimada *et al.*, 2018] have led to a model in which opening of the water channel occurs only at the R stage of the catalytic cycle, driven by conformational changes that induce residue S382 to move out of, and hence open, the largest of the water cavities. In this open state, all four translocated protons are transferred by hydronium ions towards haem *a* and the upper region, and from there into the same proton trap site around the bound Mg^{2+} *via* a short pathway involving R438/R439. Subsequent oxygen binding to R closes the lower pathway. Protons are then released sequentially into the P phase, one with each redox cycle of haem *a* in the same four steps ($O_H \rightarrow E_H$, $E_H \rightarrow R$, P→F and F→O) [Yano *et al.*, 2016] as the bacterial model above. Their exit path into the P phase was proposed to run *via* R438/R439 through the hydrogen bonded network in the "top" part of the H channel towards the mobile region around D51 at the P phase interface between subunits I and II, with the essential gating controlled by the reduction/oxidation cycles of haem *a*.

This mechanism differs from that described for bacterial and yeast C*c*Os in several crucial features. The first is obviously the proton path between N phase and the proton trap region. In the case of bovine C*c*O, this is largely based on crystallographic identification of H channel residues and associated water molecules or potential water-accessible regions, an uncoupling double mutant (V386L/M390W) in this span [Shimokata *et al.*, 2007] and the marked structural changes that have been consistently observed around residue S382 [Shimada *et al.*, 2017a; Ishigami *et al.*, 2017]. To date, however, MD simulations have been unable to show sufficient hydration of this span that could support an H-bonded proton transfer capability unless a central H413 residue could form its imidazolium, positively charged, state, which was also disfavoured by the simulations [Sharma *et al.*, 2017]. An alternative mechanism of hydronium ion diffusion raises the question of the driving force for its directionality, particularly in the presence of a counteracting protonmotive force.

A further important way in which this mechanism differs is in electrostatic considerations in relation to which stages of the catalytic cycle are linked to substrate and translocated proton movements. All four translocated protons are proposed to be transferred into the proton trap after formation of the R intermediate and before it binds oxygen to form A. On R formation and before protonation, the trap was considered to have a net charge of −2, rising to +2 when all four protons bound. This partly addresses the issue of the energy cost of such multiple charge storage, but clearly does not follow the electroneutral 'rule' of counterbalancing BNC and trap charges. Additionally, however, it requires a revision of the steps at which substrate protons are bound in order to accommodate the net proton uptakes that are associated with the formation of each stable catalytic intermediate [Mitchell *et al.*, 1992; Mitchell and Rich, 1994]. Specifically, the four substrate protons would have to be bound in the BNC only the P→F and F→O steps, requiring a revision of the chemistry of these intermediates. Further discussion of these implications in relation to available experimental data can be found in [Rich, 2017b].

The proposed exit path of protons from the trap is *via* R438/R439 and through the 'top' part of the H channel towards D51 and the P phase at the subunit I/II interface (Figure 2). Some direct experimental support for importance of this region has come from two mutations (D51N and S441P) generated in hybrid bovine/human construct that cause loss of coupled proton translocation without affecting catalytic oxygen reduction rates [Tsukihara *et al.*, 2003; Shimokata *et al.*, 2007]. Measurements of this kind are not without caveats of interpretation and the fact that none caused an associated inhibition of oxygen reduction requires explanation. Nevertheless, this region could also provide a feasible exit pathway from the trap into the P phase in bacterial/yeast C*c*Os. Although amino acids in this region are not well-conserved, the overall structural organisation is similar and its extended structure may provide multiple exit routes. Indeed, similar suggestions have been made for exit routes *via* Arg438/439 into this general region from simulations of bovine [Sugitani and Stuchebrukhov, 2009] and *Rb. sphaeroides* [Cai *et al.*, 2018] C*c*Os. Any such path between the trap and the P phase must be gated in some manner to prevent short-circuiting, and gating mechanisms involving redox changes of heme *a* or Cu_A have been suggested [Yano *et al.*, 2016; Sugitani and Stuchebrukhov, 2009].

4. Emerging Mechanisms of Control of Mammalian C*c*O

The fact that C*c*O is always operating irreversibly even at the highest cellular PMFs means that modulation of its activity, rather than activities of upstream reversible complexes I, II and III that may be operating closer to equilibrium, would most influence the overall throughput of the whole respiratory chain. In human tissues, the amount of assembled C*c*O appears to be rather constant, though differing in levels in different tissues. Yet, its activity may be required to vary considerably in different physiological conditions. The clearest and most direct form of regulation is by the pH and membrane potential components of the protonmotive force [Gregory and Ferguson-Miller, 1988; Gregory and Ferguson-Miller, 1989] and, hence, by the cellular demand for ATP. Further control by competition of cellular compounds with the C*c*O substrates at their binding sites has rather limited scope, particularly for the oxygen site which is small and with its possibly selective access route through subunit III. However, one competitive inhibitor of oxygen binding is the physiologically important nitric oxide [Brown, 2001; Moncada and Erusalimsky, 2002]. A further intriguing possible substrate control mechanism is by modulation of access of substrate protons by conformational changes within the K channel and ligand binding at its entrance [Hiser *et al.*, 2013; Liu *et al.*, 2017].

Mitochondrial CcOs are considerably larger than their bacterial counterparts because they have additional supernumerary subunits. Human and bovine CcOs have at least 10 [Kadenbach and Hüttemann, 2015] and seven of these (IV, VIa, VIb, VIc, VIIa, VIIb and VIII) have encoded isoforms that may be expressed as tissue- or cellular status-dependent isoforms. Several subunits house sites that may become phosphorylated [Hüttemann *et al.*, 2012], acetylated [Liko *et al.*, 2016] or can bind nucleotides or other ligands [Kadenbach and Hüttemann, 2015; Arnold, 2012]. Specific allosteric control functions of some of these isoforms, their post-translational modifications and ligand binding sites have been suggested, though the supporting data in many cases remain inconsistent.

Several additional proteins have been identified that are associated with digitonin-extracted mitochondrial C*c*Os but are not retained in bovine C*c*O crystals produced with harsher detergents. For example, NDUFA4, originally thought to be a subunit of complex I, has been shown to be a component of mammalian complex IV [Balsa *et al.*, 2012; Zong *et al.*, 2018]. Another protein, hypoxia-inducible Higd1a can also associate with mammalian CcO and a direct effect of binding on catalytic activity has been suggested [Hayashi *et al.*, 2015], as has the equivalent yeast homologue, Rcf1 [Rydström Lundin and Brzezinski, 2017; Schäffer *et al.*, 2018]. Higd1a is proposed to bind to a site close to subunit IV and two helices of core subunit I that provide part of the H channel structure, possibly affecting catalytic parameters through allosteric effects on haem *a* [Hayashi *et al.*, 2015]. Isoforms of subunit IV have also been reported to alter core properties [Kadenbach and Hüttemann, 2015]. Recently, it has been reported that chemoresistance of glioblastoma tumours cells is associated with a switch from subunit IV-2 to subunit IV-1 in C*c*O and that C*c*O with subunit IV–1 can be selectively inhibited by chlorpromazine [Oliva *et al.*, 2017]. It may be speculated that all of these effects could be mediated by alterations of the top region of the H channel, perhaps through modulation of its dielectric properties [Rich and Maréchal, 2013].

With the explosion of new information on respiratory supercomplexes, it has become clear that some of the supernumerary subunits, and possibly some of the additional proteins that can associate with C*c*Os, are involved in formation of supercomplexes between C*c*O and complexes I and III. The question of whether the activity of C*c*O can be further modulated by incorporation into some supercomplex configurations remains an open speculation.

5. Concluding Remarks

Atomic structures of C*c*Os at high resolution have now been known for over 20 years and many of the key details of the catalytic mechanism of oxygen reduction and coupled proton translocation have been established. Nevertheless, questions still remain, in particular related to the unstable transient states that cannot be easily captured by static structural biophysical methods. The advent of time-resolved high resolution structural methods in combination with the ability to initiate full reaction cycles *via* the "flow/flash" CO photolysis method provide a future viable avenue for definitive progress on these issues. Parallel to this is the emerging area of supercomplex structures, additional protein factors that influence aspects of C*c*O and the recognition of new allosteric regulatory mechanisms that offer future medical therapeutic opportunities.

Acknowledgements

Peter R. Rich has been funded by the BBSRC with grants BB/K001094/1 and BB/L020165/1 that are directly relevant to this topic.

References

Abramson, J., Riistama, S., Larrson, G., Jasaitis, A., Svensson-Ek, M., Laakkonen, L., Puustinen, A., Iwata, S. and Wikström, M. (2000). The structure of the ubiquinol oxidase from *Escherichia coli* and its ubiquinone binding site. *Nat. Struct. Biol.*, 7, 910–917.

Alnajjar, K.S., Hosler, J. and Prochaska, L. (2014). Role of the N-terminus of subunit III in proton uptake in cytochrome *c* oxidase of *Rhodobacter sphaeroides*. *Biochemistry*, 53, 496–504.

Arnold, S. (2012). The power of life-cytochrome *c* oxidase takes center stage in metabolic control, cell signalling and survival. *Mitochondrion*, 12, 46–56.

Balsa, E., Marco, R., Perales-Clemente, E., Szklarczyk, R., Calvo, E., Landàzuri, M.O. and Enríquez, J.A. (2012). NDUFA4 is a subunit of complex IV of the mammalian electron transport chain. *Cell Metabolism*, 16, 378–386.

Barber, J. (2017). Photosynthetic water splitting by the $Mn_4Ca^{2+}O_X$ catalyst of photosystem II: its structure, robustness and mechanism. *Q. Rev. Biophys.*, 50, e13.

Blomberg, M.R.A. and Siegbahn, P.E.M. (2012). The mechanism for proton pumping in cytochrome *c* oxidase from an electrostatic and quantum chemical perspective. *Biochim. Biophys. Acta*, 1817, 495–505.

Blomberg, M.R.A. (2015). Mechanism of oxygen reduction in cytochrome *c* oxidase and the role of the active site tyrosine. *Biochemistry*, 55, 489–500.

Blomberg, M.R.A. and Siegbahn, P.E.M. (2015a). Protonation of the binuclear active site in cytochrome *c* oxidase decreases the reduction potential of Cu_B. *Biochim. Biophys. Acta*, 1847, 1173–1180.

Blomberg, M.R.A. and Siegbahn, P.E.M. (2015b). How cytochrome *c* oxidase can pump four protons per oxygen molecule at high electrochemical gradient. *Biochim. Biophys. Acta*, 1847, 364–376.

Brown, G. C. (2001). Regulation of mitochondrial respiration by nitric oxide inhibition of cytochrome *c* oxidase. *Biochim. Biophys. Acta*, 1504, 46–57.

Cai, X., Haider, K., Lu, J., Radic, S., Son, C.-Y., Cui, Q. and Gunner, M.R. (2018). Network analysis of a proposed exit pathway for protons to the P-side of cytochrome *c* oxidase. *Biochim. Biophys. Acta*, 1859, 997–1005.

DeCoursey, T.E. (2003). Voltage-gated proton channels and other proton transfer pathways. *Physiol. Rev.*, 83, 475–579.

Dodia, R.J. (2014). Structure-function relationship of mitochondrial cytochrome *c* oxidase: redox centres, proton pathways and isozymes. PhD Thesis, University College London, London, UK.

Duursma, E.K. and Boisson, M.P.R.M. (1994). Global oceanic and atmospheric oxygen stability considered in relation to the carbon cycle and to different time scales. *Oceanologica Acta*, 17, 117–141.

Faxén, K., Gilderson, G., Ädelroth, P. and Brzezinski, P. (2005). A mechanistic principle for proton pumping by cytochrome *c* oxidase. *Nature*, 437, 286–289.

Ferguson-Miller, S., Brautigan, D.L. and Margoliash, E. (1978). Definition of cytochrome *c* binding domains by chemical modification. III. Kinetics of reaction of carboxydinitrophenyl cytochromes *c* with cytochrome *c* oxidase. *J. Biol. Chem.*, 253, 149–159.

Fetter, J.R., Qian, J., Shapleigh, J., Thomas, J.W., García-Horsman, A., Schmidt, E., Hosler, J., Babcock, G.T., Gennis, R.B. and Ferguson-Miller, S. (1995). Possible proton relay pathways in cytochrome *c* oxidase. *Proc. Natl. Acad. Sci. USA*, 92, 1604–1608.

García-Horsman, J.A., Barquera, B., Rumbley, J., Ma, J. and Gennis, R.B. (1994). The superfamily of heme-copper oxidases. *J. Bacteriol.*, 176, 5587–5600.

Goyal, P., Yang, S. and Cui, Q. (2015). Microscopic basis for kinetic gating in cytochrome *c* oxidase: insights from QM/MM analysis. *Chem. Sci.*, 6, 826–841.

Gregory, L. and Ferguson-Miller, S. (1988). Control of cytochrome *c* oxidase activity by pH and the electrical potential gradient occurs at separate electron transfer steps and does not require subunit III. *Ann. N. Y. Acad. Sci.*, 550, 260–268.

Gregory, L. and Ferguson-Miller, S. (1989). Independent control of respiration in cytochrome *c* oxidase vesicles by pH and electrical gradients. *Biochemistry*, 28, 2655–2662.

Grotthuss, C.J.T. (2006). (Translated version of) Memoir on the decomposition of water and of the bodies that it holds in solution by means of galvanic electricity. *Biochim. Biophys. Acta*, 1757, 871–875.

Han, D., Namslauer, A., Pawate, A., Morgan, J.E., Nagy, A.S., Vakkasoglu, P., Brzezinski, P. and Gennis, R.B. (2006). Replacing Asn207 by aspartate at the neck of the D channel in the aa_3-type cytochrome *c* oxidase from *Rhodobacter sphaeroides* results in decoupling the proton pump. *Biochemistry*, 45, 14064–14074.

Hayashi, T., Asano, Y., Shintani, Y., Aoyama, H., Kioka, H., Tsukamoto, O., Hikita, M., Shinzawa-Itoh, K., Takafuji, K., Higo, S., Kato, H., Yamazaki, S., Matsuoka, K., Nakano, A., Asanuma, H., Asakura, M., Minamino, T., Goto, Y.-I., Ogura, T., Kitakaza, M., Komuro, I., Sakata, Y., Tsukihara, T., Yoshikawa, S. and Takashima, S. (2015). *Hig1a* is a positive regulator of cytochrome *c* oxidase. *Proc. Natl. Acad. Sci. USA*, 112, 1553–1558.

Hemp, J. and Gennis, R.B. (2008). Results probl. cell. differ. Schäfer, G. and Penefsky, H.S., eds. In *Diversity of the Heme-Copper Superfamily in Archaea: Insights from Genomics and Structural Modeling,* (Springer-Verlag), pp. 1–31.

Hiser, C., Buhrow, L., Liu, J., Kuhn, L. and Ferguson-Miller, S. (2013). A conserved amphipathic ligand binding region influences K-path-dependent activity of cytochrome *c* oxidase. *Biochemistry*, 52, 1385–1396.

Hosler, J.P. (2004). The influence of subunit III of cytochrome *c* oxidase on the D pathway, the proton exit pathway and mechanism-based inactivation in subunit I. *Biochim. Biophys. Acta*, 1655, 332–339.

Huang, J., Huang, J., Liu, X., Li, C., Ding, L. and Yu, H. (2018). The global oxygen budget and its future projection. *Sci. Bull.*, 63, 1180–1186.

Hüttemann, M., Helling, S., Sanderson, T.H., Sinkler, C., Samavati, L., Mahapatra, G., Varughese, A., Lu, G., Liu, J., Ramzan, R., Vogt, S., Grossman, L.I., Doan, J.W. and Lee, I. (2012). Regulation of mitochondrial respiration and apoptosis through cell signaling: cytochrome *c* oxidase and cytochrome *c* in ischemia/reperfusion injury and inflammation. *Biochim. Biophys. Acta*, 1817, 598–609.

Ishigami, I., Zatespin, N.A., Hikita, M., Conrad, C.E., Nelson, G., Coe, J.D., Basu, S., Grant, T.D., Seaberg, M.H., Sierra, R.G., Hunter, M.S., Fromme, P., Fromme, R., Yeh, S.-R. and Rousseau, D.L. (2017). Crystal structure of CO-bound cytochrome *c* oxidase determined by serial femtosecond X-ray crystallography at room temperature. *Proc. Natl. Acad. Sci. USA*, 114, 8011–8016.

Iwata, S., Ostermeier, C., Ludwig, B. and Michel, H. (1995). Structure at 2.8 Å resolution of cytochrome *c* oxidase from *Paracoccus denitrificans*. *Nature*, 376, 660–669.

Jünemann, S., Meunier, B., Gennis, R.B. and Rich, P.R. (1997). Effects of mutation of the conserved lysine-362 in cytochrome *c* oxidase from *Rhodobacter sphaeroides*. *Biochemistry*, 36, 14456–14464.

Kadenbach, B. and Hüttemann, M. (2015). The subunit composition and function of mammalian cytochrome *c* oxidase. *Mitochondrion*, 24, 64–76.

Kaila, V.R.I., Verkhovsky, M.I., Hummer, G. and Wikström, M. (2009). Mechanism and energetics by which glutamic acid 242 prevents leaks in cytochrome *c* oxidase. *Biochim. Biophys. Acta*, 1787, 1205–1210.

Kim, Y.C., Wikström, M. and Hummer, G. (2009). Kinetic gating of the proton pump in cytochrome *c* oxidase. *Proc. Natl. Acad. Sci. USA*, 106, 13707–13712.

Kubo, M., Nakashima, S., Yamaguchi, S., Ogura, T., Mochizuki, M., Kang, J., Tateno, M., Shinzawa-Itoh, K., Kato, K. and Yoshikawa, S. (2013). Effective pumping proton collection facilitated by a copper site (Cu_B) of bovine heart cytochrome *c* oxidase, revealed by a newly developed time-resolved infrared system. *J. Biol. Chem.*, 288, 30259–30269.

Lee, H.-M., Das, T.K., Rousseau, D.L., Mills, D., Ferguson-Miller, S. and Gennis, R.B. (2000). Mutations in the putative H-channel in the cytochrome *c* oxidase from *Rhodobacter sphaeroides* show that this channel is not important for proton conduction but reveal modulation of the properties of heme *a*. *Biochemistry*, 39, 2989–2996.

Lepp, H., Svahn, E., Faxén, K. and Brzezinski, P. (2008). Charge transfer in the K proton pathway linked to electron transfer to the catalytic site in cytochrome *c* oxidase. *Biochemistry*, 47, 4929–4935.

Lide, D.R. (1995). *Handbook of Chemistry and Physics*. (CRC Press), Boca Raton, Florida.

Liko, I., Degiacomi, M.T., Mohammed, S., Yoshikawa, S., Schmidt, C. and Robinson, C. (2016). Dimer interface of bovine cytochrome *c* oxidase is influenced by local post-translational modifications and lipid binding. *Proc. Natl. Acad. Sci. USA*, 113, 8230–8235.

Liu, J., Hiser, C. and Ferguson-Miller, S. (2017). Role of conformational change and K-path ligands in controlling cytochrome *c* oxidase activity. *Biochem. Soc. Trans.*, 45, 1087–1095.

Luo, F., Shinzawa-Itoh, K., Hagimoto, K., Shimada, A., Shimada, S., Yamashita, E., Yoshikawa, S. and Tsukihara, T. (2018). Structure of bovine cytochrome *c* oxidase in the ligand-free reduced state at neutral pH. *Acta Cryst.*, F74, 92–98.

Maréchal, A., Haraux, F., Meunier, B. and Rich, P.R. (2014). Determination of H^+/e ratios in mitochondrial yeast cytochrome *c* oxidase. *Biochim. Biophys. Acta*, 1837, e100.

Mitchell, P. (1966). Chemiosmotic coupling in oxidative and photosynthetic phosphorylation. (Glynn Research Ltd.), Bodmin.

Mitchell, R., Mitchell, P. and Rich, P.R. (1992). Protonation states of the catalytic cycle intermediates of cytochrome *c* oxidase. *Biochim. Biophys. Acta*, 1101, 188–191.

Mitchell, R. and Rich, P.R. (1994). Proton uptake by cytochrome *c* oxidase on reduction and on ligand binding. *Biochim. Biophys. Acta.*, 1186, 19–26.

Moncada, S. and Erusalimsky, J.D. (2002). Does nitric oxide modulate mitochondrial energy generation and apoptosis? *Nature Rev. Mol. Cell. Biol.*, 3, 214–220.

Moser, C.C., Keske, J.M., Warncke, K., Farid, R.S. and Dutton, P.L. (1992). Nature of biological electron transfer. *Nature*, 355, 796–802.

Nagle, J.F. and Tristram-Nagle, S. (1983). Hydrogen bonded chain mechanisms for proton conduction and proton pumping. *J. Membrane Biol.*, 74, 1–14.

Oliva, C.R., Zhang, W., Langford, C., Suto, M.J. and Griguer, C.E. (2017). Repositioning chlopromazine for treating chemoresistant glioma through the inhibition of cytochrome *c* oxidase bearing the COX4–1 regulatory subunit. *Oncotarget*, 8, 37568–37583.

Pfitzner, U., Hoffmeier, K., Harenga, A., Kannt, A., Michel, H., Bamberg, E., Richter, O.-M.H. and Ludwig, B. (2000). Tracing the D-pathway in reconstituted site-directed mutants of cytochome *c* oxidase fom *Paracoccus denitrificans*. *Biochemistry*, 39, 6756–6762.

Pilet, E., Jasaitis, A., Liebl, U. and Vos, M.H. (2004). Electron transfer between hemes in mammalian cytochrome *c* oxidase. *Proc. Natl. Acad. Sci. USA*, 101, 16198–16203.

Rich, P.R. (1982). Electron and proton transfers in chemical and biological quinone systems. *Faraday Discuss. Chem. Soc.*, 74, 349–364.

Rich, P.R. (1995). Towards an understanding of the chemistry of oxygen reduction and proton translocation in the iron-copper respiratory oxidases. *Aust. J. Plant Physiol. (now renamed Functional Plant Biology)*, 22, 479–486.

Rich, P.R. (2003). The cost of living. *Nature*, 421, 583.

Rich, P.R. and Maréchal, A. (2013). Functions of the hydrophilic channels in protonmotive cytochrome *c* oxidase. *J. R. Soc. Interface*, 10, 1–14.

Rich, P.R. (2017a). Photosynthesis and Bioenergetics eds. Barber, J. and Ruban, A., "*Cytochrome c Oxidase: Insights into Functions from Studies of the Yeast S. cerevisiae Homologue*", (World Scientific Press) pp. 65–79.

Rich, P.R. (2017b). Mitochondrial cytochrome *c* oxidase: catalysis, coupling and controversies. *Biochem. Soc. Trans.*, 45, 813–829.

Riistama, S., Puustinen, A., Garcia-Horsman, A., Iwata, S., Michel, H. and Wikström, M. (1996). Channelling of dioxygen into the respiratory enzyme. *Biochim. Biophys. Acta*, 1275, 1–4.

Rydström Lundin, C. and Brzezinski, P. (2017). Modulation of O_2 reduction in *Saccharomyces cerevisiae* mitochondria. *FEBS Lett.*, 591, 4049–4055.

Schäffer, J., Dawitz, H., Ott, M., Ädelroth, P. and Brzezinski, P. (2018). Regulation of cytochrome *c* oxidase activity by modulation of the catalytic site. *Sci. Rep.*, 8, 11397.

Schmidt, B., McCracken, J. and Ferguson-Miller, S. (2003). A discrete water exit pathway in the membrane protein cytochrome *c* oxidase. *Proc. Natl. Acad. Sci. USA*, 100, 15539–15542.

Seibold, S.A., Mills, D.A., Ferguson-Miller, S. and Cukier, R.I. (2005). Water chain formation and possible proton pumping routes in *Rhodobacter sphaeroides* cytochrome *c* oxidase: a molecular dynamics comparison of the wild type and R481K mutant. *Biochemistry*, 44, 10475–10485.

Sharma, V., Karlin, K.D. and Wikström, M. (2013). Computational study of the activated OH state in the catalytic mechanism of cytochrome *c* oxidase. *Proc. Natl. Acad. Sci. USA*, 110, 16844–16849.

Sharma, V., Ala-Vannesluoma, P., Vattulainen, I., Wikström, M. and Rog, T. (2015). Role of subunit III and its lipids in the molecular mechanism of cytochrome *c* oxidase. *Biochim. Biophys. Acta*, 1847, 690–697.

Sharma, V. and Wikström, M.K.F. (2016). The role of the K-channel and the active-site tyrosine in the catalytic mechanism of cytochrome *c* oxidase. *Biochim. Biophys. Acta*, 1857, 1111–1115.

Sharma, V., Jambrina, P.G., Kaukonen, M., Rosta, E. and Rich, P.R. (2017). Insights into the function of the H channel of cytochrome *c* oxidase from atomistic molecular dynamics simulations. *Proc. Natl. Acad. Sci. USA*, 114, E10339–E10348.

Shimada, A., Kubo, M., Baba, S., Yamashita, K., Hirata, K., Ueno, G., Nomura, T., Kimura, T., Shinzawa-Itoh, K., Baba, J., Hatano, K., Eto, Y., Miyamoto, A., Murakami, H., Kumasaka, T., Owada, S., Tono, K., Yabashi, M., Yamaguchi, Y., Yanagisawa, S., Sakaguchi, M., Ogura, T., Komiya, R., Yan, J., Yamashita, E., Yamamoto, M., Ago, H., Yoshikawa, S. and Tsukihara, T. (2017a). A nanosecond time-resolved XFEL analysis of structural changes associated with CO release from cytochrome *c* oxidase. *Sci. Adv.*, 3, e1603042.

Shimada, A., Hatano, K., Tadehara, H., Yano, N., Shinzawa-Itoh, K., Yamashita, E., Muramoto, K., Tsukihara, T. and Yoshikawa, S. (2018). X-ray structural analyses of azide-bound cytochrome *c* oxidases reveal that the H-pathway is critically important for the proton-pumping activity. *J. Biol. Chem.*, 293, 14868–14879.

Shimada, S., Shinzawa-Itoh, K., Baba, J., Aoe, S., Shimada, A., Yamashita, E., Kang, J., Tateno, M., Yoshikawa, S. and Tsukihara, T. (2017b). Complex structure of cytochrome *c*-cytochrome *c* oxidase reveals a novel protein-protein interaction mode. *EMBO J.*, 36, 291–300.

Shimokata, K., Katayama, Y., Murayama, H., Suematsu, M., Tsukihara, T., Muramoto, K., Aoyama, H., Yoshikawa, S. and Shimada, H. (2007). The proton pumping pathway of bovine heart cytochrome *c* oxidase. *Proc. Natl. Acad. Sci. USA*, 104, 4200–4205.

Shinzawa-Itoh, K., Aoyama, H., Muramoto, K., Terada, H., Kurauchi, T., Tadehara, Y., Yamasaki, A., Sugimura, T., Kurono, S., Tsujimoto, K., Mizushima, T., Yamashita, E., Tsukihara, T. and Yoshikawa, S. (2007). Structures and physiological roles of 13 integral lipids of bovine heart cytochrome *c* oxidase. *EMBO J.*, 26, 1713–1725.

Sousa, F.L., Alves, R.J., Ribeiro, M.A., Pereira-Leal, J.B., Teixeira, M. and Pereira, M.M. (2012). The superfamily of heme-copper oxygen reductases: types and evolutionary considerations. *Biochim. Biophys. Acta*, 1817, 629–637.

Sugitani, R. and Stuchebrukhov, A.A. (2009). Molecular dynamics simulation of water in cytochrome *c* oxidase reveals two water exit pathways and the mechanism of transport. *Biochim. Biophys. Acta*, 1787, 1140–1150.

Supekar, S., Gamiz-Hernandez, A.P. and Kaila, V.R.I. (2016). A protonated water cluster as a transient proton-loading site in cytochrome *c* oxidase. *Angew. Chem.*, 128, 12119–12123.

Svensson-Ek, M., Abramson, J., Larsson, G., Törnroth, S., Brzezinski, P. and Iwata, S. (2002). The X-ray crystal structures of wild-type and EQ(I-286) mutant cytochrome *c* oxidases from *Rhodobacter sphaeroides*. *J. Mol. Biol.*, 321, 329–339.

Tsukihara, T., Aoyama, H., Yamashita, E., Tomizaki, T., Yamaguchi, H., Shinzawa-Itoh, K., Nakashima, R., Yaono, R. and Yoshikawa, S. (1996). The whole structure of the 13-subunit oxidized cytochrome *c* oxidase at 2.8 Å. *Science*, 272, 1136–1144.

Tsukihara, T., Shimokata, K., Katayama, Y., Shimada, H., Muramoto, K., Aoyama, H., Mochizuki, M., Shinozawa-Itoh, K., Yamashita, E., Yao, M., Ishimura, Y. and Yoshikawa, S. (2003). The low-spin heme of cytochrome *c* oxidase as the driving element of the proton-pumping process. *Proc. Natl. Acad. Sci. USA*, 100, 15304–15309.

Wackerhage, H., Hoffmann, U., Essfeld, D., Leyk, D., Mueller, K. and Zange, J. (1998). Recovery of free ADP, Pi, and free energy of ATP hydrolysis in human skeletal muscle. *J. Appl. Physiol.*, 85, 2140–2145.

Walker, J.C.G. (1980). The handbook of environmental chemistry. Volume 1, part A. The natural environment and the biogeochemical cycles, Hutzinger, O., ed. In *The Oxygen Cycle*, (Springer-Verlag), pp. 87–104.

Watt, I.N., Montgomery, M.G., Runswick, M.J., Leslie, A.G.W. and Walker, J.E. (2010). Bioenergetic cost of making an adenosine triphosphate molecule in animal mitochondria. *Proc. Natl. Acad. Sci. USA*, 107, 16823–16827.

Wikström, M., Verkhovsky, M.I. and Hummer, G. (2003). Water-gated mechanism of proton translocation by cytochrome *c* oxidase. *Biochim. Biophys. Acta*, 1604, 61–65.

Wikström, M., Ribacka, C., Molin, M., Laakkonen, L., Verkhovsky, M. and Puustinen, A. (2005). Gating of proton and water transfer in the respiratory enzyme cytochrome *c* oxidase. *Proc. Natl. Acad. Sci. USA*, 102, 10478–10481.

Wikström, M. (2012). Active site intermediates in the reduction of O_2 by cytochrome oxidase, and their derivatives. *Biochim. Biophys. Acta*, 1817, 468–475.

Wikström, M., Sharma, V., Kaila, V.R.I., Hosler, J.P. and Hummer, G. (2015). New perspectives on proton pumping in cellular respiration. *Chem. Rev.*, 115, 2196–2221.

Wikström, M., Krab, K. and Sharma, V. (2018). Oxygen activation and energy conservation by cytochrome *c* oxidase. *Chem. Rev.*, 118, 2469–2490.

Wikström, M.K.F. (1977). Proton pump coupled to cytochrome *c* oxidase in mitochondria. *Nature*, 266, 271–273.

Wood, P.M. (1974). The redox potential of the system oxygen-superoxide. *FEBS Lett.*, 44, 22–24.

Yano, N., Muramoto, K., Shimada, A., Takemura, S., Baba, J., Fujisawa, H., Mochizuki, M., Shinzawa-Itoh, K., Yamashita, E., Tsukihara, T. and Yoshikawa, S. (2016). The Mg^{2+}-containing water cluster of mammalian cytochrome *c* oxidase collects four pumping proton equivalents in each catalytic cycle. *J. Biol. Chem.*, 291, 2382–2394.

Yoshikawa, S., Shinzawa-Itoh, K., Nakashima, R., Yaono, R., Yamashita, E., Inoue, N., Yao, M., Fei, M.J., Libeu, C.P., Mizushima, T., Yamaguchi, H., Tomizaki, T. and Tsukihara, T. (1998). Redox-coupled crystal structural changes in bovine heart cytochrome *c* oxidase. *Science*, 280, 1723–1729.

Yoshikawa, S. and Shimada, A. (2015). Reaction Mechanism of Cytochrome *c* Oxidase. *Chem. Rev.*, 115, 1936–1989.

Zaslavsky, D. and Gennis, R. (2000). Proton pumping by cytochrome oxidase: progress, problems and postulates. *Biochim. Biophys. Acta*, 1458, 164–179.

Zhen, Y., Hoganson, C.W., Babcock, G.T. and Ferguson-Miller, S. (1999). Definition of the interaction domain for cytochrome *c* on cytochrome *c* oxidase. I. Biochemical, spectral, and kinetic characterization of surface mutants in subunit II of *Rhodobacter sphaeroides* cytochrome aa_3. *J. Biol. Chem.*, 274, 38032–38041.

Zong, S., Wu, M., Gu, J., Liu, T., Guo, R. and Yang, M. (2018). Structure of the intact 14-subunit human cytochrome *c* oxidase. *Cell Res.*, 28, 1026–1034.

Chapter 9

Mimicking the Oxygen-Evolving Center in Photosystem II

Changhui Chen*,†, Yang Chen*,† and Chunxi Zhang*,‡

**Laboratory of Photochemistry, Institute of Chemistry, Chinese Academy of Sciences, Beijing 100190, China*

†University of Chinese Academy of Sciences, Beijing 100049, China

‡chunxizhang@iccas.ac.cn

The oxygen-evolving center (OEC) in photosystem II (PSII) is a unique biological catalyst that splits water into electrons, protons and O_2 by using solar energy. Recent crystallographic studies have revealed that the structure of the OEC is an asymmetric Mn_4Ca-cluster, which provides a blueprint to develop man-made water-splitting catalysts in artificial photosynthesis. Although it was a great challenge to mimic the whole structure and function of the OEC in laboratory, significant advances have recently been achieved. One of the most important advances is the synthesis of the artificial Mn_4Ca-cluster that closely mimicked both the atomic and electronic structures of the OEC in PSII. The artificial Mn_4Ca-cluster provides a structurally well-defined chemical model to understand the structure-function relationship of the natural Mn_4Ca-cluster, and sheds new insight into the mechanism of the water-splitting reaction in PSII. The artificial Mn_4Ca-cluster and its variants may open new avenues to develop new generations of artificial catalysts for photo water-splitting reaction from earth-abundant and non-toxic chemical elements in the future.

1. Introduction

Photosynthetic oxygen evolution is a unique function of photosystem II (PSII), which is a pigment-binding, multi-subunit protein complex embedded in the thylakoid membranes of plants, algae and cyanobacteria [Barber, 2009; Dau and Zaharieva, 2009; Nelson and Yocum, 2006; Satoh *et al.*, 2005; Shen, 2015]. The key cofactors in the reaction center of PSII involved to perform this function are shown in Figure 1. Upon photo excitation, the primary electron donor (P_{680}) donates one electron to the primary electron acceptor (Pheo) to produce the $P_{680}^{+\bullet}$ and $Pheo^{-\bullet}$ charge pair in a few picoseconds [Cardona *et al.*, 2012; Diner and Rappaport, 2002; Holzwarth *et al.*, 2006; Renger, 2012]. $Pheo^{-\bullet}$ then transfers the electron to the primary and secondary plastoquinones (Q_A and Q_B) in sequence through the non-heme iron at the acceptor side [Petrouleas and Crofts, 2005; Saito *et al.*, 2013]. The high redox potential of $P_{680}^{+\bullet}$ obtains one electron from the secondary electron donor, Tyr_Z, leading to the formation of a neutral radical ($Tyr_Z^\bullet$) [Diner and Britt, 2005; Styring *et al.*, 2012; Zhang, 2007], which then drives water oxidation at the Mn_4Ca-cluster, the so-called oxygen-evolving center (OEC), in milliseconds on the donor side [McEvoy and Brudvig, 2006; Perez-Navarro *et al.*, 2016; Yano and Yachandra, 2014].

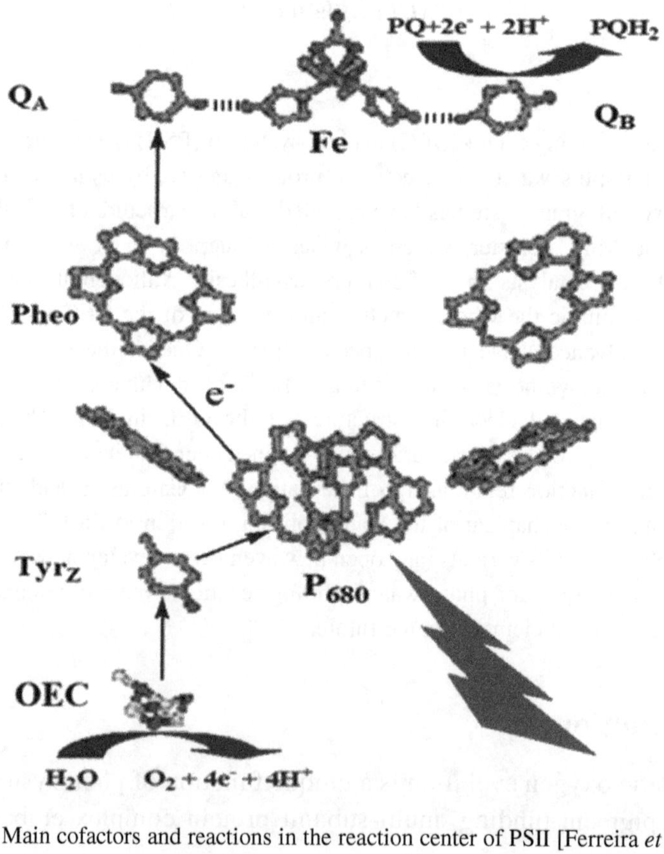

Figure 1. Main cofactors and reactions in the reaction center of PSII [Ferreira *et al.*, 2004].

The catalytic oxygen-evolving reaction (or water-splitting reaction) involves five different redox states (S_n, n = 0–4) of the OEC [Dau and Haumann, 2007; Junge *et al.*, 2002; Kok *et al.*, 1970], wherein S_0 is the initial and most reduced state and S_1 is the dark-stable state. The S_2 and S_3 states are metastable and decay eventually to the dark-stable S_1 state, whereas S_4 is a transient state that releases dioxygen and decays to the S_0 state. Changes of the redox valences of the four mangenese ions occur during the turnover. The widespread adoption of the oxidation states for the four mangenese ions are S_0 (III, III, III, IV) or (II, III, IV, IV), S_1 (III, III, IV, IV), S_2 (III, IV, IV, IV) and S_3 (IV, IV, IV, IV). This is the so-called "high-valent scheme" [Dau *et al.*, 2008; Krewald *et al.*, 2015; Sauer *et al.*, 2008; Yano and Yachandra, 2014]. However, several studies [Jin *et al.*, 2014; Pace *et al.*, 2012; Zheng and Dismukes, 1996] have proposed another "low-valent scheme" that differs by a total of two metal-unpaired electrons. Namely, the oxidation states of the four Mn ions are S_0 (II, III, III, III), S_1 (III, III, III, III) or (II, III, III, IV), S_2 (III, III, III, IV) and S_3 (III, III, IV, IV) [Krewald *et al.*, 2015], respectively. The calcium is an indispensable cofactor for the function of the OEC, and its depletion results in the complete loss of the water oxidation capability of PSII. Calcium in OEC can only be functionally replaced by strontium [van Gorkom and Yocum, 2005; Yocum, 2008]. Due to the broad interests in fundamental research, as well as potential applications in artificial photosynthesis, the structure and catalytic mechanism of the OEC have attracted extensive studies during the last two decades.

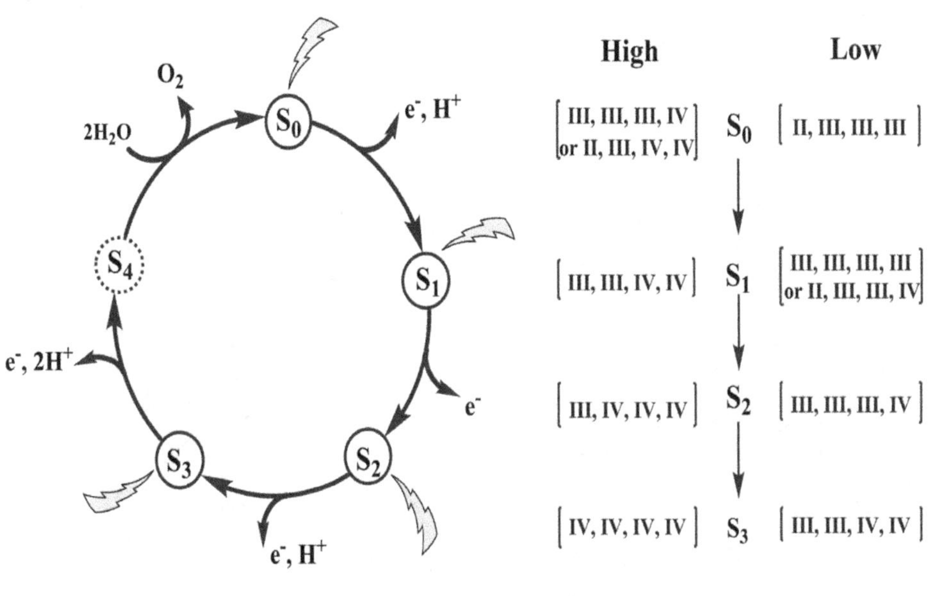

Figure 2. The turnover of the OEC [Krewald *et al.*, 2015]. The S-cycle of the OEC, indicating the oxidation and proton release events at each transition (left). The possible formal oxidation states of the four manganese ions in OEC for each S state according to the high-valent and low-valent schemes (right).

2. Structure of the OEC

The structure of the OEC has been a long-standing issue in the field of photosynthetic research. Before the appearance of the crystal structure of PSII, most structural information of the OEC came from extended X-ray absorption fine structure (EXAFS) [Dau *et al.*, 2008; Yano and Yachandra, 2014] and electron paramagnetic resonance (EPR) [Cox *et al.*, 2013; Peloquin and Britt, 2001] investigations and theoretical calculations. Different structural models were proposed to explain the properties of the OEC in PSII [Cinco *et al.*, 1998; Peloquin *et al.*, 2000; Tommos and Babcock, 1998; Vrettos *et al.*, 2001]. Figure 3A shows the structural model proposed by Zhang *et al.* in 1999 [Zhang *et al.*, 1999], in which the key component of calcium was suggested to be located in the middle of the OEC and be connected with four manganese ions through three μ-oxo and two carboxylate groups. This was the only model that predicted correctly the unique binding mode of calcium in OEC (see below).

The crystal structure data of the PSII have emerged since the beginning of this century [Ferreira *et al.*, 2004; Guskov *et al.*, 2009; Hellmich *et al.*, 2014; Kamiya and Shen, 2003; Umena *et al.*, 2011; Zouni *et al.*, 2001]. In 2001, Zouni *et al.* [Zouni *et al.*, 2001] reported the first crystal structure of PSII from thermophilic cyanobacteria at a resolution of 3.8 Å. In 2004, based on 3.5 Å resolution structure data, Ferreira *et al.* [Ferreira *et al.*, 2004], for the first time, proposed that the OEC could be comprised of a Mn_3CaO_4 cubane attached with a "dangler" Mn ion *via* one bridging oxo forming a Mn_4CaO_4-cluster (Figure 4A). However, the detailed peripheral ligands of the OEC, were still elusive due to the low resolution and possible reduction induced by X-ray radiation during the crystallographic

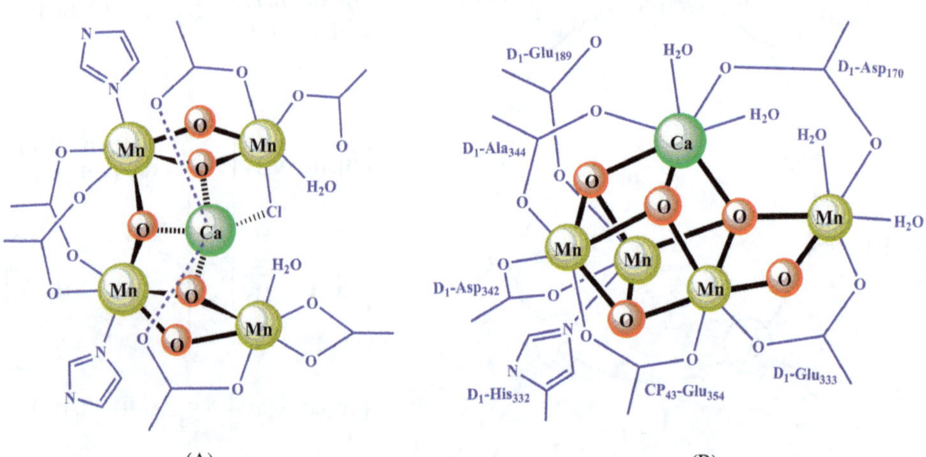

Figure 3. Scheme for the structural model [Zhang *et al.*, 1999] (A) and structure [Umena *et al.*, 2011] (B) of the OEC.

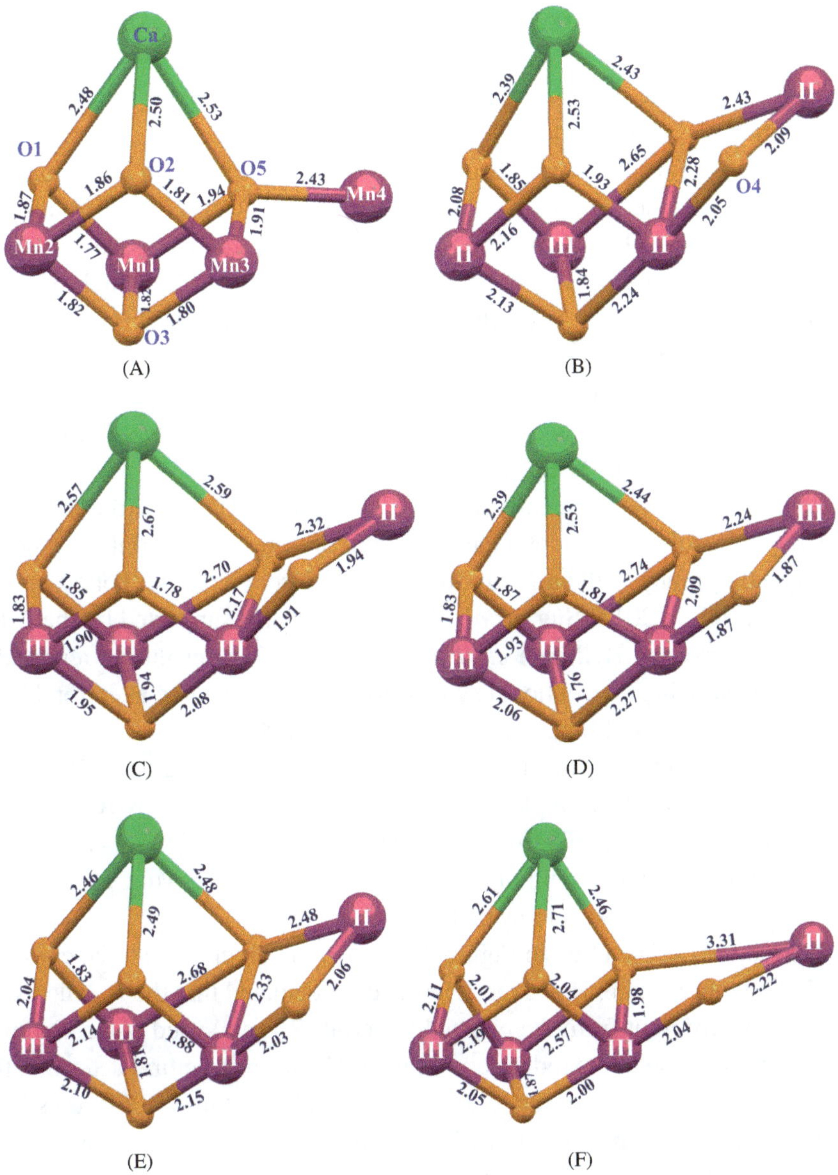

Figure 4. Core structure of the OEC at different resolutions of PSII from cyanobacteria and higher plants. (A) 3.8 Å resolution from thermocyanobacteria reported by Ferreira *et al.* [2004]; (B) 1.9 Å resolution from thermocyanobacteria reported by Umena *et al.* [2011]; (C) 1.95 Å resolution by using XFEL method reported by Suga *et al.* [2015]; (D) 1.87 Å resolution from thermocyanobacteria with extremely low X-ray doses (at 0.03 MYy) [Tanaka *et al.*, 2017]; (E) 3.2 Å resolution revealed by single-particle cryoelectron microscopy (Cryo-EM) from spinach reported by Wei *et al.* [2016]. (F) 2.7 Å resolution revealed by single-particle cryoelectron microscopy (Cryo-EM) from *Pisum sativum* reported by Su *et al.* [2017]. Mn, Ca and O are shown in purple, green, orange, respectively. Distances are given in Å. Roman numerals indicate the possible oxidation states of manganese ions based on the Bond-valence sum (BVS) calculations.

structural determination [Askerka *et al.*, 2017; Grabolle *et al.*, 2006; Yano *et al.*, 2005; Yano *et al.*, 2006].

The detailed structure of the OEC was revealed by the crystal structure of PSII at a resolution of 1.9 Å reported by Umena *et al.* in 2011 [Umena *et al.*, 2011]. All binding ligands of the OEC have been clearly resolved, including four water molecules, one imidazole group from D_1-His332 and six carboxylate groups from D_1-Asp170, D_1-Glu189, D_1-Glu333, D_1-Asp342, D_1-Ala344, CP_{43}-Glu354, respectively. One additional bridged oxygen (O4) between Mn4 and Mn3 in the OEC has been observed (Figure 4B). Thus, the whole structure of the OEC is comprised of an asymmetric Mn_4CaO_5-cluster. In this structure, Ca^{2+} is located in the middle of the OEC, and is connected to the Mn_4-cluster through three bridging oxo and two carboxylate groups, which is consistent with our previous proposal (Figure 3A) published in 1999 [Zhang *et al.*, 1999].

The aforementioned structure of the OEC (Figure 4B) was further confirmed by the 1.95 Å resolution data obtained by using time-resolved serial femtosecond X-ray crystallography with an X-ray free electron laser (XFEL) reported by Suga *et al.* (Figure 4C) in 2015 [Suga *et al.*, 2015]. It was also supported by the 2.44 Å resolution reported by Hellmich *et al.* [2014] and 1.87 Å resolution reported by Tanaka *et al.* by using conventional synchrotron radiation source at extremely low X-ray does (0.03 MGy) [Tanaka *et al.*, 2017] (Figure 4D).

Recently, the structure of the OEC in higher plants (spinach and *Pisum sativum*) was also revealed by single-particle cryoelectron microscopy (Cryo-EM) at a resolution of 3.2~2.7 Å (Figures 4E and 4F) [Su *et al.*, 2017; Wei *et al.*, 2016], which demonstrates that essentially the same structure of the Mn_4CaO_5-cluster is contained in higher plants as in cyanobacteria.

All structures of the OEC in Figure 4 were originally claimed to be correlated to the dark-stable S_1 state. However, it has been speculated that the revealed structures could be different from the original material (S_1 state) used at the beginning of the experiment due to the reduction of the OEC during structural determination. Indeed, the atom positions of the OEC in the crystal structure of PSII are still not consistent with the high-resolution EXAFS spectroscopy studies on active samples [Askerka *et al.*, 2017; Dau *et al.*, 2012].

To estimate the oxidation states of the four manganese ions of the OEC revealed by different resolution structural determinations, we have carried out bond-valence sum (BVS) calculations. The BVS method is a popular method in coordination chemistry to estimate the oxidation states of atoms [Brown, 2009; Pauling, 1929]. It is derived from the bond valence model [Pauling, 1929], which is a simple yet robust model for validating chemical structures with localized bonds and is used to predict some of their properties. This method has been used extensively to estimate the oxidation state of active site in various metalloenzymes, as well [Gatt *et al.*, 2012;

Table 1. Bond-valence sum (BVS) calculations on the structures of the OEC at different resolutions as shown in Figure 4. Roman numerals in parentheses indicate the assignment of the possible oxidation states of four manganese ions in the OEC based on BVS calculations. All atomic coordinates were taken from the first monomer of photosystem II in the crystal structure data with PDB-IDs: 3WU2, 4UB6, 5B5E, 3JCU and 5XNL, respectively.

Resolution of PSII	**1.9Å (3WU2)**	**1.95Å (4UB6)**	**1.87Å (5B5E)**	**3.2 Å (3JCU)**	**2.7Å (5XNL)**
Mn1	3.022 (III)	3.075 (III)	3.244 (III)	3.132 (III)	2.713 (III)
Mn2	2.491 (II)	3.237 (III)	3.057 (III)	2.644 (III)	2.823 (III)
Mn3	2.440 (II)	2.980 (III)	2.951 (III)	2.610 (III)	2.966 (III)
Mn4	2.009 (II)	2.318 (II)	2.603 (III)	2.008 (II)	1.690 (II)

Liu and Thorp, 1993]. Table 1 lists the results of the BVS calculations on the structures of the OEC at different resolutions. We have found that the oxidation states of the four manganese ions are likely to be (III, II, II, II), (III, III, III, II), (III, III, III, III) for the 1.9 Å [Umena *et al.*, 2011], 1.95 Å [Suga *et al.*, 2015] and 1.87 Å [Tanaka *et al.*, 2017] resolution structures, respectively (Figure 4). Whereas, the same (III, III, III, II) states are estimated for OEC in higher plants at both 3.2 Å [Wei *et al.*, 2016] and 2.7 Å [Su *et al.*, 2017] resolution. Surprisingly, the oxidation states of the four manganese ions in all structures are closer (but not the same) to that in the low-valence scheme (Figure 2), but remarkably far from that of the high-valence scheme, which has been widely adopted in the field of photosynthesis research. One possibility could be that reduction of the OEC occurred from the X-ray or electron beam during the structural determination. If this was the case, one would expect that all methods (including conventional synchrotron radiation, advanced XFEL and Cryo-EM) could give rise to a similar effect of the reduction of the OEC. Interestingly, the highest oxidation state (III, III, III, III) were seen from the highest resolution (1.87 Å) structure obtained by using extremely low X-ray doses (at 0.03 MGy). Not only does it confirm once again the crucial role of low X-ray dose in revealing the precise structure of the OEC [Tanaka *et al.*, 2017], but it also highlights the sensitive and dynamic nature of the OEC during its structural determination.

Due to the complexity of the huge protein environment and the dynamic structural changes of the OEC during the water-splitting reaction, the detailed mechanism for the water-splitting reaction in PSII is still elusive. Various mechanisms for the water-splitting reaction have been proposed based on the available structure data of the OEC reported [Barber, 2017; Chen *et al.*, 2015; Isobe *et al.*, 2012; Kawashima *et al.*, 2018; Krewald *et al.*, 2016; Siegbahn, 2013; Vinyard *et al.*, 2015]. In this regard, therefore, precise structural information for the different S states of the OEC will be crucial in the future [Suga *et al.*, 2017; Young *et al.*, 2016; Zhang and Kuang, 2018].

3. Challenge for the Synthesis of the OEC in Laboratory

In order to better understand the structure and properties of the OEC, as well as to develop highly efficient and cheap artificial catalysts for the water-splitting reaction to overcome the bottleneck of artificial photosynthesis, many groups have tried to synthesize the OEC in the laboratory since the end of last century. However, it has been a great challenge for chemists to synthesize the entire structure of the OEC in the laboratory due to several reasons [Zhang, 2015, 2016]: (i) It is very difficult to incorporate Ca^{2+} into the Mn_4-cluster through μ-oxo bridges, because the affinity of Ca^{2+} to μ-oxo is significantly weaker than that of the Mn ion. In general, only a multi-manganese cluster, instead of the manganese-calcium heterometallic cluster, can be synthesized. (ii) The core structure of the OEC is an asymmetric Mn_4Ca-cluster, which has been suggested to play crucial roles for the function of the OEC [Shen, 2015]. It was completely unknown whether such a structure could be synthesized in a chemical system. (iii) The ligands of the biological OEC are mainly composed of carboxylate groups and water molecules, which are drastically different from the multi-pyridine ligands used in most previous chemical model systems [Kärkäs *et al.*, 2014; Limburg *et al.*, 1999; Mullins and Pecoraro, 2008; Tsui *et al.*, 2013]. (iv) The redox potential of the OEC is very high (+0.8~+1.0 V *versus* NHE) [Dau and Zaharieva, 2009; Vass and Styring, 1991] due to the presence of the high oxidation states of the manganese ions.

A large number of artificial Mn complexes has been reported in the literature [Dismukes *et al.*, 2009; Gerey *et al.*, 2016; Limburg *et al.*, 1999; Mukhopadhyay *et al.*, 2004; Mullins and Pecoraro, 2008; Najafpour *et al.*, 2016; Paul *et al.*, 2017]. Among them, tetra-manganese complexes containing a Mn_4O_4-cubane [Chakov *et al.*, 2003; Dismukes *et al.*, 2009; Kanady *et al.*, 2011; Ruettinger *et al.*, 1997] (Figure 5) are attractive, because the oxidation states of the four manganese ions are (III, III, IV, IV) similar to that of the high valence scheme for OEC in the S_1 state (Figure 2). The peripheral ligands of these artificial Mn_4O_4-clusters were provided mostly by aromatic organic ligands (aromatic organic phosphate or multi-pyridine). These types of ligands are known to contribute significantly to stabilization of the high oxidation states of the manganese ions in these complexes. However, it should be pointed out that this is remarkably different to the peripheral ligands of the OEC, which are composed mainly of carboxylate groups [Umena *et al.*, 2011].

Significant advances for the synthesis of the OEC have emerged since 2011. Agapie's group reported the first artificial Mn_3CaO_4-cluster using a multi-pyridylalkoxide ligand (Figures 6A and 6B) [Kanady *et al.*, 2011]. The same group reported a series of analogues or derivatives [Tsui and Agapie, 2013], for example, the artificial Mn_3CaAgO_4-complex [Kanady *et al.*, 2014] (Figures 6E and 6F). In 2012, Christou's group reported the $Mn_3Ca_2O_4$-complex with one Ca^{2+} attached to

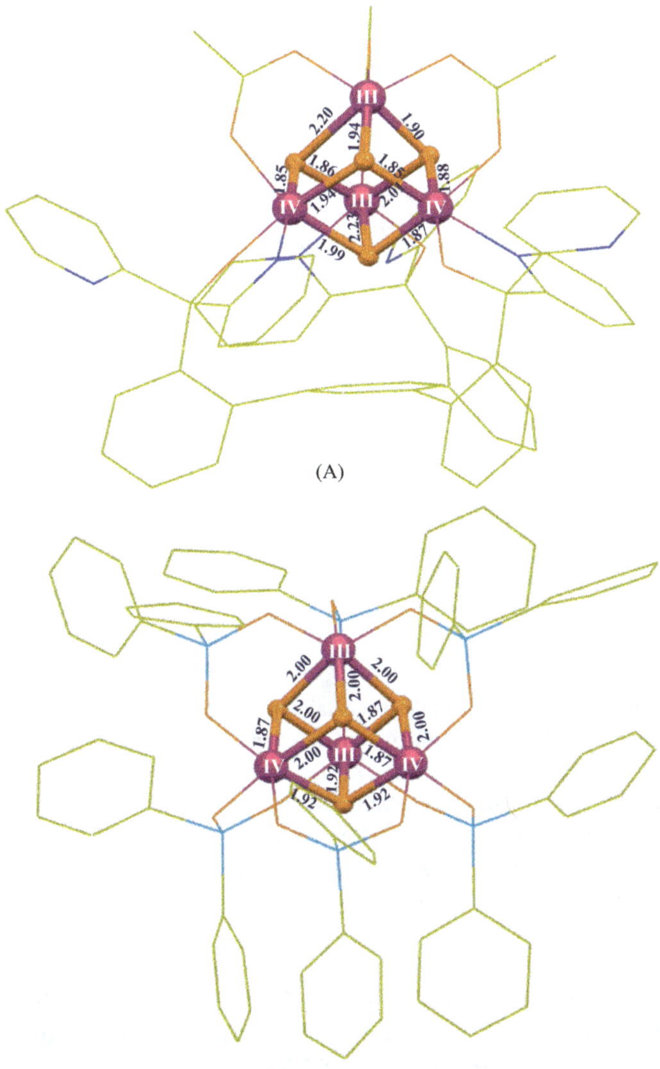

(A)

(B)

Figure 5. Structures of two artificial complexes containing $[Mn_4O_4]$ cubane. (A) $Mn_4O_4[(C_6H_5)_2PO_2]_6$ complex [Chakov *et al.*, 2003; Ruettinger *et al.*, 1997]. (B) $Mn_4O_4L(CH_3CO_2)_3$ complex (L = 1,3,5-tris(2-di(2'-pyridyl)oxymethylphenyl)benzene) [Kanady *et al.*, 2011]; Distances are given in Å units. Mn, O, N, P and C are shown in purple, orange, blue, cyan and yellow, respectively. For clarity, all hydrogen atoms are not shown.

the Mn_3CaO_4 cubane [Mukherjee *et al.*, 2012] (Figures 6C and 6D). Distinct from previous Mn_3CaO_4-complexes, all the ligands in the $Mn_3Ca_2O_4$-complex are pivalic anions or neutral pivalic acid, which closely mimicks the peripheral carboxylate ligands of the OEC in PSII. In these artificial complexes, all the manganese

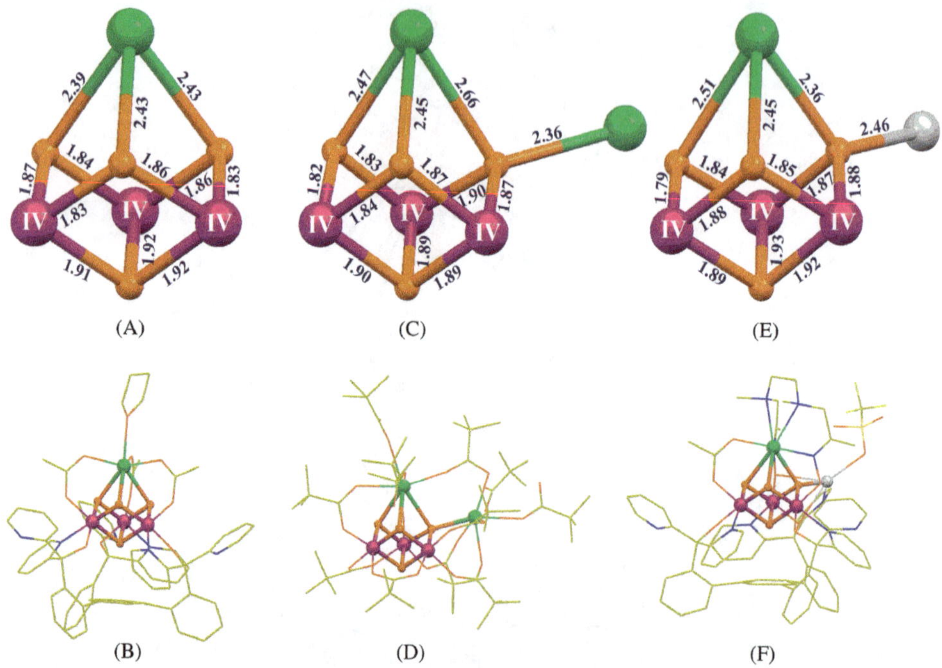

Figure 6. Structures of artificial complexes containing [Mn_3CaO_4] cubane. (A) Core structure of the Mn_3CaO_4-complex; (B) Whole structure of the Mn_3CaO_4-complex [Kanady *et al.*, 2011]; (C) Core structure of the $Mn_3Ca_2O_4$-complex; (D) Whole structure of the $Mn_3Ca_2O_4$-complex [Mukherjee *et al.*, 2012]; (E) Core structure of the Mn_3CaAgO_4-complex; (F) Whole structure of the Mn_3CaAgO_4-complex [Kanady *et al.*, 2014]. Distances are given in Å units; Mn, Ca, Ag, O, N, F, S and C are shown in purple, green, white, orange, blue, green yellow, bright yellow and yellow, respectively. For clarity, all hydrogen atoms are not shown.

ions are in a +4 oxidation state, and the typical bond lengths for Mn–O and Ca–O are in the range of 1.8–1.9 Å and 2.4–2.7 Å, respectively. The distances of the Mn…Mn and Mn…Ca are in the range of 2.7–2.8 Å and 3.2–3.5 Å, respectively.

Although all these complexes shown in Figure 6 have mimicked some aspects of the key structural motifs of the Mn_3CaO_4 cubane in the OEC, until recently, it remained a great challenge to synthesize the entire Mn_4Ca-cluster with similar ligands to those seen in PSII.

4. Closer Mimicking of the OEC

After more than 16 years of work on the natural OEC [Wang *et al.*, 2014; Zhang, 2007; 2016; Zhang *et al.*, 1999], we have started since 2012 to synthesize the artificial OEC using simple chemicals (*i.e.* $Mn(CH_3CO_2)_2$, $Ca(CH_3CO_2)_2$, $KMnO_4$ and pivalic acid). During this period, we have synthesized a series of multi-manganese [Chang *et al.*, 2017; Wang *et al.*, 2013], manganese-calcium [Chen *et al.*, 2015] and manganese-strontium heterometallic complexes [Chen *et al.*, 2014] to mimic the metal cluster in the OEC.

Inspired by the assembly processes of the OEC in PSII [Dasgupta *et al.*, 2008; Zhang *et al.*, 2017], we successfully prepared the first artificial Mn_4Ca-cluster [Zhang *et al.*, 2015] (Figures 7C and 7D). It was synthesized through a two-step procedure using inexpensive commercial chemicals. The first step was to synthesize a precursor through a reaction of $Bu^n_4NMnO_4$ (Bu^n = *n*-butyl), $Mn(CH_3CO_2)_2 \cdot (H_2O)_4$ and $Ca(CH_3CO_2)_2 \cdot H_2O$ (molar ratio of 4:1:1) in boiling acetonitrile in the presence of an excess of pivalic acid. The second step was to treat the precursor with organic base (pyridine) in ethyl acetate, leading to the formation of the final product.

The core structure of the artificial Mn_4CaO_4-complex contains a Mn_3CaO_4 cubane attached with a dangler Mn ion forming an asymmetric Mn_4CaO_4-cluster, which is exactly the same as the OEC's structural model proposed by Ferreira *et al.* in 2004 [Ferreira *et al.*, 2004] (Figure 4A). The surrounding ligands of the

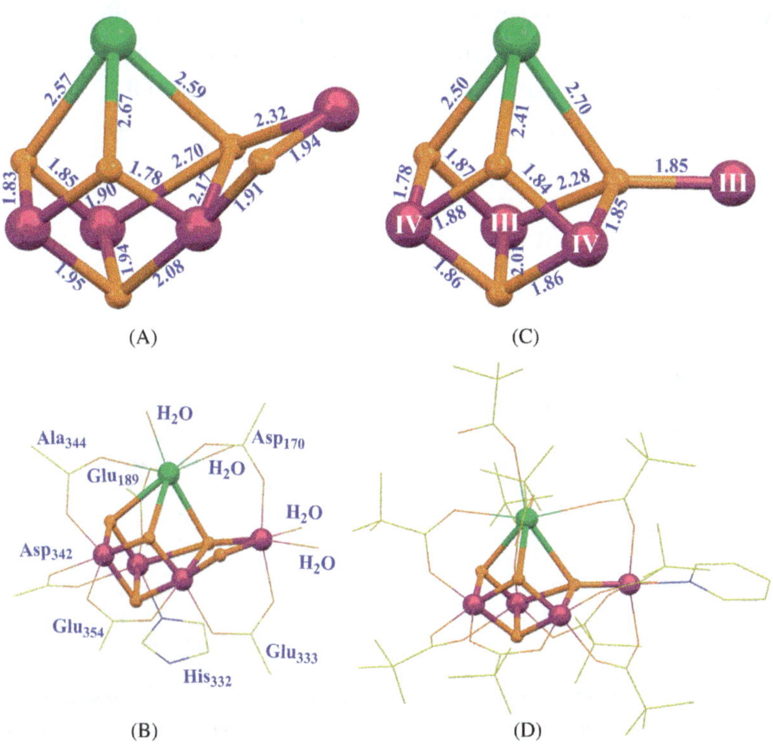

Figure 7. Structures of the OEC [Suga *et al.*, 2015] (A, B) and one artificial Mn_4Ca-cluster [Zhang *et al.*, 2015] (C, D). Distances are given in Å units; Mn, Ca, O, N and C are shown in purple, green, orange, blue and yellow, respectively. For clarity, all hydrogen atoms are not shown.

Mn_4CaO_4-cluster are provided by eight $(CH_3)_3CCO_2^-$ anions and three neutral ligands (two pivalic acid and one pyridine molecules), similar to the peripheral ligands of the OEC (Figure 7B).

It should be pointed out that the structure of the artificial Mn_4CaO_4-cluster is well-defined and effects from X-ray radiation-induced reduction is limited, mainly due to the absence of water as solvent in the crystal of the artificial Mn_4CaO_4-complex. BVS calculations have clearly shown that the oxidation states of the four manganese ions of the artificial Mn_4CaO_4-complex are (III, III, IV, IV), which are essentially the same as that proposal for the S_1 state in the high valence scheme (Figure 2). Considering the uncertainty of the oxidation states of four manganese ions in the current structure data of the OEC, the structurally well-defined artificial Mn_4CaO_4-cluster could provide strong evidence for the assignment of the S_1 state oxidation states of (III, III, IV, IV) for the four manganese ions in the OEC.

Interestingly, we have found that the neutral ligands on both the dangler manganese and calcium ions of the artificial Mn_4CaO_4-complex are exchangeable and can be replaced by other ligands. Figure 8 shows two new complexes containing one Mn_4CaO_4-cluster after the replacement of these ligands. The central structure of these two new Mn_4CaO_4-complexes is essential the same as that in Figures 7C and 7D. It is noticed that the binding sites of these exchangeable neutral ligands closely mimic the binding sites of the water molecules in the OEC [Nilsson *et al.*, 2014; Umena *et al.*, 2011].

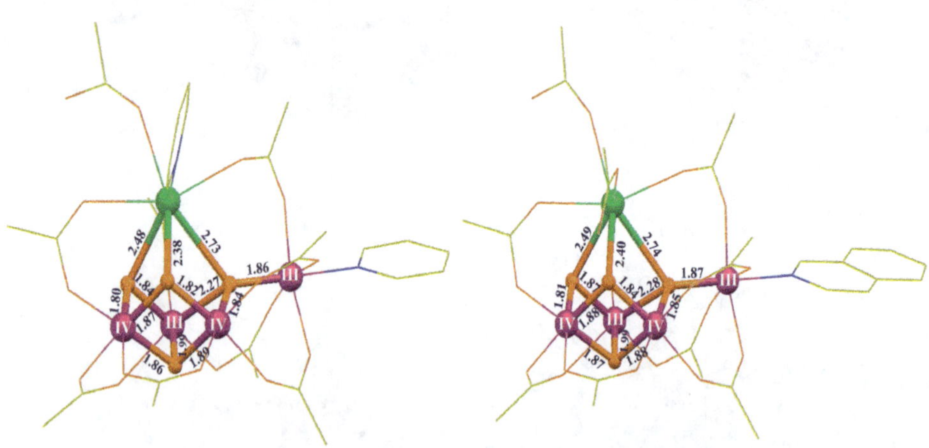

(A) (B)

Figure 8. Structures of artificial model complexes containing Mn_4CaO_4-cluster with different ligands on the calcium and the dangler manganese ions. (A) Mn_4CaO_4-complex containing two pyridine molecules; (B) Mn_4CaO_4-complex containing one isoquinine. Distances are given in Å units; Mn, Ca, O, N and C are shown in purple, green, orange, blue and yellow, respectively. For clarity, all the CH_3 and H are not shown.

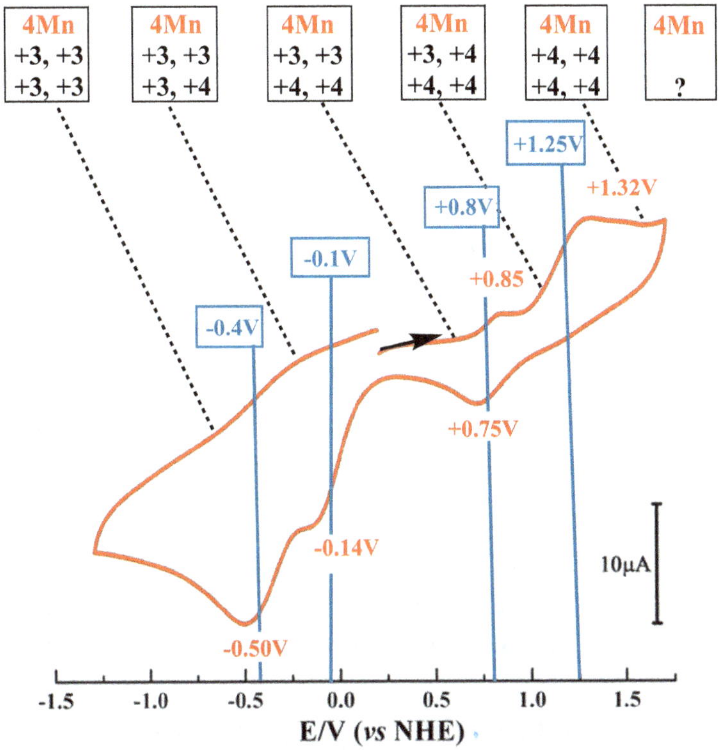

Figure 9. Cyclic voltammogram (CV) of artificial Mn_4CaO_4-complex in Figure 7 [Zhang *et al.*, 2015]. Potentials *versus* NHE; scan rate of 100 mV/s. Peak positions of CV waves are in red and estimated midpoint potentials are in blue. The likely oxidation states of the four Mn ions in various oxidation states of artificial complex are indicated using black numbers.

The similarity between the artificial Mn_4CaO_4-cluster and OEC was further supported by the observation of four redox transitions revealed by cyclic voltammogram (CV) measurements (Figure 9). The $S_1 \rightarrow S_2$ transition of the artificial Mn_4CaO_4-complex was detected at a redox potential of ~0.8 eV (*versus* NHE), which is close to the estimated potential of the corresponding OEC redox transition (≥ 0.9 V) [Dau and Zaharieva, 2009; Vass and Styring, 1991], but is remarkably different from that of a previous Mn_3CaO_4-complex without a dangling Mn ion (–940 mV *versus* ferrocene/ferrocenium) [Kanady *et al.*, 2011]. This result indicates that the dangler manganese ion could play a crucial role in tuning the redox potential of the Mn_4CaO_4-cluster.

Moreover, the one-electron oxidation of the artificial Mn_4Ca-complex gave rise to two distinct EPR signals (g = 4.9 and g = 2.0) [Zhang *et al.*, 2015] (Figure 10), similar to the $g \approx 4$ and g = 2.0 EPR signals observed in PSII for the OEC in the S_2 state [Boussac and Rutherford, 2000; Boussac *et al.*, 2018; Dismukes and Siderer, 1981; Pantazis *et al.*, 2012; Peloquin and Britt, 2001]. In the field of photosynthesis,

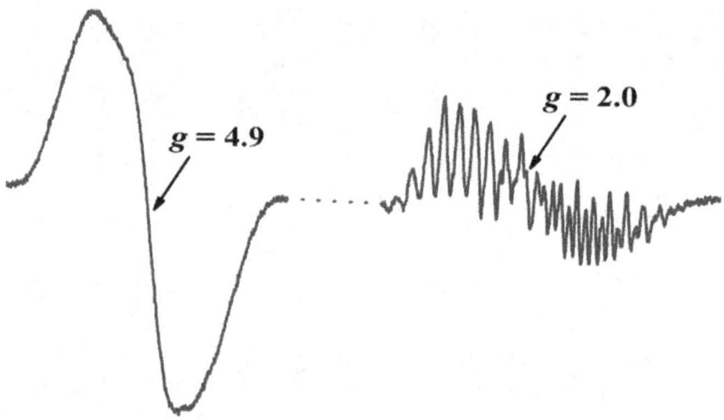

Figure 10. EPR signals induced from the one-electron oxidation of the artificial Mn_4CaO_4-cluster [Zhang *et al.*, 2015].

the latter two EPR signals have been considered as fingerprint spectroscopic characteristics to evaluate the structure and function of the OEC. Therefore, the experimental observation of two EPR signals strongly suggested that the artificial Mn_4Ca-cluster could have similar electronic structures to that of the OEC.

5. Catalytic Activity of the Mn_4CaO_4-Cluster

Considering the close mimicking of both the atomic and electronic structures as well as the redox properties of the OEC, we expect that the artificial Mn_4CaO_4-cluster should be able to serve as a catalyst for the water-splitting reaction. Indeed, a remarkable catalytic current was observed during the CV measurement in the presence of the artificial Mn_4CaO_4-complex and 1% water in acetonitrile [Zhang *et al.*, 2015] (Figure 11, left). The artificial Mn_4CaO_4-complexes can also efficiently catalyze the oxygen-evolving reaction by using $(CH_3)_3COOH$ (*tert*-butyl hydroperoxide) as an oxidant in acetonitrile in the presence of 4.5% water [Chen *et al.*, 2017] (Figure 11, right). However, these artificial complexes are very sensitive to the experimental conditions. Quantification of the catalytic reaction has been difficult due to the rapid degradation of the catalyst in the presence of water in solution. The calcium in the artificial complex has been found to be easily dissociated, leading to the formation of multi-manganese complexes. Another important issue is how to promptly remove protons and electrons that are formed during the water-splitting reaction. In PSII, the OEC is well protected by a large protein matrix, and the electrons and protons formed during the water-splitting reaction can be removed very efficiently from the OEC. Therefore, proper environmental conditions need to be developed in the future, so that the Mn_4CaO_4-cluster can be protected from the large amount of water in the solution and electrons and protons can be efficiently removed from the catalytic site during the water-splitting reaction in the artificial system.

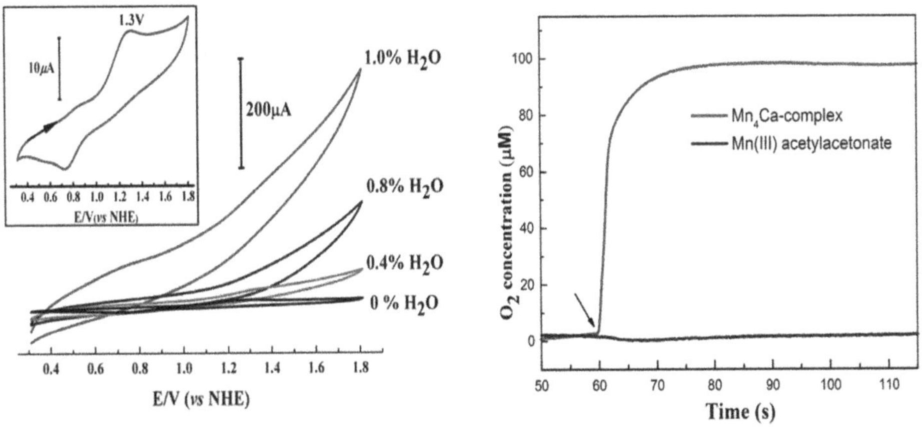

Figure 11. Catalytic properties of the artificial Mn_4CaO_4-cluster. (Left) CV measurement of artificial Mn_4CaO_4-complex in acetonitrile with different amounts of H_2O [Zhang *et al.*, 2015]. The inset shows the CV without H_2O (0% H_2O). Potentials are referenced to NHE; scan rate, 100 mV/s. (Right) Oxygen-evolving reaction catalyzed by artificial Mn_4Ca-complex (50 μM) in acetonitrile with the presence of $(CH_3)_3COOH$ (1 M) and 4.5% water in acetonitrile at 25 °C [Chen *et al.*, 2017]. A Clark-type O_2 electrode was used to detect the release of dioxygen.

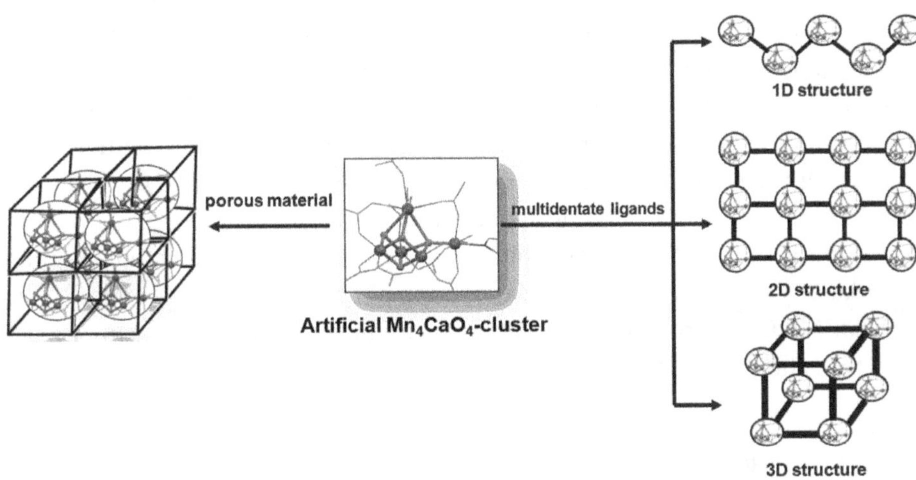

Figure 12. Strategies for building stable catalytic materials for the water-splitting reaction by using artificial Mn_4CaO_4-cluster [Chen *et al.*, 2017].

One possible strategy to tackle the stability issue and to enhance the catalytic efficiency is to incorporate the artificial Mn_4CaO_4-cluster into a porous material, or alternatively, to build 1D/2D/3D materials by linking the artificial Mn_4CaO_4-cluster *via* multi-dentate ligands as shown in Figure 12. We are currently working in this direction, *i.e.*, to build stable and efficient catalytic material by using the artificial Mn_4CaO_4-cluster as building block, aiming to develop a new generation of artificial catalysts for the photocatalytic water-splitting reaction from earth-abundant and non-toxic chemical elements in the future [Kuang, 2016].

6. Implications for the Mechanism of the Water-Splitting Reaction in OEC

The artificial Mn_4CaO_4-clusters shown in Figures 7 and 8 have precisely mimicked the main structural motifs of the OEC, which could provide distinct structural insights into understanding the princple of the OEC. Firstly, the successful isolation of different artificial Mn_4CaO_4-complexes (Figures 7 and 8) demonstrates that the Mn_4CaO_4-cluster is thermodynamically stable, supporting the previous proposal by Barber that the Mn_4CaO_4-cluster could be an evolutionary origin of the natural OEC [Barber, 2016]. Secondly, from the structure of artificial Mn_4CaO_4-cluster, one can clearly see that the μ_4-O^{2-} bridge (O5) is tightly bound to four metal ions (one calcium and three manganese ions). This is a position less prone to be removed or replaced, indicating that the similar μ_4-O^{2-} bridge (O5) in the OEC could not be directly involved in serving as an oxygen source to form the O–O bond. Interestingly, Kawashima *et al.* recently have proposed that the μ_2-O^{2-} bridge (O4) in OEC, instead, may play a role as the substrate binding site to form the O–O bond [Kawashima *et al.*, 2018]. In this proposal and some previous suggestions [Barber, 2017; Chen *et al.*, 2015; Vinyard *et al.*, 2015], the Mn_4CaO_4 fragment does not undertake significant changes during the water-splitting reaction, which is consistent with the isolation of the stable artificial Mn_4CaO_4-cluster described here. It should be pointed out that the μ_2-O^{2-} (O4) in the OEC is absent in the artifical Mn_4CaO_4-cluster, which is replaced by a bridging carboxyl group in the latter. Obviously, the future investigation of this missing μ_2-O^{2-} bridge in the artificial Mn_4Ca-cluster would provide new insights into the mechanism of O–O bond formation during turnover of the OEC.

7. Conclusion

Atomic level high resolution structures of the natural OEC have been reported with increasing accuracy. They provide a blueprint to develop artificial water-splitting catalysts in artificial photosynthesis. Although it remains a great challenge to precisely mimic the entire structure of the OEC in the laboratory, recently synthesized Mn_4CaO_4 model complexes have closely mimicked both the structure and properties of the catalytic metal cluster in the OEC. In-depth understanding of the structure and mechanisms of the water-splitting reaction both in natural and artificial photosynthesis would be greatly facilitated by using these artificial Mn_4CaO_4-clusters in the near future. To achieve the goal of generating stable and highly efficient photocatalytic water-splitting reaction, novel strategies are required to develop next generation catalytic materials by using the artificial Mn_4CaO_4-cluster as building blocks in the future.

Acknowledgement

This work was supported by the National Key Research and Development Program of China (No. 2017YFA0503704), the Strategic Priority Research Program of Chinese Academy of Sciences (No. XDB17030600) and National Natural Science Foundation of China (No. 31770258). CZ thanks Dr. Jie Pan for constructive suggestions and helps during the preparation of the manuscript.

References

Askerka, M., Brudvig, G.W. and Batista, V.S. (2017). The O_2-evolving complex of photosystem II: Recent insights from quantum mechanics/molecular mechanics (QM/MM), extended X-ray absorption fine structure (EXAFS), and femtosecond X-ray crystallography data. *Acc. Chem. Res.*, 50, 41–48.

Barber, J. (2009). Photosynthetic energy conversion: natural and artificial. *Chem. Soc. Rev.*, 38, 185–196.

Barber, J. (2016). Mn_4Ca cluster of photosynthetic oxygen-evolving center: structure, function and evolution. *Biochemistry*, 55, 5901–5906.

Barber, J. (2017). A mechanism for water splitting and oxygen production in photosynthesis. *Nat. Plant.*, 3, 17041.

Boussac, A. and Rutherford, A.W. (2000). Comparative study of the *g*=4.1 EPR signals in the S_2 state of photosystem II. *Biochim. Biophys. Acta.*, 1457, 145–156.

Boussac, A., Ugur, I., Marion, A., Sugiura, M., Kaila, V.R.I. and Rutherford, A.W. (2018). The low spin-high spin equilibrium in the S_2-state of the water oxidizing enzyme. *Biochim. Biophys. Acta.*, 1859, 342–356.

Brown, I.D. (2009). Recent developments in the methods and applications of the bond valence model. *Chem. Rev.*, 109, 6858–6919.

Cardona, T., Sedoud, A., Cox, N. and Rutherford, A.W. (2012). Charge separation in Photosystem II: A comparative and evolutionary overview. *Biochim. Biophys. Acta.*, 1817, 26–43.

Chakov, N.E., Abboud, K.A., Zakharov, L.N., Rheingold, A.L., Hendrickson, D.N. and Christou, G. (2003). Reaction of $[Mn_{12}O_{12}(O_2CR)_{16}(H_2O)_4]$ single-molecule magnets with non-carboxylate ligands. *Polyhedron.*, 22, 1759–1763.

Chang, W., Chen, C., Dong, H. and Zhang, C. (2017). Artificial Mn_4-oxido complexes mimic the oxygen-evolving center in photosynthesis. *Sci. Bull.*, 62, 665–668.

Chen, C., Li, Y., Zhao, G., Yao, R. and Zhang, C. (2017). Natural and artificial Mn_4Ca cluster for the water splitting reaction. *Chem. Sus. Chem.*, 10, 4403–4408.

Chen, C., Zhang, C., Dong, H. and Zhao, J. (2014). A synthetic model for the oxygen-evolving complex in Sr^{2+}-containing photosystem II. *Chem. Commun.*, 50, 9263–9265.

Chen, C., Zhang, C., Dong, H. and Zhao, J. (2015). Artificial synthetic Mn^{IV}Ca-oxido complexes mimic the oxygen-evolving complex in photosystem II. *Dalton Trans.*, 44, 4431–4435.

Cinco, R.M., Robblee, J.H., Rompel, A., Fernandez, C., Yachandra, V.K., Sauer, K. and Klein, M.P. (1998). Strontium EXAFS reveals the proximity of calcium to the manganese cluster of oxygen-evolving photosystem II. *J. Phys. Chem. B.*, 102, 8248–8256.

Cox, N., Pantazis, D.A., Neese, F. and Lubitz, W. (2013). Biological water oxidation. *Acc. Chem. Res.*, 46, 1588–1596.

Dasgupta, J., Ananyev, G.M. and Dismukes, G.C. (2008). Photoassembly of the water-oxidizing complex in photosystem II. *Coord. Chem. Rev.*, 252, 347–360.

Dau, H., Grundmeier, A., Loja, P. and Haumann, M. (2008). On the structure of the manganese complex of photosystem II: extended-range EXAFS data and specific atomic-resolution models for four S-states. *Phil. Trans. R. Soc. Lond. B.*, 363, 1237–1244.

Dau, H. and Haumann, M. (2007). Eight steps preceding O–O bond formation in oxygenic photosynthesis — A basic reaction cycle of the Photosystem II manganese complex. *Biochim. Biophys. Acta.*, 1767, 472–483.

Dau, H. and Zaharieva, I. (2009). Principles, efficiency and blueprint character of solar-energy conversion in photosynthetic water oxidation. *Acc. Chem. Res.*, 42, 1861–1870.

Dau, H., Zaharieva, I. and Haumann, M. (2012). Recent developments in research on water oxidation by photosystem II. *Curr. Opin. Chem. Biol.*, 16, 3–10.

Diner, B.A. and Britt, R.D. (2005). The redox-active tyrosines Y_Z and Y_D. In *Photosystem II: The Light-Driven water: Plastoquinone Oxidoreductase.* T.J. Wydrzynski. and K. Satoh, eds. (The Netherlands: Springer), pp. 207–233.

Diner, B.A. and Rappaport, F. (2002). Structure, dynamics, and energetics of the primary photochemistry of photosystem II of oxygenic photosynthesis. *Annu. Rev. Plant Biol.*, 53, 551–580.

Dismukes, G.C., Brimblecombe, R., Felton, G.A.N., Pryadun, R.S., Sheats, J.E., Spiccia, L. and Swiegers, G.F. (2009). Development of bioinspired Mn_4O_4-cubane water oxidation catalysts:Lessons from photosynthesis. *Acc. Chem. Res.*, 42, 1935–1943.

Dismukes, G.C. and Siderer, Y. (1981). Intermediates of a polynuclear manganese center involved in photosynthetic oxidation of water. *Proc. Natl. Acad. Sci. USA*, 78, 274–278.

Ferreira, K.N., Iverson, T.M., Maghlaoui, K., Barber, J. and Iwata, S. (2004). Architecture of the photosynthetic oxygen-evolving center. *Science*, 303, 1831–1838.

Gatt, P., Petrie, S., Stranger, R. and Pace, R.J. (2012). Rationalizing the 1.9 Å crystal structure of Photosystem II — A remarkable Jahn-Teller balancing act induced by a single proton transfer. *Angew. Chem. Int. Ed.*, 51, 12025–12028.

Gerey, B., Goure, E., Fortage, J., Pecaut, J. and Collomb, M.N. (2016). Manganese-calcium/strontium heterometallic compounds and their relevance for the oxygen-evolving center of photosystem II. *Coord. Chem. Rev.*, 319, 1–24.

Grabolle, M., Haumann, M., Müller, C., Liebisch, P. and Dau, H. (2006). Rapid loss of structural motifs in the manganese complex of oxygenic photosynthesis by X-ray irradiation at 10-300K. *J. Biol. Chem.*, 281, 4580–4588.

Guskov, A., Kern, J., Gabdulkhakov, A., Broser, M., Zouni, A. and Saenger, W. (2009). Cyanobacterial photosystem II at 2.9-Å resolution and the role of quinones, lipids, channels and chloride. *Nat. Struct. Mol. Biol.*, 16, 334–342.

Hellmich, J., Bommer, M., Burkhardt, A., Ibrahim, M., Kern, J., Meents, A., F. Müh, Dobbek, H. and Zouni, A. (2014). Native-like photosystem II superstructure at 2.44Å resolution through detergent extraction from the protein crystal. *Structure*, 22, 1607–1615.

Holzwarth, A.R., Müller, M.G., Reus, M., Nowaczyk, M., Sander, J. and Rögner, M. (2006). Kinetics and mechanism of electron transfer in intact photosystem II and in the isolated reaction center: Pheophytin is the primary electron acceptor. *Proc. Natl. Acad. Sci. USA*, 103, 6895–6890.

Isobe, H., Shoji, M., Yamanaka, S., Umena, Y., Kawakami, K., Kamiya, N., Shen, J.R. and Yamaguchi, K. (2012). Theoretical illumination of water-inserted structures of the $CaMn_4O_5$-cluster in the S_2 and S_3 states of oxygen-evolving complex of photosystem II:Full geometry optimizations by B3LYP hybrid density functional. *Dalton. Trans.*, 41, 13727–13740.

Jin, L., Smith, P., Noble, C.J., Stranger, R., Hanson, G.R. and Pace, R.J. (2014). Electronic structure of the oxygen evolving complex in photosystem II, as revealed by 55Mn Davies ENDOR studies at 2.5 K. *Phys. Chem. Chem. Phys.*, 16, 7799–7812.

Junge, W., Haumann, M., Ahlbrink, R., Mulkidjanian, A. and Clausen, J. (2002). Electrostatics and proton transfer in photosynthetic water oxidation. *Phil. Trans. R. Soc. Lond. B.*, 357, 1407–1417.

Kamiya, N. and Shen, J.R. (2003). Crystal structure of oxygen-evolving photosystem II from *Thermosynechococcus vulcanus* at 3.7 Å resolution. *Proc. Natl. Acad. Sci. USA*, 100, 98–103.

Kanady, J.S., Lin, P.H., Carsch, K.M., Nielsen, R.J., Takase, M.K., Goddard, W.A. and Agapie, T. (2014). Toward models for the full oxygen-evolving complex of photosystem II by ligand coordination to lower the symmetry of the Mn_3CaO_4 cubane: Demonstration that electronic effects facilitate binding of a fifth metal. *J. Am. Chem. Soc.*, 136, 14373–14376.

Kanady, J.S., Tsui, E.Y., Day, M.W. and Agapie, T. (2011). A synthetic model of the Mn_3Ca subsite of the oxygen-evolving complex in photosystem II. *Science*, 333, 733–736.

Kärkäs, M.D., Verho, O., Johnston, E.V. and Åkermark, B. (2014). Artificial photosynthesis: molecular systems for catalytic water oxidation. *Chem. Rev.*, 114, 11863–12001.

Kawashima, K., Takaoka, T., Kimura, H., Saito, K. and Ishikita, H. (2018). O_2 evolution and recovery of the water-oxidizing enzyme. *Nat. Commun.*, 9, 1247.

Kok, B., Forbush, B. and McGloin, M. (1970). Cooperation of charges in photosynthetic O_2 evolution. I. A linear four step mechanism. *Photochem. Photobiol.*, 11, 457–475.

Krewald, V., Retegan, M., Cox, N., Messinger, J., Lubitz, W., DeBeer, S., Neese, F. and Pantazis, D.A. (2015). Metal oxidation states in biological water splitting. *Chem. Sci.*, 6, 1676–1695.

Krewald, V., Retegan, M., Neese, F., Lubitz, W., Pantazis, D.A. and Cox, N. (2016). Spin state as a marker for the structural evolution of nature's water splitting satalyst. *Inorg. Chem.*, 55, 488–501.

Kuang, T. (2016). A breakthrough of artificial photosynthesis. *Nat. Sci. Rev.*, 3, 2–3.

Limburg, J., Vrettos, J.S., Liable-Sands, L.M., Rheingold, A.L., Crabtree, R.H. and Brudvig, G.W. (1999). A functional model for O-O bond formation by the O_2-evolving complex in photosystem II. *Science*, 283, 1524–1527.

Liu, W. and Thorp, H.H. (1993). Bond valence sum analysis of metal-ligand bond lengths in metalloenzymes and model complexes. 2. Refined distances and other enzymes. *Inorg. Chem.*, 32, 4102–4105.

McEvoy, J.P. and Brudvig, G.W. (2006). Water-splitting chemistry of photosystem II. *Chem. Rev.*, 106, 4455–4483.

Mukherjee, S., Stull, J.A., Yano, J., Stamatatos, T.C., Pringouri, K., Stich, T.A., Abboud, K.A., Britt, R.D., Yachandra, V.K. and Christou, G. (2012). Synthetic model of the asymmetric [Mn_3CaO_4] cubane core of the oxygen-evolving complex of photosystem II. *Proc. Natl. Acad. Sci. USA*, 109, 2257–2262.

Mukhopadhyay, S., Mandal, S.K., Bhaduri, S. and Armstrong, W.H. (2004). Manganese clusters with relevance to photosystem II. *Chem. Rev.*, 104, 3981–4026.

Mullins, C.S. and Pecoraro, V.L. (2008). Reflections on small molecule manganese models that seek to mimic photosynthetic water oxidation chemistry. *Coord. Chem. Rev.*, 252, 416–443.

Najafpour, M.M., Renger, G., Hołyńska, M., Moghaddam, A.N., Aro, E.M., Carpentier, R., Nishihara, H., Eaton-Rye, J.J., Shen, J.R. and Allakhverdiev, S.I. (2016). Manganese compounds as water-oxidizing catalysts: from the natural water-oxidizing complex to nanosized manganese oxide structures. *Chem. Rev.*, 116, 2886–2936.

Nelson, N. and Yocum, C.F. (2006). Structure and function of photosystem I and II. *Annu. Rev. Plant. Biol.*, 57, 521–565.

Nilsson, H., Rappaport, F., Boussac, A. and Messinger, J. (2014). Substrate-water exchange in photosystem II is arrested before dioxygen formation. *Nat. Commun.*, 5, 4305.

Pace, R.J., Jin, L. and Stranger, R. (2012). What spectroscopy reveals concerning the Mn oxidation levels in the oxygen evolving complex of photosystem II: X-ray to near infra-red. *Dalton. Trans.*, 41, 11145–11160.

Pantazis, D.A., Ames, W., Cox, N., Lubitz, W. and Neese, F. (2012). Two interconvertible structures that explain the spectroscopic properties of the oxygen-evolving complex of photosystem II in the S_2 state. *Angew. Chem. Int. Ed.*, 51, 9935–9940.

Paul, S., Neese, F. and Pantazis, D.A. (2017). Structural models of the biological oxygen-evolving complex: achievements, insights, and challenges for biomimicry. *Green. Chem.*, 19, 2309–2325.

Pauling, L. (1929). The principles determining the structure of complex ionic crystals. *J. Am. Chem. Soc.*, 51, 1010–1026.

Peloquin, J.M. and Britt, R.D. (2001). EPR/ENDOR characterization of the physical and electronic structure of the OEC Mn-cluster. *Biochim. Biophys. Acta.*, 1503, 96–111.

Peloquin, J.M., Campbell, K.A., Randall, D.W., Evanchik, M.A., Pecoraro, V.L., Armstrong, W.H. and Britt, R.D. (2000). ^{55}Mn ENDOR of the S_2-state multiline EPR signal of photosystem II: implications on the structure of the tetranuclear Mn cluster. *Journal of the American Chemical Society*, 122, 10926–10942.

Perez-Navarro, M., Neese, F., Lubitz, W., Pantazis, D.A. and Cox, N. (2016). Recent developments in biological water oxidation. *Curr. Opin. Chem. Biol.*, 31, 113–119.

Petrouleas, V. and Crofts, A.R. (2005). The iron-quinone accetor complex. In Photosystem II: The Light-Driven water: Plastoquinone Oxidoreductase, T.J. Wydrzynski, and K. Satoh, eds. (The Netherlands: Springer), pp. 177–206.

Renger, G. (2012). Mechanism of light induced water splitting in Photosystem II of oxygen evolving photosynthetic organisms. *Biochim. Biophys. Acta.*, 1817, 1164–1176.

Ruettinger, W.F., Campana, C. and Dismukes, G.C. (1997). Synthesis and characterization of $Mn_4O_4L_6$ complexes with cubane-like core structure: a new class of models of the active site of the photosynthetic water oxidase. *J. Am. Chem. Soc.*, 119, 6670–6671.

Saito, K., Rutherford, A.W. and Ishikita, H. (2013). Mechanism of proton-coupled quinone reduction in photosystem II. *Proc. Nat. Acad. Sci., USA*, 110, 954–959.

Satoh, K., Wydrzynski, T.J. and Govindjee (2005). Introduction to photosystem II. In Photosystem II: The Light-Driven Water: Plastoquinone Oxidoreductase, T.J. Wydrzynski, and K. Satoh, eds. (The Netherlands: Springer), pp. 11–22.

Sauer, K., Yano, J. and Yachandra, V.K. (2008). X-ray spectroscopy of the photosynthetic oxygen-evolving complex. *Coord. Chem. Rev.*, 252, 318–335.

Shen, J.R. (2015). The structure of photosystem II and the mechanism of water oxidation in photosynthesis. *Annu. Rev. Plant. Biol.*, 66, 23–48.

Siegbahn, P.E.M. (2013). Water oxidation mechanism in photosystem II, including oxidations, proton release pathways, O-O bond formation and O_2 release. *Biochim. Biophys. Acta.*, 1827, 1003–1019.

Styring, S., Sjöholm, J. and Mamedov, F. (2012). Two tyrosines that changed the world: Interfacing the oxidizing power of photochemistry to water splitting in photosystem II. *Biochim. Biophys. Acta.*, 1817, 76–87.

Su, X., Ma, J., Wei, X., Cao, P., Zhu, D., Chang, W., Liu, Z., Zhang, X. and Li, M. (2017). Structure and assembly mechanism of plant $C_2S_2M_2$-type PSII-LHCII supercomplex. *Science*, 357, 815–820.

Suga, M., Akita, F., Hirata, K., Ueno, G., Murakami, H., Nakajima, Y., Shimizu, T., Yamashita, K., Yamamoto, M., Ago, H., *et al.* (2015). Native structure of photosystem II at 1.95Å resolution revealed by a femtosecond X-ray laser. *Nature*, 517, 99–103.

Suga, M., Akita, F., Sugahara, M., Kubo, M., Nakajima, Y., Nakane, T., Yamashita, K., Umena, Y., Nakabayashi, M., Yamane, T., *et al.* (2017). Light-induced structural changes and the site of O=O bond formation in PSII caught by XFEL. *Nature*, 543, 131–135.

Tanaka, A., Fukushima, Y. and Kamiya, N. (2017). Two different structures of the oxygen-evolving complex in the same polypeptide frameworks of photosystem II. *J. Am. Chem. Soc.*, 139, 1718–1721.

Tommos, C. and Babcock, G.T. (1998). Oxygen production in nature: A light-driven metalloradical enzyme process. *Acc. Chem. Res.*, 31, 18–25.

Tsui, E.Y. and Agapie, T. (2013). Reduction potentials of heterometallic manganese–oxido cubane complexes modulated by redox-inactive metals. *Proc. Natl. Acad. Sci. USA*, 110, 10084–10088.

Tsui, E.Y., Kanady, J.S. and Agapie, T. (2013). Synthetic cluster models of biological and heterogeneous manganese catalysts for O_2 evolution. *Inorg. Chem.*, 52, 13833–13848.

Umena, Y., Kawakami, K., Shen, J.R. and Kamiya, N. (2011). Crystal structure of oxygen-evolving photosystem II at a resolution of 1.9Å. *Nature*, 473, 55–60.

van Gorkom, H.J. and Yocum, C.F. (2005). The calcium and chloride cofactor. In Photosystem II: The Light-Driven Water: Plastoquinone Oxidoreductase, T.J. Wydrzynski, and K. Satoh, eds. (The Netherlands: Springer), pp. 307–328.

Vass, I. and Styring, S. (1991). pH-dependent charge equilibria between tyrosine-D and the S states in photosystem II. Estimation of relative midpoint redox potentials. *Biochemistry*, 30, 830–839.

Vinyard, D.J., Khan, S. and Brudvig, G.W. (2015). Photosynthetic water oxidation: Binding and activation of substrate water for O-O bond formation. *Faraday. Discuss.*, 185, 37–50.

Vrettos, J.S., Limburg, J. and Brudvig, G.W. (2001). Mechanism of photosynthetic water oxidation: combining biophysical studies of photosystem II with inorganic model chemistry. *Biochim. Biophys. Acta.*, 1503, 229–245.

Wang, L., Zhang, C. and Zhao, J. (2014). Location and function of the high-affinity chloride in the oxygen-evolving complex — Implications from comparing studies on $Cl^-/Br^-/I^-$-substituted photosystem II prepared using two different methods. *J. Photochem. Photobiol. B.*, 138, 249–255.

Wang, Y., Zhang, C., Wang, L. and Zhao, J. (2013). Assignment of the μ_4-O5 atom in catalytic center for water oxidation in photosystem II. *Chin. Sci. Bull.*, 58, 3213–3216.

Wei, X., Su, X., Cao, P., Liu, X., Chang, W., Li, M., Zhang, X. and Liu, Z. (2016). Structure of spinach photosystem II–LHCII supercomplex at 3.2Å resolution. *Nature*, 534, 69–74.

Yano, J., Kern, J., Irrgang, K.D., Latimer, M.J., Bergmann, U., Glatzel, P., Pushkar, Y., Biesiadka, J., Loll, B., Sauer, K. *et al.* (2005). X-ray damage to the Mn_4Ca complex in single crystals of photosystem II: A case study for metalloprotein crystallography. *Proc. Natl. Acad. Sci. USA*, 102, 12047–12052.

Yano, J., Kern, J., Sauer, K., Latimer, M.J., Pushkar, Y., Biesiadka, J., Loll, B., Saenger, W., Messinger, J., Zouni, A., *et al.* (2006). Where water is oxidized to dioxygen: Structure of the photosynthetic Mn_4Ca cluster. *Science*, 314, 821–825.

Yano, J. and Yachandra, V. (2014). Mn_4Ca-cluster in photosynthesis: Where and how water is oxidized to dioxygen. *Chem. Rev.*, 114, 4175–4205.

Yocum, C.F. (2008). The calcium and chloride requirements of the O_2 evolving complex. *Coord. Chem. Rev.*, 252, 296–305.

Young, I.D., Ibrahim, M., Chatterjee, R., Gul, S., Fuller, F.D., Koroidov, S., Brewster, A.S., Tran, R., Alonso-Mori, R., Kroll., T. *et al.* (2016). Structure of photosystem II and substrate binding at room temperature. *Nature*, 540, 453–457.

Zhang, C. (2007). Low-barrier hydrogen bond plays key role in active photosystem II — a new model for photosynthetic water oxidation. *Biochim. Biophys. Acta.*, 1767, 493–499.

Zhang, C. (2015). The first artificial Mn_4Ca-cluster mimicking the oxygen-evolving center in photosystem II. *Sci. Chin. Life. Sci.*, 58, 816–817.

Zhang, C. (2016). From natural photosynthesis to artificial photosynthesis. *Sci. Sin. Chim.*, 46, 1101–1109.

Zhang, C., Chen, C., Dong, H., Shen, J.R., Dau, H. and Zhao, J. (2015). A synthetic Mn_4Ca-cluster mimicking the oxygen-evolving center of photosynthesis. *Science*, 348, 690–693.

Zhang, C. and Kuang, T. (2018). A new milestone for photosyntehsis. *Nat. Sci. Rev.*, 5, 444–445.

Zhang, C., Pan, J., Li, L. and Kuang, T. (1999). New structure model of oxygen-evolving center and mechanism for oxygen evolution in photosynthesis. *Chin. Sci. Bull.*, 44, 2209–2215.

Zhang, M., Bommer, M., Chatterjee, R., Hussein, R., Yano, J., Dau, H., Kern, J., Dobbek, H. and Zouni, A. (2017). Structural insights into the light-driven auto-assembly process of the water-oxidizing Mn_4CaO_5-cluster in photosystem II. *eLife*, 6, e26933.

Zheng, M. and Dismukes, G.C. (1996). Orbital configuration of the valence electrons, ligand field symmetry, and manganese oxidation states of the photosynthetic water oxidizing complex: analysis of the S_2 state multiline EPR signals. *Inorg. Chem.*, 35, 3307–3319.

Zouni, A., Witt, H.T., Kern, J., Fromme, P., Kraub, N., Saenger, W. and Orth, P. (2001). Crystal structure of photosystem II from Synechococcus elongatus at 3.8Å resolution. *Nature*, 409, 739–743.

Chapter 10

Protein Environment that Facilitates Proton Transfer and Electron Transfer in Photosystem II

Hiroshi Ishikita

Research Center for Advanced Science and Technology
The University of Tokyo 4-6-1 Komaba,
Meguro-ku, Tokyo 153-8904, Japan
hiro@appcuem.t.u-tokyo.ac.jp

The water splitting reaction $2H_2O \rightarrow O_2 + 4H^+ + 4e^-$ requires release of four protons and four electrons from two substrate water molecules. In photosystem II (PSII), this reaction is catalyzed at the Mn_4CaO_5 moiety in a protein environment. In contrast to the bulk region, the product H^+ cannot be easily diffused from the catalytic sites in protein environments. To remove protons, PSII uses proton transfer pathways that proceed from the substrate water molecules at the Mn_4CaO_5 moiety toward the protein bulk surface. The proton transfer pathways are an H-bond network, which comprises polar groups, *e.g.*, polar residues, charged residues, and water molecules in the protein electrostatic environment. On the other hand, to remove electrons from the substrate water molecules, PSII uses electron transfer pathways, which are mainly formed by the redox-active cofactors. The redox potential values of the cofactors are strongly associated with the energetics of electron transfer along the pathways and are affected by the protein electrostatic environment. Thus, the same protein electrostatic environment determines the energetics of both proton transfer pathways and electron transfer pathways in PSII. This chapter provides an overview of how the same protein electrostatic environment tunes the energetics of positively charged H^+ and negatively charged e^- simultaneously in PSII.

1. Proton Transfer in PSII

1.1. *Formation of low-barrier hydrogen bonds are required for proton transfer*

To facilitate proton transfer, formation of a hydrogen bond (H-bond) is necessary. In the H-bond geometry, the distance, which is required for proton transfer, is smallest. In a standard H-bond, the pK_a value is higher in the donor moiety than in the acceptor moiety, *i.e.*, the H-bond donor moiety is protonated, whereas the H-bond acceptor moiety is deprotonated (Figure 1, left). Thus, the proton is energetically stable at the donor moiety and will never transfer to the acceptor moiety because of the presence of the energy barrier.

This, in turn, suggests that proton transfer occurs most efficiently when the pK_a values of both the donor and acceptor moieties are nearly equal [Eigen, 1964], *i.e.*, the two moieties can be classified to neither proton donors nor acceptors. In this case, a low-barrier H-bond (LBHB; Figure 1, middle) or a single-well H-bond (LBHB; Figure 1, right) can form between the two moieties [Perrin and Nielson, 1997; Schutz and Warshel, 2001; Ishikita and Saito, 2014]. Thus, the formation of low-barrier or single-well H-bonds is a prerequisite for proton transfer. The difference between low-barrier and single-well H-bonds is somewhat vague, and both H-bonds are called low-barrier H-bonds in this chapter. According to Schutz and Warshel, low-barrier H-bonds cannot be defined solely by their length or strength; only energy-based evaluations, *e.g.*, shapes of the potential energy profiles (Figure 2) can be used to judge the type of H-bond [Schutz and Warshel, 2001].

Low-barrier H-bonds can be unambiguously identified by analyzing the potential-energy profile at the density functional theory (DFT) level, *e.g.*, without solving the nuclear Schrödinger equation [Ikeda *et al.*, 2017] (Figure 2). This fact

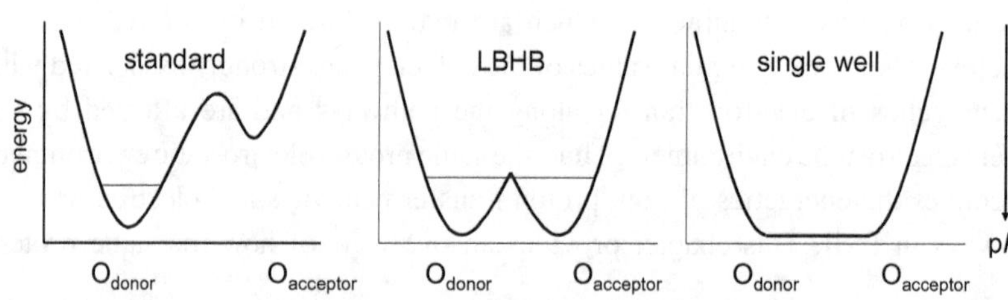

Figure 1. Potential-energy profiles of standard (left), low-barrier (LBHB; middle), and single-well (right) H-bonds ($O_{donor}\cdots O_{acceptor}$ H-bonds).

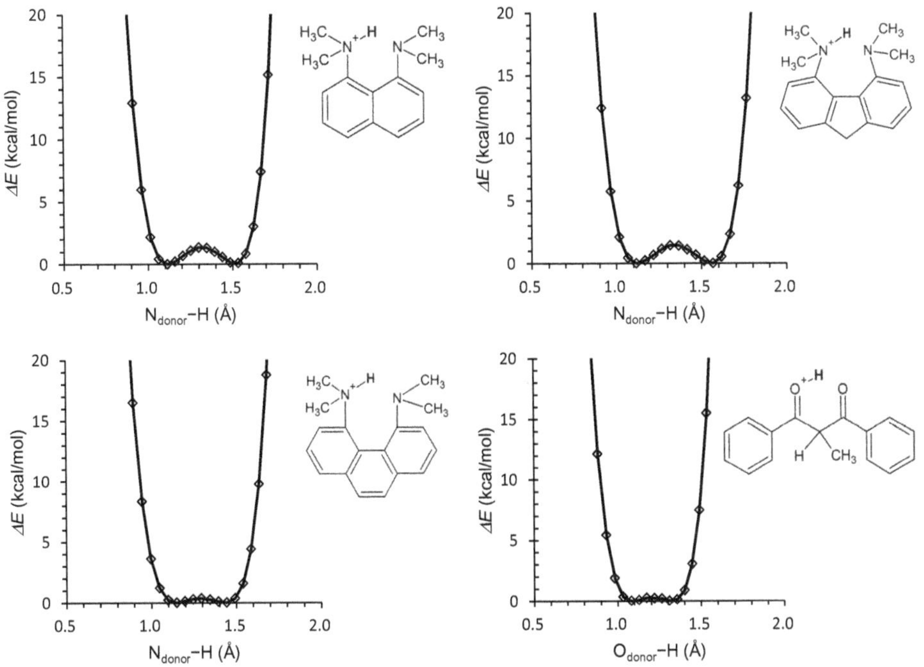

Figure 2. Potential-energy profiles of low-barrier H-bonds in organic molecules, which were analyzed at the DFT level.

strongly indicates that matching the pK_a values between the two H-bond moieties is the sole requirement for formation of low-barrier H-bonds.

1.2. *Low-barrier hydrogen bonds in PSII*

Formation of low-barrier H-bonds is strongly associated with proton transfer [Ishikita and Saito, 2014]. The first finding of a low-barrier H-bond in PSII was reported by Saito *et al.* in 2011 [Saito *et al.*, 2011]. They analyzed a remarkably short H-bond between TyrZ and D1-His190 in the PSII crystal structure ($O_{TyrZ}\cdots N\varepsilon_{D1\text{-}His190}$ = 2.46 Å [Umena *et al.*, 2011]) and revealed that the shape of the potential energy profile resembles that of a typical single-well H-bond [Saito *et al.*, 2011; Kawashima *et al.*, 2018] (Figure 1, right). pK_a of tyrosine (~10) is higher than pK_a of Nε of histidine (~7) in water. In PSII, pK_a of TyrZ is lowered to pK_a of D1-His190 because of the donation of an H-bond from a cluster of well-ordered water molecules (*i.e.*, with low *B*-factor) near the Mn_4CaO_5 cluster, resulting in the formation of a low-barrier H-bond [Saito *et al.*, 2011; Kawashima *et al.*, 2018]. The low-barrier H-bond in the TyrZ...D1-His190 pair is likely to facilitate proton-coupled electron transfer.

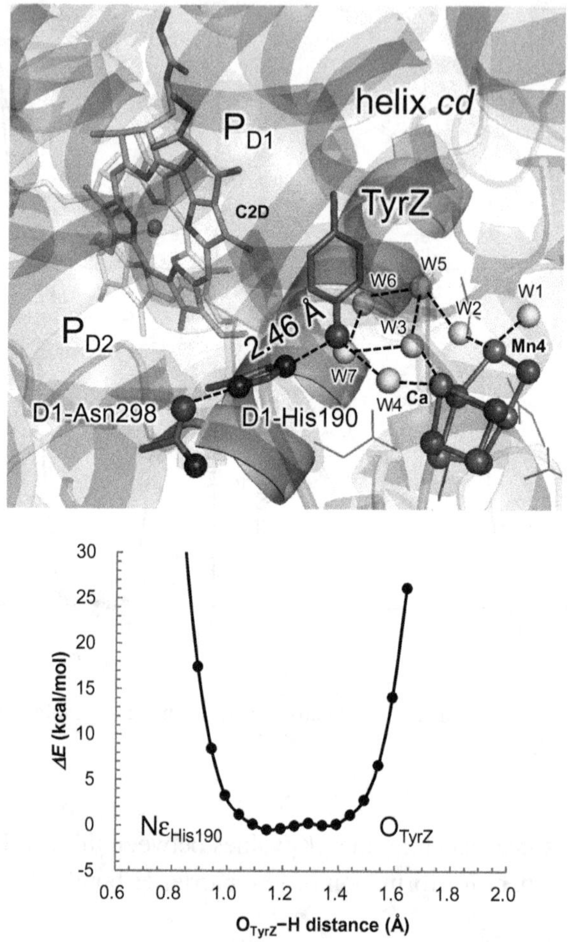

Figure 3. H-bond network (upper) and the potential-energy profile of the low-barrier H-bond between TyrZ and D1-His190 (bottom).

Formation of a low-barrier H-bond is also observed in proton-coupled electron transfer to Q_B. The initial proton donation occurs from D1-His252 *via* D1-Ser264 to the Q_B distal carbonyl O site (with respect to the non-heme Fe complex) in an energetically downhill process [Ishikita and Knapp, 2005; Saito *et al.*, 2013]. Formation of a low-barrier H-bond between Q_BH^- and D1-His215 ($O_{QB}\ldots N\delta_{D1\text{-}His215}$ = 2.47 Å) (Figure 3) suggests that the second protonation occurs from Nδ of D1-His215 directly to the Q_B proximal carbonyl O site [Saito *et al.*, 2013]. D1-His215 is only singly protonated (at Nδ), but it can release the proton due to a decrease in pK_a by the presence of Fe^{2+} at Nε [Saito *et al.*, 2013]. Indeed, Fourier transform infrared spectroscopic (FTIR) studies have indicated that deprotonation of D1-His215 occurs in response to pH changes [Berthomieu and Hienerwadel, 2001].

1.3. *Release of protons from substrate water molecules in PSII*

Formation of low-barrier H-bonds was observed at TyrZ...D1-His190 and Q_B, wherever proton transfer occurs in PSII. Then, it seems probably more reasonable to consider that release of protons from substrate water molecules at the Mn_4CaO_5 cluster is no exception. In particular, the S_0-to-S_1 transition is the process where not release of the proton but electron transfer is rate-limiting and the activation energy is specifically low [Dau and Haumann, 2008; Shimizu *et al.*, 2018]. In this case, it is highly likely that a low-barrier H-bond will form with the proton-releasing site in the S_0-to-S_1 transition.

The 1.9-Å crystal structure [Umena *et al.*, 2011] shows that the Mn_4CaO_5 cluster has two water molecules as H-bond partners (O1...O_{W923} and O4...O_{W539}). Remarkably, the H-bond between O4 and W539 is unusually short, 2.43–2.50 Å. From the short distance, there is no doubt that O4 and W539 form an H-bond. Quantum mechanical/molecular mechanical (QM/MM) calculations reproduced a short H-bond distance (~2.5 Å) for O4–O_{W539} only when O4 was OH^- and donated an H-bond to W539 in S_0 and S_1 [Saito *et al.*, 2015]. In S_0, the H-bond was short (2.55 Å), but the shape of the H-bond potential-energy profile was asymmetric, which suggests that O4–O_{W539} was a standard H-bond (Figure 4). In contrast, the shape of the H-bond potential-energy profile was symmetric in S_1, which suggests

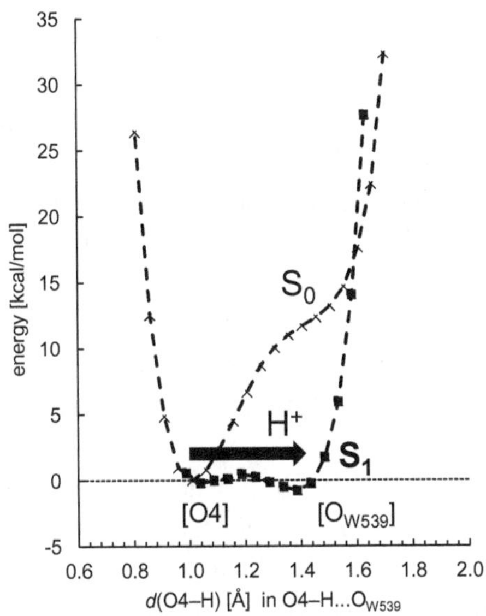

Figure 4. Potential-energy profiles of the H-bond between O4 of the Mn_4CaO_5 cluster and W539 in S_0 (standard H-bond, dotted curve) and S_1 (low-barrier H-bond, solid curve).

that a low-barrier H-bond forms between O4 and W539 and proton transfer can occur along the barrier-less O4–O_{W539} (Figure 4).

The O4 site of the Mn_4CaO_5 cluster is connected with a chain of ~8 H-bonding water molecules, the O4-water chain, *via* W539. The presence of the O4-water chain not only serves as a Grotthuss-like proton conduit [Saito *et al.*, 2015] but also decreases $pK_a(OH^-/O^{2-})$ of O4, because the positive charge of the released proton from O4 can be delocalized over the water molecules, in particular, the initial three H-bonding water molecules W539…W538…W393 [Takaoka *et al.*, 2016]. Thus, in addition to an increase of the net charge of the Mn_4CaO_5 cluster due to oxidation of Mn3(III) to Mn3(IV) [Saito *et al.*, 2015], $pK_a(OH^-/O^{2-})$ of O4 matches $pK_a(H_2O/H_3O^+)$ of W539, resulting in the formation of a low-barrier H-bond between O4 and W539 and proton transfer in the S_0-to-S_1 transition. Note that H_3O^+ is not stabilized at W539, but the proton is transferred to the terminus of the O4-water chain concertedly [Saito *et al.*, 2015].

The O4-water chain, serving as a proton transfer pathway in the S_0-to-S_1 transition, is consistent with the fact that the S_0-to-S_1 is less inhibited by Cl^- depletion [Wincencjusz *et al.*, 1997]. Note that Cl^- is not involved in the H-bond network of the O4-water chain but is involved in the proton transfer pathway *via* D1-Asp61 [Kawashima *et al.*, 2018].

1.4. *Can O5 be the deprotonation site?*

O5, which is interpreted as forming a covalent bond with O6 in a higher S state in the recent two-flash illuminated PSII structures [Suga *et al.*, 2017; Kern *et al.*, 2018] and being a candidate for substrate water molecule, has no H-bond partner in all PSII crystal structures so far reported [Umena *et al.*, 2011; Suga *et al.*, 2015; 2017]. W2 is located at 3.1 Å away from O5, but QM/MM calculations suggest that the water molecule cannot accept an H-bond from OH^- at O5 in the PSII protein environment [Pal *et al.*, 2013; Saito *et al.*, 2015]; this is because W2 is ligated to Mn4 and the H atoms of W2 cannot orient toward O5. Since the *B*-factor of the Mn_4CaO_5 moiety, including the O5 site, is low, the absence of a water molecule at the O5 moiety in the crystal structure is predominantly due to the presence of the hydrophobic D1-Val185 side-chain at O5 ($C_{D1\text{-}Val185}$…O5 = 3.7 Å [Umena *et al.*, 2011]).

Regardless of such a protein environment, however, some theoretical studies suggested that OH^- at O5 could release a proton, probably because O5 is located at the interesting position near Ca^{2+}, which is a prerequisite for water oxidation. Siegbahn tried to demonstrate that deprotonation of OH^- at O5 occurred in the S_0-to-S_1 transition in DFT calculations [Siegbahn, 2013]. To overcome the energetic problem of the absence of an H-bond partner of O5 in the original PSII crystal structure, he assumed an additional water molecule near O5, a water molecule that is not visible not only in the original 1.9-Å structure [Umena *et al.*, 2011] but also in the radiation damage-free PSII crystal structure [Suga *et al.*, 2015]. However, it seems impossible to place a water molecule at O5; the two crystal structures show that there is no space to place a water molecule at O5, because the hydrophobic D1-Val185 side-chain ($C\gamma_{D1\text{-}Val185}$...O5 = 3.7 Å [Umena *et al.*, 2011]) and the ligand D1-His332 side-chain ($C\delta_{D1\text{-}His332}$...O5 = 3.6 Å [Umena *et al.*, 2011]) are too close to O5. Hence the question: How was it possible to place a water molecule at O5 in his theoretical model?

We found [Saito *et al.*, 2015] that in DFT calculations by Siegbahn [Siegbahn, 2013], (i) the D1-Val185 side-chain was removed and (ii) the D1-His332 side-chain was twisted by ~90° (Figure 5) with respect to the crystal structure. In general DFT calculations, the electrostatic influence of the valine side-chain may be ignored. However, in the case of PSII, the D1-Val185 side-chain must be included in calculations, because it plays a role in providing a hydrophobic environment to the O5 site (i.e., preventing access of water molecules and decreasing the chance of solvation). In addition, twisting the D1-His332 side-chain cannot be energetically allowed, because it is a ligand of Mn1 and the Nδ of D1-His332 forms an H-bond with D1-Glu329 ($N\delta_{D1\text{-}His332}$...$O_{D1\text{-}Glu329}$ = 2.65 Å [Umena *et al.*, 2011]). These drastic manipulations of the atomic coordinates of the original crystal structure would fail to provide support for O5 releasing a proton in the S_0-to-S_1 transition or O5 being a substrate water molecule.

Theoretical studies by Pal *et al.* [2013] were sometimes cited as a theoretical basis for O5 being a substrate water molecule. However, caution must be paid; they clearly stated that deprotonation of OH^- at O5 was not observed in the S_0-to-S_1 transition because of the absence of the H-bond partner of O5 when the PSII protein environment was considered in calculations, *i.e.*, "the QM/MM S_0 model [Pal *et al.*, 2013]." They observed deprotonation of OH^- at O5 *only* when they switched from "the *QM/MM* S_0 model" to "*a small model* of the oxidized S_0' state". Neither the details nor the atomic coordinates of "a small model of the oxidized S_0' state" are clearly shown in the Supporting Information (even in Figure S7 in Pal *et al.* [2013]); thus, theoretical studies by Pal *et al.* [2013] can confirm only that deprotonation of OH^- at O5 was not observed in the S_0-to-S_1 transition.

In view of the preceding discussion, to the best of our knowledge, the story of deprotonation of O5 occurring in the S_0-to-S_1 transition can be made up only when removing the hydrophobic D1-Val185 side-chain and rotating the ligand D1-His332 intentionally [Siegbahn, 2013] or using a small model [Pal *et al.*, 2013]; *i.e.*, so far, in theoretical studies, deprotonation of OH^- at O5 has never been demonstrated using the original geometry of the PSII crystal structures.

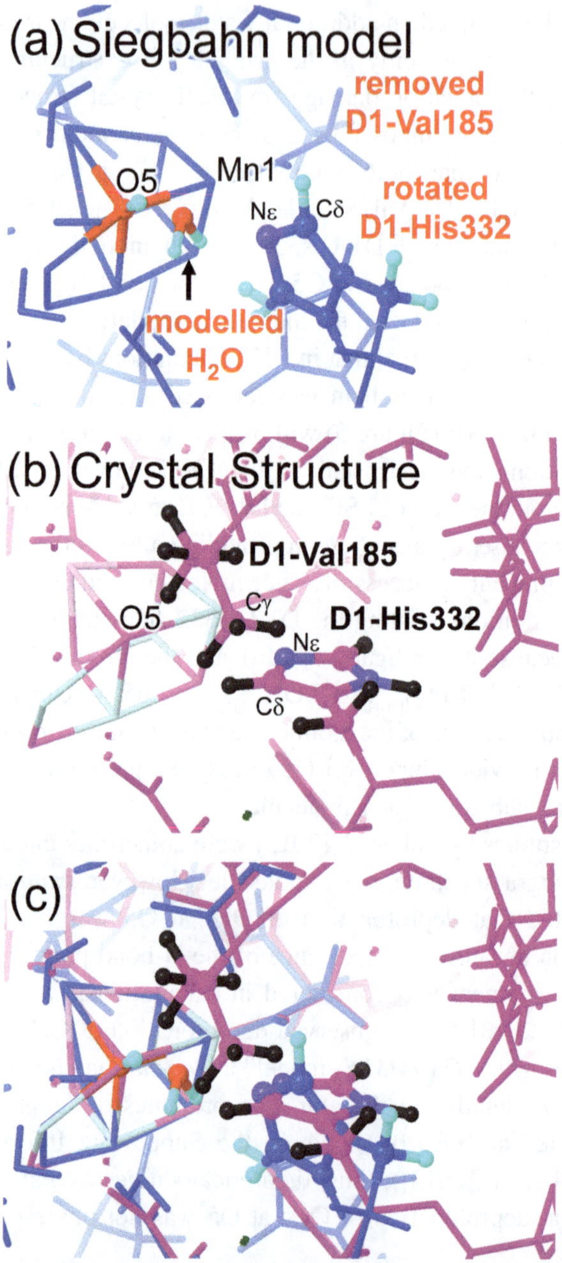

Figure 5. Protein environment at the O5 moiety of the Mn_4CaO_5 cluster: (a) QM model, which was used by Siegbahn [2013] to propose the release of the proton from OH^- at O5 in the S_0-to-S_1 transition. Compared to its position in the 1.9-Å structure [Umena *et al.*, 2011], (i) he twisted the side chain of D1-His332 (by ~90°) along the Mn1-$N\varepsilon_{His332}$ axis [Siegbahn, 2013] to create the water binding site at O5, and (ii) he removed the hydrophobic D1-Val185 side chain to stabilize the water molecule added at the O5 moiety. (b) Original geometry of the PSII crystal structure analyzed at 1.9-Å resolution [Umena *et al.*, 2011]. (c) Superposition of the two geometries.

To sufficiently demonstrate that deprotonation of O5 occurs in the S_0-to-S_1 transition, one needs to answer, on the basis of the original PSII crystal structure: (i) What is the H-bond partner of OH^- at O5 (W539 for O4), (ii) what is the shape of the potential-energy profile of the H-bond between O5 and the H-bond partner (low-barrier H-bond for $O4 \ldots O_{W539}$), and (iii) what is the proton-conducing H-bond network that is connected with the H-bond partner of O5 (the O4-water chain for O4)?

1.5. *Water channels near the Mn_4CaO_5 cluster*

If a water molecule is the deprotonation site in the S_0-to-S_1 transition, it can approach the binding site at the Mn_4CaO_5 moiety from the bulk region along the water channel before S_0. Molecular dynamics simulations indicated that water molecules can approach the W539 site, as well as the O4 site, from the bulk region along the D1-Glu65/D2-Glu312 channel [Sakashita *et al.*, 2017]. It seems plausible that both W539 and O4, which have a low-barrier H-bond in the S_0-to-S_1 transition and serve as a proton-releasing site [Saito *et al.*, 2015; Takaoka *et al.*, 2016], satisfy the requirement for a substrate water molecule.

It should be noted that the O4-water chain is too narrow [Takaoka *et al.*, 2016; Sakashita *et al.*, 2017] and all water molecules except for W539 and W538 are practically not exchangeable [Sakashita *et al.*, 2017], *i.e.*, the O4-water chain cannot serve as a water intake channel.

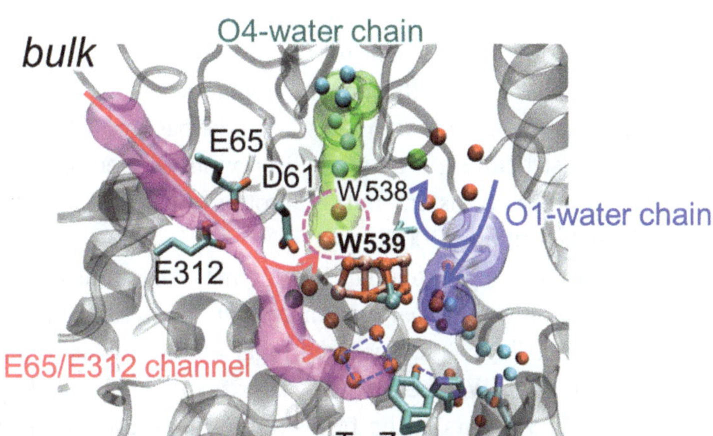

Figure 6. Water channels/chains near the Mn_4CaO_5 cluster. Red and cyan balls indicate exchangeable and less-exchangeable water molecules, respectively.

1.6. *Summary of release of a proton from the substrate water molecule in the low S state transition*

In summary,

- Solely from the original geometry of the crystal structure, O4 is the only site in the Mn_4CaO_5 cluster that has a delocalized H-bond network [Umena *et al.*, 2011; Suga *et al.*, 2015].
- One should not intentionally ignore the fact that the H-bond between O4 and W539 is remarkably short (~2.5 Å) in the original PSII crystal structure [Umena *et al.*, 2011; Suga *et al.*, 2015], before drastically manipulating/modifying the original atomic coordinates of the crystal structures to make a great effort for supporting the idea of O5 being the deprotonation site in the S_0-to-S_1 transition. If the O5 deprotonation could occur, one could have demonstrated it without drastically manipulating/modifying the original atomic coordinates of the crystal structures.
- Deprotonation of OH^- at O5 needs approach of a water molecule, i.e., "conformational dynamics." Proton transfer that occurs with help of conformational dynamics is slow, whereas proton transfer that occurs in the pre-organized protein electrostatic environment (without the help of conformational dynamics) is fast [Stuchebrukhov, 2009]. For this issue, O4 already has "the pre-organized protein electrostatic environment" [Schutz and Warshel, 2004; Adamczyk *et al.*, 2011], and is ready for proton transfer; *i.e.*, O4 already forms a remarkably short H-bond with W539 (~2.5 Å) and the pre-organized H-bonding O4-water chain. The corresponding pre-organized water chain or H-bond network is obviously absent for O5 in the PSII crystal structure.
- Some researchers like to use the phrase "dynamics helps proton transfer" when they cannot explain the proton transfer on the basis of the original geometry of the crystal structure. As mentioned previously, there are cases where "dynamics helps proton transfer," but the proton transfer is slow in this case (see below) [Stuchebrukhov, 2009].
- Warshel and coworkers suggested that in enzymatic reactions (including proton transfer), not "conformational dynamics" but "preorganized protein electrostatic environments" (or "preorganized polar environments") play a major role in enzyme catalysis [Warshel, 1998; Adamczyk *et al.*, 2011]. In particular, the S_0-to-S_1 transition occurs at the fastest rate among all S-state transitions [Dekker *et al.*, 1984; Haumann *et al.*, 2005; Shimizu *et al.*, 2018] and the activation energy is specifically low [Dau and Haumann, 2008]; this feature is likely to fit more to proton transfer in "preorganized protein electrostatic environments" along the O4-water chain (Figure 7) than proton transfer from

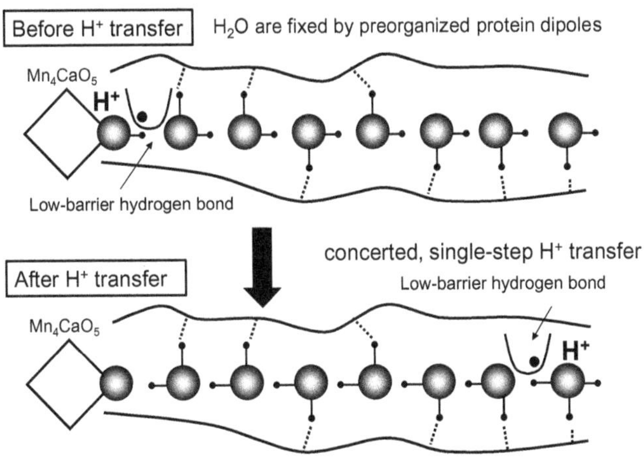

Figure 7. Overview of proton transfer from O4 along the O4-water chain in the preorganized protein electrostatic environment.

O5 in a "conformational dynamics" manner. Recent analysis using time-resolved infrared spectroscopy by Shimizu *et al.* successfully identified the pure contribution of the S_0 to S_1 transition to the signal; the S_0-to-S_1 transition is the highest among the four S-state transitions with a relatively small KIE value of 1.2. They concluded that their results are more consistent with deprotonation of O4 [Saito *et al.*, 2015] than deprotonation of O5 in the S_0-to-S_1 transition [Shimizu *et al.*, 2018].

2. Electron Transfer in PSII

2.1. *Redox potential values of cofactors in the electron transfer branches*

In PSII, electron-transfer branches (D1-, D2-branches) proceed from a pair of chlorophyll *a* (Chl*a*) (P_{D1}, P_{D2}) *via* accessory Chl*a* (Chl_{D1}, Chl_{D2}), pheophytin *a* (Pheo*a*) ($Pheo_{D1}$, $Pheo_{D2}$), and plastoquinone (Q_A, Q_B). The redox potential value for one-electron reduction (E_m) of Chl_{D1} ($E_m(Chl_{D1})$) is lower than $E_m(Chl_{D2})$, irrespective of the presence of the Mn_4CaO_5 cluster on the D1 side [Kawashima and Ishikita, 2018], which suggests that the Mn_4CaO_5 cluster does not increase the E_m values of cofactors on the D1 site electrostatically (Figure 8). Indeed, the E_m values of the cofactors obtained using the Mn-depleted PSII crystal structure (PDB codes 5MX2 [Zhang *et al.*, 2017]) did not significantly differ from the E_m values obtained using the 1.9-structure (PDB codes 3ARC) [Kawashima and Ishikita,

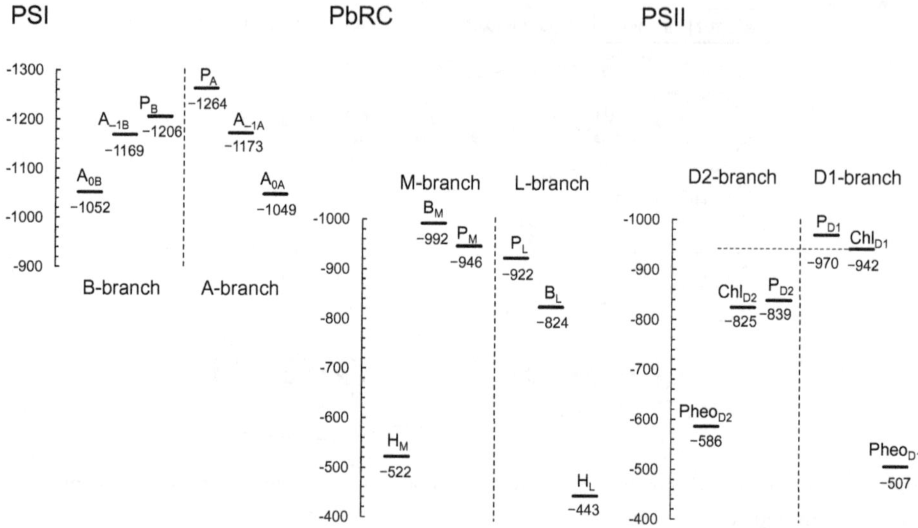

Figure 8. E_m values of cofactors in both electron transfer branches in photosystem I (PSI, left), the photosynthetic reaction centers from purple bacteria (PbRC, middle), and PSII (right).

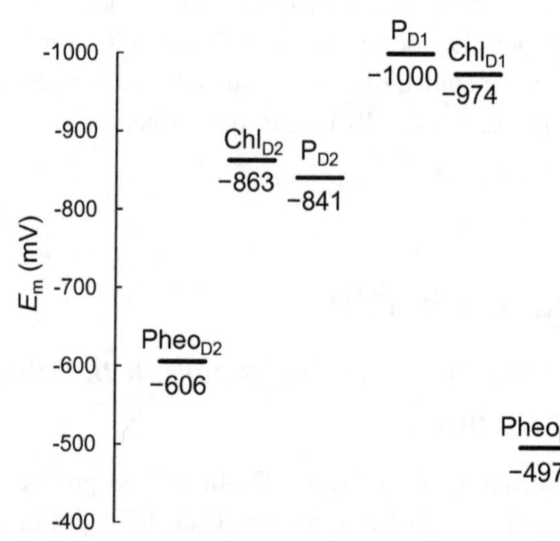

Figure 9. E_m values of cofactors in the Mn-depleted PSII (PDB codes 5MX2 [Zhang *et al.*, 2017]).

2018] (Figure 9). Changes in the protonation states of the ligand residues upon depletion of the Mn_4CaO_5 cluster could partially compensate for changes in the E_m values [Kawashima and Ishikita, 2018].

$E_m(Chl_{D1}) < E_m(Chl_{D2})$ in PSII is in sharp contrast to $E_m(B_L) > E_m(B_M)$ in the photosynthetic reaction centers from purple bacteria (PbRC; where B_L and B_M are accessory bacteriochlorophyll in active L and inactive M branches, respectively) [Kawashima and Ishikita, 2018]. The energy profile in PSII implies that electron transfer along the D1 branch is possible only when Chl_{D1} has the lowest excitation energy. The energy profile in PSII suggests that in the excited state, the electron at the lowest unoccupied molecular orbital (LUMO) chlorophyll in D1 (*e.g.*, Chl_{D1}) transfers to $Pheo_{D1}$ [Kawashima and Ishikita, 2018]. If the hole at the highest occupied molecular orbital (HOMO) of Chl_{D1} transfers to P_{D1}, it would result in $P_{D1}^{\bullet+}Pheo_{D1}^{\bullet-}$ [Kawashima and Ishikita, 2018] (Figure 8). Thus, $E_m(Chl_{D1}) < E_m(Chl_{D2})$ is not inconsistent with the view that Chl_{D1} is the primary electron donor in PSII [Diner and Rappaport, 2002].

2.2. *Residues that are crucial to electron transfer are also crucial to proton transfer*

The D1/D2 pairs that decrease $E_m(Chl_{D1})$ with respect to $E_m(Chl_{D2})$ are D1-Asp61/D2-His61 (98 mV), D1-Asn181/D2-Arg180 (73 mV), Cl-1/– (53 mV), and [D1-Tyr161…D1-His190]/[D2-Tyr160…D2-His189] (39 mV) [Kawashima and Ishikita, 2018]. These groups, which play a crucial role in electron transfer, have been suggested to also play a key role in proton transfer.

(i) D1-Asp61 forms an H-bond with the water ligand, W1 [Singh *et al.*, 2008; Umena *et al.*, 2011], and mediates release of a proton from W1 [Ferreira *et al.*, 2004; Iwata and Barber, 2004; Ishikita *et al.*, 2006; Kawashima *et al.*, 2018]. D1-Asn181 is a chloride (Cl-1) binding site. Cl^- is required to progress through the S_2-to-S_3 transition [Boussac *et al.*, 1992; van Vliet and Rutherford, 1996] in terms of proton transfer [Saito *et al.*, 2013]. All D1-Asp61, D1-Asn181, and Cl-1 seem to be located along the same proton transfer pathway [Saito *et al.*, 2013].

(ii) The corresponding residues of D1-Asp61/D1-Asn181…Cl-1 in D1 are D2-His61 and D2-Arg180 in D2. When TyrD is oxidized to TyrD-O$^\bullet$, the released proton is transferred along the proton transfer pathway, which proceeds from D2-Arg180 toward D2-His61 (Figure 10) [Saito *et al.*, 2013; Nakamura and Noguchi, 2015; Saito *et al.*, 2016]. Indeed, the EPR signal from TyrD is low in the D2-Arg180 mutant PSII [Manna *et al.*, 1998].

Notably, Kawashima *et al.* pointed out that the proton transfer pathway from TyrD in PSII corresponds to the cytochrome binding interface in PbRC [Kawashima *et al.*, 2018]. From the observation of the PbRC cytochrome c_2 co-crystal structure, Axelrod *et al.* concluded that Glu-M95 and Asp-M184 provide the largest electrostatic interaction between the protein subunit M and cytochrome c_2 [Axelrod *et al.*, 2002]. D2-His61 in loop *a–b* and D2-Arg180

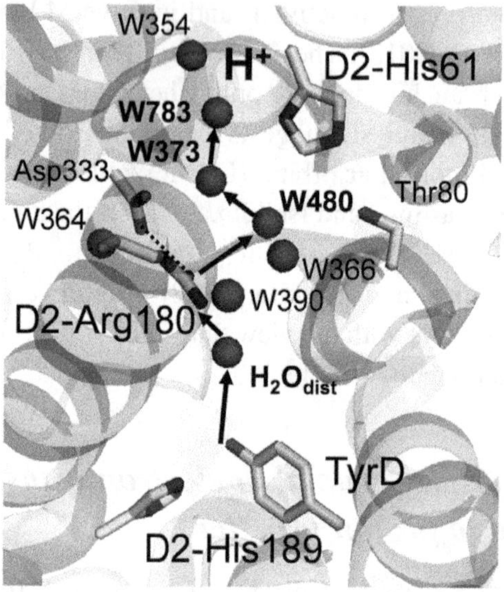

Figure 10. Proton transfer pathway that proceeds from TyrD toward the protein bulk surface. D2-Arg180 and D2-His61 in the proton-transfer pathway from TyrD also contribute to increases in E_m of the redox-active cofactors in the D2 branch.

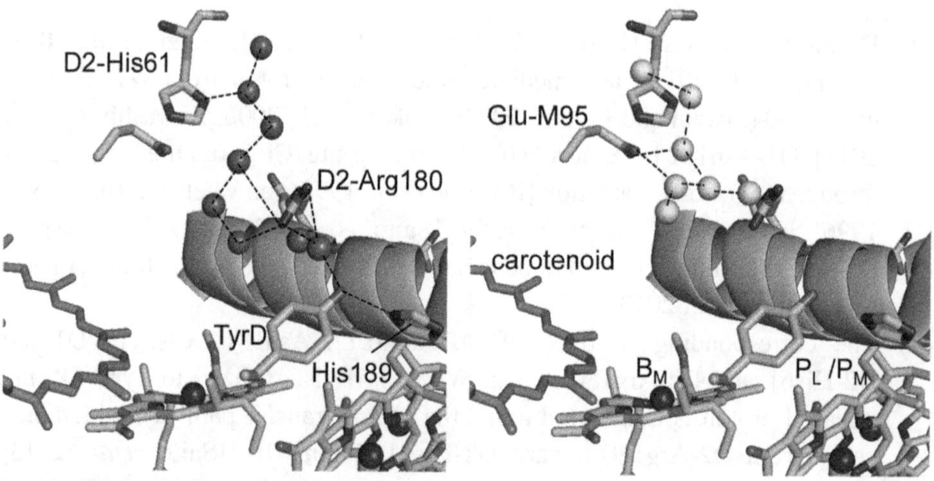

Figure 11. Arrangements of the water molecules in the PSII crystal structure (left) and PbRC (right). In both panels, the protein environments of PSII and PbRC are shown. Dotted lines indicate H-bonds.

in helix *cd* in PSII are replaced with Glu-M95 in loop *a–b* and Asp-M184 in helix *cd* in PbRC from *Rb. sphaeroides*, respectively. Conservation of even arrangement of the water molecules (Figure 11) in the proton transfer pathway from TyrD (PDB codes 3ARC [Umena *et al.*, 2011]) and the cytochrome binding interface (PDB codes 3I4D) implies that they have the same origin [Kawashima *et al.*, 2018].

(iii) D1-Asn181/D2-Arg180 (49 mV) and D1-Asp61/D2-His61 (40 mV) are the most crucial residue pairs that differentiate between $E_m(P_{D1})$ and $E_m(P_{D2})$ for one-electron oxidation, leading to the larger $P_{D1}^{\bullet+}$ population than $P_{D2}^{\bullet+}$ with the ratio of $P_{D1}^{\bullet+}$:$P_{D2}^{\bullet+}$ = 80:20 [Saito *et al.*, 2011, 2012]. The charge recombination rate between Q_A and P680$^{\bullet+}$ was increased in the D2-Arg180 mutant PSII [Manna *et al.*, 1998], which suggests that D2-Arg180 significantly contributes to the chlorophyll cationic states.

Acknowledgments

I would like to acknowledge JST CREST (JPMJCR1656), JSPS KAKENHI (JP26105012, JP18H05155, and JP18H01937), Japan Agency for Medical Research and Development (AMED; JP18AK0101084H0002), Materials Integration for engineering polymers of Cross-ministerial Strategic Innovation Promotion Program (SIP), and Interdisciplinary Computational Science Program in CCS, University of Tsukuba.

References

Adamczyk, A. J., Cao, J., Kamerlin, S. C. and Warshel, A. (2011). Catalysis by dihydrofolate reductase and other enzymes arises from electrostatic preorganization, not conformational motions, *Proc. Natl. Acad. Sci. USA*, 108, 14115–14120.

Axelrod, H. L., Abresch, E. C., Okamura, M. Y., Yeh, A. P., Rees, D. C. and Feher, G. (2002). X-ray structure determination of the cytochrome c_2: reaction center electron transfer complex from *Rhodobacter sphaeroides*, *J. Mol. Biol.*, 319, 501–515.

Berthomieu, C. and Hienerwadel, R. (2001). Iron coordination in photosystem II: Interaction between bicarbonate and the Q_B pocket studied by Fourier transform infrared spectroscopy, *Biochemistry*, 40, 4044–4052.

Boussac, A., Setif, P. and Rutherford, A. W. (1992). Inhibition of tyrosine Z photooxidation after formation of the S_3-state in Ca^{2+}-depleted and Cl^--depleted photosystem II, *Biochemistry*, 31, 1224–1234.

Dau, H. and Haumann, M. (2008). The manganese complex of photosystem II in its reaction cycle–Basic framework and possible realization at the atomic level, *Coord. Chem. Rev.*, 252, 273–295.

Dekker, J. P., Plijter, J. J., Ouwehand, L. and Van Gorkom, H. J. (1984). Kinetics of manganese redox transitions in the oxygen-evolving apparatus of photosynthesis, *Biochim. Biophys. Acta.*, 767, 176–179.

Diner, B. A. and Rappaport, F. (2002). Structure, dynamics, and energetics of the primary photochemistry of photosystem II of oxygenic photosynthesis, *Annu. Rev. Plant. Biol.*, 53, 551–580.

Eigen, M. (1964). Proton transfer, acid-base catalysis, and enzymatic hydrolysis. Part I: Elementary processes, *Angew. Chem. Int. Ed.*, 3, 1–19.

Ferreira, K. N., Iverson, T. M., Maghlaoui, K., Barber, J. and Iwata, S. (2004). Architecture of the photosynthetic oxygen-evolving center, *Science*, 303, 1831–1838.

Haumann, M., Liebisch, P., Muller, C., Barra, M., Grabolle, M. and Dau, H. (2005). Photosynthetic O_2 formation tracked by time-resolved X-ray experiments, *Science*, 310, 1019–1021.

Ikeda, T., Saito, K., Hasegawa, R. and Ishikita, H. (2017). The existence of an isolated hydronium ion in the interior of proteins, *Angew. Chem. Int. Ed.*, 56, 9151–9154.

Ishikita, H. and Knapp, E.-W. (2005). Control of quinone redox potentials in photosystem II: electron transfer and photoprotection, *J. Am. Chem. Soc.*, 127, 14714–14720.

Ishikita, H., Saenger, W., Loll, B., Biesiadka, J. and Knapp, E.-W. (2006). Energetics of a possible proton exit pathway for water oxidation in Photosystem II, *Biochemistry*, 45, 2063–2071.

Ishikita, H. and Saito, K. (2014). Proton transfer reactions and hydrogen-bond networks in protein environments, *J. R. Soc. Interface.*, 11, 20130518.

Iwata, S. and Barber, J. (2004). Structure of photosystem II and molecular architecture of the oxygen-evolving centre, *Curr. Opin. Struct. Biol.*, 14, 447–453.

Kawashima, K. and Ishikita, H. (2018). Energetic insights into two electron transfer pathways in light-driven energy-converting enzymes, *Chem. Sci.*, 9, 4083–4092.

Kawashima, K., Saito, K. and Ishikita, H. (2018). Mechanism of radical formation in the H-Bond network of D1-Asn298 in photosystem II, *Biochemistry*, 57, 4997–5004.

Kawashima, K., Takaoka, T., Kimura, H., Saito, K. and Ishikita, H. (2018). O_2 evolution and recovery of the water-oxidizing enzyme, *Nat. Commun.*, 9, 1247.

Kern, J., Chatterjee, R., Young, I. D., Fuller, F. D., Lassalle, L., Ibrahim, M., Gul, S., Fransson, T., Brewster, A. S., Alonso-Mori, R., Hussein, R., Zhang, M., Douthit, L., de Lichtenberg, C., Cheah, M. H., Shevela, D., Wersig, J., Seuffert, I., Sokaras, D., Pastor, E., Weninger, C., Kroll, T., Sierra, R. G., Aller, P., Butryn, A., Orville, A. M., Liang, M., Batyuk, A., Koglin, J. E., Carbajo, S., Boutet, S., Moriarty, N. W., Holton, J. M., Dobbek, H., Adams, P. D., Bergmann, U., Sauter, N. K., Zouni, A., Messinger, J., Yano, J., and Yachandra, V. K. (2018). Structures of the intermediates of Kok's photosynthetic water oxidation clock, *Nature*, 563, 421-425.

Manna, P., LoBrutto, R., Eijckelhoff, C., Dekker, J. P. and Vermaas, W. (1998). Role of Arg180 of the D2 protein in photosystem II structure and function, *Eur. J. Biochem.*, 251, 142–154.

Nakamura, S. and Noguchi, T. (2015). Infrared detection of a proton released from tyrosine Y_D to the bulk upon its photo-oxidation in photosystem II, *Biochemistry*, 54, 5045–5053.

Pal, R., Negre, C. F., Vogt, L., Pokhrel, R., Ertem, M. Z., Brudvig, G. W. and Batista, V. S. (2013). S-state model of the oxygen-evolving complex of photosystem II, *Biochemistry*, 52, 7703–7706.

Perrin, C. L. and Nielson, J. B. (1997). "Strong" hydrogen bonds in chemistry and biology, *Annu. Rev. Phys. Chem.*, 48, 511–544.

Saito, K., Ishida, T., Sugiura, M., Kawakami, K., Umena, Y., Kamiya, N., Shen, J.-R. and Ishikita, H. (2011). Distribution of the cationic state over the chlorophyll pair of photosystem II reaction center, *J. Am. Chem. Soc.*, 133, 14379–14388.

Saito, K., Rutherford, A. W. and Ishikita, H. (2013). Mechanism of tyrosine D oxidation in Photosystem II, *Proc. Natl. Acad. Sci. USA*, 110, 7690–7695.

Saito, K., Rutherford, A. W. and Ishikita, H. (2013). Mechanism of proton-coupled quinone reduction in Photosystem II, *Proc. Natl. Acad. Sci. US*, 110, 954–959.

Saito, K., Rutherford, A. W. and Ishikita, H. (2015). Energetics of proton release on the first oxidation step in the water-oxidizing enzyme, *Nat. Commun.*, 6, 8488.

Saito, K., Sakashita, N. and Ishikita, H. (2016). Energetics of the proton transfer pathway for tyrosine D in photosystem II, *Aust. J. Chem.*, 69, 991–998.

Saito, K., Shen, J.-R., Ishida, T. and Ishikita, H. (2011). Short hydrogen-bond between redox-active tyrosine Y_Z and D1-His190 in the photosystem II crystal structure, *Biochemistry*, 50, 9836–9844.

Saito, K., Shen, J. R. and Ishikita, H. (2012). Influence of the axial ligand on the cationic properties of the chlorophyll pair in photosystem II from *Thermosynechococcus vulcanus*, *Biophys. J.*, 102, 2634–2640.

Sakashita, N., Watanabe, H. C., Ikeda, T. and Ishikita, H. (2017). Structurally conserved channels in cyanobacterial and plant photosystem II, *Photosynth. Res.*, 133, 75–85.

Sakashita, N., Watanabe, H. C., Ikeda, T., Saito, K. and Ishikita, H. (2017). Origins of water molecules in the photosystem II crystal structure, *Biochemistry*, 56, 3049–3057.

Schutz, C. N. and Warshel, A. (2001). What are the dielectric constants of proteins and how to validate electrostatic models?, *Proteins*, 44, 400–417.

Schutz, C. N. and Warshel, A. (2004). The low barrier hydrogen bond (LBHB) proposal revisited: the case of the Asp... His pair in serine proteases, *Proteins*, 55, 711–723.

Shimizu, T., Sugiura, M., and Noguchi, T. (2018). Mechanism of proton-coupled electron transfer in the S0-to-S1 transition of photosynthetic water oxidation as revealed by time-resolved infrared spectroscopy, *J. Phys. Chem. B*, 122, 9460–9470.

Siegbahn, P. E. (2013). Water oxidation mechanism in photosystem II, including oxidations, proton release pathways, O-O bond formation and O_2 release, *Biochim. Biophys. Acta.*, 1827, 1003–1019.

Singh, S., Debus, R. J., Wydrzynski, T. and Hillier, W. (2008). Investigation of substrate water interactions at the high-affinity Mn site in the photosystem II oxygen-evolving complex, *Phil. Trans. R. Soc. Lond. B.*, 363, 1229–1235.

Stuchebrukhov, A. A. (2009) Mechanisms of proton transfer in proteins: localized charge transfer versus delocalized soliton transfer, *Phys. Rev. E Stat. Nonlin. Soft Matter Phys.*, 79, 031927.

Suga, M., Akita, F., Hirata, K., Ueno, G., Murakami, H., Nakajima, Y., Shimizu, T., Yamashita, K., Yamamoto, M., Ago, H. and Shen, J. R. (2015). Native structure of photosystem II at 1.95 Å resolution viewed by femtosecond X-ray pulses, *Nature*, 517, 99–103.

Suga, M., Akita, F., Sugahara, M., Kubo, M., Nakajima, Y., Nakane, T., Yamashita, K., Umena, Y., Nakabayashi, M., Yamane, T., Nakano, T., Suzuki, M., Masuda, T., Inoue, S.,

Kimura, T., Nomura, T., Yonekura, S., Yu, L. J., Sakamoto, T., Motomura, T., Chen, J. H., Kato, Y., Noguchi, T., Tono, K., Joti, Y., Kameshima, T., Hatsui, T., Nango, E., Tanaka, R., Naitow, H., Matsuura, Y., Yamashita, A., Yamamoto, M., Nureki, O., Yabashi, M., Ishikawa, T., Iwata, S. and Shen, J. R. (2017). Light-induced structural changes and the site of O=O bond formation in PSII caught by XFEL, *Nature*, 543, 131–135.

Takaoka, T., Sakashita, N., Saito, K. and Ishikita, H. (2016) pK_a of a proton-conducting water chain in photosystem II, *J. Phys. Chem. Lett.*, 1925–1932.

Umena, Y., Kawakami, K., Shen, J.-R. and Kamiya, N. (2011) Crystal structure of oxygen-evolving photosystem II at a resolution of 1.9 Å, *Nature*, 473, 55–60.

van Vliet, P. and Rutherford, A. W. (1996). Properties of the chloride-depleted oxygen-evolving complex of photosystem II studied by electron paramagnetic resonance, *Biochemistry*, 35, 1829–1839.

Warshel, A. (1998). Electrostatic origin of the catalytic power of enzymes and the role of preorganized active sites, *J. Biol. Chem.*, 273, 27035–27038.

Wincencjusz, H., van Gorkom, H. J. and Yocum, C. F. (1997). The photosynthetic oxygen evolving complex requires chloride for its redox state $S_2 \rightarrow S_3$ and $S_3 \rightarrow S_0$ transitions but not for $S_0 \rightarrow S_1$ or $S_1 \rightarrow S_2$ transitions, *Biochemistry*, 36, 3663–3670.

Zhang, M., Bommer, M., Chatterjee, R., Hussein, R., Yano, J., Dau, H., Kern, J., Dobbek, H. and Zouni, A. (2017). Structural insights into the light-driven auto-assembly process of the water-oxidizing Mn_4CaO_5-cluster in photosystem II, *eLife*, 6, e26933.

Chapter 11

Key Molecular Ru Complexes that have Fostered the Development and Understanding of Water Oxidation Catalysis

Carolina Gimbert Suriñach* and Antoni Llobet*,†

*Institute of Chemical Research of Catalonia (ICIQ), Barcelona Institute of Science and Technology, Av. Països Catalans 16, E-43007 Tarragona, Spain

†Universitat Autònoma de Barcelona, Departament de Química, Cerdanyola del Vallès, E-08193 Barcelona, Spain

In this chapter, we describe the key molecular coordination compounds related to water oxidation catalysis based on Ruthenium. The particularities of a few mononuclear and polynuclear Ruthenium complexes that are of interest from a mechanistic point of view are disclosed. In addition, we also discuss the properties of the most significant catalytic intermediates and transition states through which these catalytic reactions occur. Finally, some examples of how these complexes can achieve water oxidation catalysis induced by sunlight and how they can be incorporated in water splitting devices is also briefly presented.

1. Introduction

Energy is directly related to the degree of wellness of humans in planet earth. As such the prosperity of humans in the planet is linked to the access of cheap and abundant energy resources. In this way the importance of fossil fuels has been largely projected in our society dictating the world economics and politics [Kaldor, 2007]. The high rate of fossil fuel utilization, the realization that these fossil fuels are not limitless, as well as the consequences of global warming due to the increase of CO_2 concentration in the atmosphere, clearly indicate the need for a change of energy paradigm. Sunlight is the best energy to be used since it is limitless and free [Balmaseda *et al.*, 2013; Stocker *et al.*, 2013; Edenhofer *et al.*, 2014; Steffen *et al.*, 2018].

Nature has been using sunlight as a means of energy for a more than a billion years through photosynthesis in green plants and algae. For this purpose, Nature uses light induced chemical reactions to store solar energy into chemical bonds. In particular the oxidation of H_2O to molecular dioxygen and the reduction of carbon dioxide to carbohydrates as shown in the equation below [Wydrzynski and Satoh, 2005].

$$6H_2O + 6CO_2 \xrightarrow{h\upsilon} C_6H_{12}O_6 + 6O_2 \tag{1}$$

This reaction can be regarded as two redox half-reactions from a chemical standpoint involving the reduction of carbon dioxide to carbohydrates and the oxidation of H_2O to dioxygen. Both reactions involve proton-coupled electron transfer (PCET) processes as indicated in Table 1. The oxidation of water to dioxygen in the oxygen evolving complex of photosystem II (OEC-PSII), takes place *via* four protons and four electrons which is actually the lowest energy route from a thermodynamic perspective,

$$2H_2O \rightarrow O_2 + 4H^+ + 4e^- \tag{2}$$

The comprehension of this reaction at a molecular level has been the focus of many research endeavors recently. A significant input in this field come from the resolution of the crystal X-ray structures of the OEC-PSII at 1.9 Å resolution by

Table 1. Selected redox potentials for the oxidation of water and the reduction of CO_2 and H^+.

Water oxidation	**E°, V (vs. NHE), pH = 7.0**		**CO_2 Reduction**
$\cdot OH + 1H^+ + 1e^- \rightarrow H_2O$	2.39	−1.9	$CO_2 + 1e^- \rightarrow CO_2$
$HO\text{-}OH + 2H^+ + 2e^- \rightarrow 2H_2O$	1.37	−0.61	$CO_2 + 2H^+ + 2e^- \rightarrow HCO_2H$
$HO\text{-}O\cdot + 3H^+ + 3e^- \rightarrow 2H_2O$	1.26	−0.53	$CO_2 + 2H^+ + 2e^- \rightarrow CO + H_2O$
$O{=}O + 4H^+ + 4e^- \rightarrow 2H_2O$	0.82	−0.49	$2CO_2 + 2H^+ + 2e^- \rightarrow H_2C_2O_4 + H_2O$
		−0.48	$CO_2 + 4H^+ + 4e^- \rightarrow HCHO + H_2O$
H^+ Reduction		−0.38	$CO_2 + 6H^+ + 6e^- \rightarrow CH_3OH + H_2O$
$2H^+ + 2e^- \rightarrow H_2$	−0.41	−0.24	$CO_2 + 8H^+ + 8e^- \rightarrow CH_4 + 2H_2O$

the Barber group. In recent times further refinement with higher quality crystals and by using a number of spectroscopic techniques and theoretical methodologies have generated a wealth of exceptional information on the subject [Ferreira *et al.*, 2004; Haumann *et al.*, 2005; Yano *et al.*, 2006; Sproviero *et al.*, 2008; Umena *et al.*, 2011; Suga *et al.*, 2015; Barber, J., 2016]. Thanks to this there is a good understanding of the reaction mechanisms involved although there are still some caveats that will need to be sorted out in the near future.

Low molecular weight metal complexes as models for the OEC-PSII can become crucial in order to check out the potential models proposed. They can also show the critical parameters needed to fully understand the system and thus learn how to manipulate it.

A model reaction that shares some of the commonalities with the OEC-PSII is water splitting induced by sunlight,

$$2H_2O \xrightarrow{h\upsilon} 2H_2 + O_2 \tag{3}$$

As in photosynthesis this reaction can be divided into two half reactions. The oxidation of water to dioxygen, which is exactly the same as it happens in the OEC-PSII. In water splitting the other half reaction is the reduction of protons to dihydrogen (see Equation 4), instead of the reduction of CO_2 to carbohydrates. The former is obviously much simpler than the latter.

$$2H^+ + 2e^- \rightarrow H_2 \tag{4}$$

The development of good catalysts for water oxidation to dioxygen (WOC) as well as proton reduction to hydrogen (PRC) shown in Equations 2 and 4 is a requirement to increase reaction rates and is one of the keys to build technologically useful devices for new energy conversion schemes. An important subject in this topic is the complexity for a catalyst to manage multiple protons and electrons at the same time. The latter is fundamental in order to design technologically useful devices based on water splitting with sunlight.

At present there is a substantial number of complexes based on transition metals that have been described to be active for the water oxidation to oxygen reaction [Garrido-Barros *et al.*, 2014], mainly based on Ru [Sala *et al.*, 2009; Romain *et al.*, 2009] but also containing other first and third row metals such as Ir [Hull *et al.*, 2009; McDaniel *et al.*, 2008; Lalrempuya *et al.*, 2010], Co [Yin *et al.*, 2010; Wasylenko *et al.*, 2011; Rigsby *et al.*, 2012; Gimbert-Suriñach *et al.*, 2016], Fe [Ellis *et al.*, 2010], Cu [Barnett *et al.*, 2012; Gerlach *et al.*, 2014; Zhang *et al.*, 2014; Fisher *et al.*, 2017; Garrido-Barros *et al.*, 2015] or Mn [Chen *et al.*, 2004]. The main purpose of the present account is not to make a thorough description of all reported catalysts but to show the examples that are of fundamental importance for the understanding of the water oxidation reaction. Those that allow to understand the potential reaction intermediates and transition states potentially involved in the catalytic cycles.

We will always use formal oxidation states for simplicity reasons since that will give us an easy way of book-keeping the number of electron removed or gained. In addition, the auxiliary ligands that do not undergo any change in protonation or oxidation state will not be written. All redox potentials reported are referred vs. NHE.

2. Mononuclear Mono-Aqua Ru Complexes

Today there are dozens of mononuclear Ru(II)-aqua type of complexes that have been thoroughly characterized because of their capacity to act as catalysts for organic oxidation reactions [Meyer and Huynh, 2003]. Ruthenium complexes at oxidation sate II are generally low spin d^6 ions with a $t_{2g}^{6}e_{g}^{0}$ configuration and possess octahedral type of coordination. An example of this type of complexes is the recently reported *out*-$[Ru^{II}(Hppz)(trpy)(H_2O)]^{2+}$, $\mathbf{1}^{2+}$ [Mognon *et al.*, 2015], where Hppz is the monoprotonated didentate ligand ethyl 3-(pyridin-2-yl)-1H-pyrazole-5-carboxylate and trpy is 2,2′;6′:2″-terpyridine. A drawing of this ligand and all the ligands discussed in this chapter is provided in Scheme 1. Also a drawing of the molecular structure of complex $\mathbf{1}^{2+}$, emphasizing the atoms in the first coordination sphere is exhibited in Figure 1.

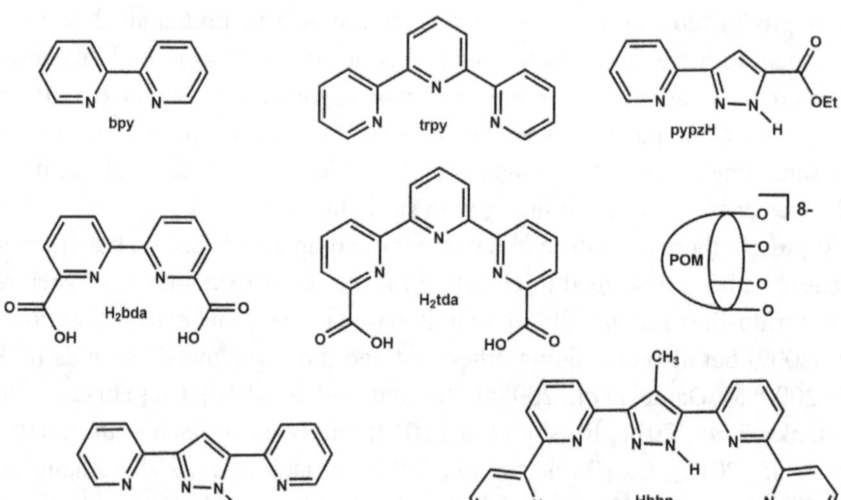

Scheme 1

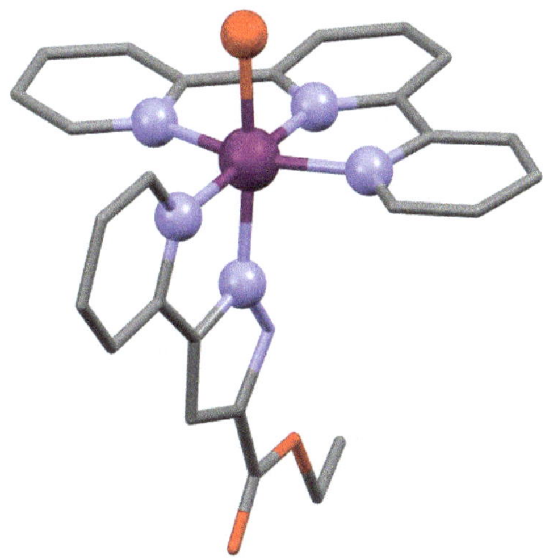

Figure 1. Drawing of the molecular structure of complex $\mathbf{1}^{2+}$, $[Ru^{II}(Hppz)(trpy)(H_2O)]^{2+}$. Color codes: Ru, cyan; O, red; N, blue; C, gray. H atoms are omitted for clarity.

2.1. *Thermodynamic considerations*

The interest in Ru-OH_2 complexes is mainly due to their capacity to undergo PCET processes at relatively low redox potentials. That is the access to reactive high oxidation states by sequential loss of protons and electrons taking place at relatively low energies [Llobet, 1994; Huynh and Meyer, 2007]. For the particular case of $\mathbf{1}^{2+}$ this is outlined in Equations 5 and 6 at pH 7 [Mognon *et al.*, 2015],

$$Ru^{III}\text{-}OH + 1H^+ + 1e^- \rightarrow Ru^{II}\text{-}OH_2; \quad E_{1/2}(Ru^{III/II} \text{ at pH} = 7) = 0.52\text{ V} \tag{5}$$

$$Ru^{IV}{=}O + 1H^+ + e^- \rightarrow Ru^{III}\text{-}OH; \quad E_{1/2}(Ru^{IV/III} \text{ at pH} = 7) = 0.88\text{ V} \tag{6}$$

Further one more redox process is observed in this case that does not involve protons and thus is an electron transfer only process,

$$Ru^{V}{=}O + 1e^- \rightarrow Ru^{IV}{=}O; \quad E_{1/2}(Ru^{V/IV} \text{ at pH} = 7) = 1.59\text{ V} \tag{7}$$

The complete thermodynamic characterization as a function of pH is presented as a Pourbaix diagram that is shown in Figure 2. This diagram provides a precise information about the different redox potentials of complex $\mathbf{1}^{2+}$ as a function of pH and the zone of predominance of the different species depending of their degree of protonation and oxidation states is also shown. The slope brakes of the redox potentials coincide with pK_a's that are indicated as vertical lines. For

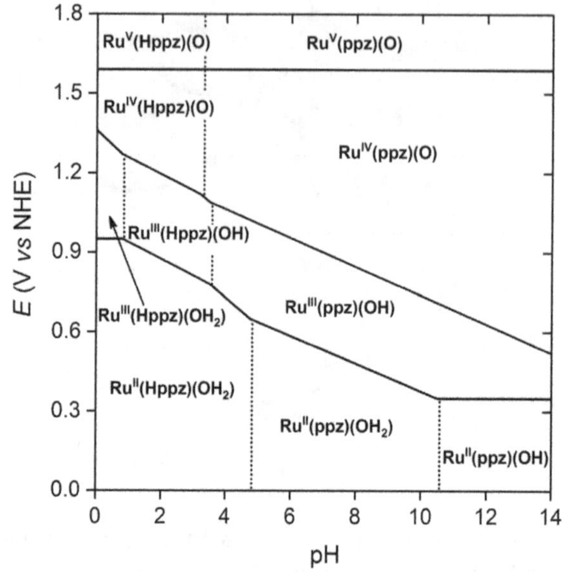

Figure 2. Pourbaix diagram for 1^{2+}, $[Ru^{II}(Hppz)(trpy)(H_2O)]^{2+}$. The zones of stability of the different species generated by proton and electron loss or gain are indicated.

instance, the $Ru\text{-}OH_2$ species at oxidation state II and III have pK_a of 10.5 and 0.8 respectively as shown in the equation below,

$$Ru^{II}\text{-}OH_2 \rightarrow Ru^{II}\text{-}OH + H^+; \quad pK_a,II = 10.5 \tag{8}$$

$$Ru^{III}\text{-}OH_2 \rightarrow Ru^{III}\text{-}OH + H^+; \quad pK_a,III = 0.8 \tag{9}$$

Further the pK_a's due to the deprotonation of the Hppz ligand can also be identified and appear at 4.75, 3.5 and 3.2 for oxidation states II, III and IV respectively.

It is also worth noticing that at a sufficiently high pH the redox potentials for the IV/III and III/II couples could end up merging and thus a two electron process would take place,

$$Ru^{IV} = O + 2e^- + 2H^+ \rightarrow Ru^{II}\text{-}OH_2 \tag{10}$$

Although this is not occurring for complex 1^{2+}, there are other examples where this can be clearly observed [Llobet, 1994]. On the other hand, at a sufficiently low pH the potentials for the V/IV and the IV/III redox couples could also fuse generating a two-electron wave as indicated in the following equation,

$$Ru^{V} = O + 2e^- + 2H^+ \rightarrow Ru^{III}\text{-}OH_2 \tag{11}$$

The high oxidation state species Ru(IV)=O and Ru(V)=O are energetically stabilized thanks to the σ and π donating character of the oxido ligands that supply electron density to the Ru center. In addition the access to PCET pathways allow to reach high oxidation state without increasing the overall charge of the metal complex and thus is energetically viable [Huynh and Meyer, 2007].

2.2 *Access to high oxidation states and O–O bond formation*

From the water oxidation catalysis perspective, the access to high oxidation states derived from the Ru-aqua complexes at low energy cost is important because it can promote a series of reactions that end up making oxygen–oxygen bonds. One of the main pathways that has been reported for the O–O bond formation is the so-called water nucleophilic attack of a solvent water molecule (WNA) to an active high oxidation state Ru=O [Concepcion *et al.*, 2008; Romain *et al.*, 2009] species as indicated below,

$$Ru^{V}{=}O + H_2O \rightarrow Ru^{III}\text{-}OOH + H^+ \tag{12}$$

The interaction of water with the Ru(V)=O in this case forms an O-O bond generating the corresponding hydroperoxido species, Ru(III)-OOH, that can be further oxidized to Ru(IV) and end up evolving dioxygen as indicated in Equations 13–14,

$$Ru^{III}\text{-}OOH - 1e^- \rightarrow Ru^{IV}\text{-}OOH \tag{13}$$

$$Ru^{IV}\text{-}OOH \rightarrow Ru^{II} + O\text{-}O + 1H^+ \tag{14}$$

Access to Ru^V species can also be potentially achieved by disproportion of Ru^{IV} to give Ru^{III} and Ru^V. This is of interest because of the high reactivity of Ru^V towards O–O bond formation as just indicated above. An example of a water oxidation catalyst that reach high oxidation states *via* disproportionation is the mononuclear complex *out*-$[Ru^{II}(Hbpp)(trpy)(H_2O)]^{2+}$, **2**$^{2+}$ (see Scheme 1 for Hbpp ligand structure) [Roeser *et al.*, 2011].

Another mechanistic pathway that leads to O–O bond formation is the so called I2M (interaction between two metal-oxo groups), that can happen in an intra or inter-molecular fashion [Romain *et al.*, 2009].

A number of other potential reaction pathways complementary to the ones just described here have also been reported mainly based on kinetic considerations [Wasylenko *et al.*, 2010; Concepcion *et al.*, 2010]. Other pathways involving radical CeO· have been proposed when CAN (Cerium ammonium nitrate) is used as a chemical oxidant, that is the O–O bond formation is proposed to occur between CeO· and a Ru(V)=O species [Yoshida *et al.*, 2010]. Further in one instance the detection of trace amount of NO_2 is also found indicating the degree of potential complexity of the catalytic water oxidation reaction [Wasylenko *et al.*, 2010]. In addition, several complexes have also been reported where the auxiliary ligand can be redox active and participate in the overall water oxidation in combination with the Ru-aqua group [Muckerman *et al.*, 2008].

Finally, Iridium complexes containing an aqua group have also been reported as analogues of the Ru counterparts and are proposed to release dioxygen *via* a WNA pathway [Blakemore *et al.*, 2010].

3. Mononuclear di-Aqua Ruthenium Complexes

Ruthenium complexes that have two aqua groups are of interest because they can potentially lose up to $4H^+$ and $4e^-$. Nevertheless, there are only a few complexes containing two aqua groups implying that the metal center has to cycle within 5 different oxidation states and thus has to display certain stability at each of these states. One of the complexes that comply these requirements is *cis*-$[Ru^{II}(bpy)_2(H_2O)_2]^{2+}$, $\mathbf{3}^{2+}$ (bpy is 2,2′-bipyridine) that can lose $4H^+$ and $4e^-$ consecutively and generate $Ru^{VI}(O)_2$, as can be observed in the Pourbaix diagram exhibited in Figure 3.

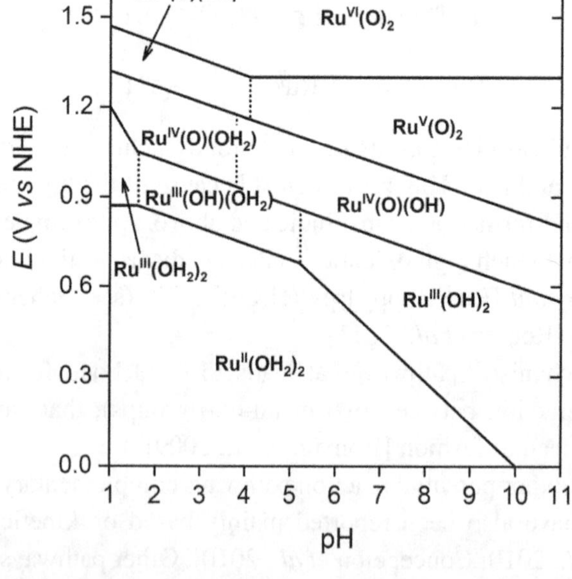

Figure 3. Pourbaix diagram for *cis*-$[RuII(bpy)_2(H_2O)_2]^{2+}$, 3^{2+}.

$$Ru^{II}(H_2O)_2 - 4\,H^+ - 4e^- \rightarrow Ru^{VI}(O)_2 \tag{15}$$

The diaqua complex $\mathbf{3}^{2+}$ shows a Pourbaix diagram, which is much more complicated than its mono-aqua counterpart due to the increased number of oxidation states available and also because there are more protonation degrees as a function of pH. Complex $\mathbf{3}^{2+}$ is certainly of interest from a mechanistic point of view but it is not a spectacular catalyst. It can only make a few catalytic cycles before it isomerizes to its trans isomer, *trans*-$[Ru^{II}(bpy)_2(H_2O)_2]^{2+}$. The latter does not have the sufficiently high redox potentials to carry out the water oxidation reaction. For $\mathbf{3}^{2+}$, the active species that fosters the formation of molecular dioxygen is the *cis*-Ru-dioxo, $Ru^{VI}(O)_2$. For this complex an elegant labelling experiment with O^{18} has been performed showing that once the $Ru^{VI}(O)_2$ species is formed it follows a WNA mechanism that ends up making the corresponding Ru-hydroperoxido intermediate that finally releases oxygen [Sala *et al.*, 2010] as pointed out in Equations 16 and 17,

$$[Ru^{VI}(O)_2] + H_2O \rightarrow [Ru^{IV}(OH)(OOH)] \tag{16}$$

$$[Ru^{IV}(OH)(OOH)] + H_2O \rightarrow [Ru^{II}(H_2O)_2] + O\text{–}O \tag{17}$$

4. Seven Coordinated Complexes

The Ruthenium complexes described up to this point have a pseudo octahedral coordination geometry. However, there are a few examples where the active species increase their coordination number to seven. The most significant ones are the Ru complexes based on the H_2bda (bda^{2-} is [2,2′-bipyridine]-6,6′-dicarboxylato; See Scheme 1) and H_2tda (tda^{2-} is [2,2′:6′,2″-terpyridine]-6,6″-dicarboxylato) ligands, $[Ru^{IV}(O)(bda)(pic)_2]^{2+}$, $\mathbf{4}^{2+}$ [Duan *et al.*, 2009] and $[Ru^{IV}(O)(tda)(py)_2]^{2+}$, $\mathbf{5}^{2+}$ [Matheu *et al.*, 2015], respectively where pic is 4-methyl picolinate and py is pyridine.

There is now mounting evidence of the seven coordination reaction intermediates derived for both $\mathbf{4}^{2+}$ and $\mathbf{5}^{2+}$. In the case of $\mathbf{4}^{2+}$ it includes a X-ray crystal structure that contain a proton that bridges two units of Ru^{IV} and two hydrogen bonded water molecules, $\{[Ru^{IV}(OH)(bda)(pic)_2{\cdot}(H_2O)]_2(\mu\text{-}H)\}^{3+}$. For $\mathbf{5}^{2+}$, evidence include a diamagnetic behavior due to the pentagonal bipyramid geometry that leads to a spin pared high field d^4, $(\boldsymbol{e}1'')^4(\boldsymbol{e}2')^0(\boldsymbol{a}1')^0$, electronic configuration for the Ru metal center.

From a mechanistic perspective $\mathbf{4}^{2+}$ follows a I2M mechanism [Duan *et al.*, 2012]. That is when the $Ru^V(O)$ species are generated they dimerize to generate the corresponding peroxido intermediate as indicated in the equation below,

$$2\,Ru^V\text{-}O \rightarrow Ru^{IV}\text{-}OO\text{-}Ru^{IV} \tag{18}$$

The peroxido group then suffers an intramolecular electron transfer leading to Ru^{III} and releasing dioxygen as shown in Equation 19.

$$Ru^{IV}\text{-OO-}Ru^{IV} \rightarrow 2\ Ru^{III} + O_2 \tag{19}$$

An interesting complex that further supports this mechanism is $[Ru^{IV}(O)(bda)(MeO\text{-}isoq)_2]^{2+}$, **6**$^{2+}$ [Richmond *et al.*, 2014], where MeO-isoq is 6-methoxyisoquinoline, see Figure 4 (left). The presence of the pi-extended system in the axial ligand MeO-isoq, provides a platform for significant π-π stacking interaction that decreases the energy of activation to bring together two M-O units and as a consequence of this interaction, the turnover frequency increases by two orders of magnitude under exactly the same conditions.

Finally, complex **5**$^{2+}$, that is nowadays the fastest water oxidation catalyst reported, follows a WNA mechanism thanks to the dangling carboxylate that reduces the energy of activation at the rate determining step *via* an intramolecular proton transfer (see Figure 4, right).

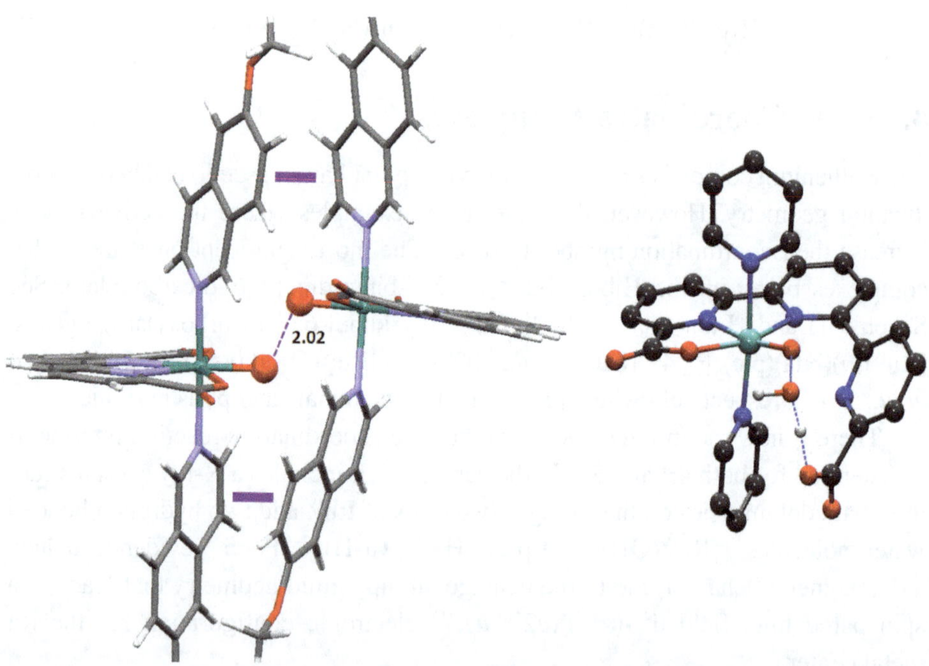

Figure 4. (Left) Transition state associated with the O–O bond formation step for catalyst $[Ru^{IV}(bda)(MeO\text{-}isoq)_2(O)]^{2+}$, **6**$^{2+}$ at the $Ru^{V}=O$ species (I2M). Right) Transition state associated with the nucleophilic attach of a water molecule to the $Ru^{V}=O$ group of catalyst $[Ru^{IV}(tda)(py)_2(O)]^{2+}$, **5**$^{2+}$ (WNA). Color code: Ru, turquoise; C, gray; N, blue; O, red; H, white. Hydrogen atoms on the ligands are omitted for clarity.

5. Multiple Ru-Aqua Complexes

Several multinuclear complexes containing various Ru-OH_2 groups have been reported to be capable of catalytically oxidize H_2O to dioxygen [Gilbert *et al.*, 1985; Concepcion *et al.*, 2008; Romain *et al.*, 2009; Bozoglian *et al.*, 2009; Mola *et al.*, 2011; Sartorel *et al.*, 2008; Geletii *et al.*, 2008; Sens *et al.*, 2004]. These complexes contain two or more Ru-OH_2 moieties that are linked together by means of a bridging ligand (BL). The BL has two very important roles: (a) it is responsible for the extent of electronic coupling between the Ruthenium centers and (b) it controls the relative geometry of the Ru-aqua groups as well as their relative situation in space.

Compared to the mononuclear *cis*-$Ru^{II}(H_2O)_2$ complex where two aqua groups are situated at the same metal center, the dinuclear complexes containing two Ru-OH_2 groups with a bridging ligand can distribute the burden of several PCET steps between the two metal centers, as long as there is a sufficient electronic coupling through the BL. An example of this type of complex is $\{[Ru^{II}(pyC)_2(H_2O)]_2(\mu\text{-bbp})\}^{3+}$, $\mathbf{7}^{3+}$ (pyC is 4-carboxypyridine, see Scheme 1 for bbp structure) [Neudeck *et al.*, 2014; Berardi *et al.*, 2015]. As can be observed in the Pourbaix diagram in Figure 5, at pH 7 this complex undergoes three PCET steps followed by a single electron transfer step to reach the reactive species [O=Ru^{IV}(μ-BL)Ru^{V}= O],

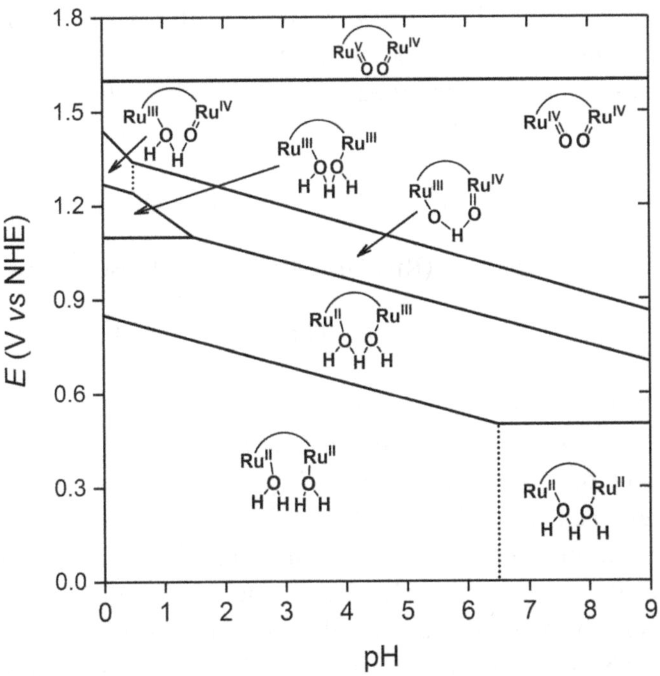

Figure 5. Pourbaix diagram for $\{[Ru^{II}(pyC)_2(H_2O)]_2(\mu\text{-bbp})\}^{3+}$, $\mathbf{7}^{3+}$.

where $4H^+$ and $5e^-$ have been removed from the original complex. This means that one of the centers will cycle through 3 oxidation sates whereas the other will cycle through 4 oxidation states, as indicated in the equation below,

$$[H_2O\text{-}Ru^{II}(\mu\text{-BL})Ru^{II}\text{-}OH_2] - 4H^+ - 5e^- \rightarrow [O{=}Ru^{IV}(\mu\text{-BL})Ru^{V}{=}O] \quad (20)$$

The mechanism found for this complex is WNA. That is once the reactive $[O{=}Ru^{IV}(\mu\text{-BL})Ru^{V}{=}O]$ species is generated it will undergo a water nucleophilic attack by a water solvent molecule at the $Ru^V{=}O$ moiety as shown in Equation 21,

$$[O{=}Ru^{IV}(\mu\text{-BL})Ru^{V}{=}O] + H_2O \rightarrow [HO\text{-}Ru^{III}(\mu\text{-BL})Ru^{IV}(OOH)] \quad (21)$$

That will generate a terminal hydroperoxo that will end up releasing dioxygen and generating the $[H_2O\text{-}Ru^{II}(\mu\text{-BL})Ru^{III}\text{-}OH]$ species.

The WNA mechanism has been also proposed for the oxido-bridged dinuclear complex known as the blue dimer [Gilbert *et al.*, 1985], $\{[Ru^{III}(bpy)_2]_2(\mu\text{-}O)\}^{4+}$, $\mathbf{8}^{4+}$. Ru polyoxometallates have also been shown to be capable of catalyzing the water oxidation reaction. An example is the complex $[Ru^{IV}{}_4(\mu\text{-}O)_4(\mu\text{-}OH)_2(H_2O)_4(\gamma\text{-}SiW_{10}O_{36})_2]^{10-}$, $\mathbf{9}^{10-}$ [Sartorel *et al.*, 2008], (POM in Scheme 1 is the polyoxometalate ligand $\gamma\text{-}SiW_{10}O_{36}{}^{8-}$), that contains 4 $Ru\text{-}OH_2$ groups connected *via* several oxido and hydroxido type of ligands. Thanks to the presence of four Ru-aqua groups, here each metal only needs to access two different oxidation states, to be capable of releasing dioxygen as described in the Equations 22–24. In this case a WNA mechanism is also proposed based on electrochemistry, kinetics and complemented with DFT calculations.

$$[(Ru^{IV}\text{-}OH_2)_4] - 4H^+ - 4e^- \rightarrow [(Ru^{V}\text{-}OH)_4] \quad (22)$$

$$[(Ru^{V}\text{-}OH)_4] + H_2O \rightarrow [(Ru^{V}\text{-}OH)_2(Ru^{IV}\text{-}OH_2)(Ru^{IV}\text{-}OOH)] + H^+ \quad (23)$$

$$[(Ru^{V}\text{-}OH)_2(Ru^{IV}\text{-}OH_2)(Ru^{IV}\text{-}OOH)] + H_2O + H^+ \rightarrow [(Ru^{IV}\text{-}OH_2)_4] + O\text{–}O \quad (24)$$

Several dinuclear Ru-aqua complexes have been described recently in the literature that most likely operate through multiple $Ru\text{-}OH_2$ groups as described above. However the degree of understanding in each case varies substantially [Zong and Thummel, 2005]. The possibility of using many different type of bridging ligands could bring a large family of Ru WOCs that could set up the mechanistic base for the complex water oxidation catalysis reaction. However, from a synthetic perspective the preparation and purification of these complexes is certainly not straightforward [Catalano and Craig, 2003]. Given the large number of bridging ligands available the potential for complex design is basically limitless.

6. Water Oxidation Induced by Light and Devices

One of the main goals in the field is to use molecular catalysts for water oxidation into water splitting devices that work with just sunlight. For this reason, it is important to carry out water oxidation catalysis induced by light. Since most of the water oxidation catalysts do not efficiently absorb light at the desired wavelength one of the most used strategies consists of using complexes or materials that efficiently capture light (photosensitizers, PS) and couple them to the water oxidation catalyst. One of the most used molecular PS in combination with WOCs in homogeneous phase is $[Ru^{II}(bpy)_3]^{2+}$ and its derivatives. The $[Ru^{II}(bpy)_3]^{2+}$ molecule absorbs light at the visible range and has a long lasting excited state in the microseconds timescale that is long enough to interact with other molecules and thus carry out chemical reactions. In addition, the redox potential for the III/II couple for this family of molecules is in the range of 1.3–17 V, which is ideal for the water oxidation reaction. Thus a typical arrangement involves $[Ru^{II}(bpy)_3]^{2+}$ as PS, a molecular WOC and a sacrificial electron acceptor (SEA) that interacts with the $[Ru^{II}(bpy)_3]^{2+}$ excited state. Typical SEAs include persulfate and Co(III) salts.

Scheme 2 displays the combination of desired reactions that occur when visible light is shone to a mixture of PS, WOC and SEA. In this process, a number of potential reactions that either are not productive or that can lead to decomposition products can also occur. Such undesired reactions include the excited state decay or the back electron transfers from the excited or reduced photosensitizer (PS* or PS_{red}) to the oxidized catalyst or intermediate species (WOC_{ox}). Although these processes will not lead to decomposition products, they are non-productive reactions that will lead to low quantum yields and low solar to fuel conversion efficiencies. It is thus important to understand the different thermodynamic and kinetic parameters that will be involved in such a complex system in order to be able to build efficient light induced water splitting devices.

Several examples using mononuclear [Li *et al.*, 2010; Duan *et al.*, 2010; Roeser *et al.*, 2011; Francàs *et al.*, 2017], dinuclear [Berardi *et al.*, 2015] and tetranuclear [Orlandi *et al.*, 2010; Puntoriero *et al.*, 2010; Geletii *et al.*, 2009] Ruthenium water oxidation catalysts have been shown to successfully promote this combination of reactions in a congruent way and therefore generate O_2 when shining light. However, the performance of the system under photochemically induced conditions still needs a significant improvement. More work has to be focused on the optimization of all the parameters involved namely, pH, type of the photosensitizer, sacrificial electron acceptor, etc. In addition, the different pathways that lead to deactivation need to be understood.

The best systems developed today involve the combination of different molecules responsible for the light absorption process and the catalysis. However,

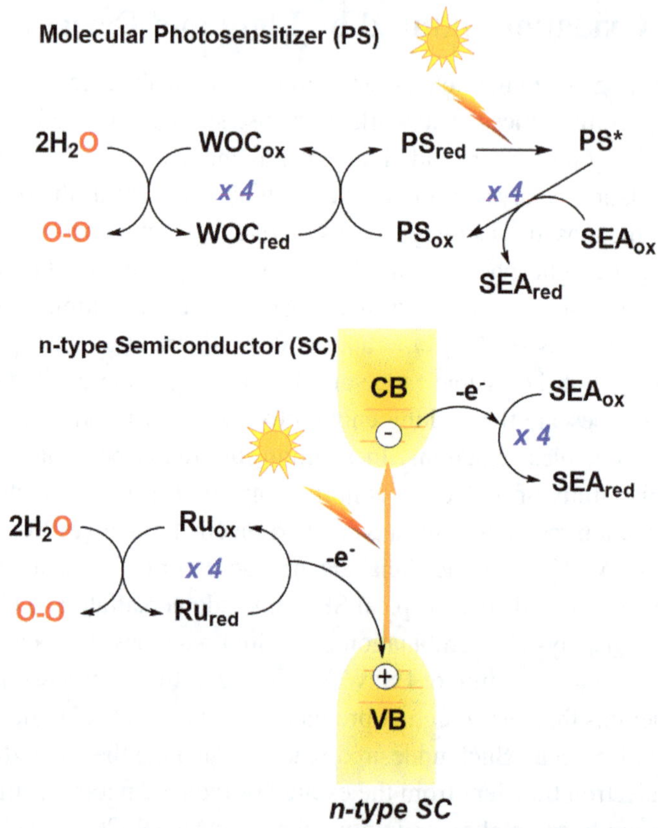

Scheme 2. Light driven water oxidation using a molecular photosensitizer (top) and a *n*-type semiconductor (bottom). WOC: water oxidation catalyst; PS: photosensitizer; SEA: sacrificial electron acceptor; SC: semiconductor; CB: conduction band; VB: valence band; ox: oxidized form of the species; red: reduced form of the species.

there are some cases where the two roles have been combined in a single molecule named *diad* where two active sites (light absorber and catalyst) are connected by a bridging ligand. For this type of systems one of the key parameters that will dictate its success is the degree of electronic coupling provided by the bridging ligand that will be responsible for avoiding non-productive reactions.

Up to now *diads*, including both Ru based PS and Ru based WOC have been synthesized but have not been shown capable of performing water oxidation catalysis photochemically, although there are some examples described as electron relays [Concepcion *et al.*, 2009].

From a technological perspective, the building of devices is much simpler if the different parts needed can be assembled in the solid state. From this point of view, it is thus very important to be able to attached molecular catalysts into

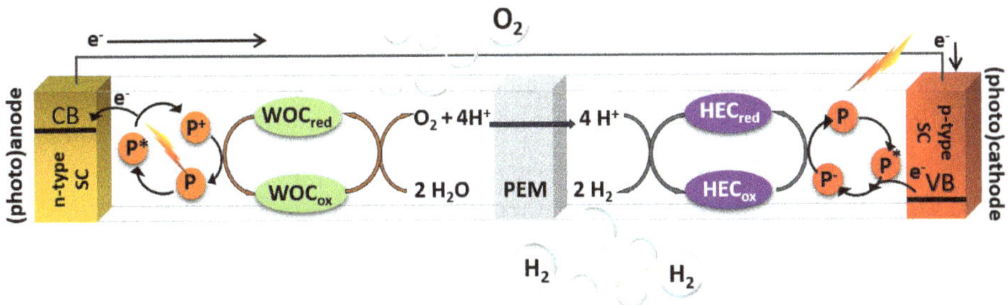

Figure 6. Water-splitting photo-electrochemical cell with four components: (i) light-harvesting molecules, **P**, attached to a semiconductor (**SC**), (ii) a water-oxidation catalyst labeled as **WOC**, (iii) a hydrogen evolution catalyst labeled as **HEC** and (iv) a proton exchange membrane to separate the anodic and cathodic compartments (**PEM**). CB: conduction band; VB: valence band.

solid electrodes or photoanodes and show that their activity remains intact or is even improved [Chen *et al.*, 2009; Mola *et al.*, 2008; Francàs *et al.*, 2009; Toma *et al.*, 2010; Llobet, 2012; Creus *et al.*, 2016; Matheu *et al.*, 2017; Grau *et al.*, 2018].

Today there is already a family of Ru water oxidation catalysts that have been shown to be very efficient both in homogeneous and heterogeneous phase. It is thus the right time to start building integrated light induced water splitting devices based on anchored molecular complexes [Matheu *et al.*, 2017; Grau *et al.*, 2018]. A generic example of such devices is provided in Figure 6 where a tandem cell with two absorbers one for the anode and the other for the cathode is illustrated. The photonanode is inspired in Grätzel cells where the dye or PS is attached to a n-type semiconductor, like TiO_2, *via* a phosophonate derivatized $[Ru(bpy)_3]^{2+}$. A similar strategy can be followed in the cathode using a *p*-type semiconductor, the adequate dye and a proton reduction catalyst. A proton exchange membrane such as nafion is used here to separate the anode from the cathode. In these type of devices, it is very important not only that the thermodynamics for the water oxidation and proton reduction are adequately met, but that all the kinetics of all the reaction occur in a harmonious manner. A successful example of a related device containing a Ru based WOC but with the photosensitizer directly attached at the surface of an ITO electrode has been reported [Li *et al.*, 2010].Other devices have been reported using solid oxides of Ir [Youngblood, Lee, Kobayashi *et al.*, 2009; Youngblood, Lee, Maeda *et al.*, 2009] and Mn [Brimblecombe *et al.*, 2010] as WOCs.

The construction of such devices will allow to start identifying the problems that will need to be solved to be able to generate efficient devices.

7. Conclusions

Over the last decade there has been an impressive development in the field of molecular water oxidation catalysis. Notably seven coordination Ru complexes now achieve turnover frequencies that are two orders of magnitude better than that of the OEC-PSII in nature. While the mechanistic understanding has significantly improved there are still many caveats that need to be sorted out in order to be able to rationally design effective and efficient molecular water oxidation catalysts. Nevertheless, the WOCs that we have nowadays are sufficiently rugged and efficient that they can be incorporated in water splitting devices using just sunlight. That will allow to start tackling the enormous complexity of these devices due to the large number of reactions that have to work synchronically. The task ahead is certainly colossal but the potential payback to our society and the very promising recent developments strongly encourages us to keep ahead in this direction.

8. Numbering Scheme

out-$[Ru^{II}(Hppz)(trpy)(H_2O)]^{2+}$, $\mathbf{1}^{2+}$
out-$[Ru^{II}(Hbpp)(trpy)(H_2O)]^{2+}$, $\mathbf{2}^{2+}$
cis-$[Ru^{II}(bpy)_2(H_2O)_2]^{2+}$, $\mathbf{3}^{3+}$
$\{[Ru^{IV}(O)(bda)(pic)_2]^{2+}$, $\mathbf{4}^{2+}$
$[Ru^{IV}(O)(tda)(py)_2]^{2+}$, $\mathbf{5}^{2+}$
$[Ru^{IV}(bda)(MeO\text{-}isoq)_2(O)]^{2+}$, $\mathbf{6}^{2+}$,
$\{[Ru^{II}(pyC)_2(H_2O)]_2(\mu\text{-}bbp)\}^{3+}$, $\mathbf{7}^{3+}$
$\{[Ru^{III}(bpy)_2]_2(\mu\text{-}O)\}^{4+}$, $\mathbf{8}^{4+}$
$[Ru^{IV}{}_4(\mu\text{-}O)_4(\mu\text{-}OH)_2(H_2O)_4(\gamma\text{-}SiW_{10}O_{36})_2]^{10-}$, $\mathbf{9}^{10-}$

Acknowledgments

Sustained support from MINECO, FEDER and AGAUR are gratefully acknowledged. Recent grants include CTQ2016-80058-R, CTQ2015-73028-EXP, SEV 2013-0319, ENE2016-82025-REDT, CTQ2016-81923-REDC, and 2017-SGR-1631. We also thank the European Commission for the eSCALED project with ID number 7653 within the H2020-MSCA-ITN-2017 program.

References

Balmaseda, M. A., Trenberth, K. E., and Källén, W. (2013) Distinctive climate signals in reanalysis of global ocean heat content, *Geophys. Res. Lett.*, 40, 1754–1759.

Barber, J. (2016) Mn_4Ca Cluster of photosynthetic oxygen-evolving center: structure, function and evolution, *Biochemistry*, 55, 5901—5906.

Barnett, S. M., Goldberg, K. I., and Mayer, J. M. (2012). A soluble copper–bipyridine water-oxidation electrocatalyst, *Nat. Chem.*, 4, 498–502.

Berardi, S., Francàs, L., Neudeck, S., Maji, S., BenetBuchholz, J., Meyer, F., and Llobet, A. (2015). Efficient light-driven water oxidation catalysis by dinuclear ruthenium complexes, *ChemSusChem*, 8, 3688–3696.

Blakemore, J. D., Schley, N. D., Balcells, D., Hull, J. F., Olack, G. W.; Incarvito, C. D., Eisenstein, O., Brudvig, G. W., and Crabtree, R. H. (2010). Half-sandwich iridium complexes for homogeneous water-oxidation catalysis, *J. Am. Chem. Soc.*, 132, 16017–16029.

Bozoglian, F, Romain, S., Ertem, M. Z., Todorova, T.K., Sens, C., Mola, J., Rodriguez, M., Romero, I., Benet-Buchholz, J., Fontrodona, S., Cramer, C. J., Gagliardi, L., and Llobet, A. (2009). The Ru-Hbpp water oxidation catalyst, *J. Am. Chem. Soc.*, 131, 15176–15187.

Brimblecombe, R., Koo, A., Dismukes, G.C., Swiegers, G. F., and Spiccia, L. (2010). Solar driven water oxidation by a bioinspired manganese molecular catalyst, *J. Am. Chem. Soc.*, 132, 2892–2894.

Catalano, V. J., and Craig, T. J. (2003). Monometallic and dimetallic ruthenium(II)–terpyridine complexes employing the tetradentate ligands dipyridylpyrazolyl, dipyridyloxadiazole, and their dimethyl derivatives, *Inorg. Chem.*, 42, 321–334.

Chen, J., Faller, J. W., Crabtree, R. H., and, Brudvig, G. W. (2004). Dimer-of-dimers model for the oxygen-evolving complex of photosystem II. Synthesis and properties of $[Mn^{IV}_4O_5(terpy)_4(H_2O)_2](ClO_4)_6$, *J. Am. Chem. Soc.*, 126, 7345–7349.

Chen, Z., Concepcion, J. J., Jurss, J. W., and Meyer, T. J. (2009). Single-site, catalytic water oxidation on oxide surfaces, *J. Am. Chem. Soc.*, 131, 15580–15581.

Concepcion, J. J., Jurss, J. W., Hoertz, P. G., and Meyer, T. J. (2009). Catalytic and surface-electrocatalytic water oxidation by redox mediator-catalyst assemblies, *Angew. Chem. Int. Ed.*, 48, 9473–9476.

Concepcion, J. J., Jurss, J. W., Templeton, J. L., and Meyer, T. J. (2008). One site is enough. Catalytic water oxidation by $[Ru(tpy)(bpm)(OH_2)]^{2+}$ and $[Ru(tpy)(bpz)(OH_2)]^{2+}$, *J. Am. Chem. Soc.*, 49, 16462–16463.

Concepcion, J. J., Jurss, J. W., Templeton, J. L., and Meyer, T. J. (2008). Mediator-assisted water oxidation by the ruthenium "blue dimer" *cis,cis*-$[(bpy)_2(OH_2)RuORu(OH_2)(bpy)_2]^{4+}$, *Proc. Nat. Acad. Sci.*, 105, 17632–17635.

Concepcion, J. J., Tsai, M.-K., Muckerman, J. T., and Meyer, T. J. (2010). Mechanism of water oxidation by single-site ruthenium complex catalysts, *J. Am. Chem. Soc.*, 132, 1545–1557.

Creus, J., Matheu, R., Peñafiel, I., Moonshiram, D., Blondeau, P., Benet-Buchholz, J., García-Antón, J., Sala, X., Godard, C., and Llobet, A. (2016). A million turnover molecular anode for catalytic water oxidation, *Angew. Chem. Int. Ed.*, 55, 15382–15386.

Duan, L., Bozoglian, F., Mandal, S., Stewart, B., Privalov, T., Llobet, A., and Sun. L. (2012). A molecular ruthenium catalyst with water-oxidation activity comparable to that of photosystem II, *Nat. Chem.*, 4, 418–423.

Duan, L., Fischer, A., Xu, Y., and Sun, L. (2009). Isolated seven-coordinate Ru(IV) dimer complex with $[HOHOH]_-$ bridging ligand as an intermediate for catalytic water oxidation, *J. Am. Chem. Soc.*, 131, 10397–10399.

Duan, L., Xu, Y., Gorlov, M, Tong, L., Andersson, S., and Sun, L. (2010). Chemical and photochemical water oxidation catalyzed by mononuclear ruthenium complexes with a negatively charged tridentate ligand, *Chem. Eur. J.*, 16, 4659–4668.

Edenhofer, O., Pichs-Madruga, R., Sokona, Y.. Minx, J. C., Farahani, E., Kadner, S., Seyboth, K., Adler, A., Baum, I., Brunner, S., Schlömer, S., von Stechow, C., and Zwickel, T. (2014). *IPCC Climate Change 2014: Mitigation of Climate Change*, Cambridge Univ. Press.

Ellis, W. C., McDaniel, N. D., Bernhard, S., and Collins, T. J. (2010). Fast water oxidation using iron, *J. Am. Chem. Soc.*, 132, 10990–10991.

Ferreira, K. N., Iverson, T. M., Maghlaoui, K., Barber, J., and Iwata, S. (2004). Architecture of the photosynthetic oxygen-evolving center, *Science*, 303, 1831–1838.

Fisher, K. J., Materna, K. L., Mercado, B. Q., Crabtree, R. H., and Brudvig, G. W. (2017) Electrocatalytic water oxidation by a copper(II) complex of an oxidation-resistant ligand, *ACS Catal.*, 7, 3384–3387.

Francàs, L., Matheu, R., Pastor, E., Reynal, A., Berardi, S., Sala, X., Llobet, A., and Durrant, J. (2017). Kinetic analysis of an efficient molecular light-driven water oxidation system, *ACS Catal.*, 7, 5142–5150.

Francàs, L., Sala, X., Benet-Buchholz, J., Escriche, L., and Llobet, A. (2009). A Ru-Hbpp-based water-oxidation catalyst anchored on rutile TiO_2, *Chem. Sus. Chem.*, 2, 321–329.

Garrido-Barros, P., Funes-Ardoiz, I., Drouet, S., Benet-Buchholz, J., Maseras, F., and Llobet, A. (2015). Redox non-innocent ligand controls water oxidation overpotential in a new family of mononuclear Cu-based efficient catalysts, *J. Am. Chem. Soc.*, 137, 6758–6761.

Garrido-Barros, P., Gimbert-Suriñach, C., Matheu, R., Sala, X., and Llobet, A. (2017). How to make an efficient and robust molecular catalyst for water oxidation. *Chem. Soc. Rev.*, 46, 6088–6098.

Geletii, Y. V., Huang, Z., Hou, Y., Musaev, D. G., Lian, T., and Hill, C. L. (2009). Homogeneous light-driven water oxidation catalyzed by a tetraruthenium complex with all inorganic ligands, *J. Am. Chem. Soc.*, 131, 7522–7523.

Geletii, Y. V., Botar, B., Kogerler, P., Hillesheim, D. A., Musaev, D. G., and Hill, C. L. (2008). An all-inorganic, stable, and highly active tetraruthenium homogeneous catalyst for water oxidation, *Angew. Chem. Int. Ed.*, 47, 3896–3899.

Gerlach, D. L., Bhagan, S., Cruce, A. A., Burks, D. B., Nieto, I., Truong, H. T., Kelley, S. P., Herbst-Gervasoni, C. J., Jernigan, K. L., Bowman, M. K., Pan, S., Zeller, M., and Papish, E. T. (2014). Studies of the pathways open to copper water oxidation catalysts containing proximal hydroxy groups during basic electrocatalysis, *Inorg. Chem.*, 53, 12689–12698.

Gilbert, J. A., Eggleston, D. A., Murphy Jr. W. A., Geselowitz, D. A., Gersten, S. W., Hodgson, D. J., and Meyer, T. J. (1985). Structure and redox properties of the water-oxidation catalyst $[(bpy)_2(OH_2)RuORu(OH_2)(bpy)_2]^{4+}$, *J. Am. Chem. Soc.*, 107, 3855–3864.

Gimbert-Suriñach, C., Moonshiram, D., Francàs, L., Planas, N., Bernales, V., Bozoglian, F., Guda, A., Mognon, L., López, I., Hoque, Md A., Gagliardi, L., Cramer, C. J., and

Llobet, A. (2016) Structural and spectroscopic characterization of reaction intermediates involved in a dinuclear Co-Hbpp water oxidation catalyst, *J. Am. Chem. Soc.*, 138, 15291–15294.

Grau, S., Berardi, S., Moya, A., Matheu, R., Vito, C., Vilatela, J., Bignozzi, C. A., Caramori, S., Gimbert-Suriñach, C., and Llobet, A. (2018). A hybrid molecular photoanode for efficient light-induced water oxidation, *Sustainable Energy & Fuels,* 2, 1979–1985.

Haumann, M., Liebisch, P., Müller, C., Barra, M., and Grabolle, M. (2005). Photosynthetic O_2 formation tracked by time-resolved x-ray experiments, *Science*, 310, 1019–1021.

Hull, J. F., Balcells, D., Blakemore, J. D., Incarvito, C. D., Eisenstein, O., Brudvig, G. W., and Crabtree, R. H. (2009). Highly active and robust Cp* iridium complexes for catalytic water oxidation, *J. Am. Chem. Soc.*, 131, 8730–8731.

Huynh, M. H.-V., and Meyer, T. J. (2007). Proton-coupled electron transfer, *Chem. Rev.*, 107, 5004–5064.

Kaldor, M., Karl, T. L., and Said, Y. (2007). *Oil Wars*, Pluto Press.

Lalrempuya, R., McDaniel, M. D., Müller-Bunz, H., Bernhard, S., and Albrecht, M. (2010). Water oxidation catalyzed by strong carbenetype donorligand complexes of iridium, *Angew. Chem. Int. Ed.*, 49, 9765–9768.

Li, L., Duan, L., Xu, Y., Gorlov, M., Hagfeld, A., and Sun, L. (2010). A photoelectrochemical device for visible light driven water splitting by a molecular ruthenium catalyst assembled on dye-sensitized nanostructured TiO_2, *Chem. Commun.*, 46, 7307–7309.

Llobet, A. (1994). Synthesis, spectral and redox properties of a new series of aqua complexes of ruthenium (II), *Inorg. Chim. Acta*, 221, 125–131.

Llobet, A. (2012). Catalytic water oxidation: Rugged water-oxidation anodes, *Nature Chem.*, 2, 804–805.

Matheu, R., Ertem, M. Z., Benet-Buchholz, J., Coronado, E., Batista, V. S., Sala, X., and Llobet, A. (2015). Intramolecular proton transfer boosts water oxidation catalyzed by a Ru complex, *J. Am. Chem. Soc.*, 137, 10786–10795.

Matheu, R., Moreno-Hernandez, I. A., Sala, X., Gray, H. B., Brunschwig, B. S., Llobet, A., and Lewis, N. S. (2017). Photoelectrochemical behavior of a molecular Ru-based water-oxidation catalyst bound to TiO_2-protected Si photoanodes, *J. Am. Chem. Soc.*, 139, 11345–11348.

McDaniel, N. D., Coughlin, F. J., Tinker, L. L., and Bernhard, S. (2008). Cyclometalated iridium(III) Aquo complexes: efficient and tunable catalysts for the homogeneous oxidation of water, *J. Am. Chem. Soc.*, 130, 210–217.

Meyer, T. J., Huynh, and M. H.-V. (2003). The remarkable reactivity of high oxidation state ruthenium and osmium polypyridyl complexes, *Inorg. Chem.*, 42, 8140–8160.

Mognon, L., Benet-Buchholz, J., and Llobet, A. (2015). Single site isomeric Ru WOCs with an electron-withdrawing group: Synthesis, electrochemical characterization, and reactivity, *Inorg. Chem.*, 54, 11948–11957.

Mola, J., Dinoi, C., Sala, X., Rodriguez, M., Romero, I., Parella, T., Fontrodona, X., and Llobet, A. (2011). A new dinuclear Ru-Hbpp based water oxidation catalyst with a trans-disposition of the Ru-OH, *Dalton Trans.*, 40, 3640–3646.

Mola, J., Mas-Marza, E., Sala, X., Romero, I., Rodriguez, M., Viñas, C., and Llobet, A. (2008). Ru-hbpp-based water-oxidation catalysts anchored on conducting solid supports, *Angew. Chem. Int. Ed.*, 47, 5830–5832.

Muckerman, J. T., Polyansky, D. E., Wada, T., Tanaka, K., and Fujita, E. (2008). Water oxidation by a ruthenium complex with noninnocent quinone ligands: possible formation of an O--O bond at a low oxidation state of the metal, *Inorg. Chem.*, 47, 1787–1802.

Neudeck, S., Maji, S., López, I., Meyer, S., Meyer, F., and Llobet, A. (2014). New powerful and oxidatively rugged dinuclear Ru water oxidation catalyst: control of mechanistic pathways by tailored ligand design, *J. Am. Chem. Soc.*, *136*, 24–27.

Orlandi, M., Argazzi, R., Sartorel, A., Carraro, M., Scorraro, G., Bonchio, M., and Scandola, F. (2010). Ruthenium polyoxometalate water splitting catalyst: very fast hole scavenging from photogenerated oxidants, *Chem. Commun.*, 46, 3152–3154.

Puntoriero, F., La Ganga, G., Sartorel, A., Carraro, M., Scorrano, G., Bonchio, M., and Campagna, S. (2010). Photo-induced water oxidation with tetra-nuclear ruthenium sensitizer and catalyst: A unique 4 × 4 ruthenium interplay triggering high efficiency with low-energy visible light, *Chem. Commun.*, 46, 4725–4727.

Richmond, C. J., Matheu, R., Poater, A., Falivene, L., Benet-Buchholz, J., Sala, X., Cavallo, L., and Llobet, A. (2014). Supramolecular water oxidation with Ru-bda-based catalysts, *Chem. Eur. J.*, 22, 17282–17286.

Rigsby, M. L., Mandal, L., Spencer, L. C., Nam, W., Llobet, A., and Stahl, S. S. (2012). Cobalt analogs of Ru-based water oxidation catalysts: overcoming thermodynamic in,stability and kinetic lability to achieve electrocatalytic O_2 evolution, *Chem. Sci.*, 3, 3058–3062.

Roeser, S., Farràs, P., Bozoglian, F., Martínez-Belmonte, M., Benet-Buchholz, J., and Llobet, A. (2011). Chemical, electrochemical, and photochemical catalytic oxidation of water to dioxygen with mononuclear ruthenium complexes, *Chem. Sus. Chem.*, 4, 197–207.

Romain, S., Bozoglian, F., Sala, X., and Llobet, A. (2009). Oxygen-oxygen bond formation by the Ru-Hbpp water oxidation catalyst occurs solely *via* an intramolecular reaction pathway, *J. Am. Chem. Soc.*, 131, 2768–2769.

Romain, S., Vigara, L., and Llobet, A. (2009). Oxygen-oxygen bond formation pathways promoted by ruthenium complexes, *Acc. Chem. Res.*, 42, 1944–1953.

Sala, X., Ertem, M. Z., Vigara, L., Todorova, T. K., Chen, W., Rocha, R. C., Aquilante, F., Cramer, C. J., Gagliardi, L., and Llobet, A. (2010). The cis-$[Ru(II)(bpy)_2(H_2O)_2]^{2+}$ water-oxidation catalyst revisited, *Angew. Chem. Int. Ed.*, 49, 7745.

Sala, X., Rodriguez, M., Romero, I., Escriche, L., and Llobet, A. (2009). Molecular catalysts that oxidize water to dioxygen, *Angew. Chem. Int. Ed.*, 48, 2842–2852.

Sartorel, A., Carraro, M., Scorrano, G., De Zorzi, R., Geremia, S., McDaniel, N.D., Bernhard, S., and Bonchio, M. (2008). Polyoxometalate embedding of a tetraruthenium(IV)-oxo-core by template-directed metalation of [gamma-$SiW_{10}O_{36}]^{8-}$: a totally inorganic oxygen-evolving catalyst, *J. Am. Chem. Soc.*, 130, 5006–5007.

Sens, C., Romero, I., Rodriguez, M., Llobet, A., Parella, T., and Benet-Buchholz, J. (2004). A new Ru complex capable of catalytically oxidizing water to molecular dioxygen, *J. Am. Chem. Soc.*, 126, 7798–7799.

Sproviero, E. M., Gascon, J. A., McEvoy, J. P., Brudvig, G. W., and Batista, V. S. (2008). A model of the oxygen-evolving center of photosystem II predicted by structural refinement based on EXAFS simulations, *J. Am. Chem. Soc.*, 130, 6728–6730.

Steffen, W., Rockström, J., Richardson, K., Lenton, T. M., Folke, C., Liverman, D., Summerhayes, C. P., Barnosky, A. D., Cornell, S. E., Crucifix, M., Donges, J. F., Fetzer, I., Lade, S. J., Scheffer, M., Winkelmann, R. and Schellnhuber, H. J. (2018). Trajectories of the Earth System in the Anthropocene, *Proc. Nat. Acad. Sci.*, 115, 8252–8259.

Stocker, T. F., Qin, D., Plattner, G.-K., Tignor, M. M. B., Allen, S. K., Boschung, J., Nauels, A., Xia, Y., Bex, V., and Midgley, P. M. (2013). *IPCC Climate Change 2013: The Physical Science Basis*, Cambridge Univ. Press.

Suga, M., Akita, F., Hirata, K., Ueno, G., Murakami, H., Nakajima, Y., Shimizu, T., Yamashita, K., Yamamoto, M., Ago, H., and Shen, J.-R. (2015) Native structure of photosystem II at 1.95 Å resolution viewed by femtosecond X-ray pulses, *Nature*, 517, 99–103.

Toma, F. M., Sartorel, A., Iurlo, M., Carraro, M., Parisse, P., Maccato, C., Rapino, S., Gonzalez, B. R., Amenitsch, H., Ros, T. D., Casalis, L., Goldoni, A., Marcaccio, M., Scorrano, G., Scoles, G., Paolucci, F., Prato, M., and Bonchio, M. (2010). Efficient water oxidation at carbon nanotube-polyoxometalate electrocatalytic interfases, *Nature Chem.*, 2, 826–831.

Umena, Y., Kawakami, K., Shen, J.-R., and Kamiya, N. (2011) Crystal structure of oxygen-evolving photosystem II at a resolution of 1.9 Å, *Nature*, 473, 55–60.

Wasylenko, D. J., Ganesamoorthy, C., Borau-Garcia, J., and Berlinguette, C. P. (2011) Electrochemical evidence for catalytic water oxidation mediated by a high-valent cobalt complex, *Chem. Commun.*, 47, 4249–4251.

Wasylenko, D. J., Ganesamoorthy, C., Henderson, M. A., Koivisto, B. D., Osthoff, H. D., and Berlinguette, C. P. (2010). Electronic modification of the $[Ru^{II}(tpy)(bpy)(OH_2)]^{2+}$ scaffold: effects on catalytic water oxidation, *J. Am. Chem. Soc.*, 132, 16094–16106.

Wydrzynski, T., and Satoh, K. (2005). *Photosystem II: The Light-Driven Water: Plastoquinone Oxidoreductase*, Springer, Dordrecht.

Yano, J., Kern, J., Sauer, K., Latimer, M. J., Pushkar, Y., Biesiadka, J., Loll, B., Saenger, W., Messinger, J., Zouni, A., and Yachandra, V. K. (2006). Where water is oxidized to dioxygen: structure of the photosynthetic Mn_4Ca cluster, *Science*, 314, 821–825.

Yin, Q., Tan, J.M., Besson, C., Geletii, Y.V., Musaev, D. G., Kuznetsov, A.E., Luo, Z., Hardcastle, K.I., and Hill, C.L. (2010). A fast soluble carbon-free molecular water oxidation catalyst based on abundant metals, *Science*, 328, 342–345.

Yoshida, M., Masaoka, S., Abe, J., and Sakai, K (2010). Catalysis of mononuclear aquaruthenium complexes in oxygen evolution from water: a new radical coupling path using hydroxocerium(IV) species, *Chem. Asian.*, 5, 2369–2378.

Youngblood, W. J., Lee, S.-H. A., Kobayashi, Y., Hernandez-Pagan, E. A., Hoertz, P. G., Moore, T. A., Moore, A. L., Gust, D., and Mallouk, T. E. (2009). Photoassisted overall water splitting in a visible light-absorbing dye-sensitized photoelectrochemical cell, *J. Am. Chem. Soc.*, 191, 926–927.

Youngblood, W. J., Lee, S.-H. A., Maeda, K., and Mallouk, T. E. (2009). Visible light water splitting using dye-sensitized oxide semiconductors, *Acc. Chem. Res.*, 42, 1966–1973.

Zhang, T., Wang, C., Liu, S., Wang, J.-L., and Lin, W. (2014) A biomimetic copper water oxidation catalyst with low overpotential, *J. Am. Chem. Soc.*, 136, 273–281.

Zong, and R., Thummel, R. P. (2005). A new family of Ru complexes for water oxidation, *J. Am. Chem. Soc.*, 127, 12802–12803.

Chapter 12

Material Design for Artificial Photosynthesis using Photoelectrodes for Hydrogen Production

Wilman Septina[*,†], James Barber[*,‡] and Lydia Helena Wong[*,§]

[*]*School of Materials Science and Engineering Nanyang Technological University, Singapore*

[†]*Hawaii Natural Energy Institute, University of Hawaii, Honolulu, Hawaii 96822, USA*

[‡]*Division of Molecular Biosciences, Imperial College London, London SW7 2AZ, UK*

[§]*lydiawong@ntu.edu.sg*

Artificial photosynthesis system for hydrogen production from water involves multiple components that work in harmony converting light energy into chemical fuel. In a photoelectrochemical (PEC) system, this typically consists of light absorber, charge separator, and catalyst. Configuration of each of the components within the system as well as their material's properties (optical, electronic, or structural) influences greatly the solar to hydrogen efficiency. In this chapter, we provide a brief review of multiple configurations in the artificial photosynthesis system found in the literature, especially those using solid-state photoelectrodes for water splitting. In particular, we focus on properties of the materials which make-up the artificial photosynthesis system and discuss how they affect the performance of the PEC cell. We then follow up by giving some examples from our works and the literature which employ semiconductor photoelectrodes.

1. Introduction

Natural photosynthesis involves multiple electron/proton transfers (electron/hole separation) comparable to that observed in semiconductors upon absorption of light. This feature has inspired physicists, chemists, and materials scientists to design an "artificial photosynthesis" system with the purpose of converting solar energy to chemical energy with a much higher efficiency than natural photosynthesis (nature hardly converts more than 1% of the solar energy into chemical fuel [Barber 2009]. The process of photosynthesis involves light absorption, charge generation and separation, and catalysis to produce chemical fuels. The primary reactions are initiated by absorption of photons by photosystems one and two (PSI and PSII) which are coupled together in series by an electron/proton transfer chain and often referred to as a Z-scheme. This tandem system involves two separate charge separations P700+Chl+ in PSI and P680+Pheo- in PSII. These two charge separation states are comparable to light induced charge separation in semiconductors between the valence and conduction bands. In photosynthesis the charge transferred states provide the energy to split water into O_2 and "hydrogen" in the form of protons and high energy electron. These reducing equivalents are then used to convert CO_2 into organic molecules (see Chapter by Barber). In this way solar energy is stored in the chemical bonds of food, global biomass and fossil fuels.

These past decades, scientists have been exploring various effective artificial photosynthesis strategies involving semiconductors to drive the water splitting reaction. Early on, the Z-scheme water splitting process involving semiconductor photocatalyst powder is the most comparable with those observed in the natural photosynthesis [Barber and Tran, 2013]. Systems involving photoelectrochemical (PEC) cells consisting of semiconductor photoelectrodes are also intensively investigated nowadays to obtain a far more efficient system than the photocatalytic powder approach. Recent technoeconomic analyses suggest that the cost of hydrogen produced by PEC cell could be competitive with that from fossil fuels provided that the PEC cell could achieve a solar to hydrogen efficiency of 15% with a lifetime of 10 years and cost of less than 200 USD/m^2 [Pinaud *et al.,* 2013].

To achieve efficient artificial photosynthesis, the system has to be designed in such a way that the electron/hole pairs generated upon absorption of light by the semiconductor are able to drive the water splitting reaction with minimal losses. For this purpose, extensive research efforts have been carried out to improve the properties of the materials that are responsible for the light absorption, the charge generation and separation, and the final catalytic reaction. This chapter will review material design requirement for the artificial photosynthesis system based on semiconductor photoelectrochemical (PEC) devices. It should be noted, however, that this chapter is meant as an introduction for those new to the field of artificial photosynthesis and is not meant to be a comprehensive review of the materials used in PEC. Whenever possible, good review papers will be cited in this chapter so readers could refer to them for more detailed and current progress of a specific material system.

2. System Design for Artificial Photosynthesis

As mentioned above, artificial photosynthesis system involves sequential light absorption, charges generation and separation, and catalytic processes to oxidize water to hydrogen. To do so, various photoelectrodes designs have been proposed and intensively studied by the scientific community. Figure 1 shows those designs

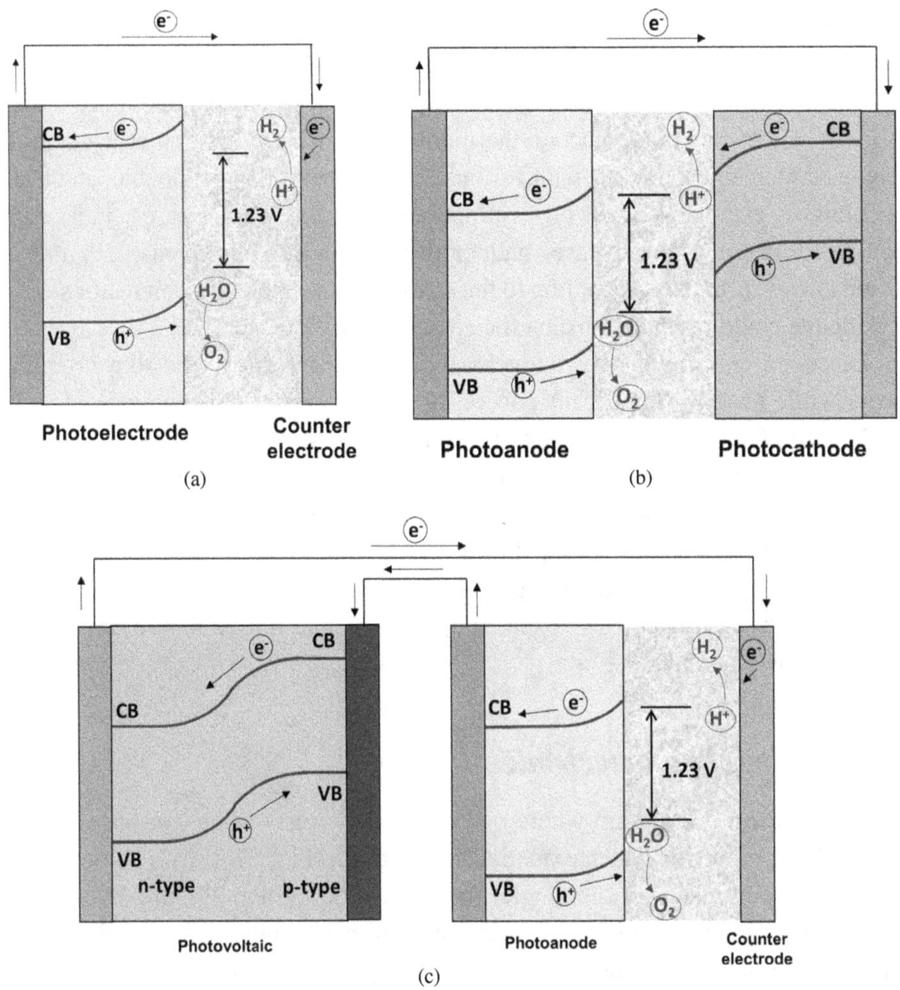

Figure 1. System designs for artificial photosynthesis using photoelectrodes, (a) single photoelectrode, (b) two photoelectrodes (tandem of photocathode and photoanode), (c) tandem of photovoltaic with a photoelectrode.

namely a single photoelectrode system, a tandem photoelectrodes (photocathodes and photoanodes), and a tandem of photovoltaic and a photoelectrode.

2.1. *Single photoelectrode*

The single photoelectrode system was the first explored system for the artificial photosynthesis (Figure 1a). The famous Fujishima-Honda configuration where an illuminated photoelectrode using TiO_2 film coupled with a Pt counter-electrode generating H_2 bubbles is an example of this system [Fujishima & Honda, 1972]. In this system, a single semiconductor (or a single junction of semiconductors) has to be able to perform all the required processes from light absorption, charge separation, and catalysis. In addition, the position of their conduction band minimum (CBM) and valence band maximum (VBM) has to straddle the water reduction/oxidation potential (~1.23 eV or 1.7 to 2.1 eV considering overpotential and resistance losses) to drive the water splitting reaction thermodynamically. The limiting efficiencies of photoelectrochemical devices with various configurations (single, double, and triple junctions) is estimated based on assumptions of three limiting cases: ideal case, high-performance realistic case, and earth-abundance realistic case (Figure 2) [Fountaine *et al.*, 2016]. According to the estimation, a single semiconductor system in principle could reach a realistic efficiency up to 15% (Figure 2a).

Discovering a single semiconductor that could conform all the above requirements while absorbs significant portion of the solar spectrum has proved to be challenging. TiO_2 and $SrTiO_3$ are two intensively-studied oxide materials that could satisfy the thermodynamic requirements. However, their band gaps are too large (>3 eV) so that they only harvest the UV portion of the solar spectrum resulted in STH efficiency <1% [Murphy *et al.*, 2006]. GaN is a non-oxide material that has been demonstrated to perform unassisted water splitting [Waki *et al.*, 2007], although photocorrosion under operating condition still remains a challenge for this material.

2.2. *Tandem photoelectrodes*

Two photoelectrodes system where one electrode is responsible for water reduction, and the other for water oxidation has been developed to ease the complex materials requirements in the single photoelectrode system (Figure 1b). In this configuration, each photoelectrodes can be optimized independently allowing a wider choice of semiconductors from the library of materials. Typically, the system is mimicking a tandem configuration rather like that adopted in natural photosynthesis where one electrode is using a wide-gap semiconductor, and the other is connected in series and has a narrower gap to absorb the rest of the incoming

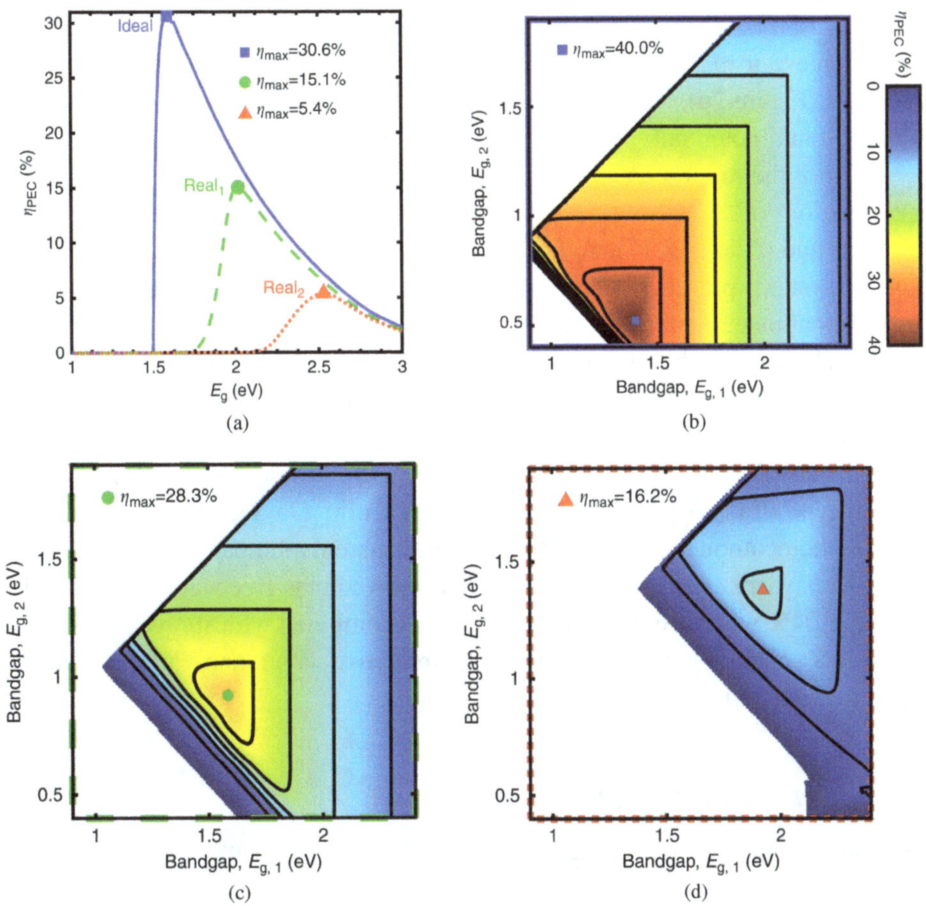

Figure 2. Photoelectrochemical water splitting efficiencies (η_{PEC}) *versus* band gap of semiconductor in a photoelectrode using (a) single junction within ideal case, high performance realistic case (Real$_1$), and earth-abundance realistic case (Real$_2$), and for the tandem photoelectrodes in (b) ideal, (c) Real$_1$, and (d) Real$_2$ cases. Reprinted from [Fountaine *et al.*, 2016] under a creative commons license.

solar radiation. The thermodynamic requirement to straddle the water reduction/ potential is basically divided between those two semiconductors, where the conduction band (CB) energy level of one semiconductor has to be above the water reduction potential (photocathode) and the maximum potential of the valence band (VB) of the other electrode has to be below the water oxidation potential (photoanode). In this configuration, the current flowing between them will be matched, and the voltages are added to satisfy the thermodynamic over potential needed to split water. The estimation by Fountaine *et al.* for double junction suggested that in an ideal case, a tandem of semiconductors with bandgaps of 1.40 eV and 0.52 eV could achieve a maximum solar to hydrogen efficiency of 40% (Figure 2b) [Fountaine *et al.*, 2016]. In a realistic case using high-performance materials

(Figure 2c), a tandem of semiconductors with bandgaps of 1.59 eV and 0.92 eV could achieve a maximum efficiency of 28.3%. Using available earth abundant materials (Figure 2d), a tandem of semiconductors with bandgaps of 1.93 eV and 1.38 eV could potentially achieve 16.2% maximum efficiency.

Most of the reports on PEC water splitting have been largely focusing on the optimization of either the photoanode or the photocathode independently [Septina and Tilley, 2017]. A few groups demonstrated the actual un-assisted tandem system, primarily using photovoltaic-quality p-type semiconductors as the photocathode, combined with $BiVO_4$ as the photoanode. $BiVO_4$ has been very popular choices for the photoanode due to its relatively negative onset potential compared to other metal oxides [Abdi and Berglund, 2017]. Pan *et al.* utilized Cu_2O/Ga_2O_3 buried-junction photocathode coupled with $BiVO_4$ photoanode and showed 3% of STH efficiency [Pan *et al.*, 2018]. Kobayashi *et al.* reported a tandem-type PEC containing a co-evaporated $CuIn_{1-x}Ga_xSe_2$ modified with CdS and Pt photocathode and a $BiVO_4$ semi-transparent photoanode with STH of 3.7% [Kobayashi *et al.*, 2018]. Huang *et al.* showed bias-free water splitting with STH of 1.04% when a $BiVO_4$ photoanode was combined with sprayed-sulfurized Cu_2ZnSnS_4 coated with CdS and HfO_2-protective layer and Pt as a catalyst [Huang *et al.*, 2018].

2.3. *Photovoltaic-photoelectrode*

To attain a high solar to hydrogen efficiency, photovoltaic (PV) can also be coupled with a photoelectrode (Figure 1c). In this configuration, the PV supplies the required over potential to split water, when the photoelectrode alone does not produce sufficient voltage. Since most of the commercial PV have a relatively good photocurrent and can also be assembled in parallel with the PEC, higher solar-to-hydrogen conversion efficiency could be achieved. The photovoltaic is normally placed outside the aqueous solution chamber and wire-connected to an immersed metal cathode/anode such as Pt or Ru. In this way, the photovoltaic is relatively protected from the harsh condition of the solution.

With the meteoric rise in the development of perovskite solar cells, there have been substantially growing reports on the demonstration of PEC water splitting using PV-photoelectrode. State-of-the-art perovskite solar cells typically have an open circuit voltage of over 1 V with a high photocurrent of more than 15 mA/cm^2 [Seok *et al.*, 2018]. These remarkable photovoltaic properties would benefit PV-photoelectrode configuration since the current density-potential curves between the solar cell and the photoelectrode could cross-over at a high photocurrent. For example, Gurudayal *et al.* combined a perovskite photovoltaic with hematite (Fe_2O_3) photoanode (Figure 4a) [Gurudayal *et al.*, 2017]. The hematite

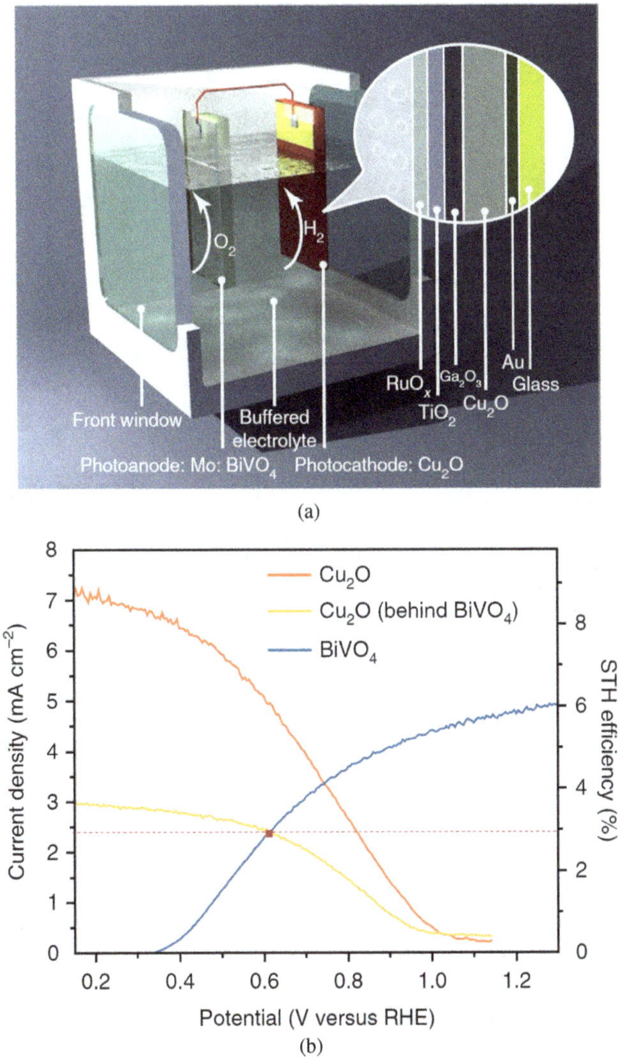

Figure 3. (a) Examples of tandem photoelectrode systems where Cu_2O photocathode is partnered with $BiVO_4$ photoanode. (b) Current density and its respective STH efficiency *versus* potential curves show that the Cu_2O photocathode (behind $BiVO_4$) curve and $BiVO_4$ curve intersect at STH efficiency around 3%. Reprinted from [Pan *et al.*, 2018] with permission from Springer Nature.

is well known to suffer from a poor photovoltage despite its wide band gap (2.2 eV) [Sivula *et al.*, 2011]. However, due to the high open circuit voltage of the perovskite solar cell (V_{OC}: 1.15 V) (Figure 4b), the current density-potential curves could cross-over at a relatively positive voltage and produced an operation current at around 2.8 mA/cm^2 (Figure 4c). Qiu *et al.* also used perovskite solar cells to drive water splitting on $BiVO_4$ photoanode and reported an STH efficiency

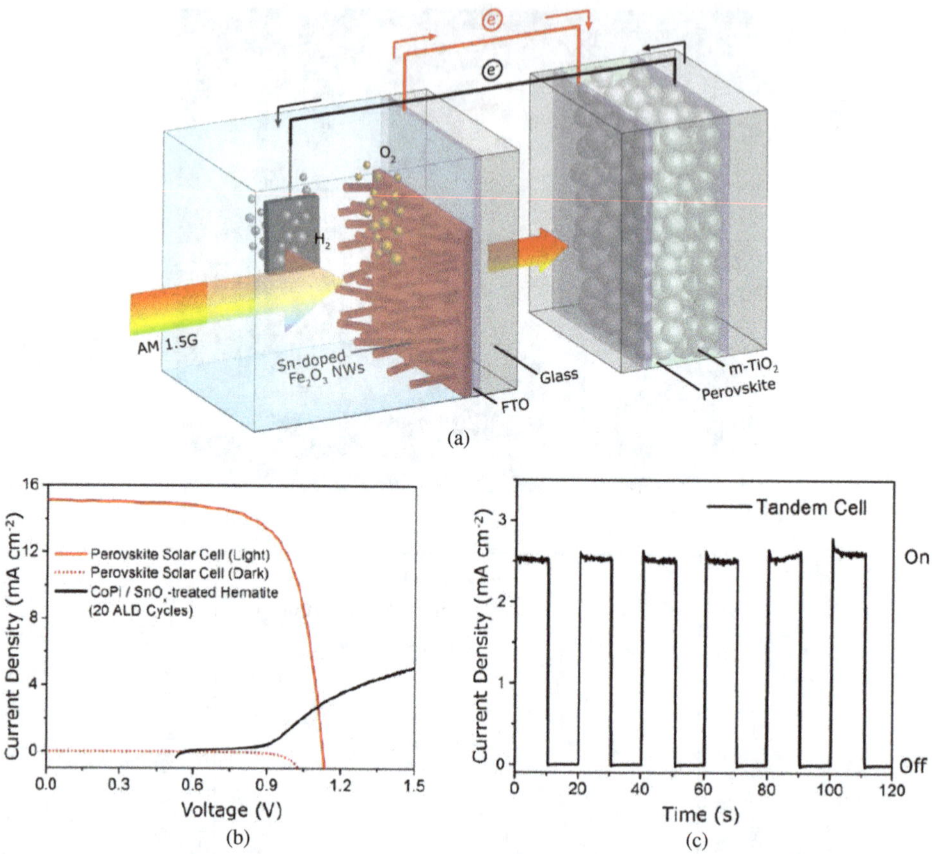

Figure 4. (a) Example of photovoltaic-photoelectrode system for water splitting [Gurudayal *et al.*, 2017]. In this case, perovskite photovoltaic was paired up with Fe_2O_3 photoanode. (b) Current density *versus* voltage curves showing the intersection between perovskite and hematite curves and (c) their operation current at around 2.8 mA/cm^2 which correspond to STH efficiency of 3.4%. Reprinted with permission from John Wiley and Sons.

of 6.2% [Qiu *et al.*, 2016]. Apart from perovskite solar cells, various groups also demonstrated water splitting using other solar cell technologies such as silicon [Abdi *et al.*, 2013] and dye-sensitized solar cells [Brillet *et al.*, 2012].

The efficiency in PV-photoelectrode system is optimized when the photocurrent of the PV and the photoelectrode are well matched. Since the photocurrent of the photoelectrode is usually much lower than that of the PV, the performance of this system is still limited by the photocurrent of the photoelectrode. In this configuration, one could also argue that a total photovoltaic-assisted electrolysis system can be obtained by replacing the photoelectrode with a metal electrode. Therefore, the majority of the cost of this system would come from the cost of the photovoltaic.

Table 1. Advantage and limitation of each artificial photosynthesis design based on photoelectrode

System	Advantage	Limitation
Single photoelectrode	• Simple • Lowest material cost	• Limited materials choices • Harness only limited portion of solar spectrum
Tandem photoelectrodes	• Each photoelectrodes can be optimized independently • Extensive library of materials can be investigated • Wide solar spectrum can be harnessed • Medium material cost	• More complex arrangement of the electrodes • The top electrode requires sufficient light transmittance • The less performing photoelectrode could limit the overall performance of the system
PV-photoelectrode	• Most efficient system • Commercial PV is generally robust under practical condition	• The photoelectrode requires sufficient light transmittance • Cost of the PV would dominate the total cost of the system • The highest materials cost

The advantages and disadvantages of the 3 systems discussed above are summarized in Table 1. Reader can also refer to [Walter *et al.*, 2010] for a good review on various approaches for solar water splitting using photoelectrodes.

3. Materials Requirements

The performance of the artificial photosynthesis system is highly dependent on the materials used, regardless of the device configuration. For example, given similar optical properties, semiconductor with a higher charge transport efficiency (i.e. mobility, carrier lifetime) would result in a higher photocurrent and result in a higher solar to hydrogen efficiency. For materials with a short carrier diffusion length, nanostructuring would enhance the distance for the collection of charge from only few nanometers to tens of nanometer. This section will explain the key materials features used in the artificial photosynthesis system as the light absorbers, charge separation, and catalyst.

3.1. *Light absorber*

Most of the investigation to find an efficient artificial photosynthesis system usually starts with synthesis and optimization of the light absorber semiconductors as it is the most important component of an artificial photosynthesis system. Reader can refer to [Sivula and van de Krol, 2016], [Septina and Tilley, 2017], and [Abdi and Berglund, 2017] for reviews on selection of semiconducting materials for PEC application. The properties of a light absorber that would greatly affect the photocatalytic efficiency are energetic band alignments, optical and electronic properties, as well as morphology. These will be discussed below.

3.1.1. *Energetic band alignment*

As discussed previously in Section 2, the first perquisite for a material to be used as a photoelectrode for splitting water is its favorable inherent position of their energy levels with respect to the levels of water reduction/oxidation. In acidic electrolyte, the redox level of water (in general proton) reduction to H_2 is located at 0 V *versus* NHE, and the water oxidation to O_2 is located at 1.23 V *versus* NHE. Figure 5 shows the energetic band alignment of various semiconductors relative to the water oxidation/reduction potentials. As mentioned in Section 2.1, if the Conduction Band Minima (CBM) and the Valence Band Maxima (VBM) of a semiconductor could straddle the redox levels of water, the semiconductor would

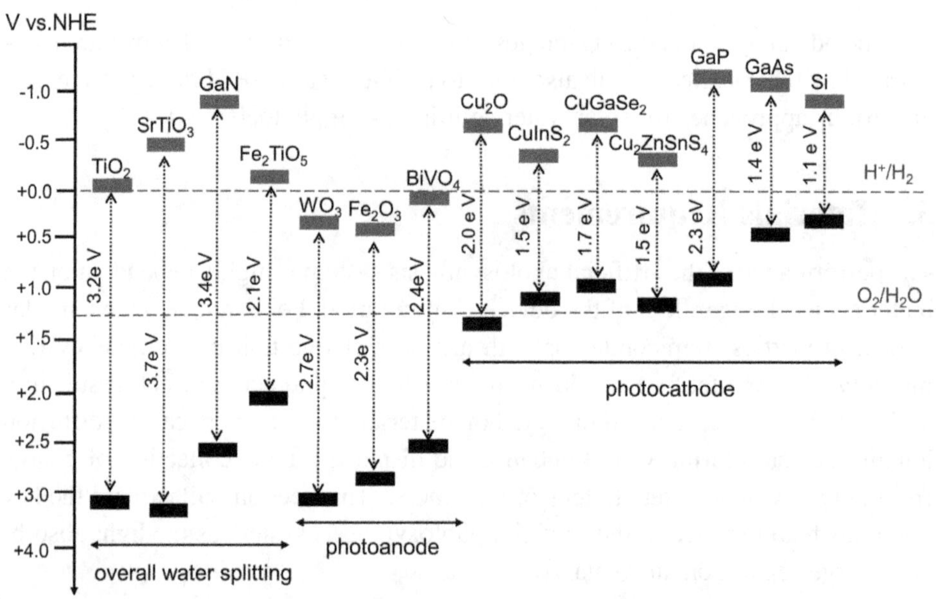

Figure 5. Band positions of various semiconductors in NHE scale relative to the redox levels for water reduction (H^+/H_2) and water oxidation (O_2/H_2O) at pH 0. The band positions for TiO_2, $SrTiO_3$, GaN, WO_3, Fe_2O_3, GaP, GaAs and Si are obtained from [Ager, 2016]. The band positions for Fe_2TiO_5 are obtained from [Bassi *et al.*, 2016]. The band positions for $BiVO_4$, Cu_2O, and Cu_2ZnSnS_4 are obtained from [Chen and Wang, 2012]. The band positions for $CuInS_2$ and $CuGaSe_2$ are adapted from [Klein, 2015].

be able to split water into hydrogen upon illumination without additional power, as in the case of the single photoelectrode system. If the CBM is above the water reduction level and the VBM is above the water oxidation level, the semiconductor is able to reduce water but needs additional power, for example, from potentiostat or PV to complete the full reaction of water splitting. Such semiconductors are commonly referred to as ***the photocathode***. On the other hand, if the CBM of the semiconductor is below the water reduction level, and the VBM is below the water oxidation level, the semiconductor is able to oxidize the water but needs the additional power to complete full water splitting reaction. Such semiconductor is usually referred to as ***the photoanode***.

Altering the band energetics of semiconductors to improve their photocatalytic activity have been pursued extensively. The goals are either to move the CBM or VBM so that they straddle the water reduction/oxidation levels, or to generate a higher photovoltage. This can be attained by cation/anion substitution to the lattice of the semiconductors. For instance, Bassi *et al.* incorporated Ti into Fe-O system via hydrothermal method to form Fe_2TiO_5. Its CBM and VBM straddle the water reduction/oxidation levels [Bassi *et al.*, 2014]. Septina *et al.* showed gradual deepening of the VBM of $CuInS_2$ as a function of the level Ga added, which resulted in a higher photovoltage when making buried junction with CdS [Septina *et al.*, 2015]. Similar trend of deepening the VBM was also observed when Cu in $Cu(In,Ga)S_2$ was partially substituted with Ag [Septina *et al.*, 2017].

3.1.2. *Optical properties*

One of the first properties in the library of materials used to screen for photoactivity, is the optical bandgap necessary to absorb visible light as efficiently as possible. This translates into materials with a bandgap of lower than 2.0 eV. There is a wide selection of narrow gap materials (<1.5 eV) for PEC water splitting, as shown in Figure 5 above. Most of these materials are usually good PV materials, such as Si, GaAs, and CIGS [Green *et al.*, 2018]. Since these materials are p-type semiconductors (electron as minority carrier), they are generally used as a photocathode. The challenge for these narrow gap materials as a photocathode in water splitting application, however, is their poor corrosion resistance in an aqueous solution especially in strong acidic solution. Metal oxides protection layers have been used to enhance the stability of these materials under PEC operating condition [Bae *et al.*, 2017].

For materials with bandgap around 2 eV, Fe_2O_3 (hematite) is widely considered as a frontrunner as a photoanode. With this bandgap, its theoretical STH efficiency is 15%, which is required for commercial possibility [Gurudayal *et al.*, 2018]. Readers can refer to [Gurudayal *et al.*, 2018] for more comprehensive review on recent development of the hematite. For materials with bandgap >2eV, $BiVO_4$ has been widely studied due to cheapness and its relatively negative CB position close to the thermodynamic level of H_2. Readers can refer to [Abdi and Berglund, 2017] and [Park *et al.*, 2013] for more detailed review on this material. However, its relatively wide gap of 2.4 eV means that the maximum photocurrent is only ~7.5 mA/cm^2 [Abdi and Berglund, 2017], translating to overall STH efficiency <10%. This limitation has motivated the search for narrower gap metal oxide materials within the vanadate class. For instance, Iron Vanadate ($FeVO_4$) has been recently demonstrated as promising metal oxides for a photoanode with its relatively narrower band gap (2.1 eV) [Zhang *et al.*, 2018]. Copper vanadate also recently emerged via high throughput combinatorial synthesis and shown to have a band gap near 2 eV [Zhou *et al.*, 2015]. Although both vanadate materials have a narrower band gap than $BiVO_4$, the short carrier diffusion length of both materials results in a low photoelectron currant of less than 1 mA/cm^2.

3.1.3. *Electronic properties*

Although the semiconductors with appropriate energy properties with respect to the water reduction/oxidation levels, in order to split water efficiently it depends strongly on the inherent electronic properties of the semiconductors. A good photoelectrode should have sufficient carrier diffusion length, which is determined by the carrier lifetime and mobility. Ideally the carrier diffusion length should be longer than the absorption depth so all photogenerated carriers (electrons and holes) can be efficiently transported to the semiconductor/liquid interface. The semiconductor should also have good conductivity so the majority carriers could be extracted effectively to the external circuit without recombining with the minority carriers needed for water redox reactions.

Introduction of metal dopants had been performed by numerous groups to alter the electronic properties of a poor performing semiconductor. Hematite, for instance, has both low electrical conductivity of $10^{-14}\ \Omega^{-1}cm^{-1}$ and electron mobility of $10^{-2}\ cm^2\ V^{-1}\ s^{-1}$ [Liao *et al.*, 2011, Liao *et al.*, 2012]. Addition of tetravalent dopants (Mn, Sn, Ti, Zr and Si) to replace some of the Fe^{3+} cations can increase their electrical conductivity due to the extra electrons donated to the crystal [Gurudayal *et al.*, 2014, 2017; Deng *et al.*, 2012; Kumar *et al.*, 2012; Cesar *et al.*, 2008]. Gurudayal *et al.* showed an enhancement of the oxygen evolution reactions kinetics of hematite by doping with Mn^{4+}, which has a similar ionic radius (53 pm) with Fe^{3+} (55 pm) [Gurudayal *et al.*, 2014]. Intensity-modulated photocurrent spectroscopy (IMPS) measurements reveal that the Mn doping not only increases the rate constant of hole transfer but also reduces the rate constant for surface recombination [Gurudayal *et al.*, 2017]. Undoped $BiVO_4$ also suffers from poor charge transfer properties due to its low electrical conductivity [Abdi and Berglund, 2017]. Small amount of Mo^{6+} or W^{6+} binding at the V^{5+} site of $BiVO_4$ is shown to increase the photocurrent significantly [Antony *et al.*, 2016]. The doping strategy is also effective in improving the PEC performance of photocathode

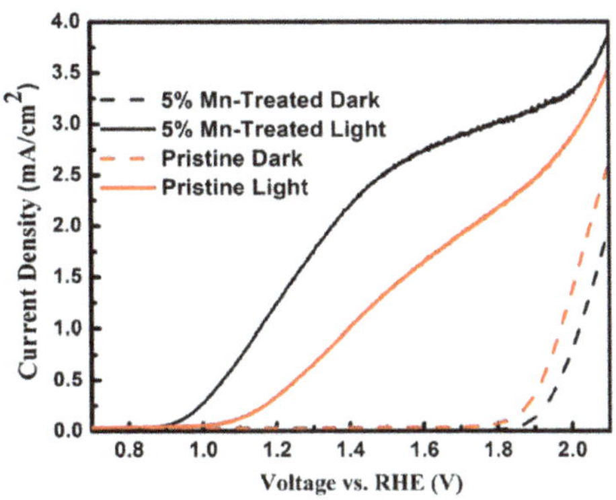

(a)

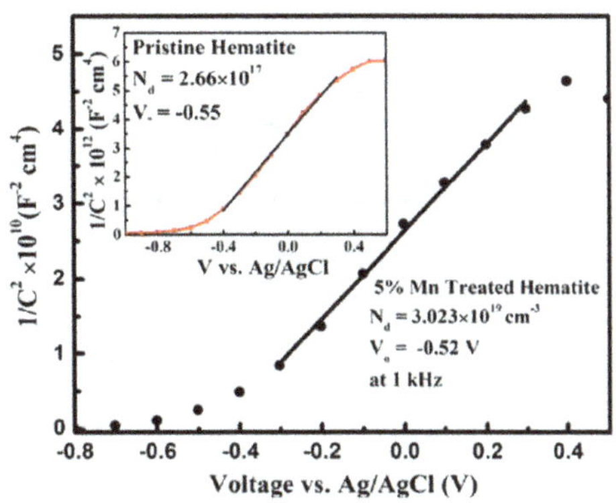

(b)

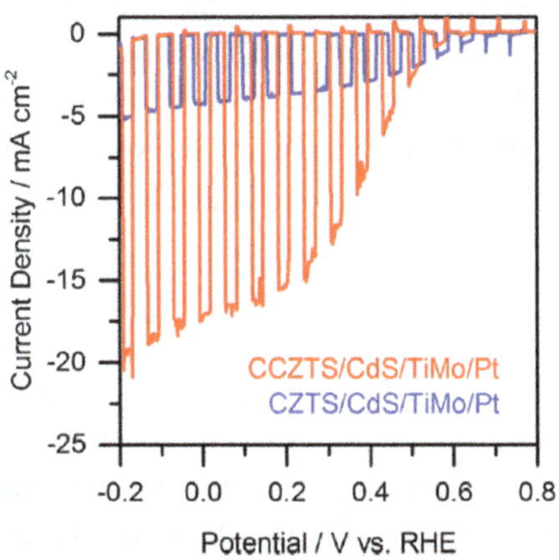

(c)

Figure 6. (a) Current density-potential curves for hematite with and without addition of 5%Mn under light illumination and dark which shows the beneficial effect of addition of Mn [Gurudayal *et al.*, 2014] (b) Capacitance-voltage curves for undoped and Mn-doped hematite which shows a higher doping density for Mn-doped hematite. Reprinted with permission from American Chemical Society. (c) Current density — potential curves for Cu_2ZnSnS_4 (CZTS) and Cd-doped CZTS (CCZTS) photocathodes modified with CdS/TiMo/Pt showing significantly higher photocurrent for the CCZTS photocathode [Tay *et al.*, 2018]. (d) Femtosecond transient reflectivity decay dynamics of CZTS and CCZTS at their photobleaching wavelength and their parameters obtained through fitting. The fitting indicated that carrier dynamics in CZTS were dominated by trapping and recombination indicated a large amount of surface and bulk defects, while the CCZTS had lower density of these defects. Reprinted with permission from Elsevier.

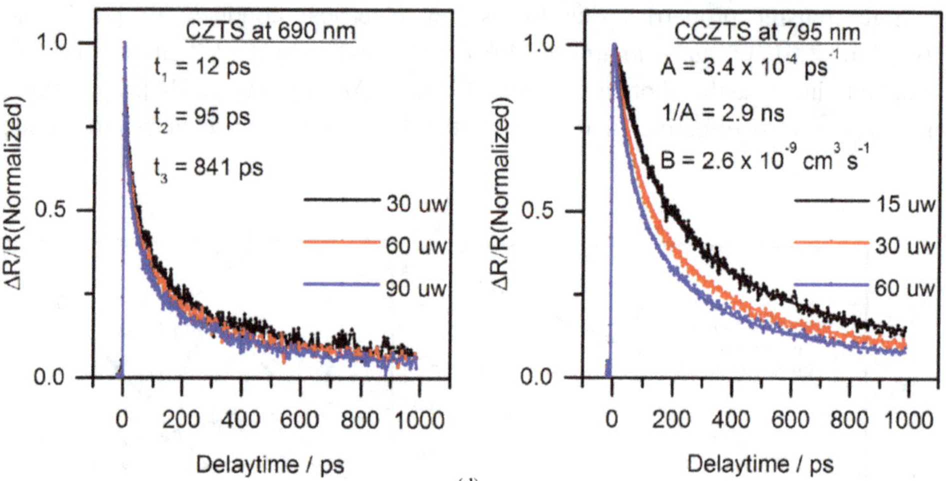

(d)

Figure 6. (*Continued*)

based on p-type semiconductors. Substitution of Cd into Cu_2ZnSnS_4 (CZTS) to form $Cu_2Cd_{0.4}Zn_{0.6}SnS_4$ (CCZTS) increases the majority carrier mobility of the CZTS which resulted in enhancement of the photocurrent of the PEC water splitting [Tay *et al.*, 2018]. Partial substitution of Ag into Cu lattice on $Cu(In,Ga)S_2$ (CIGS) was demonstrated not only increasing the grain sizes of the CIGS but also altering its electronic defects [Septina *et al.*, 2017].

3.1.4. *Morphology*

One method to improve carrier transport properties of materials with a short diffusion length is to do nanostructuring. Most of the metal oxides for photoanodes has a diffusion length only a few to tens of nanometer [Abdi *et al.*, 2013; Sivula *et al.*, 2011, Hong *et al.*, 2009]. Therefore, the majority of the reports for photoanodes showing a photocurrent higher than 1 mA/cm^2 typically used nanostructure [Huang *et al.*, 2014; Lin *et al.*, 2014]. The reader can refer to [Osterloh, 2013] for detailed review on inorganic nanostructures for water splitting. A good example of this is hematite, with its low hole diffusion length (2–4nm). Figure 7 shows the effect of nanostructuring hematite on the photoelectrochemical performance [Lin *et al.*, 2011]. Lin and colleagues used $TiSi_2$ as a structural support for the nanostructure as well as a charge collector. The nanostructured hematite photoanode achieved a higher photocurrent than its planar equivalent (Figure 7b), which was confirmed by the higher EQE (Figure 7c).

In the case where the semiconductor absorber has an appreciable diffusion length (>100 nm), as in the case of many photovoltaic materials for photocathode

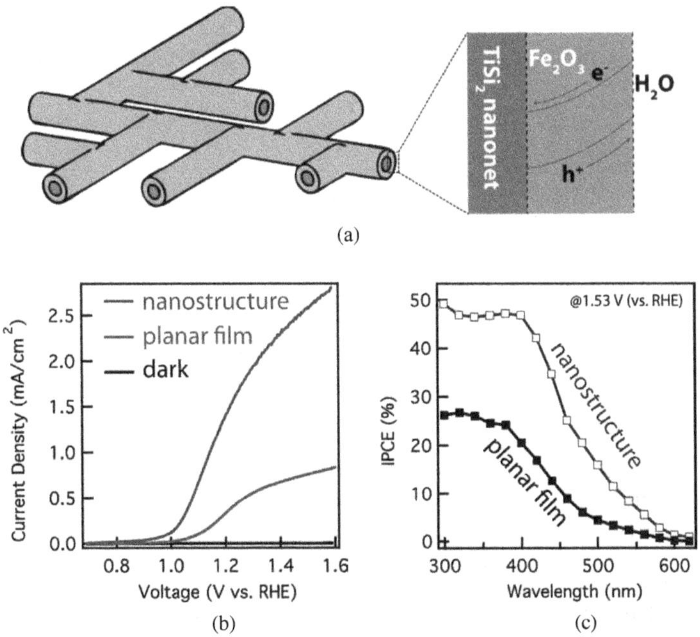

Figure 7. (a) Hematite nanostructure with $TiSi_2$ nanonet, (b) the comparison of photoelectrochemical performance between nanostructure and planar hematite showing higher photocurrent from the nanostructure hematite, and (c) external quantum efficiency comparison between them showing better collection of the nanostructure across the wavelength spectrum. Reprinted with permission from [Lin *et al.*, 2011] with permission from American Chemical Society.

such CIGS, CZTS, and GaAs, thin film structure is more appropriate to take advantage of collection of charges in the deeper part of the semiconductor films.

3.2. *Charge separation*

Upon the formation of photogenerated electron-hole pairs in the semiconductor, the charges have to be separated so that the minority carrier (electron in p-type semiconductor, hole in p-type semiconductor) can travel to the reaction sites at the semiconductor/electrolyte interface while the majority carriers move toward the back-contact of the device. When a semiconducting electrode is in contact with an electrolyte, a potential difference between them creates an electric field inside a depletion layer which can drive the movement of these charges toward the respective sites. This electric field, as well as the depletion layer width, varies depending on the doping level of the semiconductor and also condition of the semiconductor/electrolyte interface (energy alignment, surface state). Typically, due to a non-ideal energy alignment between the semiconductor and the electrolyte, the photovoltage generated from the system is far from what is expected from their band gap. Some strategies in overcoming this issue are explained below.

3.2.1. *Heterojunction*

Imposing asymmetry by depositing another semiconductor layer on top of the main semiconductor absorber can often enhance the charge separation and transport to the respective contact. The formation of p-n heterojunction semiconductors will enhance the charge separation due to a more optimal band alignment compared to just merely a semiconductor/electrolyte junction. For example, pairing of Cu_2O photocathode with n-type semiconductors such as ZnO or Ga_2O_3 enhanced the photocurrent greatly compared to bare Cu_2O [Paracchino *et al.*, 2011; Niu *et al.*, 2018]. Depositions of n-type CdS to CIGS [Ikeda *et al.,* 2013], CZTS [Jiang *et al.*, 2015], CuO [Septina *et al.*, 2017] were also demonstrated to be effective in boosting the photocurrent and the photovoltage.

When considering suitable pairing of semiconductors to form heterojunction, band alignment matching should first be considered. Figure 8 shows various types

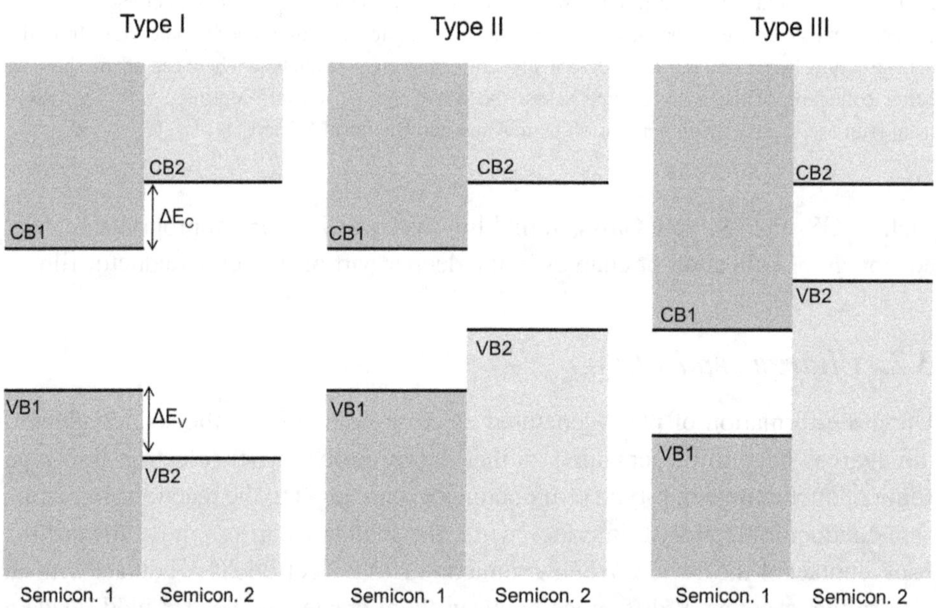

Figure 8. Different types of band alignments in semiconductor heterojuctions. In type 1, the conduction and valence bands of semiconductor 1 (CB1 and VB1) are straddled between the conduction band and valence bands of semiconductor 2 (CB2 and VB2). In this case, the semiconductor 2 has a wider band gap than the semiconductor 1. In type 2, the CB1 and VB1 are staggered in between the CB2 and VB2. In type 3, the bands of semiconductor 1 are overlapped with those that semiconductor 2. ΔE_C and ΔE_V are conduction band and valence band offsets, respectively.

of band alignments. In a photocathode system, to promote an efficient electron transfer from the conduction band of the p-type semiconductor to the n-type semiconductor, the conduction band offset (ΔE_C) should be as minimal as possible to prevent a barrier for electron transfer. In addition, the ΔE_C and ΔE_V should be positive to maximize the photovoltage. Therefore, type I band alignment with relatively small ΔE_C is typically desirable for photocathode (Figure 8). As an example, Gunawan *et al.* showed that a photocathode based on $CuInS_2$ coated with n-type In_2S_3 had better photoelectrochemical performance compared with the one coated with CdS (Figure 9a). The former demonstrated half-cell solar-to-hydrogen efficiency (HC-STH) of

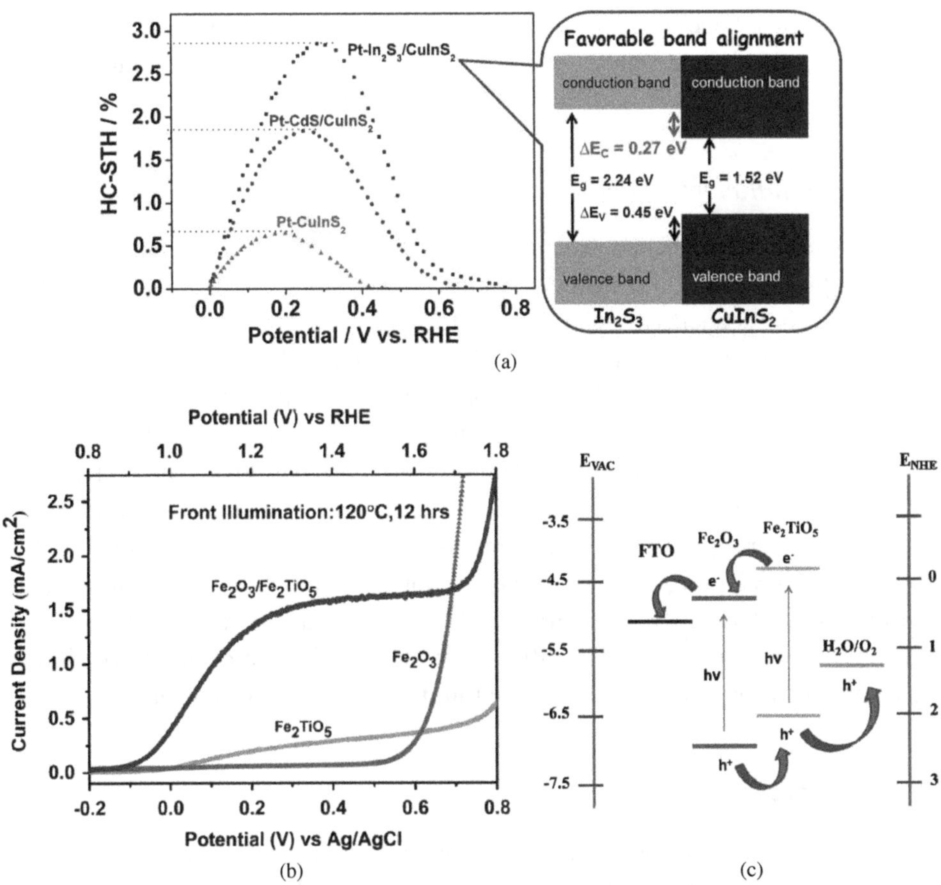

Figure 9. Example of the effect of formation of buried junction. (a) Half-cell solar-to-hydrogen (HC-STH) efficiencies of $CuInS_2$ coated with In_2S_3/Pt, CdS/Pt, and only Pt (co-catalyst). $CuInS_2$ coated with In_2S_3/Pt (Pt-In_2S_3/$CuInS_2$) had the highest HC-STH due to favorable band alignment between $CuInS_2$ and In_2S_3. Reprinted from [Gunawan *et al.*, 2015] with permission from American Chemical Society. (b) Photoelectrochemical performance of photoanode based on Fe_2O_3/Fe_2TiO_5 junction and (c) band alignment between Fe_2O_3 and Fe_2TiO_5. Reprinted from [Bassi *et al.*, 2016] with permission from Elsevier.

2.9% while the latter showed HC-STH of 1.8% [Gunawan *et al.,* 2013, 2015]. Evaluation of the conduction band offset of $CuInS_2/In_2S_3$ and $CuInS_2/CdS$ using x-ray photoelectron spectroscopy revealed a favorable positive offset for the former junction while the latter junction had an unfavorable negative conduction band offset [Gunawan *et al.,* 2015]. For photoanode, the VBM of the upper layer should be lower in energy than the underlying semiconductor to promote hopping transfer of the holes. Type II band alignment is generally considered for the photoanode system (Figure 8). For example, Bassi *et al.* reported an enhancement of the photocurrent of Fe_2O_3 nanorods when coated with Fe_2TiO_5 nanoporous approaching 1.4 mA/cm^2 at 1.23 V *versus* RHE compared to only 0.01 mA/cm^2 when only pristine Fe_2O_3 was used (Figure 9b) [Bassi *et al.,* 2016]. Fe_2TiO_5 is shown to have a favorable conduction and valence band positions with Fe_2O_3 (Figure 9c) which enhanced the charge separation and hole injection to the electrolyte. The reader can also refer to [Kaneko *et al.*, 2018] for review on surface modification in photoelectrodes.

3.2.2. *Internal electric field in the bulk materials*

Another strategy to enhance the charge separation process is by creating an intentional doping gradient throughout the film. Wang *et al.* showed a concentration gradient of Cu vacancies in a $CuBi_2O_4$ photocathode via two-step diffusion assisted spray pyrolysis process (Figure 10a) [Wang *et al.,* 2017]. The doping profiles introduced an electric field that enhanced the separation of charge carriers. The $CuBi_2O_4$ photocathode with forward gradient showed higher photocurrent than the homogenous and reverse gradient when measured using H_2O_2 as electron scavengers. Similar doping strategies have also been adopted for metal oxide photoanodes by many groups. Antony *et al.* demonstrated enhancement of photocurrent of nanoporous $BiVO_4$ through a gradient Mo-doping in the bulk of the film (Figure 10b) [Antony *et al.,* 2018]. The gradient Mo-doping was found to enhance charge separation efficiency, which is confirmed by the observation of ~200 mV shift of onset potential toward negative potential in its current density-potential curve.

3.3. *Catalyst*

Even if the band positions of the semiconductors could satisfy the thermodynamic requirement to reduce/oxidize water, generally the hydrogen/oxygen evolution reactions do not occur efficiently due to kinetic issues. A catalyst deposited on the photoabsorber reduces the overpotential necessary to perform the hydrogen and oxygen evolution reactions. Readers can refer to benchmark of various catalysts for hydrogen evolution and oxygen evolution reactions by McCrory *et al.* [McCrory *et al.*, 2015] (Figure 11). They give an overview of the required overpotential to support the reactions. The actual role of the

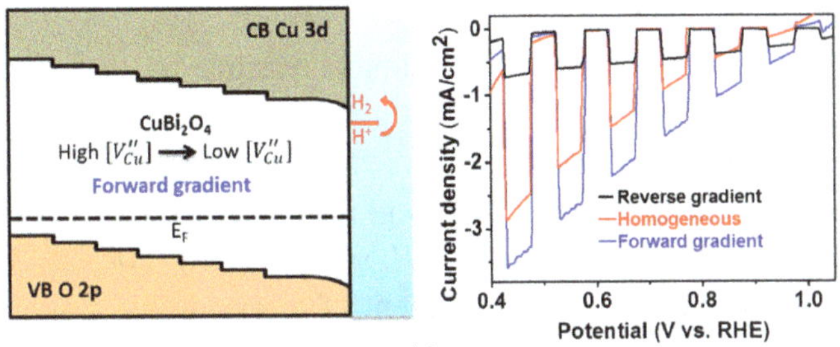

(a)

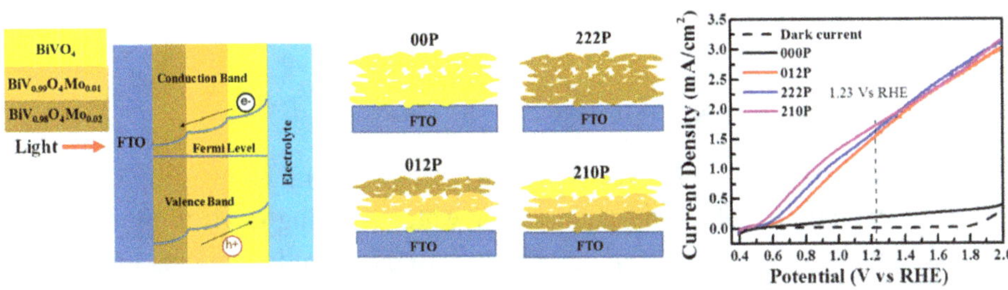

(b)

Figure 10. (a) Band diagram of $CuBi_2O_4$ with forward gradient of Cu vacancies and its photoelectrochemical performance compared to homogenous and reverse gradient measured with H_2O_2 as electron scavenger. Reprinted from [Wang *et al.,* 2017] with permission from American Chemical Society. (b) Band diagram of various Mo-gradient doped $BiVO_4$ and their photoelectrochemical performance. Reprinted from [Antony *et al.,* 2018] with permission from American Chemical Society.

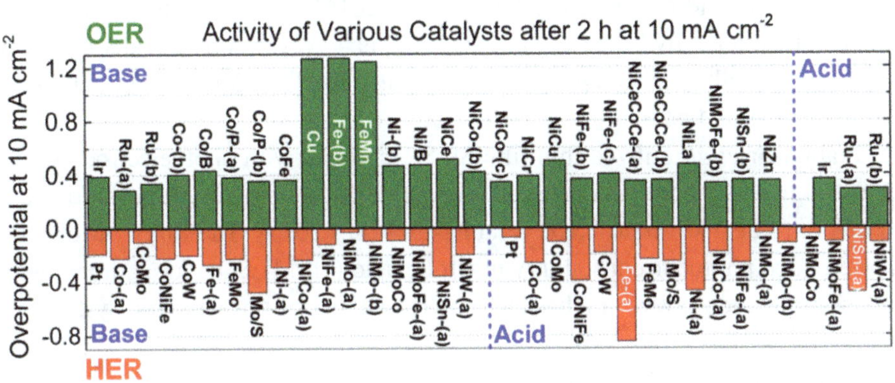

Figure 11. Overpotential at 10 mA cm^{-2} of various catalysts for hydrogen and oxygen evolution reactions. Reprinted from [McCrory *et al.*, 2015] with permission from American Chemical Society.

catalyst upon integration with a semiconductor surface can serve one of the following [Montoya *et al.,* 2017]:

(i) catalyzing the bond-making and bond-breaking reactions,
(ii) passivizing recombination sites,

(iii) tuning band structure,
(iv) protecting the semiconductor from photocorrosion.

Since there have been many extensive reviews exclusively for the electrocatalyst in photoelectrochemical water splitting, readers can refer to such articles by [Zou and Zhang, 2015], [Suen *et al.*, 2017], and [Vesborg *et al.*, 2015]. Below we summarize the snapshot of the development of emerging catalysts for hydrogen evolution and oxygen evolution reactions, which are integrated into photoelectrodes for water splitting.

3.3.1. *Hydrogen evolving catalyst*

Platinum is the most widely used hydrogen evolving catalyst to date. According to recent benchmark of catalysts for solar water splitting devices by McCrory *et al.* (Figure 11) [McCrory *et al.*, 2015], Pt based electrodes showed the highest activity in acidic solution with overpotential of ~ –40 mV to achieve 10 mA cm^{-2}. Therefore, it is usually used as a benchmark when finding a more cost-effective catalyst using earth-abundant materials. Among the myriad of non-noble metals catalyst for hydrogen evolution reaction, MoS_2 has been the most studied for its integration into the photocathode system. MoS_2 has been applied to several photocathode materials such as Cu_2O [Morales *et al.*, 2014], Si [Seger *et al.*, 2012], and Sb_2Se_3 [Prabhakar *et al.*, 2017]. MoS_2 typically has an overpotential of ~200 mV at acidic condition [Benck *et al.*, 2014]. Although not as efficient as Pt in catalyzing the hydrogen evolution, MoS_2 is however, robust under operation condition as wellbeing straight forward to synthesis. Thus MoS_2 is very attractive as a low-cost catalyst.

Prabhakar *et al.* utilized thin film Sb_2Se_3 coupled with photoelectrodeposited MoS_x (Figure 12) and demonstrated relatively high photocurrent densities up to 16 mA cm^{-2} at 0 V *versus* RHE [Prabhakar *et al.*, 2017]. The Sb_2Se_3-MoS_x also showed good photocurrent stability in strong acid (1 M H_2SO_4) for 20 hours without any protection layers emphasizing the robustness of this catalytic system.

3.3.2. *Oxygen evolving catalyst*

The water oxidation reaction is a more of a challenge compared with water reduction since it involves four electron transfers contrary to two electron transfers for the water reduction reaction. Due to this reason, exploration of catalyst for oxygen evolution reaction (OER) has been exclusively performed in basic solution. Ruthenium and Iridium-based materials are the most commonly used catalyst for OER, which is shown to have the lowest overpotential of around 300 mV (for Ru)

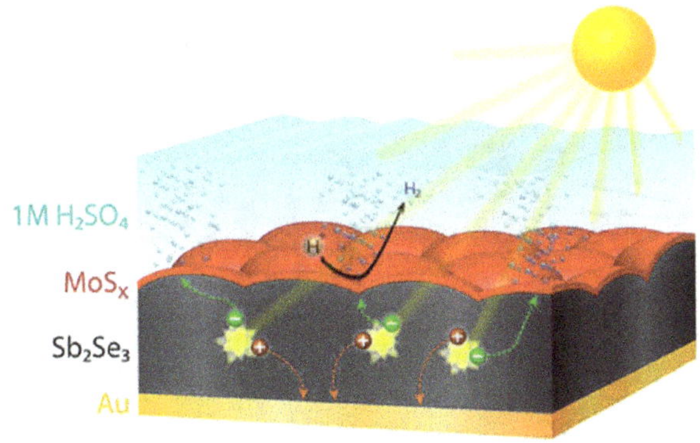

(a)

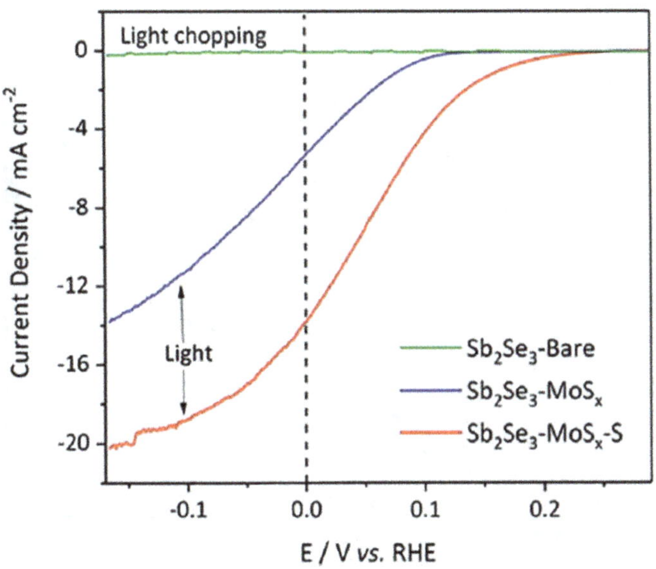

(b)

Figure 12. (a) Structure of Sb_2Se_3-MoS$_x$ photocathode, (b) photoelectrochemical performance of Sb_2Se_3 bare, Sb_2Se_3-MoS$_x$ and those that with post-sulfurization treatment Sb_2Se_3-MoS$_x$-S. Reproduced from [Prabhakar *et al.*, 2017] with permission from The Royal Society of Chemistry.

in the basic condition [McCrory *et al.*, 2015], as can be seen from Figure 11. The rarity of the noble metals as well as the need to search for an earth-abundant catalyst with much lower overpotential for oxygen evolution prompted researchers to explore other alternative catalysts, especially one that can be integrated into a PEC device.

Hematite is one of the most studied material platforms for the exploration and integration of the OER into a PEC device. Hematite itself has quite high

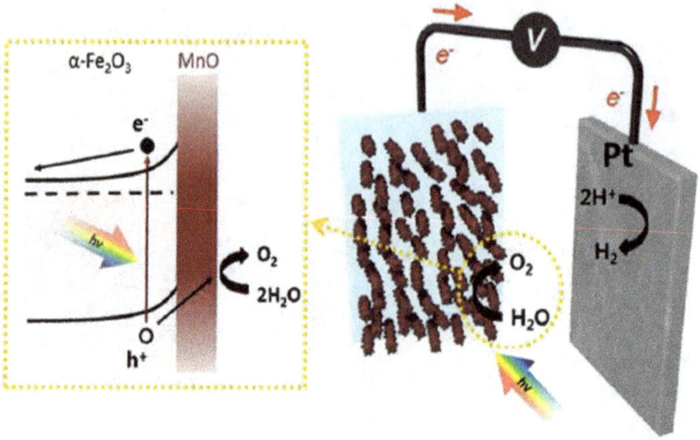

(a)

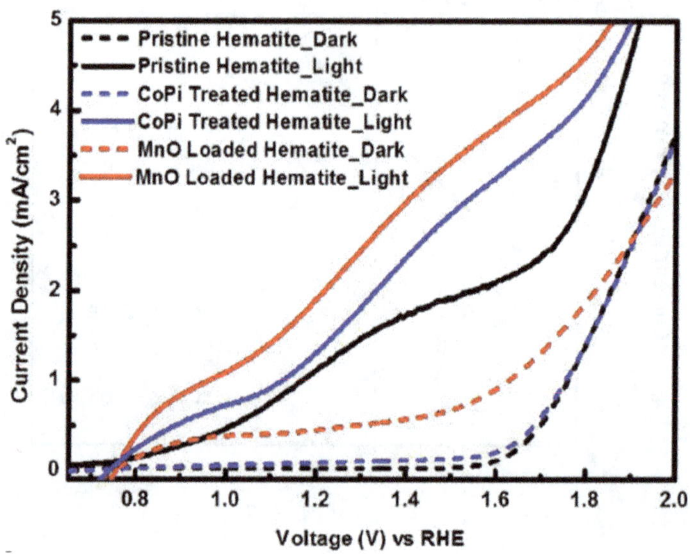

(b)

Figure 13. (a) Illustration of hematite nanorod loaded with MnO nanoparticles, and (b) its photoelectrohemical performance compared with bare and Co-Pi treated hematite. Reprinted from [Gurudayal *et al.,* 2016] with permission from John Wiley and Sons.

overpotential for oxygen production from water. Therefore, it is usually considered necessary to use co-catalyst to enhance the oxygen evolution kinetics on the surface of the hematite. Cobalt-based catalyst such as cobalt oxides or cobalt-phosphates (CoPi) have been extensively used as the co-catalyst for the hematite [Roger *et al.*, 2017; Artero *et al.*, 2011]. They can be deposited on hematite with various mild methods such as electrodeposition [Zong *et al.,* 2013] and hydrothermal [Xi *et al.*, 2012]. Their overpotential is typically around 400–500 mV (at 10 mA cm^{-2})

depending on the preparation methods [Roger *et al.*, 2017]. Another emerging catalyst for oxygen evolution reaction is Mn-based oxides (Figure 13a). [Gurudayal *et al.*, 2016]. In this study, it was postulated that the MnO layer improved the charge injection from the hematite to the electrolyte, while the beneficial effect of the CoPi layer originated from surface passivation effect.

4. Summary

In this chapter, we presented an overview of the design requirements of an artificial photosynthesis system based on semiconductor photoelectrodes. Three photoelectrochemical (PEC) device designs are considered: (i) a single semiconductor/electrolyte junction, (ii) tandem system involving a photocathode and a photoanode and (iii) a photovoltaic/photoelectrode tandem system. We overview the material being explored to optimize the basic processes in the PEC/Artificial Photosynthesis systems such as light absorption, charge separation, and electrochemical catalysis. For optimized light absorption, the intrinsic thermodynamic energy levels, electronic properties, and band gaps of the semiconductors are crucial for material selection. To promote a more efficient charge separation and transfer from the semiconductor to the reaction site, the formations of heterojunction has been widely adopted. Deposition of catalyst becomes a definite prerequisite to reduce the required overpotential to split water, and there is a drive in the scientific community to find earth abundant catalysts with similar catalytic activities as the noble metals.

References

Abdi, F.F., Han, L., Smets, A.H., Zeman, M., Dam, B. and van de Krol, R. (2013). Efficient solar water splitting by enhanced charge separation in a bismuth vanadate-silicon tandem photoelectrode. *Nat. Commun.*, 4, 2195.

Abdi, F.F., Savenije, T.J., May, M.M., Dam, B. and van de Krol, R. (2013). The origin of slow carrier transport in $BiVO_4$ thin film photoanodes: a time-resolved microwave conductivity study. *J. Phys. Chem. Lett.*, 4(16), 2752–2757.

Abdi, F.F. and Berglund, S.P. (2017). Recent developments in complex metal oxide photoelectrodes. *J. Phys. D: Appl. Phys.*, 50(19), 193002.

Ager J.W. (2016). Photoelectrochemical Approach for Water Splitting. In: Sugiyama M., Fujii K., Nakamura S. (eds) Solar to Chemical Energy Conversion. Lecture Notes in Energy, Vol 32. Springer, Cham.

Antony, R.P., Bassi, P.S., Abdi, F.F., Chiam, S.Y., Ren, Y., Barber, J. and Wong, L.H. (2016). Electrospun Mo-$BiVO_4$ for efficient photoelectrochemical water oxidation: direct evidence of improved hole diffusion length and charge separation. *Electrochim. Acta*, 211, 173–182.

Antony, R.P., Zhang, M., Zhou, K., Loo, S.C.J., Barber, J. and Wong, L.H. (2018). Synergistic effect of porosity and gradient doping in efficient solar water oxidation of catalyst-free gradient Mo: $BiVO_4$. *ACS Omega*, 3(3), 2724–2734.

Artero, V., Chavarot-Kerlidou, M. and Fontecave, M. (2011). Splitting water with cobalt. *Angew. Chem. Int. Ed.*, 50(32), 7238–7266.

Bae, D., Seger, B., Vesborg, P.C., Hansen, O. and Chorkendorff, I. (2017). Strategies for stable water splitting via protected photoelectrodes. *Chem. Soc. Rev.*, 46(7), 1933–1954.

Barber, J. and Tran, P.D. (2013). From natural to artificial photosynthesis. *J. Roy. Soc. Interface*, 10(81), 20120984.

Barroso, M., Pendlebury, S.R., Cowan, A.J. and Durrant, J.R. (2013). Charge carrier trapping, recombination and transfer in hematite (α-Fe_2O_3) water splitting photoanodes. *Chem. Sci.*, 4(7), 2724–2734.

Bassi, P.S., Chiam, S.Y., Barber, J. and Wong, L.H. (2014). Hydrothermal grown nanoporous iron based titanate, Fe_2TiO_5 for light driven water splitting. *ACS Appl. Mater. Inter.*, 6(24), 22490–22495.

Bassi, P.S., Antony, R.P., Boix, P.P., Fang, Y., Barber, J. and Wong, L.H. (2016). Crystalline Fe_2O_3/Fe_2TiO_5 heterojunction nanorods with efficient charge separation and hole injection as photoanode for solar water oxidation. *Nano Energy*, 22, 310–318.

Benck, J.D., Hellstern, T.R., Kibsgaard, J., Chakthranont, P. and Jaramillo, T.F. (2014). Catalyzing the hydrogen evolution reaction (HER) with molybdenum sulfide nanomaterials. *ACS Catal.*, 4(11), 3957–3971.

Brillet, J., Yum, J.H., Cornuz, M., Hisatomi, T., Solarska, R., Augustynski, J. and Sivula, K. (2012). Highly efficient water splitting by a dual-absorber tandem cell. *Nat. Photonics*, 6(12), 824.

Chen, S. and Wang, L.W. (2012). Thermodynamic oxidation and reduction potentials of photocatalytic semiconductors in aqueous solution. *Chem. Mater.*, 24(18), 3659–3666.

Cesar, I., Sivula, K., Kay, A., Zboril, R. and Grätzel, M. (2008). Influence of feature size, film thickness, and silicon doping on the performance of nanostructured hematite photoanodes for solar water splitting. *J. Phys. Chem. C*, 113(2), 772–782.

Deng, J., Zhong, J., Pu, A., Zhang, D., Li, M., Sun, X. and Lee, S.T. (2012). Ti-doped hematite nanostructures for solar water splitting with high efficiency. *J. Appl. Phys.*, 112(8), 084312.

Fountaine, K.T., Lewerenz, H.J. and Atwater, H.A. (2016). Efficiency limits for photoelectrochemical water-splitting. *Nat. Commun.,* 7, 13706.

Fujishima, A. and Honda, K. (1972) Electrochemical photolysis of water at a semiconductor electrode. *Nature,* 238, 37–38.

Gunawan, Septina, W., Ikeda, S., Harada, T., Minegishi, T., Domen, K. and Matsumura, M. (2014). Platinum and indium sulfide-modified $CuInS_2$ as efficient photocathodes for photoelectrochemical water splitting. *Chem. Comm.*, 50(64), 8941–8943.

Gunawan, Septina, W., Harada, T., Nose, Y. and Ikeda, S. (2015). Investigation of the electric structures of heterointerfaces in Pt-and In_2S_3-modified $CuInS_2$ photocathodes used for Sunlight-induced hydrogen evolution. *ACS Appl. Mater. Inter.*, 7(29), 16086–16092.

Gurudayal, John, R.A., Boix, P.P., Yi, C., Shi, C., Scott, M.C., Veldhuis, S.A. and Mathews, N. (2017). Atomically altered hematite for highly efficient perovskite tandem water-splitting devices. *ChemSusChem*, 10(11), 2449–2456.

Gurudayal, Chiam, S.Y., Kumar, M.H., Bassi, P.S., Seng, H.L., Barber, J. and Wong, L.H. (2014). Improving the efficiency of hematite nanorods for photoelectrochemical water splitting by doping with manganese. *ACS Appl. Mater. Inter.*, 6(8), 5852–5859.

Gurudayal, Peter, L.M., Wong, L.H. and Abdi, F.F. (2017). Revealing the influence of doping and surface treatment on the surface carrier dynamics in hematite nanorod photoanodes. *ACS Appl. Mater. Inter.*, 9(47), 41265–41272.

Gurudayal, Jeong, D., Jin, K., Ahn, H.Y., Boix, P.P., Abdi, F.F., Mathews, N. and Wong, L.H. (2016). Highly active MnO catalysts integrated onto Fe_2O_3 nanorods for efficient water splitting. *Adv. Mater. Inter.*, 3(15), 1600176.

Gurudayal, Bassi, P.S., Sritharan, T. and Wong, L.H. (2018). Recent progress in iron oxide based photoanodes for solar water splitting. *J. Phys. D: Appl. Phys.*, in press https://doi.org/10.1088/1361–6463/aae138.

Green, M.A., Hishikawa, Y., Dunlop, E.D., Levi, D.H., Hohl-Ebinger, J. and Ho-Baillie, A.W. (2018). Solar cell efficiency tables (version 52). *Prog. Photovoltaics*, 26(7).

Huang, Z.F., Pan, L., Zou, J.J., Zhang, X. and Wang, L. (2014). Nanostructured bismuth vanadate-based materials for solar-energy-driven water oxidation: a review on recent progress. *Nanoscale*, 6(23), 14044–14063.

Huang, D., Wang, K., Yu, L., Nguyen, T.H., Ikeda, S. and Jiang, F. (2018). Over 1% efficient unbiased stable solar water splitting based on a sprayed Cu_2ZnSnS_4 photocathode protected by a HfO_2 photocorrosion-resistant film. *ACS Energy Lett.*, 3(8), 1875–1881.

Hong, S.J., Jun, H., Borse, P.H. and Lee, J.S. (2009). Size effects of WO_3 nanocrystals for photooxidation of water in particulate suspension and photoelectrochemical film systems. *Int. J. Hydrogen Energy*, 34(8), 3234–3242.

Ikeda, S., Nonogaki, M., Septina, W., Gunawan, G., Harada, T. and Matsumura, M. (2013). Fabrication of $CuInS_2$ and $Cu(In,Ga)S_2$ thin films by a facile spray pyrolysis and their photovoltaic and photoelectrochemical properties. *Catal. Sci. Technol.*, 3(7), 1849–1854.

Jiang, F., Harada, T., Kuang, Y., Minegishi, T., Domen, K. and Ikeda, S. (2015). Pt/In2S3/CdS/Cu2ZnSnS4 thin film as an efficient and stable photocathode for water reduction under sunlight radiation. *J. Am. Chem. Soc.*, 137(42), 13691–13697.

Kaneko, H., Minegishi, T. and Domen, K. (2018). Recent progress in the surface modification of photoelectrodes toward efficient and stable overall water splitting. *Chem. Eur. J.*, 24(22), 5697–5706.

Klein, A. (2015). Energy band alignment in chalcogenide thin film solar cells from photoelectron spectroscopy. *J. Phys.: Condens. Matter*, 27(13), 134201.

Kobayashi, H., Sato, N., Orita, M., Kuang, Y., Kaneko, H., Minegishi, T. and Domen, K. (2018). Development of highly efficient $CuIn_{0.5}Ga_{0.5}Se_2$-based photocathode and application to overall solar driven water splitting. *Energy Environ. Sci.*, DOI: 10.1039/C8EE01783B.

Kumar, P., Sharma, P., Shrivastav, R., Dass, S. and Satsangi, V.R. (2011). Electrodeposited zirconium-doped α-Fe_2O_3 thin film for photoelectrochemical water splitting. *Int. J. Hydrogen Energy*, 36(4), 2777–2784.

Liao, P., Toroker, M.C. and Carter, E.A. (2011). Electron transport in pure and doped hematite. *Nano Lette.*, 11(4), 1775–1781.

Liao, P. and Carter, E.A. (2012). Hole transport in pure and doped hematite. *J. Appl.Phys.*, 112(1), 013701.

Lin, Y., Zhou, S., Sheehan, S.W. and Wang, D. (2011). Nanonet-based hematite heteronanostructures for efficient solar water splitting. *J. Am. Chem. Soc.*, 133(8), 2398–2401.

Lin, Y., Yuan, G., Liu, R., Zhou, S., Sheehan, S.W. and Wang, D. (2011). Semiconductor nanostructure-based photoelectrochemical water splitting: A brief review. *Chem. Phys. Lett.*, 507(4–6), 209–215.

McCrory, C.C., Jung, S., Ferrer, I.M., Chatman, S.M., Peters, J.C. and Jaramillo, T.F. (2015). Benchmarking hydrogen evolving reaction and oxygen evolving reaction electrocatalysts for solar water splitting devices. *J. Am. Chem. Soc.*, 137(13), 4347–4357.

Montoya, J.H., Seitz, L.C., Chakthranont, P., Vojvodic, A., Jaramillo, T.F. and Nørskov, J.K. (2017). Materials for solar fuels and chemicals. *Nat. Mater.*, 16(1), 70.

Morales-Guio, C.G., Tilley, S.D., Vrubel, H., Grätzel, M. and Hu, X. (2014). Hydrogen evolution from a copper (I) oxide photocathode coated with an amorphous molybdenum sulphide catalyst. *Nat. Commun.*, 5, 3059.

Murphy, A.B., Barnes, P.R.F., Randeniya, L.K., Plumb, I.C., Grey, I.E., Horne, M.D. and Glasscock, J.A. (2006). Efficiency of solar water splitting using semiconductor electrodes. *Int. J. Hydrogen Energy,* 31, 1999–2017.

Niu, W., Moehl, T., Cui, W., Wick-Joliat, R., Zhu, L. and Tilley, S.D. (2018). Extended light harvesting with dual Cu_2O-Based photocathodes for high efficiency water splitting. *Adv. Energy Mater.*, 8(10), 1702323.

Osterloh, F.E. (2013). Inorganic nanostructures for photoelectrochemical and photocatalytic water splitting. *Chem. Soc. Rev.*, 42(6), 2294–2320.

Pan, L., Kim, J.H., Mayer, M.T., Son, M.K., Ummadisingu, A., Lee, J.S. and Grätzel, M. (2018). Boosting the performance of Cu_2O photocathodes for unassisted solar water splitting devices. *Nat. Catal.*, 1(6), 412.

Paracchino, A., Laporte, V., Sivula, K., Grätzel, M. and Thimsen, E. (2011). Highly active oxide photocathode for photoelectrochemical water reduction. *Nat. Mater.*, 10(6), 456.

Park, Y., McDonald, K.J. and Choi, K.S. (2013). Progress in bismuth vanadate photoanodes for use in solar water oxidation. *Chem. Soc. Rev.*, 42(6), 2321–2337.

Pinaud, B.A., Benck, J.D., Seitz, L.C., Forman, A.J., Chen, Z., Deutsch, T.G. and Wang, H. (2013). Technical and economic feasibility of centralized facilities for solar hydrogen production via photocatalysis and photoelectrochemistry. *Energy Environ. Sci.*, 6(7), 1983–2002.

Prabhakar, R.R., Septina, W., Siol, S., Moehl, T., Wick-Joliat, R. and Tilley, S.D. (2017). Photocorrosion-resistant Sb_2Se_3 photocathodes with earth abundant MoS_x hydrogen evolution catalyst. *J. Mat. Chem. A*, 5(44), 23139–23145.

Qiu, Y., Liu, W., Chen, W., Zhou, G., Hsu, P.C., Zhang, R. and Cui, Y. (2016). Efficient solar-driven water splitting by nanocone $BiVO_4$-perovskite tandem cells. *Sci. Adv.*, 2(6), e1501764.

Roger, I., Shipman, M.A. and Symes, M.D. (2017). Earth-abundant catalysts for electrochemical and photoelectrochemical water splitting. *Nat. Rev. Chem.*, 1(1), 0003.

Seger, B., Laursen, A.B., Vesborg, P.C., Pedersen, T., Hansen, O., Dahl, S. and Chorkendorff, I.B. (2012). Hydrogen production using a molybdenum sulfide catalyst on a titanium-protected n+ p-silicon photocathode. *Angew. Chem. Inter. Ed.*, 51(36), 9128–9131.

Seok, S.I., Grätzel, M. and Park, N.G. (2018). Methodologies toward highly efficient perovskite solar cells. *Small*, 14(20), 1704177.

Septina, W., Ikeda, S., Harada, T., Higashi, M., Abe, R. and Matsumura, M. (2015). Photosplitting of water from wide-gap Cu(In, Ga)S_2 thin films modified with a CdS layer and Pt nanoparticles for a high-onset-potential photocathode. *J. Phys. Chem. C*, 119(16), 8576–8583.

Septina, W., Sugimoto, M., Chao, D., Shen, Q., Nakatsuka, S., Nose, Y. and Ikeda, S. (2017). Photoelectrochemical water reduction over wide gap (Ag,Cu)(In,Ga)S_2 thin film photocathodes. *Phys. Chem. Chem. Phys.*, 19(19), 12502–12508.

Septina, W., Prabhakar, R.R., Wick, R., Moehl, T. and Tilley, S.D. (2017). Stabilized solar hydrogen production with CuO/CdS heterojunction thin film photocathodes. *Chem. Mater.*, 29(4), 1735–1743.

Septina, W. and Tilley, S.D. (2017). Emerging Earth-abundant materials for scalable solar water splitting. *Curr. Opin. Electrochem.*, 2(1), 120–127.

Sivula, K. and van de Krol, R. (2016). Semiconducting materials for photoelectrochemical energy conversion. *Nat. Rev. Mater.*, 1(2), 15010.

Sivula, K., Le Formal, F. and Grätzel, M. (2011). Solar water splitting: progress using hematite (α-Fe_2O_3) photoelectrodes. *ChemSusChem*, 4(4), 432–449.

Suen, N.T., Hung, S.F., Quan, Q., Zhang, N., Xu, Y.J. and Chen, H.M. (2017). Electrocatalysis for the oxygen evolution reaction: recent development and future perspectives. *Chem. Soc. Rev.*, 46(2), 337–365.

Tay, Y.F., Kaneko, H., Chiam, S.Y., Lie, S., Zheng, Q., Wu, B. and Sum, T.C. (2018). Solution-processed Cd-substituted CZTS photocathode for efficient solar hydrogen evolution from neutral water. *Joule*, 2(3), 537–548.

Vesborg, P.C., Seger, B. and Chorkendorff, I.B. (2015). Recent development in hydrogen evolution reaction catalysts and their practical implementation. *J. Phys. Chem. Lett.*, 6(6), 951–957.

Waki, I., Cohen, D., Lal, R., Mishra, U., DenBaars, S.P. and Nakamura, S. (2007). Direct water photoelectrolysis with patterned n-GaN. *Appl. Phys. Lett.*, 91(9), 093519.

Walter, M.G., Warren, E.L., McKone, J.R., Boettcher, S.W., Mi, Q., Santori, E.A. and Lewis, N.S. (2010). Solar water splitting cells. *Chem. Rev.*, 110(11), 6446–6473.

Wang, F., Septina, W., Chemseddine, A., Abdi, F.F., Friedrich, D., Bogdanoff, P. and Berglund, S.P. (2017). Gradient self-doped $CuBi_2O_4$ with highly improved charge separation efficiency. *J. Am. Chem. Soc.*, 139(42), 15094–15103.

Xi, L., Tran, P.D., Chiam, S.Y., Bassi, P.S., Mak, W.F., Mulmudi, H.K. and Wong, L.H. (2012). Co_3O_4-decorated hematite nanorods as an effective photoanode for solar water oxidation. *J. Phys. Chem. C*, 116(26), 13884–13889.

Zhang, M., Ma, Y., Friedrich, D., van de Krol, R., Wong, L.H. and Abdi, F.F. (2018). Elucidation of the opto-electronic and photoelectrochemical properties of $FeVO_4$ photoanodes for solar water oxidation. *J. Mater. Chem. A*, 6(2), 548–555.

Zhou, L., Yan, Q., Shinde, A., Guevarra, D., Newhouse, P.F., Becerra-Stasiewicz, N. and Gregoire, J.M. (2015). High throughput discovery of solar fuels photoanodes in the CuO–V_2O_5 system. *Adv. Energy Mater.*, 5(22), 1500968.

Zong, X., Thaweesak, S., Xu, H., Xing, Z., Zou, J., Lu, G.M. and Wang, L. (2013). A scalable colloidal approach to prepare hematite films for efficient solar water splitting. *Phys. Chem. Chem. Phys.*, 15(29), 2314–12321.

Zou, X. and Zhang, Y. (2015). Noble metal-free hydrogen evolution catalysts for water splitting. *Chem. Soc. Rev.*, 44(15), 5148–5180.

Chapter 13

Synthesis, Electronic Structure, and Spectroscopy of Multinuclear Mn Complexes Relevant to the Oxygen Evolving Complex of Photosystem II

Heui Beom Lee, Paul H. Oyala and Theodor Agapie

California Institute of Technology
1200 E California Blvd, Pasadena, CA 91125, USA
agapie@caltech.edu

Biochemical, spectroscopic, structural, and computational studies have provided insight into the biological water oxidation by the Oxygen Evolving Complex [OEC] of Photosystem II. Nevertheless, the atomic level mechanism of O_2 formation remains a matter of debate. Mechanistic proposals have been guided by the spectroscopic characterization of reaction intermediates. A key intermediate in the catalytic cycle, the S_2 state, features two characteristic EPR signals at $g = 2$ and $g = 4.1$, with the relative intensity of the two signals being condition-dependent. Two nearly isoenergetic structural isomers have been proposed as the source of these distinct signals, but relevant structure — electronic structure studies remain rare. In general, studies that probe the effect of small structural changes on the spectroscopic and magnetic properties of synthetic model clusters of S-states are rare, likely due to the synthetic challenges of accessing a series of isolable clusters that are suitable for comparisons. In this chapter, structurally characterized tetranuclear complexes with Mn oxidation states relevant to S_0–S_3 states are surveyed. In particular, magnetic and spectroscopic studies of a series of $[Mn^{III}Mn_3^{IV}O_4]$ cuboidal complexes show that the electronic structure of tetranuclear Mn complexes is highly sensitive to small geometric changes and the nature of the bridging ligands. Thus, the deduction of the geometry of OEC S-state intermediates based on spectroscopic features needs to be complemented with appropriate structural determination.

1. Introduction: S-state Intermediates in the Oxygen Evolving Complex of Photosystem II

Despite extensive biochemical, spectroscopic, and computational studies, the mechanism of biological water oxidation by the Oxygen Evolving Complex (OEC) of Photosystem II remains a subject of significant debate [Cox, 2013; Pantazis, 2018; Shen, 2015; Yano and Yachandra, 2014]. Mechanistic studies are performed in the context of the Joliot-Kok cycle of S_n (n = 0–4) states [Joliot, 1965; Kok, 1970]. Starting from the most reduced S_0 form, light-induced sequential one-electron oxidations lead to progression to higher S_n states, culminating in the formation of the elusive S_4 state, where O_2 formation is proposed to occur (Figure 1) [Cox, 2014]. Importantly, four oxidizing equivalents are stored in the OEC prior to O–O bond formation, precluding the generation and release of harmful products such as peroxide or superoxide [Rutherford, 2012]. Thus, precise spatial and temporal resolution in the sequence of electron-transport, proton-transport, and substrate binding events is necessary to fully understand the mechanism of biological water oxidation. Structural determination and spectroscopic characterization of S_n state intermediates heavily influence mechanistic proposals for O–O bond formation [Boussac, 1998; Boussac and Rutherford, 2000; Britt,

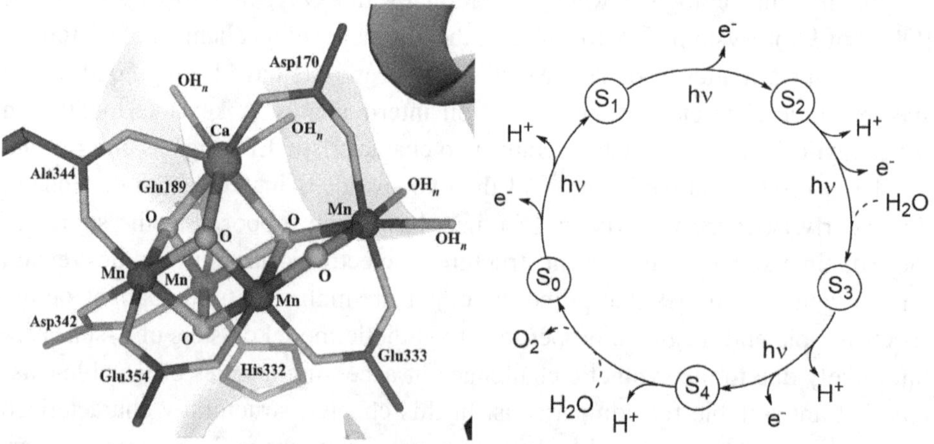

Figure 1. (a) The OEC as described by the 1.9 Å resolution structure highlighting the $CaMn_4O_5$ core. [Umena, 2011]. (b) The Joliot-Kok S-state catalytic cycle.

1992; Dismukes and Siderer, 1981; Haddy, 2004; Kim, 1992; Oyala, 2014, 2015; Pérez Navarro, 2013; Rapatskiy, 2012; Siegbahn, 2009].

The dark-stable, resting S_1 state of the OEC has been the primary target of structural methods such as X-ray absorption spectroscopy and protein crystallography [Ferreira, 2004; Kamiya and Shen, 2003; Zouni, 2001]. The 1.9 Å, atomic resolution model indicates that the structure of the OEC consists of a $CaMn_3$ subsite with a pendant Mn center, with bridging oxo moieties (Figure 1) [Umena, 2011]. It has been subsequently shown that X-ray induced Mn reduction resulted in a structural model that must be considered as a spatially averaged representation of physiological and non-physiological states [Yano, 2005]. This problem can be mitigated by using X-ray free electron laser (XFEL) pulses [Kern, 2013, 2014; Kupitz, 2014; Suga, 2014; Young, 2016]. Although issues with S-state heterogeneity and resolution need to be addressed, XFEL studies offer the possibility of observing structural and spectroscopic changes in the OEC under dynamic conditions [Sauter, 2016].

A significant level of insight derived from EPR spectroscopy and related magnetic resonance techniques has been obtained for the S_2 state, which features two characteristic EPR signals at $g = 2$ and $g = 4.1$ [Boussac, 1998; Boussac and Rutherford, 2000; Kim, 1992; Peloquin, 2000]. The relative intensity of the two signals is condition-dependent, and each signal is proposed to originate from different structures (Figure 2): an "open cubane" structure with a low-spin [LS] $S = 1/2$ ground state and a "closed cubane" structure with a high-spin [HS] $S = 5/2$ ground state [Horner, 1998; Lohmiller, 2013, 2014; Pantazis, 2012]. The interconvertibility of the two EPR signals suggests a small energy difference between

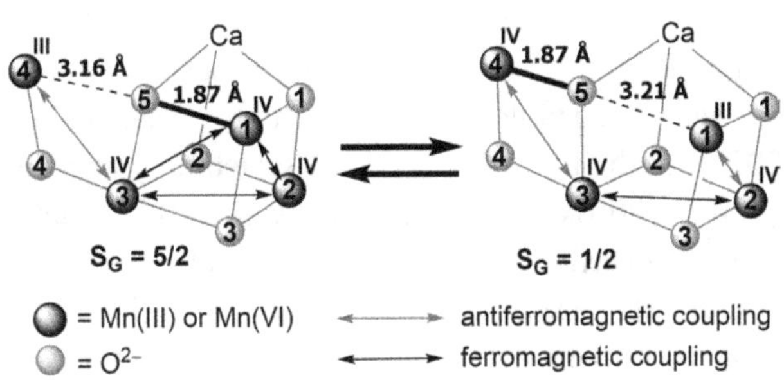

Figure 2. Computed structures for the proposed isomers of the inorganic $CaMn_4O_5$ core of the OEC in the S_2 state. The large structural changes in the Mn-oxo distances have been calculated to lead to different electronic coupling between the Mn centers and a change in the ground spin state, which explain the two observed EPR signals. Mn(4)–O(5) and Mn(1)–O(5) distances shown with bold and dashed lines, interchangeably. Nature of the computed magnetic exchange coupling interactions shown in grey (antiferromagnetic) and black (ferromagnetic) double headed arrows, respectively. [Pantazis, 2012]

these two structures, which effectively differ only by the relative position of the bridging O[5] oxygen, a proposed substrate oxygen [Bovi, 2013; Vinyard, 2017]. EXAFS studies support structural differences between the species responsible for the g = 2 and the g = 4.1 signal, but further atomistic details are unknown [Chatterjee, 2016; Liang, 1994].

Chemical changes such as Ca^{2+} removal or treatment with NH_3 or F^- inhibit the $S_2 \rightarrow S_3$ transition specifically [DeRose, 1995; Oyala, 2015], highlighting opportunities for mechanistic insight, but also the need for benchmarking with well characterized synthetic models to address the effect of structural changes on electronic structure and spectroscopy. Growing experimental data support that conversion from the LS form of the S_2 state to its HS form is an intermediate step in the $S_2 \rightarrow S_3$ transition [Boussac, 2015]. pH dependence studies indicate that deprotonation of the LS form leads to the HS form [Boussac, 2018]. Structural changes following deprotonation are unknown, but computational studies suggest that the electronic structure of the OEC is highly sensitive to small changes in structure as reported by EPR spectroscopy [Ames, 2011]. Deprotonation of a Mn-bound water and/or a reorientation of a Glu residue in the S_2 state may perturb the spin state from LS to HS, indicating that spectroscopic properties may react very sensitively to small geometric changes that do not affect significantly the total energy of the cluster.

Synthetic cluster models have been targeted to benchmark the spectroscopy and to mimic the reactivity of the OEC. In general, systematic studies that probe the effect of small structural changes on the spectroscopic and magnetic properties of S-state model clusters are very rare, likely due to the synthetic challenges of accessing a series of isolable clusters that are suitable for comparisons [Mukhopadhyay, 2004a; Mullins and Pecoraro, 2008]. Indeed, despite significant efforts to prepare tetra- and penta-nuclear clusters that are relevant to the S-state intermediates in terms of structure, redox state, or spectroscopy, accurate models for benchmarking against the biological system are rare [Kanady, 2011; Mukherjee, 2012; Zhang, 2015]. In the following sections, structurally characterized tetranuclear Mn complexes with oxidation states relevant to S_0–S_3 states ($Mn_3^{III}Mn^{IV}$–Mn_4^{IV}) are surveyed, primarily focusing on aspects of electronic structure and EPR spectroscopy. In particular, a series of structurally related ($Mn^{III}Mn_3^{IV}O_4$) cuboidal complexes as spectroscopic models of the S_2 state are highlighted [Lee, 2018].

2. Mixed-valent $Mn^{III}Mn^{IV}$ Dinuclear Complexes

Although structurally not related to the OEC, oxo-bridged, mixed-valent $(Mn^{III}Mn^{IV}O_2)^{3+}$ complexes feature EPR spectra that are highly reminiscent of the multiline signal observed in the S2 state (Figure 3) [Lohmiller, 2013; Randall, 1995, 1997; Rapatskiy, 2015; Schäfer, 1998]. The nature of the exchange coupling

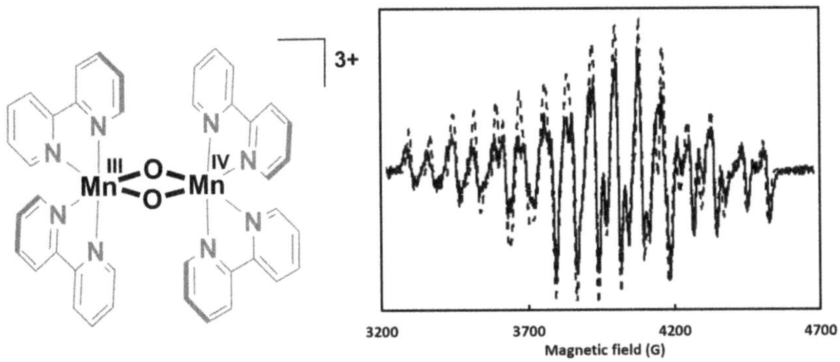

Figure 3. Structure of $(Mn_2O_2[bpy]_4)^{3+}$ and its X-band EPR spectrum. [Peloquin, 2000].

interaction in such complexes is typically antiferromagnetic, ranging from $J =$ –100 to –400 cm^{-1} [Pecoraro, 1988]. Accordingly, the ground state in these systems is $S = 1/2$, and broad EPR transitions centered at $g = 2$ featuring Mn hyperfine interactions are observed. Analysis of the ENDOR spectra of $(Mn_2O_2(bpy)_4)^{3+}$ allowed the unambiguous assignment of ^{55}Mn nuclear spin transitions of both the Mn^{III} and Mn^{IV} ions [Randall, 1995]. Parameters found for the dimeric species failed to simulate the ^{55}Mn ENDOR spectrum of the S_2 state, easily ruling out the dimeric origin of the multiline signal in the S_2 state. The deduction of a "3 + 1" arrangement of Mn ions in the OEC by ENDOR spectroscopy prior to the advent of atomic resolution crystal structures is a landmark assignment from EPR spectroscopy [Peloquin, 2000]. The interaction of ligand nuclei (*e.g.* $^{1/2}H$, ^{13}C, $^{14/15}N$, ^{17}O) with the electron spin of the Mn centers has been studied by pulsed EPR techniques to obtain structural information and determine plausible sites of substrate water binding [Oyala, 2015; Rapatskiy, 2012]. Pulse-EPR data for ^{17}O-labelled dinuclear complexes have provided a magnetic fingerprint for bridging oxo ligands similar to those found in the OEC [Rapatskiy, 2015]. In the following sections, a selective survey of structurally characterized tetranuclear Mn complexes will be presented, with an emphasis on electronic structure as probed by magnetic susceptibility and EPR studies.

3. $Mn_3^{III}Mn^{IV}$ Complexes as Models of the S_0 State of the OEC

A class of carboxylate-bridged $Mn_3^{III}Mn^{IV}$ complexes with a general $(Mn_4O_3X)^{6+}$ core ($X^- = F^-, Cl^-, Br^-, NO_3^-, OH^-, OCH_3^-, OPh^-, OAc^-, OBz^-, N_3^-$, or NCO^-) has been described [Andres, 2000; Aromí, 1998; Aubin, 1996; Cinco, 1999; Hendrickson, 1992; Wang, 1994, 1996, 2000; Wemple, 1993, 1995]. Single crystal X-ray crystal structures reveal a *pseudo*-C_3 symmetric core with the μ^3–X ligand

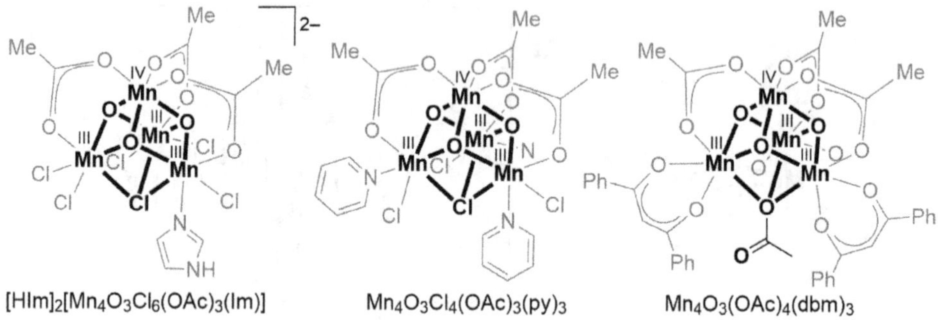

Figure 4. Examples of carboxylate bridged $Mn_3^{III}Mn^{IV}$ complexes with a $(Mn_4O_3X)^{6+}$ core.

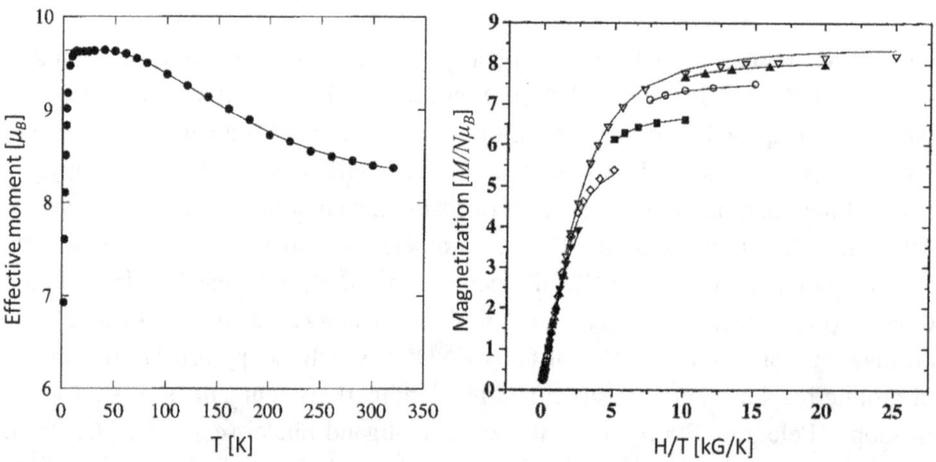

Figure 5. Temperature dependence of the effective moment (left) and reduced magnetization plot of $Mn_4O_3(OAc)_4(dbm)_3$. [Wang, 2000]

bridging the three Mn^{III} centers (Figure 4). Each metal center is six-coordinate, with the coordination sphere of the Mn^{IV} center occupied by three carboxylate and three oxo ligands. Monodentate ligands such as Cl^- or heterocycles such as pyridine or imidazole bind in a *cis* fashion to complete the coordination sphere of each Mn^{III} center. Clusters with bidentate ligands such as dibenzoylmethane (dbm) have also been reported (Figure 4).

Magnetic susceptibility studies are consistent with a well-isolated $S = 9/2$ ground state, consistent with the Curie behavior observed between 5–50 K (Figure 5). The sudden decrease in the effective magnetic moment below 5 K was attributed in part to the presence of zero field splitting. From reduced magnetization data, a positive axial zero-field splitting parameter in the range of the X-band EPR quanta (*c.a.* 0.3 cm^{-1}) has been estimated (Figure 5). To fit the susceptibility data, an exchange coupling model with two coupling constants $J(Mn^{III}–Mn^{IV})$ and $J(Mn^{III}–Mn^{III})$ has been employed. Invariably, for the series of compounds with different

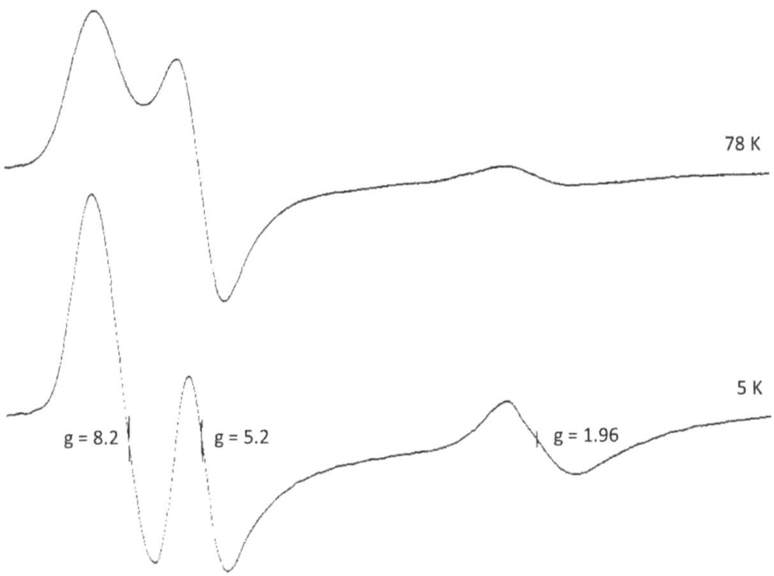

Figure 6. X-band EPR spectra of $Mn_4O_3Cl_4(OBz)_3(py)_3$. [Wemple, 1993].

X^-, bridging carboxylate, and terminal ligands, the $J(Mn^{III}–Mn^{IV})$ coupling was found to be antiferromagnetic, with values in the range of −30 to −20 cm^{-1}; the $J(Mn^{III}–Mn^{III})$ coupling was found to be ferromagnetic, with values in the range +2 to +11 cm^{-1}. The first excited state $S = 7/2$ was estimated to lie *c.a.* 200 cm^{-1} above the $S = 9/2$ ground state. Accordingly, the variable temperature EPR spectra of these complexes do not show significant population of spin excited states; aside from line broadening, the spectrum at 5 K is nearly identical to that at 78 K (Figure 6) [Wemple, 1993]. The transitions occurring within the $|\pm 1/2\rangle$ Kramers doublet were observed at $g_\perp = 8.2$ and $g_\parallel = 2$. The feature at $g = 5.2$ was tentatively assigned to the $|-1/2\rangle \rightarrow |-3/2\rangle$ transition [Hendrickson, 1992]. In contrast to the electronic structure of $(Mn_4O_3X)^{6+}$ complexes, the ground state of the S_0 state of the OEC has been shown to be $S = 1/2$, with a broad multiline EPR signal centered at $g = 2$ [Kulik, 2007]. Although not structurally characterized, *in situ* one electron reduction of $((Mn_2O_2)_2(tphpn)_2)^{4+}$ (*vide infra*]) leads to a species with an EPR spectrum reminiscent of the S_0 state [Mukhopadhyay, 2004b].

4. $Mn_2^{III}Mn_2^{IV}$ Complexes as Models of the S_1 State of the OEC

Three classes of $Mn_2^{III}Mn_2^{IV}$ complexes have been described: [1] cubane complexes with a $(Mn_4O_4)^{6+}$ core [Kanady, 2011; Ruettinger, 1997], [2] square complexes that can be described as an alkoxide bridged dimer of two $(Mn_2O_2)^{3+}$ units [Chan and

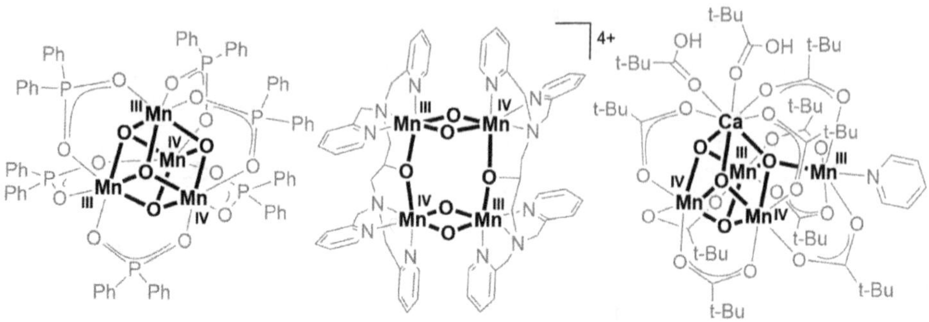

$Mn_4O_4(O_2PPh_2)_6$ $[(Mn_2O_2)_2(tphpn)_2]^{4+}$ $CaMn_4O_4(OPiv)_8(HOPiv)_2(py)$

Figure 7. Examples of $Mn_2^{III}Mn_2^{IV}$ complexes.

Armstrong, 1991], and [3] a unique $(CaMn_4O_4)^{8+}$ complex highly reminiscent of the OEC featuring a $(CaMn_3O_4)$ subsite with a dangling Mn ion (Figure 7) [Zhang, 2015]. Thus far, magnetic susceptibility studies of these complexes have not been reported. The parallel mode X-band EPR of $((Mn_2O_2)_2(tphpn)_2)^{4+}$ shows a broad signal at $g = 6$. Similarly, the EPR spectrum of the $(CaMn_4O_4)^{8+}$ complex shows a broad signal at $g = 12$ (Figure 8). For the S_1 state, parallel mode signals at $g = 4.9$ and $g = 12$ have been observed [Campbell, 1998; Dexheimer and Klein, 1992; Yamauchi, 1997]. Temperature dependence studies on the S_1 state indicate that these two signals arise from spin excited states $S = 1$ and $S = 2$, respectively, and that the S_1 state has a diamagnetic ground state [Yamauchi, 1997]. Magnetic studies of the cubane cluster $LMn_2^{III}Mn_2^{IV}O_4(OAc)_3$ [**2**] reported by our group (*vide infra*, Figure 14) indicate an $S = 1$ ground state, consistent with the observed χT value of 0.9 emu K mol^{-1} at 5 K [Kanady, 2013a]. The perpendicular EPR spectrum shows a broad signal centered at $g = 2$, and was rationalized as coming from the $S = 1$ ground state with a small zero field splitting (Figure 9) [Kanady, 2013b].

5. $Mn^{III}Mn_3^{IV}$ Complexes as Models of the S_2 State of the OEC

In addition to the extensive spectroscopic studies performed on the native S_2 state of the OEC, non-native, MeOH or NH_3 modified S_2 states have also been studied [Marchiori, 2018; Oyala, 2014, 2015; Pérez Navarro, 2013; Su, 2011]. In contrast, structural and spectroscopic studies of $Mn^{III}Mn_3^{IV}$ model complexes remain rare. Three classes of synthetic $Mn^{III}Mn_3^{IV}$ complexes have been described: [1] cubane complexes with a $(Mn_4O_4)^{7+}$ core [Ruettinger, 1999], [2] adamantane-shaped complexes with a $(Mn_4O_6)^{3+}$ core [Dubé, 1998, 1999], and [3] linear-chain complexes with a $(Mn_4O_6)^{3+}$ core [Figure 10] [Blondin, 1997]. The phosphinate-bridged, cuboidal complex $(Mn_4O_4(Ph_2PO_2)_6)^+$ shows a broad EPR spectrum with

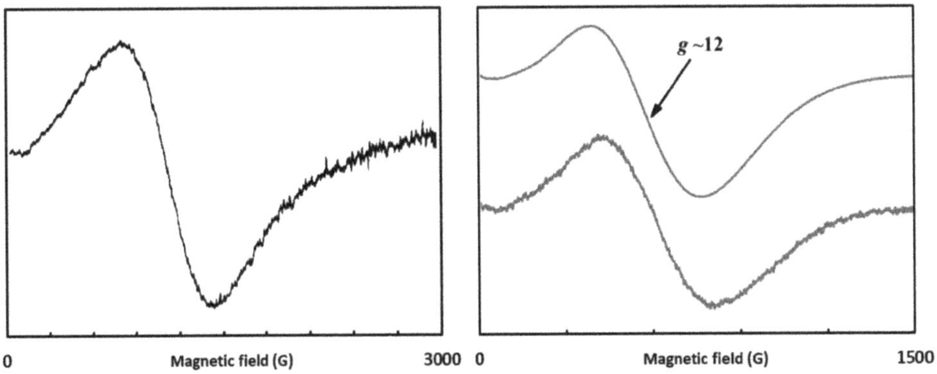

Figure 8. Parallel mode X-band EPR of $((Mn_2O_2)_2(tphpn)_2)^{4+}$ [left] and $CaMn_4O_4(OPiv)_8$ (right). [Chan and Armstrong, 1991, Zhang, 2015].

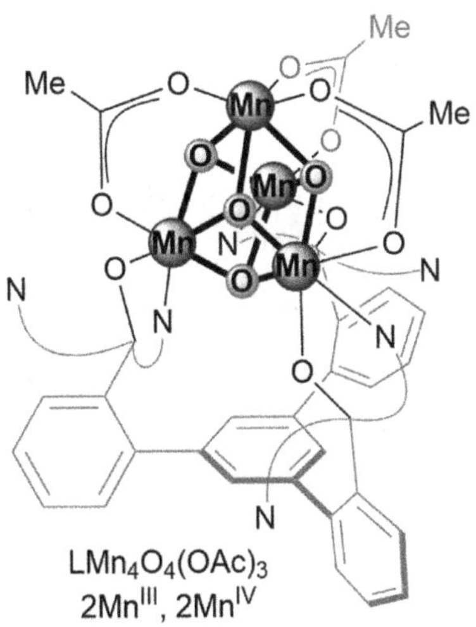

$LMn_4O_4(OAc)_3$
$2Mn^{III}$, $2Mn^{IV}$
2

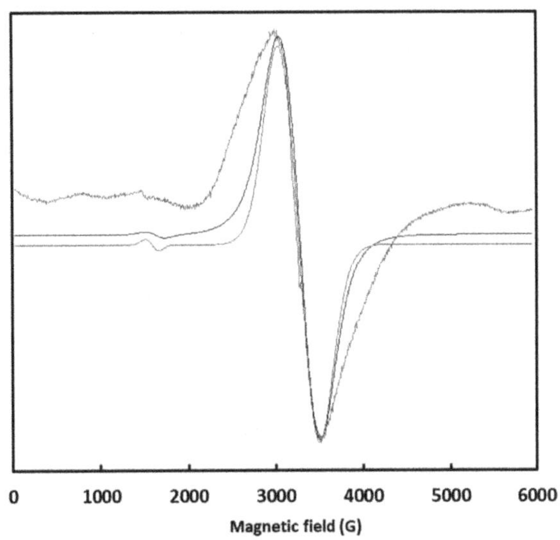

Figure 9. Perpendicular mode X-band EPR of $LMn_4O_4(OAc)_3$. [Kanady, 2013b].

transitions consistent with an $S \geq 3/2$ ground state (Figure 11) [Ruettinger, 1999]. Complementary susceptibility studies have not been provided. The EPR spectrum of the adamantane-shaped complex $(Mn_4O_6(bpea)_4)^{3+}$ features a broad signal at $g = 4.1$ that has been assigned to the transition within the first excited $|\pm 3/2\rangle$ Kramers doublet of an $S = 5/2$ ground state determined from magnetization data [Dubé, 1999]. This $g = 4.1$ signal is highly reminiscent of the high-spin form of the S_2 state [Boussac, 1998, Boussac and Rutherford, 2000]. Starting from a structurally characterized linear-chain $(Mn_4O_6(bpy)_6)^{4+}$ precursor, a putative $(Mn_4O_6(bpy)_6)^{3+}$ complex was generated by radiolysis in a frozen matrix and its EPR spectrum features a multiline signal centered at $g = 2$ consistent with an $S = 1/2$ ground state [Blondin, 1997]. *In situ* oxidation of a close OEC structural

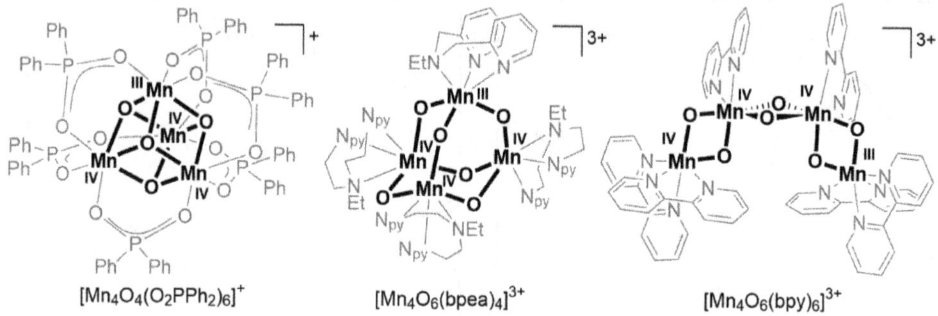

Figure 10. Examples of $Mn^{III}Mn_3^{IV}$ complexes.

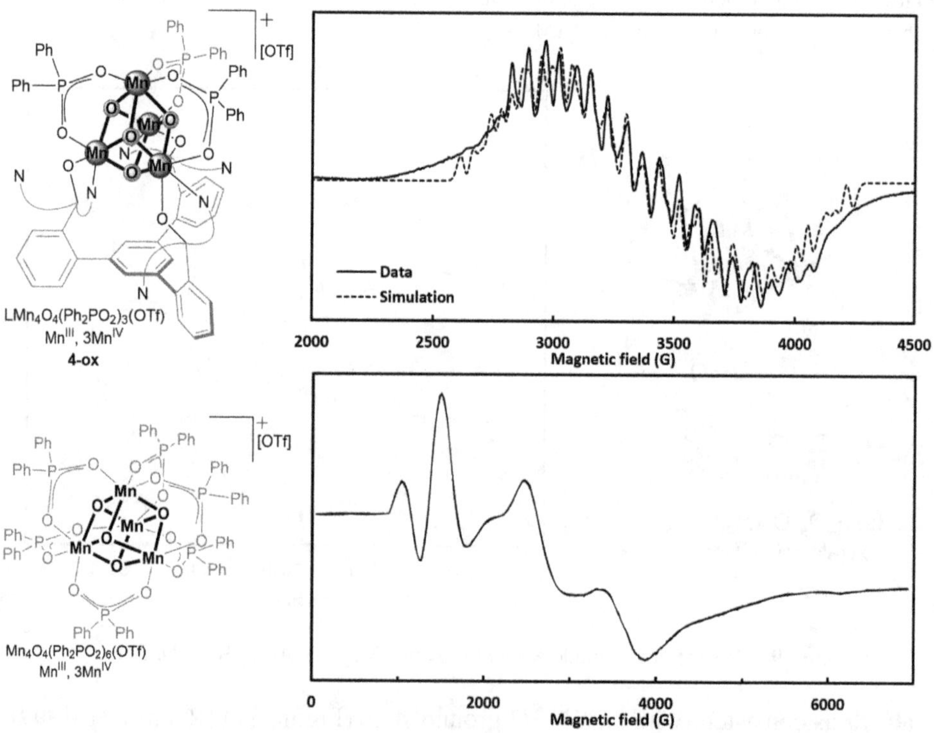

Figure 11. [Top] X-band EPR spectrum of $(LMn_4O_4(Ph_2PO_2)_3)^+$. Multiline signal centered at $g = 2$ consistent with an $S = 1/2$ ground state. [Bottom] EPR spectrum of the structurally related $(Mn_4O_4(Ph_2PO_2)_6)^+$ complex. Spectrum consistent with a higher $S \geq 3/2$ ground state. [Ruettinger, 1997].

model complex, $CaMn_4O_4(OPiv)_8$ (*vide supra*), gives rise to two EPR signals at $g = 4.9$ and $g = 2$, attributed to different spin states of the cluster corresponding to a $(CaMn_4O_4(OPiv)_8)^+$ species, although further structural, spectroscopic, and magnetic data have not been provided [Zhang, 2015]. Follow-up computational studies disagree on the assignment of the two signals, one of them suggesting that the two signals must be due to structurally very different clusters [Paul, 2017; Shoji, 2016]. In general, systematic studies akin to those performed on the S_2 state that probe the effect of small structural changes on the spectroscopic and magnetic properties of S_2 model clusters are uncommon, likely due to the synthetic challenges of accessing a series of isolable clusters that are suitable for comparisons. A series of cuboidal complexes in the relevant $Mn^{III}Mn_3^{IV}$ redox state, with systematic variation of bridging ligands has recently been reported and provides useful insight into structure-property relationships (*vide infra*) [Lee, 2018].

6. Mn_4^{IV} Complexes as Models of the S_3 State of the OEC

EPR spectroscopic studies show all Mn ions to be present as Mn(IV) in the S_3 state [Cox, 2014]. Similar to the S_2 state, the S_3 state is spectroscopically heterogeneous, with an EPR observable species having an $S = 3$ ground state [Pantazis, 2018]. The $S_2 \rightarrow S_3$ transition is highly debated, with the details of substrate water incorporation and deprotonation being largely unknown. Broadly, two general classes of synthetic tetranuclear Mn_4^{IV} complexes have been described related to S_3: linear complexes featuring $(Mn_4O_5)^{6+}$ and $(Mn_4O_6)^{4+}$ cores [Chen, 2004, 2005; Philouze, 1994], and adamantane-shaped complexes with $(Mn_4O_6)^{4+}$ cores (Figure 12) [Dubé, 1998, 1999, 2005; Hagen, 1989; Mukhopadhyay, 2002; Wieghardt, 1988]. For the linear complexes, strong pair-wise antiferromagnetic exchange coupling interactions lead to a diamagnetic ground state. An EPR spectrum has been reported at 140 K from the population of thermally accessible spin excited states, but further characterization has not been provided [Philouze, 1994]. Magnetic susceptibility studies of a series of adamantane-shaped $[Mn_4O_6]^{4+}$ complexes

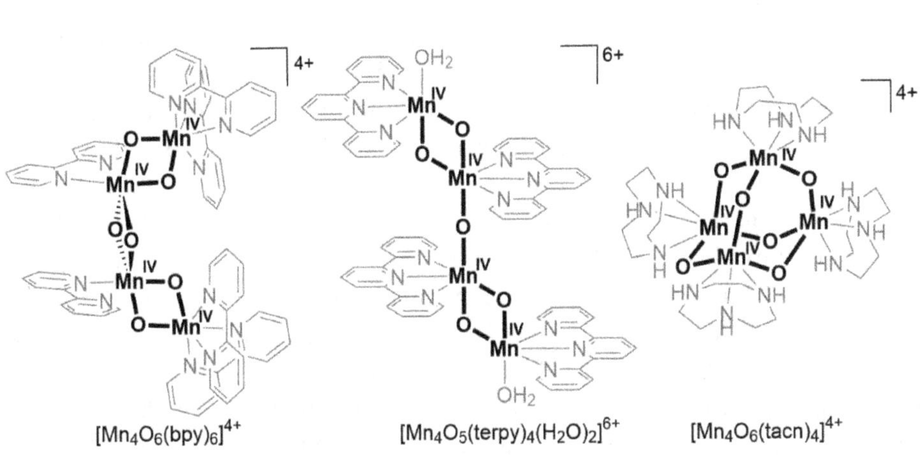

$[Mn_4O_6(bpy)_6]^{4+}$ $[Mn_4O_5(terpy)_4(H_2O)_2]^{6+}$ $[Mn_4O_6(tacn)_4]^{4+}$

Figure 12. Examples of Mn_4^{IV} complexes.

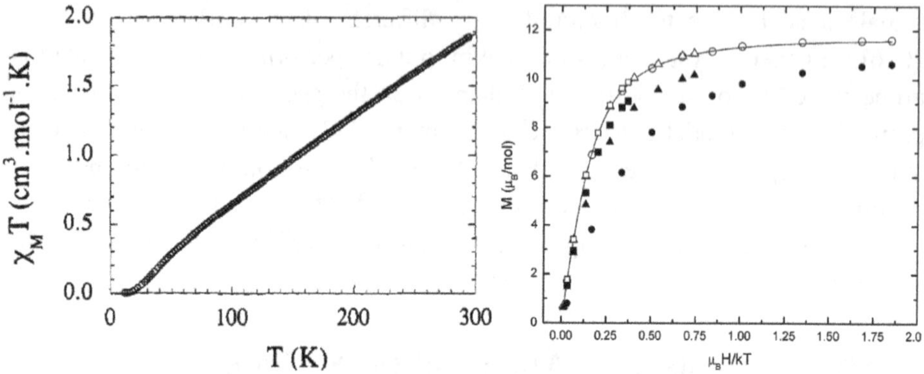

Figure 13. (Left) χT *versus* T plot of the linear $(Mn_4O_6(bpy)_6)^{4+}$ complex showing a diamagnetic ground state. (Right) Reduced magnetization plot of the adamantane-shaped $(Mn_4O_6(tacn)_4)^{4+}$ complex showing a high-spin $S = 6$ ground state. [Philouze, 1994; Dubé, 2005].

indicate overall ferromagnetic exchange coupling interactions giving rise to an $S = 6$ ground state (Figure 13) [Dubé, 2005]. EPR studies for the adamantane-shaped complexes have not been reported. Upon protonation of a bridging oxo, the overall ferromagnetic interaction in the adamantane-shaped $(Mn_4O_6)^{4+}$ complex was found to switch to antiferromagnetic [Hagen, 1989].

7. Structurally Related Mn_4 Clusters with Variable Oxido Incorporation, Ancillary Ligand Coordination, and Mn Oxidation States

Toward developing a methodology to access Mn clusters related to the OEC that can be systematically changed to provide rational access to different metal arrangement, oxo content, and heterometal content, geometrically restricted ligands were employed. Taking advantage of rigid ligand frameworks, stepwise protocols toward the synthesis of tetra- and pentanuclear complexes have been designed [Han, 2017, Kanady, 2011, 2013a, 2013b, 2014; Lee, 2017; Lin, 2016; Reed and Agapie, 2018; Tsui and Agapie, 2013; Tsui, 2013a, 2013b]. For the series of tetranuclear Mn complexes relevant to the OEC, initial metalation of H_3L with $Mn(OAc)_2$ leads to the trinuclear complex $LMn_3(OAc)_3$ [Tsui, 2011]. Oxidation of $LMn_3(OAc)_3$ with KO_2 in the presence of $Ca(OTf)_2$ leads to the formation of the Ca cubane complex $LCaMn_3O_4(OAc)_3(THF)$ [**1**] [Figure 14] [Kanady, 2011]. Substitution of Ca^{2+} with Mn^{2+} lead to the formation of $LMn_2^{III}Mn_2^{IV}O_4(OAc)_3$ [**2**]. Oxygen-atom transfer reactivity of **2** has been studied, with addition of PMe_3 resulting in the formation of $LMn_4^{III}O_3(OAc)_3$ [**3**] [Kanady, 2013a]. Reoxidation of **3** to **2** in the presence of labelled water has been

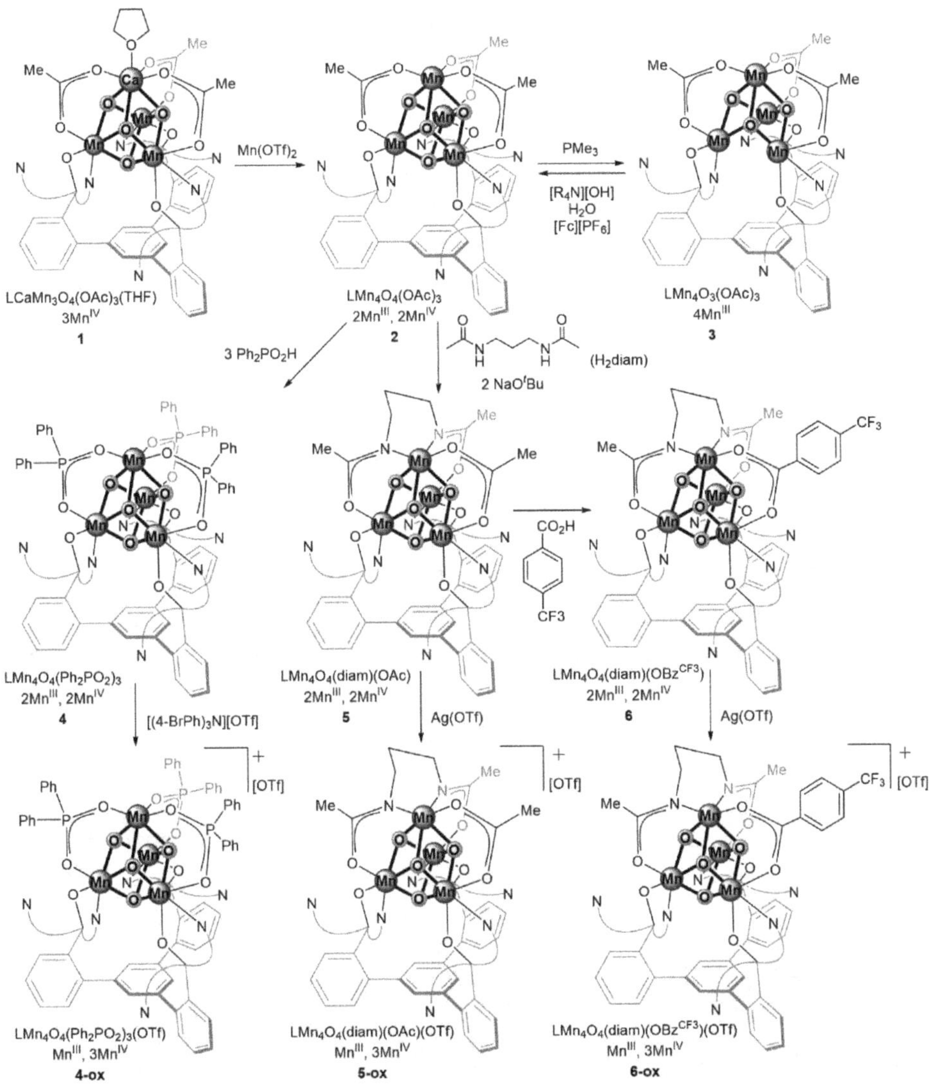

Figure 14. Synthesis of tetranuclear (Mn_4O_4) cuboidal complexes studied in our group. Ligand substitution of **2** leads to a series of structurally related complexes **4–6**. One electron oxidation leads to **4-ox–6-ox**, with Mn oxidation states mirroring those in the S_2 state.

demonstrated to involve oxide bridge movements reminiscent of the proposal for O(5) in the OEC [Kanady, 2013b]. Depending on the synthetic procedure, other tetranuclear complexes with cores $(Mn_3^{II}Mn^{III}O)^{7+}$, $(Mn_2^{II}Mn_2^{III}O)^{8+}$, and $(Mn_2^{II}Mn_2^{III}O_2)^{6+}$ have been obtained, underscoring the flexibility of the ligand scaffold in stabilizing tetranuclear Mn complexes of different oxidation states [Kanady, 2013b].

Electrochemical studies of **2** indicate the presence of an accessible one-electron oxidized species. The cyclic voltammogram (CV) of **2** shows a reversible redox process at +250 mV *versus* Fc/Fc^+ assigned to the $(Mn_2^{III}Mn_2^{IV})/(Mn^{III}Mn_3^{IV})$ couple [Kanady, 2011]. Isolation of the corresponding $(LMn_4O_4(OAc)_3)^+$ species has been unsuccessful, but freeze-quenching a mixture of **2** and $((4\text{-BrPh})_3N)(SbCl_6)$ allows the observation of an intense EPR spectrum with a broad signal centered at $g = 2$ consistent with an $S = 1/2$ ground state. This is in contrast to the EPR spectrum of $(Mn_4O_4(Ph_2PO_2)_6)^+$ with a higher spin ground state $S \geq 3/2$ [Lee, 2018].

Toward stabilizing the $Mn^{III}Mn_3^{IV}$ oxidation state, phosphinate-bridged (Mn_4O_4) cuboidal complexes were synthesized [Lee, 2018]. Treatment of **2** with 3 equiv. HO_2PPh_2 in THF leads to the formation of $LMn_4O_4(Ph_2PO_2)_3$ [**4**] via a protonolysis reaction (Figure 12). The CV of **4** shows a reversible redox process at +190 mV *versus* Fc/Fc^+ assigned to the $(Mn_2^{III}Mn_2^{IV})/(Mn^{III}Mn_3^{IV})$ couple. Treatment of **4** with 1 equiv. of $((4\text{-BrPh})_3N)(OTf)$ leads to the formation of the one-electron oxidized species **4-ox** (Figure 15). Based on Mn–oxo distances, the oxidation state of Mn(4) is assigned as Mn^{III}. The elongated Mn(4)–O(4) distance of 2.241(1) Å is consistent with this assignment. This complex represents only the second example

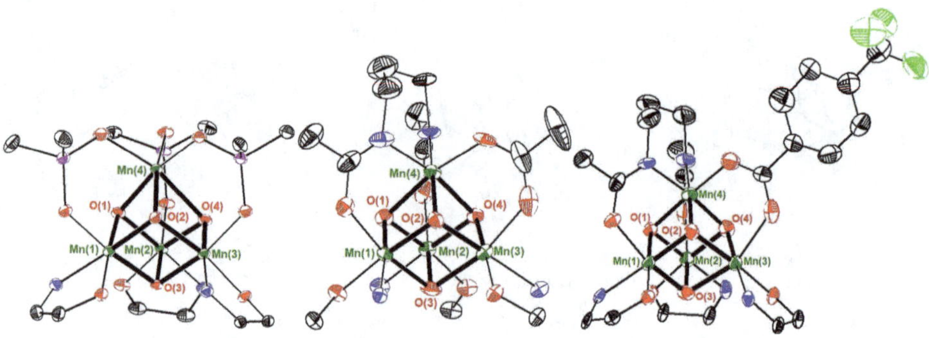

Figure 15. Truncated crystal structures of **4-ox** (left), **5-ox** (middle), and **6-ox** (right). Mn (green), O (red), N (blue), P (purple), C (black), F (light green). Bolded bonds highlight metal-oxo bonds. Selected bond distances (Å):

4-ox: Mn[1]–O[1] 1.920[1], Mn[1]–O[2] 1.862[1], Mn[1]–O[3] 1.919[1], Mn[2]–O[1] 1.898[1], Mn[2]–O[3] 1.920[1], Mn[2]–O[4] 1.858[1], Mn[3]–O[2] 1.931[1], Mn[3]–O[3] 1.929[1], Mn[3]–O[4] 1.831[1], Mn[4]–O[1] 1.932[1], Mn[4]–O[2] 1.976[2], Mn[4]–O[4] 2.241[1]

5-ox: Mn[1]–O[1] 1.909[5], Mn[1]–O[2] 1.896[3], Mn[1]–O[3] 1.937[4], Mn[2]–O[1] 1.893[4], Mn[2]–O[3] 1.931[5], Mn[2]–O[4] 1.899[3], Mn[3]–O[2] 1.911[4], Mn[3]–O[3] 2.051[4], Mn[3]–O[4] 1.926[4], Mn[4]–O[1] 1.897[4], Mn[4]–O[2] 1.909[4], Mn[4]–O[4] 1.901[4]

6-ox: Mn[1]–O[1] 1.901[2], Mn[1]–O[2] 1.873[3], Mn[1]–O[3] 1.874[3], Mn[2]–O[1] 1.880[3], Mn[2]–O[3] 1.892[3], Mn[2]–O[4] 1.902[3], Mn[3]–O[2] 1.926[3], Mn[3]–O[3] 2.143[3], Mn[3]–O[4] 1.965[3], Mn[4]–O[1] 1.894[3], Mn[4]–O[2] 1.917[3], Mn[4]–O[4] 1.883[3].

of a $(Mn_4O_4)^{7+}$ cluster, and is highly reminiscent of $(Mn_4O_4(Ph_2PO_2)_6)^+$, encouraging further magnetic and spectroscopic studies (*vide infra*).

Toward expanding the series of Mn_4 clusters with the same redox state as the S_2 state of the OEC, oxidatively stable amidate-bridged (Mn_4O_4) cuboidal complexes were synthesized [Gupta, 2015]. An *n*-propyl-linked diacetamide proligand [H_2diam] was used to replace two acetate moieties. Treatment of **2** with 1 equiv. H_2diam and 2 equiv. NaO^{t}Bu in DMF leads to the formation of LMn_4O_4[diam][OAc] [**5**] (Figure 14). The *n*-propyl-linked diamidate serves as a bridging ligand across two faces of the $[Mn_4O_4]$ cubane moiety, resulting in a *pseudo*-C_S symmetric complex. The reversible $[Mn_2^{III}Mn_2^{IV}]/[Mn^{III}Mn_3^{IV}]$ couple is observed at −150 mV *versus* Fc/Fc$^+$. Treatment of **5** with 1 equiv. Ag[OTf] leads to the formation of the one-electron oxidized species **5-ox**. The oxidation state of Mn[3] is assigned as Mn^{III} in **5-ox**. With the exception of an elongated Mn[3]–O[3] distance of 2.051[4] Å, all other Mn–oxo distances are in the range 1.893[4]–1.937[4] Å, consistent with this oxidation state assignment [Figure 15].

To further investigate the effect of small geometric changes on the electronic structure of the resulting cluster, a diamidate-benzoate complex was synthesized. Treatment of **5** with 1 equiv. of p-$CF_3C_6H_4CO_2H$ [^{CF3}BzOH] leads to the formation of LMn_4O_4[diam][OBz^{CF3}] [**6**] via a protonolysis reaction [Figure 14]. The CV of **6** shows a reversible redox process at −15 mV *versus* Fc/Fc$^+$ assigned to the $[Mn_2^{III}Mn_2^{IV}]/[Mn^{III}Mn_3^{IV}]$ couple. Treatment of **6** with 1 equiv. Ag[OTf] leads to the formation of the one-electron oxidized species **6-ox**. Similar to **5-ox**, the oxidation state of Mn[3] is assigned as Mn^{III} in **6-ox**. An elongated Mn[3]–O[3] distance of 2.143[3] Å is consistent with this assignment [Figure 15]. Overall, comparing complexes **4-ox**, **5-ox**, and **6-ox**, the elongated Mn^{III}-oxo distance varies from 2.241[1], to 2.051[4], and 2.143[3], while the other Mn-oxo distances are in the range 1.831[1]–1.976[2] Å, 1.893[4]–1.937[4] Å, and 1.873[3]–1.965[3] Å, respectively.

Magnetic susceptibility measurements for **4-ox**, **5-ox**, and **6-ox** were performed. At 300 K, the χT values of 5.64, 5.32, and 6.0 emu K mol^{-1} respectively indicate antiferromagnetic coupling between the Mn centers, deviating from the expected spin-only value of 8.62 emu K mol^{-1} [$g = 2$] for uncoupled Mn^{III} [$S = 2$] and Mn^{IV} [$S = 3/2$] centers (Figure 16). For all three complexes, χT decreases monotonically with temperature, consistent with an overall antiferromagnetic interaction between the Mn centers. For **4-ox** at 1.8 K, a χT value of 0.383 emu K mol^{-1} was observed, in good agreement with the expected χT value of 0.375 emu K mol^{-1} for an $S = 1/2$ [$g = 2$] ground state. The near-ideal Curie behavior observed between 1.8–10 K was attributed to the absence of thermally accessible spin excited states $S > 1/2$ in this temperature range. The calculated energy level diagram indicates a quartet excited state 25 cm^{-1} [equivalent temperature of 36 K] above the doublet ground state. This energy separation is comparable to the 25–35 cm^{-1} measured for MeOH-treated OEC poised in the multiline S_2 state [Su, 2011]. For **5-ox** at 1.8 K, a χT value of 0.383 emu K mol^{-1} was also observed, is in good agreement with the $S = 1/2$ [$g = 2$] ground state. Upon warming to 5 K, a χT value of 0.85 emu K mol^{-1} was observed. The increase in χT can be attributed to the presence of thermally accessible spin excited states $S > 1/2$. The calculated energy level diagram indicates the presence of low-lying quartet excited states at 5.3 cm^{-1} and 11 cm^{-1} [equivalent temperature of 7.6 K and 15.8 K] above the doublet ground state. Interestingly, an energy separation of 3–6 cm^{-1} has been reported for untreated higher plant OEC in the S_2 state [Pace, 1991; Su, 2011]. For **6-ox** at 1.8 K, a χT value of 0.82 emu K mol^{-1} was observed, deviating significantly from the expected χT value of 0.375 emu K mol^{-1} for an $S = 1/2$ [$g = 2$] ground state. This indicates a further decreased energy separation between the ground and excited states, as expected from a weaker antiferromagnetic coupling between the Mn centers. The calculated energy level diagram indicates two quartet excited states separated by 1.3 cm^{-1} and 5.7 cm^{-1} [equivalent temperature of 1.9 K and 8.2 K, respectively] above the doublet ground state. Overall, magnetic susceptibility studies indicate that the spin ladder in exchange-coupled tetranuclear Mn complexes are highly sensitive to small changes in structure and nature of the bridging ligands.

The EPR spectrum of **4-ox** at 5 K features a broad signal centered at $g = 2$ featuring Mn hyperfine interactions (Figure 11). This is consistent with the $S = 1/2$ ground state determined from susceptibility studies. As the temperature is increased, the $g = 2$ signal loses intensity until no signal is observed above 15 K. Importantly, EPR signals originating from thermally populated spin excited states are not observed, in agreement with the predicted energy separation of 25 cm^{-1} [36 K] between the ground and excited states. The spectrum can be approximated by $g = [2.053, 2.003, 1.952]$, $g_{iso} = 2.00$ and $A_{Mn} = [434, 434, 313; 293, 155, 245; 128, 146, 198; 134, 133, 72]$ MHz, $A_{Mn,iso} = [394; 231; 157; 113]$ MHz. The unique, larger Mn hyperfine coupling constant is consistent with the Mn^{III} center in **4-ox** [Randall, 1995]. The EPR spectrum of **4-ox** is in stark contrast to that of the related $[Mn_4O_4[Ph_2PO_2]_6]^+$ complex with a higher spin ground state $S \geq 3/2$ [Ruettinger, 1999], indicating that the EPR of exchange-coupled tetranuclear Mn complexes are highly sensitive to the Mn coordination environment, even if the $[Mn_4O_4]$ core is maintained (Figure 11).

For the EPR spectrum of **5-ox** at 5 K, only the multiline signal centered at $g = 2.0$ is discernible. As the sample is warmed, the signal at $g = 2$ decreases in intensity as a signal centered at $g = 4.2$ gains intensity (Figure 17). Above 20 K, both signals start to lose intensity due to relaxation. The signal at $g = 2$ is consistent with the $S = 1/2$ ground state as determined from susceptibility studies. The $g = 4.2$

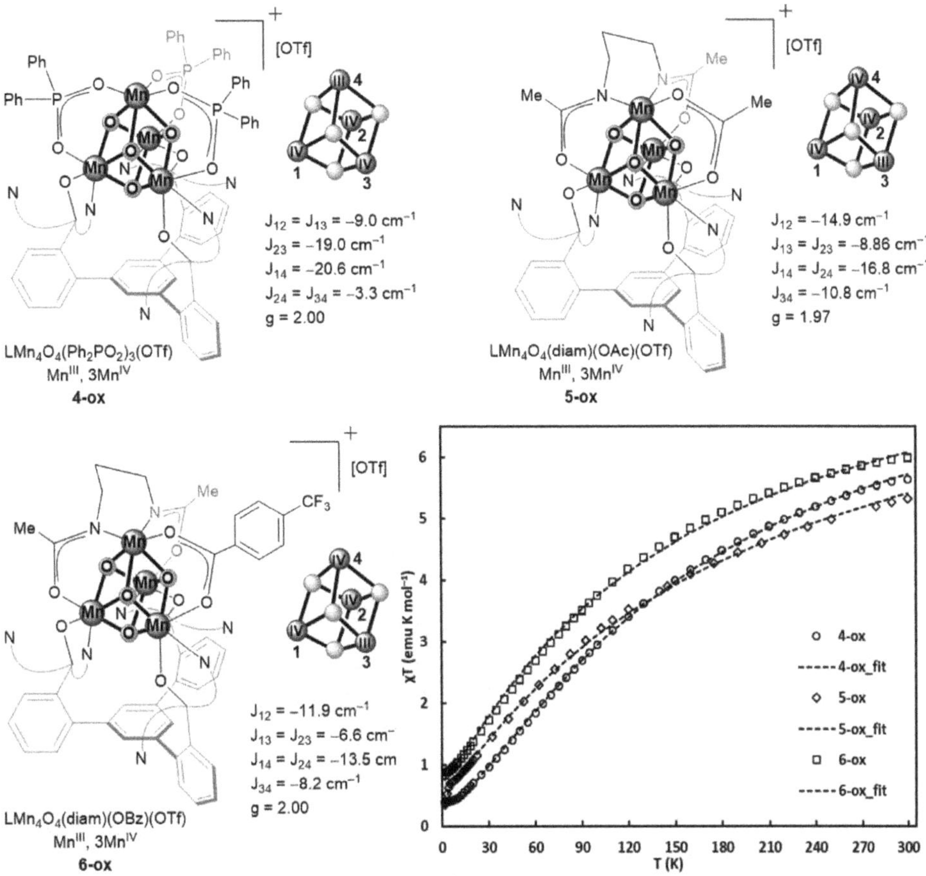

Figure 16. Exchange coupling model, fit parameters, and χT *versus* T plot of complexes **4-ox**, **5-ox**, and **6-ox**.

signal, assigned to the $S = 3/2$ excited state of **5-ox**, is highly reminiscent of the S_2 state in its HS form. The $g = 2$ and the $g = 4.2$ signals of the S_2 state arise from the ground states of structurally distinct species, the relative ratio of which is affected by external chemical stimuli such as pH [Boussac, 2018]. As such, both EPR signals can be observed at low temperatures in a ratio that reflects the relative population of the two species. In the case of **5-ox**, the two signals arise from different spin states of a single, structurally static species, as supported by solid-state and solution EXAFS measurements [Lee, 2018]. The temperature dependence of the EPR spectrum of **5-ox** can be explained in terms of small differences in the Boltzmann distribution of the ground and excited states. At 5 K, only the $S = 1/2$ ground state is significantly populated, and the $g = 2$ signal corresponds to the $|-1/2\rangle \rightarrow |1/2\rangle$ transition. As the temperature is increased, the difference in the population of the $|-1/2\rangle$ and the $|1/2\rangle$ states decrease, resulting in weaker absorption. Concurrently,

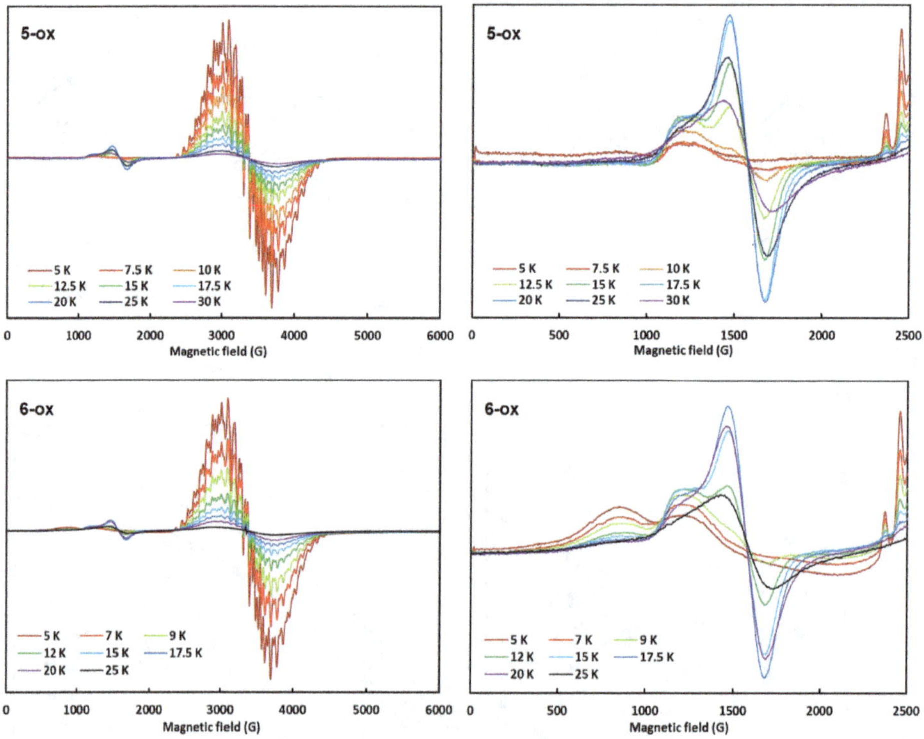

Figure 17. (Left) Variable-temperature X-band continuous-wave EPR spectra of **5-ox** and **6-ox**. (Right) Expanded view of the g > 2 region.

as the temperature is increased, the $S = 3/2$ spin excited state is populated, and the $g = 4.2$ signal corresponds to the transition within the $|\pm 3/2\rangle$ Kramers doublet. In contrast to **4-ox**, the observation of the $g = 4.2$ signal is consistent with a smaller energy separation between the ground doublet and excited quartet state in **5-ox**, in agreement with the magnetic susceptibility studies. The EPR spectrum of a weakly antiferromagnetically coupled $Mn^{III}Mn^{IV}$ dimer shows a similar temperature dependence: At 20 K, only the $g = 2$ signal is observed, but upon warming to 43 K and to 110 K, a new low-field signal at $g = 5$ gains intensity. This low-field signal has been assigned to the $S = 3/2$ excited state [Larson, 1992]. For more strongly coupled $Mn^{III}Mn^{IV}$ dimeric systems, the quartet excited state is separated from the doublet ground state by hundreds of wavenumbers and the signal corresponding to the $S = 3/2$ excited state is not observed [Bashkin, 1988; Brewer, 1989; Hagen, 1988; Pal, 1992; Schäfer, 1998; Wieghardt, 1988]. For the EPR spectrum of **6-ox**, in addition to the signal at $g = 2$, other signals assignable to $S = 3/2$ excited states were observed even at 5 K, at $g = 7.5$ and $g = 5.5$ (Figure 17). This indicates a very small energy separation between the doublet and quartet states in **6-ox**, in agreement with the susceptibility studies.

Table 1. Summary of the effective *g* and ^{55}Mn HFI tensors for **5-ox**, and the S_2 states of *T. elongatus* and spinach PSII.

			A_i [MHz]			
		g	A_1	A_2	A_3	A_4
5-ox	x	1.944	376	233	253	193
	y	1.964	297	198	283	222
	z	2.002	272	260	149	131
	iso	1.970	315	230	228	198
T. elongatus	x	1.971	350	249	202	148
	y	1.948	310	227	182	162
	z	1.985	275	278	240	263
	iso	1.968	312	251	208	191
Spinach	x	1.997	310	235	185	170
	y	1.970	310	235	185	170
	z	1.965	275	275	245	240
	iso	1.977	298	248	205	193

^{55}Mn Davies ENDOR spectra were collected for **5-ox** at Q-band. By incorporating the ESE-EPR spectrum at D-band [130 GHz], the *g* value was constrained at $g = [1.944, 1.964, 2.002]$, $g_{iso} = 1.97$. A global fit of the Mn hyperfine interactions incorporating the ENDOR spectra as well as the X-band CW spectrum yielded the parameters listed in Table 1. Similar to **4-ox**, the unique, larger Mn hyperfine coupling constant is consistent with the Mn^{III} center in **5-ox**. The Mn hyperfine coupling constants reported for **5-ox** are similar in magnitude to that of the S_2 state of *T. elongatus* [Cox, 2011; Kulik, 2007; Peloquin, 2000; Su, 2011].

Between the optimized structures of the proposed open- and closed-cubane forms of the S_2 state, Mn[4]–O[5] and Mn[1]–O[5] distances interchange from 1.87 Å to ~3.2 Å, tantamount to a bond breaking-reforming process (Figure 2). This rearrangement process is accompanied by a change in the electronic structure of the S_2 state, as explained from the computed magnetic exchange coupling constants in both open and closed forms. Our studies indicate that such large structural changes to the inorganic $CaMn_4O_5$ core of the OEC may not be necessary to perturb its electronic structure. Comparisons between the crystal structures of **4-ox**, **5-ox**, and **6-ox** indicate only small variations in Mn–oxo bond distances, with the longest Mn–oxo bond in each species varying from 2.051[4] Å to 2.241[1] Å. The remaining Mn-O distances are in the range 1.831[1]–1.976[2] Å. Yet, such small geometric changes in the $[Mn^{III}Mn_3^{IV}O_4]$ core have a substantial effect in the electronic structure, as evident from magnetic susceptibility and EPR studies. These studies indicate that spectroscopic properties of tetranuclear Mn complexes are highly dependent on the nature and magnitude of the magnetic exchange coupling interactions, which are finely tuned by the nature of not only bridging ligands but also terminal ligands in the immediate coordination sphere of each Mn ion. While the two signals observed in **5-ox** and **6-ox** are not resulting from ground states of two different isomers, as proposed in PSII, they do correspond to low- and high-spin electronic states. As in PSII, the degrees of population of the two states are affected by different coupling schemes between the Mn centers, which are connected to structural differences. Most importantly, large structural distortions are not necessary for the complete disappearance of one of the signals (the HS in this case). In an earlier reported $Mn^{III}Mn_3^{IV}O_4$ cubane with six phosphinate ligands, the LS signal completely disappears, consistent with a higher $S \geq 3/2$ ground state. Given that in the series of compounds **4-ox**, **5-ox** and **6-ox**, even a change in the nature of a single carboxylate ligand affects the state energies and EPR signals, it is expected that features such as the protonation state of aquo ligands, bridging oxos, and nearby His residues could greatly affect the electronic structure of the OEC. Therefore, interpretation of EPR signals in terms of structural implications must be done very cautiously. Ideally, complementary structural information will be obtained to corroborate spectroscopic assignments.

8. Concluding Remarks

In the absence of direct and unambiguous structural data, proposals for the mechanism of water oxidation have relied heavily on the spectroscopic characterization of S-state intermediates. Spectroscopic features were interpreted in terms of structural implications, but studies that probe the effect of structural changes on the electronic structure and spectroscopy of relevant tetranuclear Mn complexes remain rare. In this chapter, tetranuclear Mn complexes with oxidation states corresponding to the S_0–S_3 states have been surveyed. With the exception of a special $[CaMn_4O_4[OPiv]_8]$ case, most tetranuclear complexes fall within a small number of structure types: cuboidal, square, linear, or adamantane-shaped. Although magnetic susceptibility and spectroscopic studies indicate that the spin ground state correlates with the cluster geometry, with linear clusters generally having low-spin ground states while cuboidal and adamantane clusters having higher spin ground states, systematic studies with a series of structurally related complexes are rare. A versatile ligand platform that can support a variety of oxidation states and allow systematic changes to the nature of bridging and terminal ligands has been developed in our laboratories. Studies of Mn_4O_4 models of the OEC suggest that geometrical changes much smaller than the ones proposed for PSII with respect to the metal and oxo/hydroxo motifs could have substantial effects on the EPR signals. Therefore, the deduction of the geometry of OEC S-state intermediates based on EPR spectroscopic features need to be complemented with appropriate structural determination.

Spectroscopic studies have been invaluable for gaining insight into the nature of intermediates in the catalytic cycle of water oxidation by the OEC. To further inform the interpretation of these studies, benchmarking with synthetic clusters that are thoroughly characterized structurally and spectroscopically is imperative. Given progress over recent years, goals such as the synthesis of more accurate structural models of the OEC and systematic structure–electronic structures studies may be within reach. Spectroscopic characterization of Mn_4 complexes in oxidation states relevant to S_2–S_4 will prove valuable for benchmarking against the biological system. Additionally, the influence of protonation state on the electronic structure and ^{17}O labelling and spectroscopic characterization will provide further mechanistic insight.

References and Citations

Ames, W., Pantazis, D.A., Krewald, V., Cox, N., Messinger, J., Lubitz, W. and Neese, F. (2011). *J. Am. Chem. Soc.*, 133, 19743–19757.

Andres, H., Basler, R., Güdel, H.-U., Aromí, G., Christou, G., Büttner, H. and Rufflé, B. (2000). *J. Am. Chem. Soc.*, 122, 12469–12477.

Aromí, G., Wemple, M.W., Aubin, S.J., Folting, K., Hendrickson, D.N. and Christou, G. (1998). *J. Am. Chem. Soc.*, 120, 5850–5851.

Aubin, S.M.J., Wemple, M.W., Adams, D.M., Tsai, H.-L., Christou, G. and Hendrickson, D.N. (1996). *J. Am. Chem. Soc.*, 118, 7746–7754.

Bashkin, J.S., Schake, A.R., Vincent, J.B., Chang, H.R., Li, Q., Huffman, J.C., Christou, G. and Hendrickson, D.N. (1988). *J. Chem. Soc. Chem. Commun.*, 700–702.

Blondin, G., Davydov, R., Philouze, C., Charlot, M.-F., Styring, T., Akermark, B., Girerd, J.-J. and Boussac, A. (1997). *J. Chem. Soc. Dalton Trans.*, 4069–4074.

Boussac, A., Un, S., Horner, O. and Rutherford, A.W. (1998). *Biochemistry*, 37, 4001–4007.

Boussac, A. and Rutherford, A.W. (2000). *Biochim. Biophys. Acta*, 1457, 145–156.

Boussac, A., Rutherford, A.W. and Sugiura, M. (2015). *Biochim. Biophys. Acta — Bioenergetics*, 1847, 576–586.

Boussac, A., Ugur, I., Marion, A., Sugiura, M., Kaila, V.R.I. and Rutherford, A.W. (2018). *Biochim. Biophys. Acta — Bioenergetics*, 1859, 342–356.

Bovi, D., Narzi, D. and Guidoni, L. (2013). *Angew. Chem. Int. Ed.*, 52, 11744–11749.

Brewer, K.J., Calvin, M., Lumpkin, R.S., Otvos, J.W. and Spreer, L.O. (1989). *Inorg. Chem.*, 28, 4446–4451.

Britt, R.D., Lorigan, G.A., Sauer, K., Klein, M.P. and Zimmermann, J.-L. (1992). *Biochim. Biophys. Acta — Bioenergetics*, 1140, 95–101.

Campbell, K.A., Peloquin, J.M., Pham, D.P., Debus, R.J. and Britt, R.D. (1998). *J. Am. Chem. Soc.*, 120, 447–448.

Chan, M.K. and Armstrong, W.H. (1991). *J. Am. Chem. Soc.*, 113, 5055–5057.

Chatterjee, R., Han, G., Kern, J., Gul, S., Fuller, F.D., *et al.* (2016). *Chem. Sci.*, 7, 5236–5248.

Chen Collomb, M.-N., Duboc, C., Blondin, G., Rivière, E., Faller, J. W., Crabtree, R.H. and Brudvig, G.W. (2005). *Inorg. Chem.*, 44, 9567–9573.

Chen, H., Faller, J.W., Crabtree, R.H. and Brudvig, G.W. (2004). *J. Am. Chem. Soc.*, 126, 7345–7349.

Cinco, R.M., Rompel, A., Visser, H., Aromí, G., Christou, G., Sauer, K., Klein, M.P. and Yachandra, V.K. (1999). *Inorg. Chem.*, 38, 5988–5998.

Cox, N., Rapatskiy, L., Su, J.-H., Pantazis, D.A., Sugiura, M., Kulik, L., Dorlet, P., Rutherford, A.W., Neese, F., Boussac, A., Lubitz. and W., Messinger, J. (2011). *J. Am. Chem. Soc.*, 133, 3635–3648.

Cox, N., Pantazis, D.A., Neese, F. and Lubitz, W. (2013). *Acc. Chem. Res.*, 46, 1588–1596.

Cox, N., Retegan, M., Neese, F., Pantazis, D.A., Boussac, A. and Lubitz, W. (2014). *Science*, 345, 804.

DeRose, V.J., Latimer, M.J., Zimmermann, J.-L., Mukerji, I., Yachandra, V.K., Sauer, K. and Klein, M.P. (1995). *Chem. Phys.*, 194, 443–459.

Dexheimer, S.L. and Klein, M.P. (1992). *J. Am. Chem. Soc.*, 114, 2821–2826.

Dismukes, G.C. and Siderer, Y. (1981). *Proc. Nat. Acad. Sci.*, 78, 274.

Dubé, C.E., Wright, D.W., Pal, S., Bonitatebus, P.J. and Armstrong, W.H. (1998). *J. Am. Chem. Soc.*, 120, 3704–3716.

Dubé, C.E., Sessoli, R., Hendrich, M.P., Gatteschi, D. and Armstrong, W.H. (1999). *J. Am. Chem. Soc.*, 121, 3537–3538.

Dubé, C.E., Mukhopadhyay, S., Bonitatebus, P.J., Staples, R.J. and Armstrong, W.H. (2005). *Inorg. Chem.*, 44, 5161–5175.

Ferreira, K.N., Iverson, T.M., Maghlaoui, K., Barber, J. and Iwata, S. (2004). *Science*, 303, 1831.

Gupta, R., Taguchi, T., Lassalle-Kaiser, B., Bominaar, E.L., Yano, J., Hendrich, M.P. and Borovik, A.S. (2015). *Proc. Nat. Acad. Sci.*, 112, 5319–5324.

Haddy, A., Lakshmi, K.V., Brudvig, G.W. and Frank, H.A. (2004). *Biophys. J.*, 87, 2885–2896.

Hagen, K.S., Armstrong, W.H. and Hope, H. (1988). *Inorg. Chem.*, 27, 967–969.

Hagen, K.S., Westmoreland, T.D., Scott, M.J. and Armstrong, W.H. (1989). *J. Am. Chem. Soc.*, 111, 1907–1909.

Han, Z., Horak, K.T., Lee, H.B. and Agapie, T. (2017). *J. Am. Chem. Soc.*, 139, 9108–9111.

Hendrickson, D.N., Christou, G., Schmitt, E.A., Libby, E., Bashkin, J.S., Wang, S., Tsai, H.L., Vincent, J.B. and Boyd, P.D.W. (1992). *J. Am. Chem. Soc.*, 114, 2455–2471.

Horner, O., Rivière, E., Blondin, G., Un, S., Rutherford, A.W., Girerd, J.-J. and Boussac, A. (1998). *J. Am. Chem. Soc.*, 120, 7924–7928.

Joliot, P. (1965). *Biochim. Biophys. Acta*, 102, 116–134.

Kamiya, N. and Shen, J.-R. (2003). *Proc. Nat. Acad. Sci.*, 100, 98.

Kanady, J.S., Tsui, E.Y., Day, M.W. and Agapie, T. (2011). *Science*, 333, 733.

Kanady, J.S., Mendoza-Cortes, J.L., Tsui, E.Y., Nielsen, R.J., Goddard, W.A. and Agapie, T. (2013a). *J. Am. Chem. Soc.*, 135, 1073–1082.

Kanady, J.S., Tran, R., Stull, J.A., Lu, L., Stich, T.A., Day, M.W., Yano, J., Britt, R.D. and Agapie, T. (2013b). *Chem. Sci.*, 4, 3986–3996.

Kanady, J.S., Lin, P.-H., Carsch, K.M., Nielsen, R.J., Takase, M.K., Goddard, W.A. and Agapie, T. (2014). *J. Am. Chem. Soc.*, 136, 14373–14376.

Kern, J., Alonso-Mori, R., Tran, R., Hattne, J., Gildea, R.J., *et al.* (2013). *Science*, 340, 491.

Kern, J., Tran, R., Alonso-Mori, R., Koroidov, S., Echols, N., *et al.* (2014). *Nat. Commun.*, 5, 4371.

Kim, D.H., Britt, R.D., Klein, M.P. and Sauer, K. (1992). *Biochemistry*, 31, 541–547.

Kok, B., Forbush, B. and McGloin, M. (1970). *Photochem. Photobiol.*, 11, 457–475.

Kulik, L.V., Epel, B., Lubitz, W. and Messinger, J. (2007). *J. Am. Chem. Soc.*, 129, 13421–13435.

Kupitz, C., Basu, S., Grotjohann, I., Fromme, R., Zatsepin, N.A., *et al.* (2014). *Nature*, 513, 261.

Larson, E., Haddy, A., Kirk, M.L., Sands, R.H., Hatfield, W.E. and Pecoraro, V.L. (1992). *J. Am. Chem. Soc.*, 114, 6263–6265.

Lee, H.B., Tsui, E.Y. and Agapie, T. (2017). *Chem. Commun.*, 53, 6832–6835.

Lee, H.B., Shiau, A.A., Oyala, P.H., Marchiori, D.A., Gul, S., Chatterjee, R., Yano, J., Britt, R.D. and Agapie, T. (2018). *J. Am. Chem. Soc.*, 140, 17175–17187.

Liang, W., Latimer, M.J., Dau, H., Roelofs, T.A., Yachandra, V.K., Sauer, K. and Klein, M.P. (1994). *Biochemistry*, 33, 4923–4932.

Lin, P.-H., Tsui, E.Y., Habib, F., Murugesu, M. and Agapie, T. (2016). *Inorg. Chem.*, 55, 6095–6099.

Lohmiller, T., Ames, W., Lubitz, W., Cox, N. and Misra, S.K. (2013). *Appl. Magn. Reson.*, 44, 691–720.

Lohmiller, T., Krewald, V., Navarro, M.P., Retegan, M., Rapatskiy, L., Nowaczyk, M.M., Boussac, A., Neese, F., Lubitz, W., Pantazis, D.A. and Cox, N. (2014). *Phys. Chem. Chem. Phys.*, 16, 11877–11892.

Marchiori, D.A., Oyala, P.H., Debus, R.J., Stich, T.A. and Britt, R.D. (2018). *J. Phys. Chem. B*, 122, 1588–1599.

Mukherjee, S., Stull, J.A., Yano, J., Stamatatos, T.C., Pringouri, K., Stich, T.A., Abboud, K.A., Britt, R.D., Yachandra, V.K. and Christou, G. (2012). *Proc. Nat. Acad. Sci.*, 109, 2257.

Mukhopadhyay, S., Staples, R.J. and Armstrong, W.H. (2002). *Chem. Commun.*, 864–865.

Mukhopadhyay, S., Mandal, S.K., Bhaduri, S. and Armstrong, W.H. (2004a). *Chem. Rev.*, 104, 3981–4026.

Mukhopadhyay, S., Mok, H.J., Staples, R.J. and Armstrong, W.H. (2004b). *J. Am. Chem. Soc.*, 126, 9202–9204.

Mullins, C.S. and Pecoraro, V.L. (2008). *Coord. Chem. Rev.*, 252, 416–443.

Oyala, P.H., Stich, T.A., Stull, J.A., Yu, F., Pecoraro, V.L. and Britt, R.D. (2014). *Biochemistry*, 53, 7914–7928.

Oyala, P.H., Stich, T.A., Debus, R.J. and Britt, R.D. (2015). *J. Am. Chem. Soc.*, 137, 8829–8837.

Pace, R.J., Smith, P., Bramley, R. and Stehlik, D. (1991). *Biochim. Biophys. Acta — Bioenergetics*, 1058, 161–170.

Pal, S., Gohdes, J.W., Wilisch, W.C.A. and Armstrong, W.H. (1992). *Inorg. Chem.*, 31, 713–716.

Pantazis, D.A., Ames, W., Cox, N., Lubitz, W. and Neese, F. (2012). *Angew. Chem. Int. Ed.*, 51, 9935–9940.

Pantazis, D.A. (2018). *ACS Catal.*, 9477–9507.

Paul, S., Cox, N. and Pantazis, D.A. (2017). *Inorg. Chem.*, 56, 3875–3888.

Pecoraro, V.L. (1988). *Photochem. Photobiol.*, 48, 249–264.

Peloquin, J.M., Campbell, K.A., Randall, D.W., Evanchik, M.A., Pecoraro, V.L., Armstrong, W.H. and Britt, R.D. (2000). *J. Am. Chem. Soc.*, 122, 10926–10942.

Pérez Navarro, M., Ames, W.M., Nilsson, H., Lohmiller, T., Pantazis, D.A., Rapatskiy, L., Nowaczyk, M.M., Neese, F., Boussac, A., Messinger, J., Lubitz, W. and Cox, N. (2013). *Proc. Nat. Acad. Sci.*, 110, 15561.

Philouze, C., Blondin, G., Girerd, J.-J., Guilhem, J., Pascard, C. and Lexa, D. (1994). *J. Am. Chem. Soc.*, 116, 8557–8565.

Randall, D.W., Sturgeon, B.E., Ball, J.A., Lorigan, G.A., Chan, M.K., Klein, M.P., Armstrong, W.H. and Britt, R.D. (1995). *J. Am. Chem. Soc.*, 117, 11780–11789.

Randall, D.W., Gelasco, A., Caudle, M.T., Pecoraro, V.L. and Britt, R.D. (1997). *J. Am. Chem. Soc.*, 119, 4481–4491.

Rapatskiy, L., Cox, N., Savitsky, A., Ames, W.M., Sander, J., Nowaczyk, M.M., Rögner, M., Boussac, A., Neese, F., Messinger, J. and Lubitz, W. (2012). *J. Am. Chem. Soc.*, 134, 16619–16634.

Rapatskiy, L., Ames, W.M., Pérez-Navarro, M., Savitsky, A., Griese, J.J., Weyhermüller, T., Shafaat, H.S., Högbom, M., Neese, F., Pantazis, D.A. and Cox, N. (2015). *J. Phys. Chem. B*, 119, 13904–13921.

Reed, C.J. and Agapie, T. (2018). *J. Am. Chem. Soc.*, 140, 10900–10908.

Ruettinger, W.F., Campana, C. and Dismukes, G.C. (1997). *J. Am. Chem. Soc.*, 119, 6670–6671.

Ruettinger, W.F., Ho, D.M. and Dismukes, G.C. (1999). *Inorg. Chem.*, 38, 1036–1037.

Rutherford, A.W., Osyczka, A. and Rappaport, F. (2012). *FEBS Lett.*, 586, 603–616.

Sauter, N.K., Echols, N., Adams, P.D., Zwart, P.H., Kern, J., Brewster, A.S., Koroidov, S., Alonso-Mori, R., Zouni, A., Messinger, J., Bergmann, U., Yano, J. and Yachandra, V.K. (2016). *Nature*, 533, E1.

Schäfer, K.-O., Bittl, R., Zweygart, W., Lendzian, F., Haselhorst, G., Weyhermüller, T., Wieghardt, K. and Lubitz, W. (1998). *J. Am. Chem. Soc.*, 120, 13104–13120.

Shen, J.-R. (2015). *Annu. Rev. Plant Biol.*, 66, 23–48.

Shoji, M., Isobe, H., Shen, J.-R. and Yamaguchi, K. (2016). *Phys. Chem. Chem. Phys.*, 18, 11330–11340.

Siegbahn, P.E.M. (2009). *Acc. Chem. Res.*, 42, 1871–1880.

Su, J.-H., Cox, N., Ames, W., Pantazis, D.A., Rapatskiy, L., Lohmiller, T., Kulik, L.V., Dorlet, P., Rutherford, A.W., Neese, F., Boussac, A., Lubitz, W. and Messinger, J. (2011). *Biochim. Biophys. Acta — Bioenergetics*, 1807, 829–840.

Suga, M., Akita, F., Hirata, K., Ueno, G., Murakami, H., Nakajima, Y., Shimizu, T., Yamashita, K., Yamamoto, M., Ago, H. and Shen, J.-R. (2014). *Nature*, 517, 99.

Tsui, E.Y., Kanady, J.S., Day, M.W. and Agapie, T. (2011). *Chem. Commun.*, 47, 4189–4191.

Tsui, E.Y. and Agapie, T. (2013). *Proc. Nat. Acad. Sci.*, 110, 10084.

Tsui, E.Y., Kanady, J.S. and Agapie, T. (2013a). *Inorg. Chem.*, 52, 13833–13848.

Tsui, E.Y., Tran, R., Yano, J. and Agapie, T. (2013b). *Nat. Chem.*, 5, 293.

Umena, Y., Kawakami, K., Shen, J.-R. and Kamiya, N. (2011). *Nature*, 473, 55.

Vinyard, D.J., Khan, S., Askerka, M., Batista, V.S. and Brudvig, G.W. (2017). *J. Phys. Chem. B*, 121, 1020–1025.

Wang, S., Tsai, H.-L., Hagen, K.S., Hendrickson, D.N. and Christou, G. (1994). *J. Am. Chem. Soc.*, 116, 8376–8377.

Wang, S., Tsai, H.-L., Libby, E., Folting, K., Streib, W.E., Hendrickson, D.N. and Christou, G. (1996). *Inorg. Chem.*, 35, 7578–7589.

Wang, S., Wemple, M.S., Yoo, J., Folting, K., Huffman, J.C., Hagen, K.S., Hendrickson, D.N. and Christou, G. (2000). *Inorg. Chem.*, 39, 1501–1513.

Wemple, M.W., Tsai, H.L., Folting, K., Hendrickson, D.N. and Christou, G. (1993). *Inorg. Chem.*, 32, 2025–2031.

Wemple, M.W., Adams, D.M., Folting, K., Hendrickson, D.N. and Christou, G. (1995). *J. Am. Chem. Soc.*, 117, 7275–7276.

Wieghardt, K., Bossek, U., Nuber, B., Weiss, J., Bonvoisin, J., Corbella, M., Vitols, S.E. and Girerd, J.J. (1988). *J. Am. Chem. Soc.*, 110, 7398–7411.

Yamauchi, T., Mino, H., Matsukawa, T., Kawamori, A. and Ono, T.-a. (1997). *Biochemistry*, 36, 7520–7526.

Yano, J., Kern, J., Irrgang, K.-D., Latimer, M.J., Bergmann, U., Glatzel, P., Pushkar, Y., Biesiadka, J., Loll, B., Sauer, K., Messinger, J., Zouni, A. and Yachandra, V.K. (2005). *Proc. Nat. Acad. Sci.*, 102, 12047.

Yano, J. and Yachandra, V. (2014). *Chem. Rev.*, 114, 4175–4205.

Young, I.D., Ibrahim, M., Chatterjee, R., Gul, S., Fuller, F.D., *et al.* (2016). *Nature*, 540, 453–457.

Zhang, C., Chen, C., Dong, H., Shen, J.-R., Dau, H. and Zhao, J. (2015). *Science*, 348, 690–693.

Zouni, A., Witt, H.-T., Kern, J., Fromme, P., Krauss, N., Saenger, W. and Orth, P. (2001). *Nature*, 409, 739.

Chapter 14

Progress Towards Unraveling the Water-Oxidation Mechanism of Photosystem II

Hao-Li Huang, Ipsita Ghosh, Gourab Banerjee and Gary W. Brudvig

Department of Chemistry, Yale University
225 Prospect Street, New Haven, CT 06511, USA
gary.brudvig@yale.edu

Unraveling the mechanism of the water-oxidation reaction of photosystem II (PSII) is important to provide a blueprint to aid the development of artificial photosynthetic processes for global solar fuel production. The overall reaction mechanism of PSII involves a complex multistep process, and the details of many of the steps remain unclear. In this chapter, we discuss how substrate-water exchange and small-molecule water analogues have been used to probe the binding of substrate waters, which in turn can provide critical information on the mechanism of O–O bond formation. Further, we discuss the nature of the intermediates in the S-state transitions revealed by studies involving isotope effects. Finally, we discuss modification of the rate-determining step by perturbation of the hydrogen-bonding network surrounding the oxygen-evolving complex. The mechanistic details revealed by these studies will thereby assist in uncovering the mechanism of water oxidation and motivate the design of bioinspired water-oxidation catalysts.

1. Introduction

Photosystem II (PSII) is a protein complex in higher plants, algae, and cyanobacteria that exists as a homodimer, with an overall molecular weight of ~650 kDa

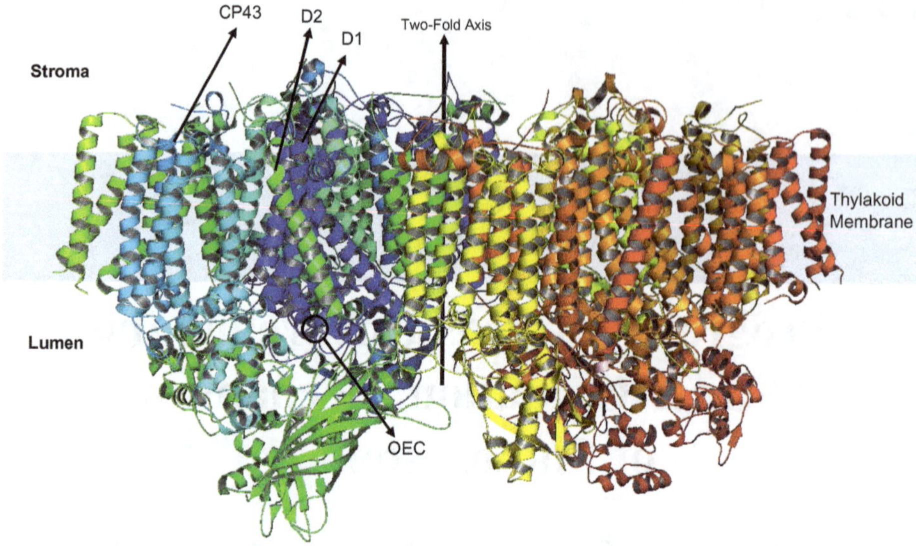

Figure 1. Crystal structure of PSII. Helices are represented as ribbons with D1 in dark blue, D2 in light green and CP43 in cyan. Figure created using Pymol software from the 3WU2 structure [Umena *et al.* 2011].

(Figure 1). The complex consists of the D1, D2, CP43, CP47, and cytochrome (cyt) *b*-559 proteins, along with a number of small protein subunits and more than 50 cofactors, which together are responsible for the water-oxidation reaction (Equation 1) in photosynthesis [McEvoy and Brudvig, 2006; Vinyard *et al.*, 2015; Vogt *et al.*, 2015]. The oxygen-evolving complex (OEC) extracts electrons from water to provide reducing equivalents for solar fuel production in later steps of photosynthesis, and releases dioxygen as a byproduct.

$$2H_2O \xrightarrow{h\nu} O_2 + 4H^+ + 4e^-. \tag{1}$$

It is an important reaction to harness solar energy, for it is light-driven and produces the electrons necessary for fuel production from water and carbon dioxide, which are abundant enough to support solar fuel production on a global scale. Figuring out how the reaction is carried out in the biological system could greatly help us design bioinspired water-oxidation catalysts.

The water-oxidation reaction in PSII is initiated upon a photon absorption at the reaction center chlorophyll P_{680}, where a charge separation is then generated. The electron is transferred to a pheophytin, a tightly-bound plastoquinone A (Q_A), and a loosely-bound plastoquinone B (Q_B). After Q_B accepts two electrons and two protons to form plastoquinol (Q_BH_2), it exchanges with another plastoquinone from the quinone pool in the thylakoid membrane. On the electron-donor side, the hole at P_{680} is filled by electrons extracted from water at the OEC, via the

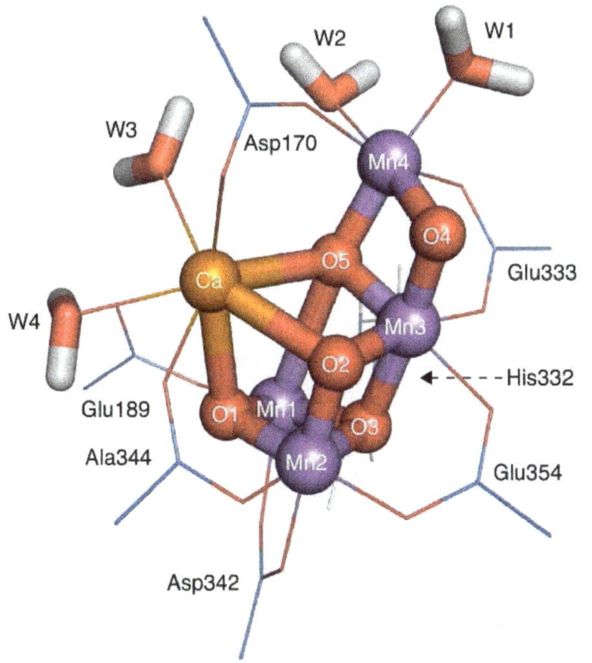

Figure 2. Structure of the OEC in the S_1 state. Reprinted (with permission) from [Vogt *et al.*, 2015].

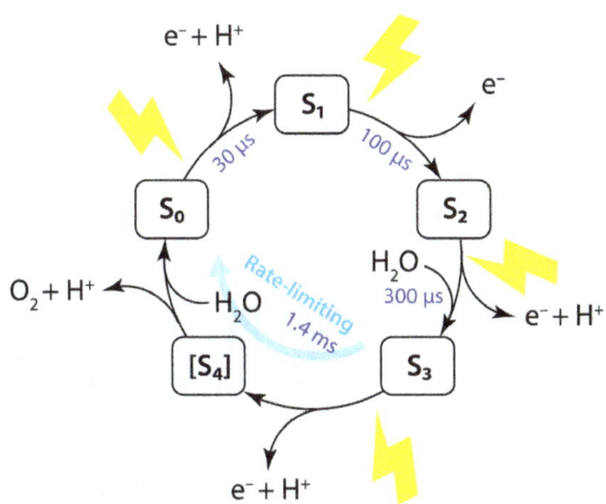

Figure 3. S-state cycle with the half-time of each transition.

redox-active tyrosine (Y_Z) [McEvoy and Brudvig, 2006; Vinyard *et al.*, 2015; Vogt *et al.*, 2015].

The OEC is a metal ion cluster composed of 1 calcium, 4 manganese, and 5 oxygen ions (Figure 2). The oxidation state of the OEC is described by the S_i states in the Kok cycle (Figure 3) [Kok *et al.*, 1970]. Following each light-driven charge separation, the OEC releases an electron and proceeds from the S_i to S_{i+1} state. Dark-adapted PSII exists in the S_0 and S_1 states.

Many of the S-states are paramagnetic and, hence, have been characterized by electron paramagnetic resonance (EPR) spectroscopy [McEvoy and Brudvig, 2006]. These studies along with extended X-ray absorption fine structure (EXAFS) measurements, X-ray crystallographic studies and computational calculations have helped to determine the oxidation states of Mn in each of the S states [Haumann *et al.*, 2005; Loll *et al.*, 2005; Luber *et al.*, 2011; Pal *et al.*, 2013; Suga *et al.*, 2015]. The S_1 state is characterized as a $(Mn^{3+})_2(Mn^{4+})_2$ state. Flash illumination results in the formation of the S_2 state, $(Mn^{3+})(Mn^{4+})_3$, which exists in two redox-isomeric forms. These redox isomers have a low-spin, S = 1/2, form (S_2^A) characterized by a g = 2 multiline EPR signal and a high-spin, S = 5/2, form (S_2^B) characterized by a g = 4 EPR signal (Figure 4) [Pantazis *et al.*, 2012; Pokhrel and Brudvig, 2014].

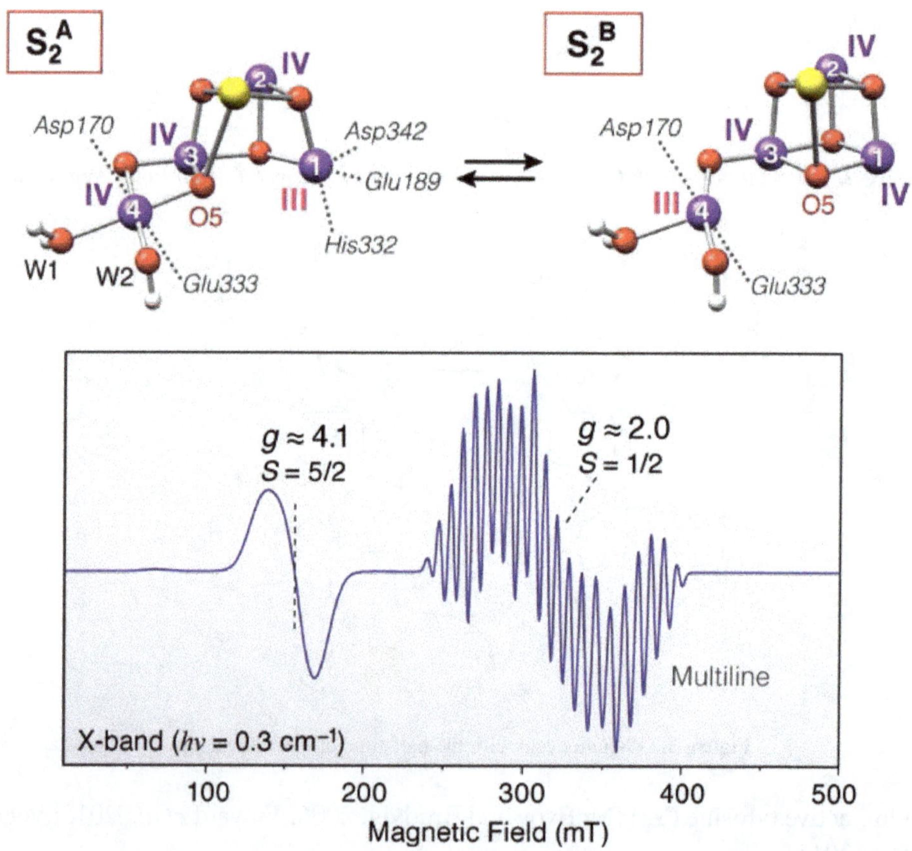

Figure 4. Structural isomers of the OEC in the S_2 state. The top figure shows the geometries of the OEC and the bottom figure shows the corresponding X-band EPR spectra. Reprinted (with permission) from [Pantazis, 2018].

A second flash results in the formation of the S_3 state accompanied by a proton loss and a substrate water entry (Figure 3). The structure of the S_3 state has recently been studied by X-ray free-electron laser (XFEL) crystallography and the $(Mn^{4+})_4$ state has been characterized by EPR spectroscopy [Cox *et al.*, 2014; Sauter *et al.*, 2016; Suga *et al.*, 2017]. However, there are ongoing debates on the mechanism of formation of the S_3 state from the S_2 state. Further, the large number of waters bound to and surrounding the OEC creates a problem for identification of the substrate waters which are crucial in understanding the O–O bond-formation reaction. Several substrate water analogues like ammonia and methanol have been employed to identify the substrate waters. Further substrate-water exchange studies have been carried out. These studies will be discussed in further detail later in this chapter. The third flash is followed by the formation of a $S_3^+Y_Z\bullet$ intermediate, which has been successfully trapped [Zahariou *et al.*, 2014]. This intermediate then undergoes a proton transfer step to form the $S_3Y_Z\bullet$ intermediate [Bao and Burnap, 2015]. The OEC then undergoes spontaneous water oxidation resulting in O–O bond formation and subsequent formation of the S_0 state (Figure 5).

Kinetics of the S-state cycling reveal that the $S_3Y_Z\bullet$ to S_0 transition is the slowest and, thus, the rate-determining step; hence, it's crucial to understand this transition in order to determine the mechanism of water oxidation [Haumann *et al.*, 2005]. The mechanism of O–O bond formation is highly debated with two leading proposals involving oxo–oxyl radical coupling and nucleophilic water attack, respectively (Figure 6) [Brudvig, 2008; Siegbahn, 2013].

In this chapter, we will focus on three key areas: (a) how isotope effects have been used to understand the nature of the rate-determining step, (b) resolving the substrate water binding sites by using substrate water exchange studies and small molecule analogues, and (c) creating perturbations in the hydrogen-bonding network surrounding the OEC by site-directed mutations and inorganic substitutions to delay the kinetics of the rate-determining step.

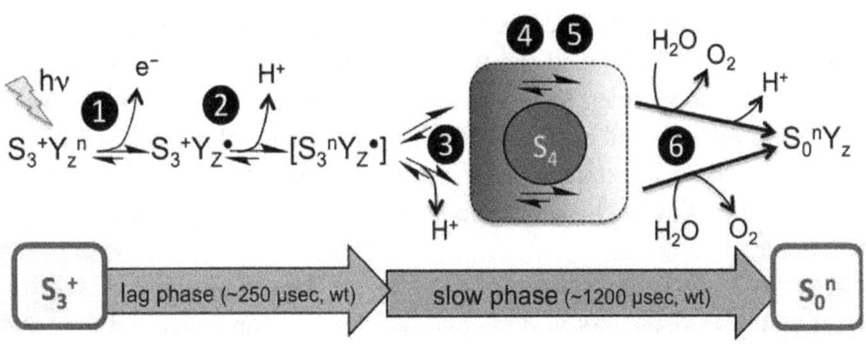

Figure 5. Intermediate steps in the S_3 to S_0 transition. Reprinted (with permission) from [Bao and Burnap, 2015].

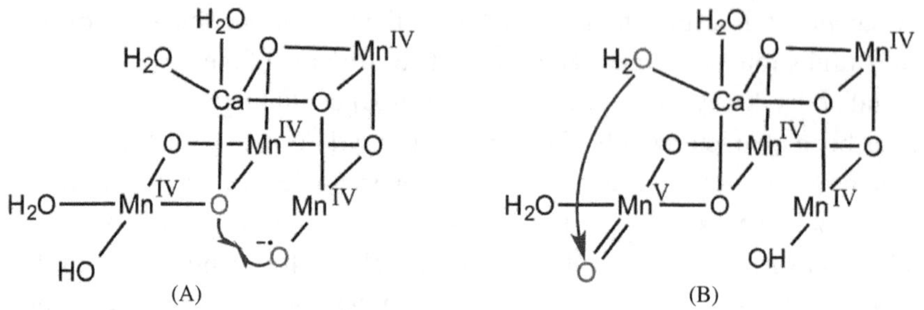

Figure 6. Proposed O–O bond formation mechanisms. (A) Oxo–oxyl mechanism and (B) Nucleophilic water attack. Reprinted (with permission) from [Vinyard *et al.*, 2015].

2. Identification of Substrate Waters

2.1. *^{18}O-labeled substrate water exchange*

A reactant can be isotope-labeled to allow for tracking along the way of the reaction, which often can help to understand the mechanism of the reaction. The labeled isotopologues can be discerned from the unlabeled ones with mass spectrometry, Fourier transform infrared (FTIR) spectroscopy, or EPR spectroscopy for isotopes with non-zero nuclear spin (*e.g.* ^{17}O). Isotope labeling is especially useful, or even necessary, for studying the substrate waters in the water-oxidation reaction carried out by PSII, where the substrate waters are overwhelmed by the solvent water molecules.

^{18}O-labeled substrate water exchange experiments have been reported [Hillier and Wydrzynski, 2000, 2004; Cox and Messinger, 2013]. Briefly, the PSII samples are poised at specific S states by applying a certain number of flashes. At the given S state, $H_2{}^{18}O$ is injected into the system and allowed to exchange with the waters bound on the OEC for various incubation times (from 4 ms to 200 s), before a series of turnover flashes are applied to drive the oxygen-evolution process. The evolved oxygen is then analyzed by time-resolved mass spectrometry, at m/z = 34 for singly-labeled ^{16}O–^{18}O, and at m/z = 36 for doubly-labeled ^{18}O–^{18}O (Figure 7). For longer incubation times, there would be more labeled oxygen evolved because more labeled water would exchange with the water molecules originally bound to the OEC to become substrate water. By fitting a plot of the labeled oxygen yield *versus* incubation time with exponential functions, the exchange rate constants can be determined.

At m/z = 34, for the evolution of ^{16}O–^{18}O, the data show a biphasic behavior, whereas at m/z = 36, the plot is mono-phasic. The slow component for the ^{16}O–^{18}O evolution matches with the exchange rate constant for ^{18}O–^{18}O evolution, suggesting a slowly-exchanging substrate water, W_s. The fast component at m/z = 34, on

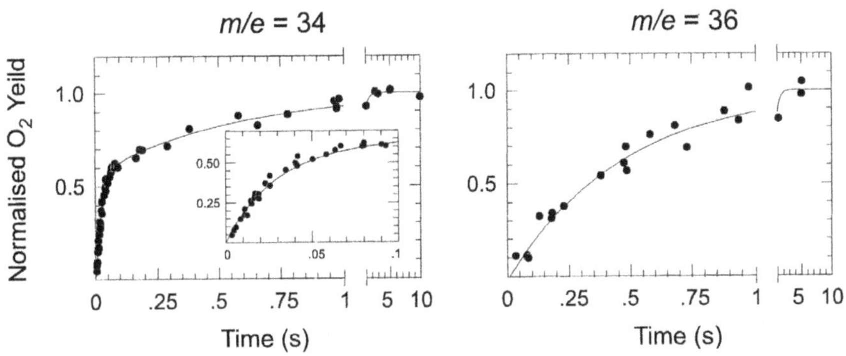

Figure 7. Normalized yield of singly-labeled ^{16}O–^{18}O (A), and that of doubly-labeled ^{18}O–^{18}O (B) plotted against various incubation times in the S_3 state of spinach thylakoids. Reprinted (with permission) from [Hillier and Wydrzynski, 2000].

Table 1. S-state dependent rate constants for substrate water exchange in spinach thylakoids and Sr^{2+}-substituted PSII membranes. Data taken from [Cox and Messinger, 2013].

	Thylakoid membranes		Sr^{2+}-substituted PSII membranes	
S_i state	**k_s /s^{-1}**	**k_f /s^{-1}**	**k_s /s^{-1}**	**k_f /s^{-1}**
S_0	~10	>120	—	—
S_1	~0.02	>120	~0.08	>120
S_2	~2.0	~120	~9.0	>120
S_3	~2.0	~40	~6.0	~23

the other hand, corresponds to a fast-exchanging substrate water, W_f. To generate a singly-labeled oxygen, one of the two substrate waters, W_s or W_f, is exchanged, and thus the biphasic behavior. For doubly-labeled oxygen to be generated, both W_s and W_f need to be exchanged; the fast-exchanging W_f would be followed by W_s exchange, and, therefore, the overall exchange kinetics for m/z = 36 match the rate of the W_s component in the m/z = 34 case. The rate constants for W_s and W_f exchange, k_s and k_f, respectively, are listed in Table 1.

The Ca^{2+} in the OEC can be exchanged with Sr^{2+} with retention of some activity [Boussac *et al.*, 2004]. The ^{18}O-labeled water exchange kinetics in Sr^{2+}-substituted PSII show a major effect on the exchange rate of W_s, but not so much of an effect on W_f (Table 1), suggesting the influence of the calcium on the binding of W_s. Furthermore, the slow exchange rate of W_s suggests that W_s may be a μ-oxo bridge, rather than a terminal ligand. Combining the results obtained from ^{17}O-labeled water exchange experiments measured by electron-electron double resonance detected nuclear magnetic resonance spectroscopy [Rapatskiy *et al.*, 2012], Cox and Messinger assigned W_s to O5 (refer to Figure 2 for the numbering) [Cox and Messinger, 2013]. Based on the assignment, they further propose that the W_f is likely either W2, or Wx, as suggested by Siegbahn [Siegbahn, 2013]. Their assignments have been used to support the oxo–oxyl coupling mechanism of O–O bond formation. An alternate model for the exchange kinetics of W_s was proposed by Vinyard *et al.*, which assigns the slowly-exchanging substrate water to a μ-oxo bridge in the S_2 state and a terminal water ligand in the S_3 state, in support of the nucleophilic attack mechanism [Vinyard *et al.*, 2015].

2.2. *Small molecules as substrate water analogues*

The OEC is surrounded by hydrogen-bonded waters that are crucial for its efficient function. However, it is difficult to distinguish and track the substrate waters that enter the S-state cycle from these surrounding waters. Hence, small molecules like ammonia and methanol have been employed as substrate water analogs. Their interaction with the OEC can be followed by using various spectroscopic techniques to reveal the possible interactions of the substrate waters with the Mn_4O_5Ca cluster.

2.1.1. *Methanol*

Methanol has been used as a spectroscopic probe to study water binding. Steady-state oxygen measurements of PSII in the presence of methanol revealed that it doesn't affect the S-state cycling or retard oxygen production [Force *et al.*, 1998; Nöring *et al.*, 2008]. However, higher concentrations of methanol (>1 M) as well as higher alcohols affect oxygen evolution, which can be due to a number of factors including modification in the water-oxidation mechanism, the details of which are yet to be investigated [Britt *et al.*, 2004]. Further, methanol stabilizes the g = 2 multiline form of the S_2 state. Hence, its interaction with the Mn_4O_5Ca cluster was probed in greater detail [Su *et al.*, 2011]. While electron spin echo envelope modulation (ESEEM) studies provide evidence for methanol ligation to one of the metal ions in the OEC, other pulsed EPR studies negate this possibility and indicate the presence of methanol in the vicinity of the Mn_4O_5Ca cluster [Force *et al.*, 1998; Oyala *et al.*, 2014; Retegan and Pantazis, 2017]. Methanol binding varies across species, and, based on this, the "narrow" protein channel leading to the OEC has been proposed to be the entry gate for methanol [Retegan and Pantazis, 2017]. Although methanol binding to the Mn_4O_5Ca cluster is yet to be conclusively demonstrated, its role as a substrate water analog to aid in elucidation of the substrate water delivery is important for future studies.

2.2.2. *Ammonia*

Ammonia has been historically used as a substrate water analogue and its interaction with the OEC has been studied extensively. Owing to its small size, hydrogen bonding capabilities and isoelectronic configuration, ammonia is an ideal substrate water analogue. Ammonia was shown to bind to the Mn_4O_5Ca cluster by Brudvig *et al.* in the S_2 state, altering the g = 2 multiline signal in the cw-EPR spectrum (Figure 8) [Beck *et al.*, 1986].

The mechanism of ammonia binding, however, has been widely debated and contrasting viewpoints exist in the literature about the mode of ammonia binding to the Mn_4O_5Ca cluster. While ESEEM studies have been used to support ammonia binding as a terminal ligand to Mn4 by substituting W1, QM/MM calculations benchmarked against EXAFS results indicate that ammonia binds to Mn4 as an additional ligand (Figure 9) [Askerka *et al.*, 2015; Oyala *et al.*, 2015].

The binding of ammonia in the higher S-states is less extensively probed. Velthuys *et al.* showed ammonia stays bound to the OEC in the S_3 state using

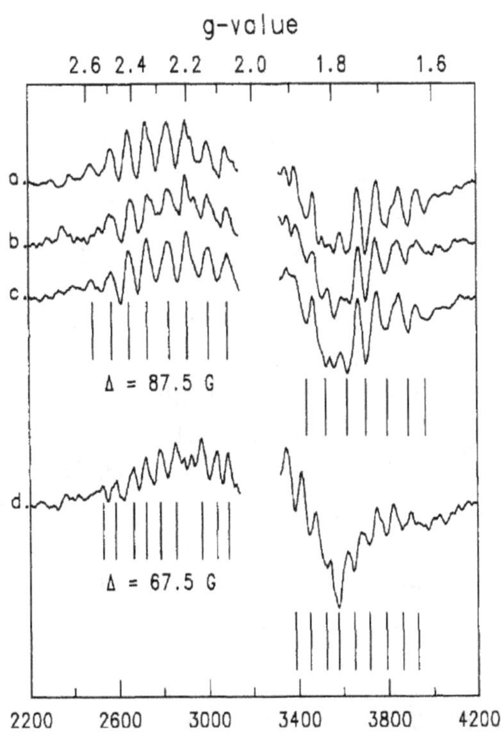

Figure 8. Comparison of S_2 state EPR spectra: (a) untreated spectra of PSII membranes illuminated at 210 K, (b) 100 mM NH_4Cl-treated PSII membranes illuminated at 210 K (c) untreated PSII membranes illuminated at 0°C (d) 100 mM NH_4Cl-treated PSII membranes illuminated at 0°C. Reprinted (with permission) from [Beck *et al.*, 1986].

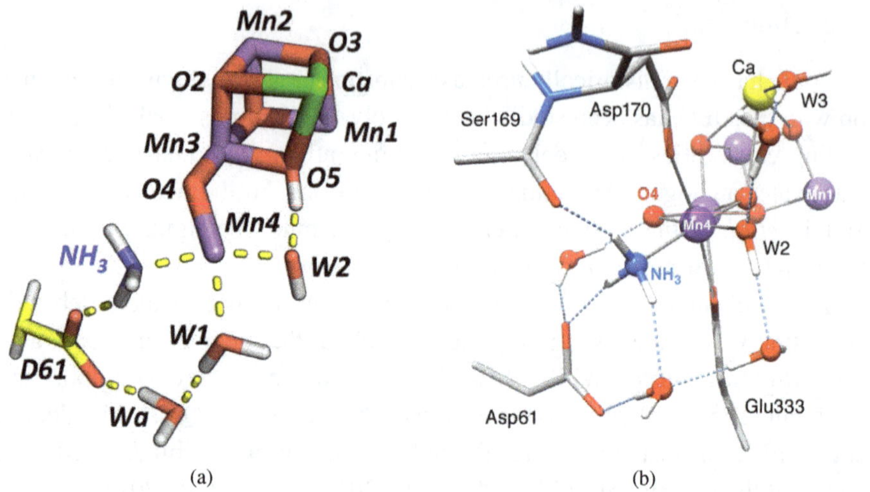

Figure 9. Comparison of models for the binding of ammonia to the OEC: (A) QM/MM structure depicting ammonia bound as an additional lidgand to Mn4, (B) model based on ESEEM showing ammonia binding by substitution of W1. Reprinted (with permission) from [Askerka *et al.*, 2015; Pantazis, 2018].

luminescence studies, whereas more recently Boussac *et al.* have used cw-EPR experiments to contradict this conclusion [Velthuys, 1975; Boussac *et al.*, 2018]. The S_3 state of ammonia-treated PSII shows modest changes to the rate of substrate water exchange underscoring the lack of significant alteration to the water-oxidation mechanism in the presence of ammonia [Young *et al.*, 2016]. However, in a recent study, Dau *et al.* have proposed a substantial delay in the oxygen-evolution rate in the presence of ammonia, contradicting the previous interpretations and proposing an important role of the ammonia-binding site in the water-oxidation chemistry [Schuth *et al.*, 2017]. Further, ammonia binding also stabilizes the S_2 state and affects S-state cycling [Vinyard and Brudvig, 2015]. However, these effects can also be due to the secondary binding site of ammonia which is chloride dependent and is known to retard oxygen evolution [Mandal *et al.*, 2017; Schuth *et al.*, 2017]. While further work is still required to fully understand the role of the ammonia-binding site in the mechanism of water oxidation, its importance to providing greater insights into the process is well-established.

3. Probing Mechanistic Details with Kinetic Isotope Effects

Isotopes are atoms with the same number of protons, but different number of neutrons. They are the same element, so they undergo chemical reactions in the same way. However, the difference in mass may lead to different reaction rates, shedding light on valuable mechanistic information at the transition state(s). The kinetic isotope effect (KIE), defined as the rate of the light isotopologue (k_L) over that of the heavy one (k_H) (Equation 2), is an effective approach for probing the transition state(s) of the rate-limiting step(s) of a reaction. The experiments can be carried out under the most physiologically-relevant conditions, without a need of trapping any intermediates.

$$\mathrm{KIE} = \frac{k_L}{k_H}. \tag{2}$$

Consider an energy diagram of the bond cleavage reaction. The electronic potential energy curve does not depend on the mass, according to the Born-Oppenheimer approximation, so different isotopologues share the same potential energy curve. The vibrational zero-point energy (ZPE) E_0, on the other hand, is dependent on mass, as is given in Equation 3, where the harmonic approximation is applied. The difference in the vibrational ZPEs often diminishes at the transition state, due to a reduced force constant k upon bond cleavage. Therefore, the activation barrier, and thus the rate of the reaction would be different for the two isotopologues.

$$E_0 = \frac{h}{4\pi}\sqrt{\frac{k}{\mu}}. \tag{3}$$

To calculate a KIE, we use the apparent rate of the overall reaction, which is dominated by the rate of the rate-limiting step. As a result, the KIEs are sensitive to the rate-limiting step. That is, if the substituted isotope is not involved in the rate-limiting step, there would be no kinetic isotope effect (KIE = 1).

3.1. *H/D Kinetic isotope effects*

To complete the Kok cycle, four electrons and four protons are released. It is known that one electron is released when proceeding from the S_i to the S_{i+1} state [Kok *et al.*, 1970]. In order to understand the proton-release pattern in the Kok cycle, H/D KIE measurements for each S state transition were made to determine whether or not a deprotonation process is involved in the rate-limiting step [Haumann *et al.*, 1997]. The results were mostly interpreted qualitatively. That is, if there is a KIE, then a proton is involved, and *vice versa*. The UV-visible transient absorption spectra at 360 nm reflect the redox state of the OEC, and were fitted with exponential functions to determine the rate constants of the S-state transitions.

The experiments were done using H_2O and D_2O as solvent, and the H/D isotope effects were calculated by taking the ratio of the rate in H_2O and that in D_2O.

A more comprehensive discussion of the H/D KIEs on the S-state transitions can be found in the review article by Dau and Haumann [Dau and Haumann, 2008]. To summarize, no H/D KIE is observed in the S_1 to S_2 transition, so it is concluded that deprotonation is not involved in the transition, consistent with the pH-independent transition rate. An H/D KIE of 2.1 is observed in the S_2 to S_3 transition, indicating a proton movement, consistent with a pH-dependent rate constant. The H/D KIEs in other transitions (S_3 to S_4 to S_0, S_0 to S_1) are between 1.0 and 1.4. The effects are too small to be used alone to confidently conclude whether there is a proton movement during the transition. Other techniques must be combined to assign the timing of deprotonation in the S-state cycle.

3.2. ^{18}O Kinetic isotope effects

Studies have shown that the water-oxidation reaction carried out by both spinach thylakoids and cyanobacterial PSII shows no ^{18}O isotope fractionation between the substrate water and the evolved oxygen; *i.e.*, the oxygen isotope ratios of the reactant and product are the same; therefore, the $^{16}O/^{18}O$ KIE = 1 and there is no ^{18}O KIE [Guy *et al.*, 1993; Helman *et al.*, 2005]. One explanation for the absence of an ^{18}O KIE is that the bottleneck of the overall water-oxidation process is quinone exchange on the electron-acceptor side, for which no chemistry involving oxygen from water occurs. The overall rate-limiting step is the exchange of reduced plastoquinol (Q_BH_2) at the very last step of turnover of the electron-acceptor side (see introduction for more details), which occurs with a half-time of ~10 ms [Cao, 1990], almost an order of magnitude slower than that of the S_3 to S_0 transition, whose half-time is only ~1.4 ms [Vinyard *et al.*, 2013]. As a result, even though there might be a difference in rate of turnover of the two oxygen isotopes at the OEC, it would be shielded by the rate-limiting step of quinone exchange, and thus no ^{18}O KIE would be observed, consistent with the literature.

Still under debate are two possible O–O bond-formation mechanisms at the S_4 intermediate: water nucleophilic attack and oxo–oxyl radical coupling. The expected ^{18}O KIE values for these mechanisms can be inferred from studies of inorganic model systems coupled with computational calculations [Sarma *et al.*, 2012; Angeles-Boza *et al.*, 2014]. If researchers were to use ^{18}O KIE measurements to directly probe the O–O bond formation chemistry at the OEC, important mechanistic information could be gained by comparing the ^{18}O KIE values for the OEC with published values for inorganic water-oxidation catalysts. Experimental data could also be supplemented with computational calculations to provide a more complete mechanistic insight.

4. Effect of the Perturbation of the Hydrogen-Bonding Network on the Rate-Determining Step

The oxygen-evolving complex is surrounded by a number of hydrogen-bonded amino-acid residues and waters (Figure 10). The amino-acid residues that are directly ligated to the Mn_4O_5Ca cluster are crucial for photoassembly of the cluster and oxygen evolution. Mutations of many of these primary-shell amino-acid residues inhibit oxygen evolution. In addition, secondary shell residues (amino-acid residues that are not directly bonded to the Mn_4O_5Ca cluster but are within a 15 Å radius) play crucial roles in proton transfer, electron transfer and substrate water delivery, and thereby promote efficient water oxidation [Vogt *et al.*, 2015].

The least well understood reactions in the water-oxidation mechanism are the intermediate steps that occur during the spontaneous conversion of the $S_3Y_Z\bullet$ state into the S_0 state with the concomitant release of oxygen. For native PSII, the slowest step in this reaction is electron transfer from the S_3 state to $Y_Z\bullet$ to form the transient S_4 state; the S_4 state decays faster than it forms and, thus, cannot be

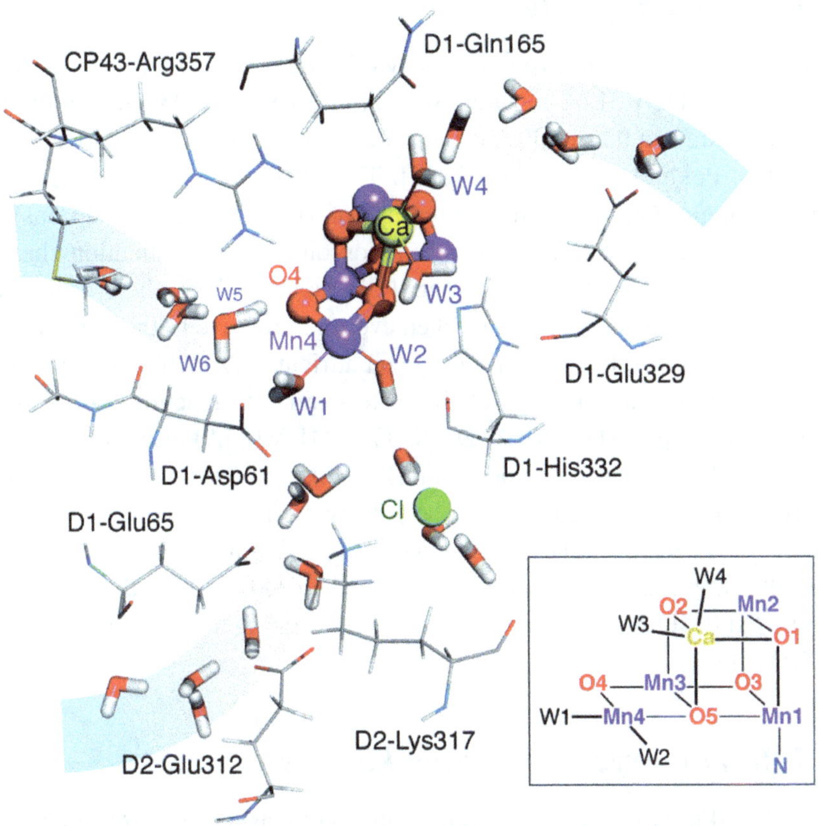

Figure 10. Structure of the OEC depicting the hydrogen-bonding network in cyanobacterial PSII based on the 1.9 Å crystal structure. Reprinted (with permission) from [Retegan and Pantazis, 2016].

observed as a kinetic intermediate. Hence, several attempts have been made to slow down this reaction by perturbing the hydrogen-bonding network by creating single point mutations of the amino-acid residues surrounding the OEC. Some of the residues that affect the oxygen-evolution chemistry are D1-D61, D1-V185, D1-N181, D2-K317, CP43-R357 and D1-N87 [Clausen *et al.*, 2004; Hwang *et al.*, 2007; Dilbeck *et al.*, 2013; Pokhrel *et al.*, 2013, 2015; Banerjee *et al.*, 2018; Kim *et al.*, 2018]. The effects of single point mutations of these residues have been studied in detail using fluorescence kinetics, FTIR spectroscopy, flash-induced polarographic oxygen-evolution measurements and EPR spectroscopy. In this chapter, we focus on the substitutions of amino-acid residues that directly affect the rate-determining step (these include the D1-D61A, D1-N181A, D1-V185N and D2-K317A mutants) and discuss how site-directed mutations of these residues can be used to probe the intermediates formed during the rate-determining conversion of the S_3Y_Z• state into the S_0 state.

4.1. *Effect on steady-state oxygen-evolution and S-state cycling*

The steady-state oxygen evolution of wild-type and D1-D61A, D1-N181A, D1-V185N and D2-K317A mutated *Synechocystis* PCC 6803 cells and the PSII protein isolated from these cells has been measured [Dilbeck *et al.*, 2013; Clausen *et al.*, 2004; Pokhrel *et al.*, 2013, 2015]. The PSII activity (amount of oxygen evolved per mg of Chl per hr in the presence of saturating light conditions) is the primary indicator of the effect of the mutation on the water-oxidation chemistry. All the single-point mutated cells and PSII isolated from these cells exhibit 20–45% reduction in steady-state oxygen evolution relative to the wild-type samples. The reduction of PSII activity has been attributed to inefficient photoassembly of the OEC or an inhibition of the water-oxidation reaction. Some of these mutated PSII samples (D1-D61A, D2-K317A, D1-V185A) also showed reduced efficiency of the S-state transitions, as reflected by a damping of the oscillations of the flash-induced oxygen yields. The perturbation in S-state cycling indicates that these residues are crucial for the effective functioning of PSII [Clausen *et al.*, 2004; Dilbeck *et al.*, 2013; Pokhrel *et al.*, 2013, 2015]. Hence, these secondary-shell mutated PSII became interesting targets for probing the rate-determining step in further detail.

4.2. *Effect on oxygen-evolution kinetics*

Flash-induced polarographic oxygen-evolution measurements have been done to study the oxygen-evolution kinetics of the site-directed mutants. As shown

Table 2. Miss parameters depicting the efficiency of S-state advancement in different single-point mutants and their oxygen-release kinetics. Reprinted (with permission) from [Pokhrel *et al.*, 2015].

	Miss factor (%)	**O_2-release kinetics ($t_{1/2}$)**
D1-D61N	26% vs 17% in WT	15 ms vs 2.7 ms in WT
D1-D61A	NA	12 ms vs 2.7 ms in WT
CP43-R357K	46% vs 10% in WT	3.2 ms vs 0.9 ms in WT
D1-V185N	18% vs 14% in WT	27 ms vs 1.2 ms in WT
D1-V185T	13% vs 14% in WT	1.5 ms vs 1.2 ms in WT
D1-K317A	22% vs 8% in WT	slow
D1-K317R	18% vs 8% in WT	slow
D1-N181A	8% vs 8% in WT	slow
D1-N181S	10% vs 8% in WT	slow

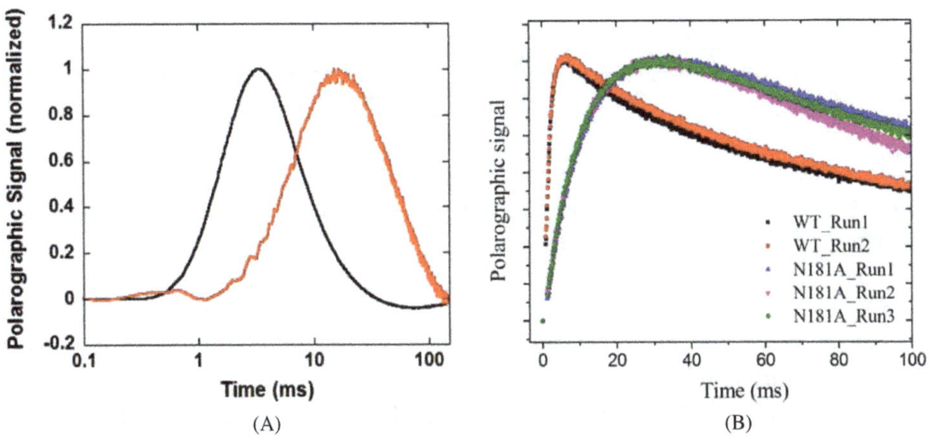

Figure 11. O_2-release kinetics depicting a delay for mutated PSII relative to wild-type PSII. (A) D1-V185N PSII (red trace) compared to wild-type PSII (black trace). (B) D1-N181S PSII compared to wild-type PSII. Reprinted (with permission) from [Dilbeck *et al.*, 2013; Pokhrel *et al.*, 2015].

in Table 2 and Figure 11, some of these mutated PSII exhibit a delay in oxygen evolution. This delay can on one hand be attributed to changes in the O–O bond-formation mechanism. On the other hand, this delay can also be caused by slower advancement of the S-state transitions as is evident from the higher miss factors (Table 2).

FT-IR studies indicated that these site-directed mutations strongly perturbed the hydrogen-bonding network [Debus, 2015]. In order to reveal the mechanism

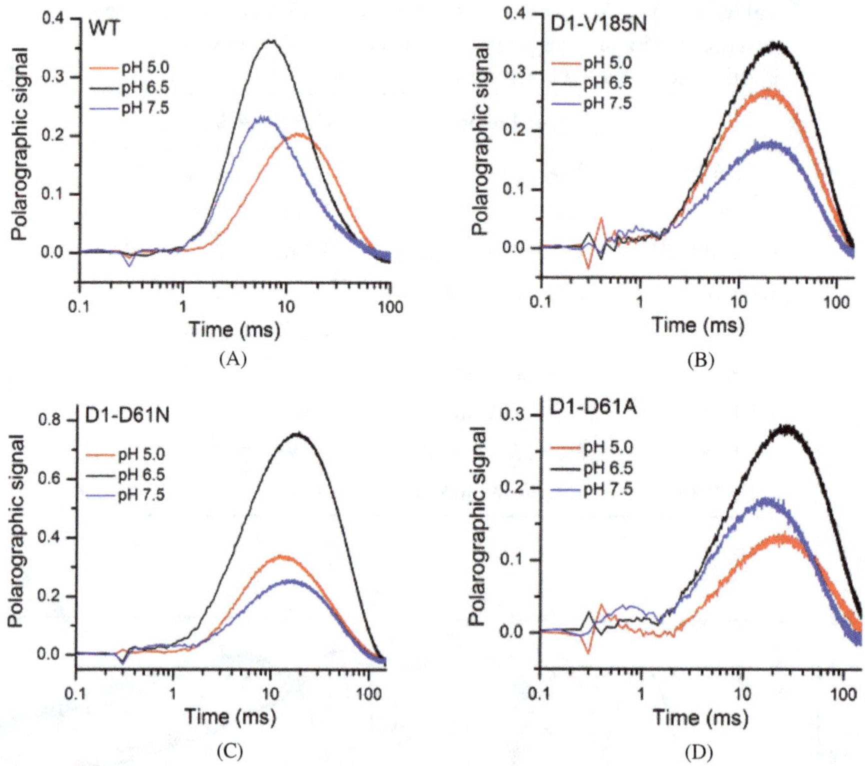

Figure 12. Dependence of oxygen-release kinetics on pH. Reprinted (with permission) from [Bao and Burnap, 2015].

behind the delay, the oxygen-evolution kinetics were studied under different pH and temperature conditions.

For D1-D61A PSII, the dependence of the kinetic phases on pH indicated that the delay was caused by a hindered proton-transfer process (Figure 12). This emphasized the importance of the D1-D61 residue in proton transfer. In the case of D1-V185N PSII, the oxygen-evolution kinetics were independent of pH and revealed an unfavorable entropy. Hence, it was proposed, in case of V185N, that a structural rearrangement may take place that affected the O–O bond formation [Bao and Burnap, 2015]. Similar proposals were made for D1-N181A PSII, for which the S-state cycling remains unaffected [Pokhrel *et al.*, 2015]. The steady-state oxygen-evolution studies in the absence of chloride revealed that the D2-K317 residue is important for chloride binding, which is an important cofactor with a possible role in proton transfer. Hence, the delay observed for D2-K317A PSII was concluded to be mainly caused by a perturbation in the chloride-binding site [Pokhrel *et al.*, 2013].

4.3. *Is the rate-determining step different across species?*

In recent years, the protein channel architecture of cyanobacterial PSII and PSII from higher plants has been compared and contrasted. It was revealed that the only significant difference between higher plant and cyanobacterial species around the OEC was present at the D1-87 position [Vogt *et al.*, 2015]. In most cyanobacteria, it is D1-N87 whereas in plants it is D1-A87. However, this difference does not affect the oxygen-evolution kinetics. *Synechocystis* PCC 6803 D1-N87A PSII was also studied to reveal potential differences between the higher plant and cyanobacterial species. It was found that S-state cycling and steady-state oxygen evolution remain unchanged in *Synechocystis* D1-N87A PSII, further emphasizing that the rate-limiting step is independent of the species [Banerjee *et al.*, 2018].

4.4. *Effect of Ca^{2+} and Cl^{-} substitution*

The crystal structure reveals the presence of inorganic cofactors like Ca^{2+} and Cl^{-} on the donor side of the PSII [Umena *et al.*, 2011]. Hence, to resolve their role in the water-oxidation chemistry, a number of studies have been carried out. Biosynthetic substitution of Ca^{2+}/Sr^{2+} and Cl^{-}/Br^{-} was studied to understand whether these inorganic mutants affect the oxygen-evolution kinetics. These substitutions slow down the rate of oxygen evolution and also result in a lower rate of oxygen evolution [Boussac *et al.*, 2015]. Other biochemical and computational studies have indicated that Cl^{-} is involved in proton transfer [Pokhrel *et al.*, 2011]. Hence, we can attribute the delayed oxygen-evolution kinetics of Cl^{-} substitution to a hindered proton-transfer process. The waters around Ca^{2+} have often been proposed as the substrate waters and its substitution could affect substrate water binding. However, it was concluded from a study of Ca^{2+}/Sr^{2+}-substituted PSII that the observation of delayed oxygen-evolution kinetics was due to a decrease in the free energy of the S_3 state [Boussac *et al.*, 2015].

5. Conclusions

Though substrate exchange kinetics and small molecule analogues have helped in identifying candidates for substrate waters, still there exist contrasting proposals. Hence, the water-oxidation mechanism remains unclear. The H/D isotope effects, S-state kinetics and ^{18}O-isotope effects have played critical roles in characterizing the nature of the S-state transitions and emphasizing the choreographed proton- and electron-transfer events necessary for efficient water oxidation. However, our capability to probe deeper into the S_3Y_Z• to S_0 transition still remains a challenge. Several site-directed mutants have been constructed to delay this transition. Some of these mutants do affect the water-oxidation mechanism; however, these mutants need to be studied in greater detail to uncover the rate-determining step and clarify the mechanism of O–O bond formation.

References

Angeles-Boza, A.M., Ertem, M.Z., Sarma, R., Ibanez, C.H., Maji, S., Llobet, A., Cramer, C.J. and Roth, J.P. (2014). Competitive oxygen-18 kinetic isotope effects expose O–O bond formation in water oxidation catalysis by monomeric and dimeric ruthenium complexes, *Chem. Sci.*, 5, 1141–1152.

Askerka, M., Vinyard, D.J., Brudvig, G.W. and Batista, V.S. (2015). NH_3 binding to the S_2 state of the O_2-evolving complex of photosystem II: Analogue to H_2O binding during the $S_2 \rightarrow S_3$ transition, *Biochemistry*, 54, 5783–5786.

Banerjee, G., Ghosh, I., Kim, C.J., Debus, R.J. and Brudvig, G.W. (2018). Substitution of the D1-Asn87 site in photosystem II of cyanobacteria mimics the chloride-binding characteristics of spinach photosystem II, *J. Biol. Chem.*, 293, 2487–2497.

Bao, H. and Burnap, R.L. (2015). Structural rearrangements preceding dioxygen formation by the water oxidation complex of photosystem II, *Proc. Natl. Acad. Sci. USA.*, 112, E6139–E6147.

Beck, W.F., de Paula, J.C. and Brudvig, G.W. (1986). Ammonia binds to the manganese site of the oxygen-evolving complex of photosystem II in the S_2 state, *J. Am. Chem. Soc.*, 108, 4018–4022.

Boussac, A., Rappaport, F., Carrier, P., Verbavatz, J.M., Gobin, R., Kirilovsky, D., Rutherford, A.W. and Sugiura, M. (2004). Biosynthetic Ca^{2+}/Sr^{2+} exchange in the photosystem II oxygen-evolving enzyme of *Thermosynechococcus elongatus*, *J. Biol. Chem.*, 279, 22809–22819.

Boussac, A., Rutherford, A.W. and Sugiura, M. (2015). Electron transfer pathways from the S_2-states to the S_3-states either after a Ca^{2+}/Sr^{2+} or a Cl^-/I^- exchange in photosystem II from *Thermosynechococcus elongatus*, *Biochim. Biophys. Acta-Bioenergetics*, 1847, 576–586.

Boussac, A., Ugur, I., Marion, A., Sugiura, M., Kaila, V.R. and Rutherford, A.W. (2018). The low spin-high spin equilibrium in the S_2-state of the water oxidizing enzyme, *Biochim. Biophys. Acta-Bioenergetics*, 1859, 342–356.

Britt, R.D., Campbell, K.A., Peloquin, J.M., Gilchrist, M.L., Aznar, C.P., Dicus, M.M., Robblee, J. and Messinger, J. (2004). Recent pulsed EPR studies of the photosystem II oxygen-evolving complex: Implications as to water oxidation mechanisms, *Biochim. Biophys. Acta*, 1655, 158–171.

Brudvig, G.W. (2008). Water oxidation chemistry of photosystem II, *Phil. Trans. R. Soc. Lond. B*, 363, 1211–1219.

Cao, J. (1990). Chlorophyll a fluorescence transient as an indicator of active and inactive photosystem II in thylakoid membranes, *Biochim. Biophys. Acta-Bioenergetics*, 1015, 180–188.

Clausen, J., Debus, R.J. and Junge, W. (2004). Time-resolved oxygen production by PSII: Chasing chemical intermediates, *Biochim. Biophys. Acta-Bioenergetics*, 1655, 184–194.

Cox, N. and Messinger, J. (2013). Reflections on substrate water and dioxygen formation, *Biochim. Biophys. Acta*, 1827, 1020–1030.

Cox, N., Retegan, M., Neese, F., Pantazis, D.A., Boussac, A. and Lubitz, W. (2014). Electronic structure of the oxygen-evolving complex in photosystem II prior to O–O bond formation, *Science*, 345, 804–808.

Dau, H. and Haumann, M. (2008). The manganese complex of photosystem II in its reaction cycle — Basic framework and possible realization at the atomic level, *Coord. Chem. Rev.*, 252, 273–295.

Debus, R.J. (2015). FTIR studies of metal ligands, networks of hydrogen bonds, and water molecules near the active site Mn_4CaO_5 cluster in photosystem II, *Biochim. Biophys. Acta-Bioenergetics*, 1847, 19–34.

Dilbeck, P.L., Bao, H., Neveu, C.L. and Burnap, R.L. (2013). Perturbing the water cavity surrounding the manganese cluster by mutating the residue D1-valine 185 has a strong effect on the water oxidation mechanism of photosystem II, *Biochemistry*, 52, 6824–6833.

Ferreira, K.N., Iverson, T.M., Maghlaoui, K., Barber, J. and Iwata, S. (2004). Architecture of the photosynthetic oxygen-evolving center, *Science*, 303, 1831–1838.

Force, D.A., Randall, D.W., Lorigan, G.A., Clemens, K.L. and Britt, R.D. (1998). ESEEM studies of alcohol binding to the manganese cluster of the oxygen evolving complex of Photosystem II, *J. Am. Chem. Soc.*, 120, 13321–13333.

Guy, R.D., Fogel, M.L. and Berry, J.A. (1993). Photosynthetic fractionation of the stable isotopes of oxygen and carbon, *Plant Physiol.*, 101, 37–47.

Haumann, M., Bögershausen, O., Cherepanov, D., Ahlbrink, R. and Junge, W. (1997). Photosynthetic oxygen evolution: H/D isotope effects and the coupling between electron and proton transfer during the redox reactions at the oxidizing side of photosystem II, *Photosynth. Res.*, 51, 193–208.

Haumann, M., Liebisch, P., Müller, C., Barra, M., Grabolle, M. and Dau, H. (2005). Photosynthetic O_2 formation tracked by time-resolved X-ray experiments, *Science*, 310, 1019–1021.

Helman, Y., Barkan, E., Eisenstadt, D., Luz, B. and Kaplan, A. (2005). Fractionation of the three stable oxygen isotopes by oxygen-producing and oxygen-consuming reactions in photosynthetic organisms, *Plant Physiol.*, 138, 2292–2298.

Hillier, W. and Wydrzynski, T. (2000). The affinities for the two substrate water binding sites in the O_2 evolving complex of photosystem II vary independently during S-state turnover, *Biochemistry*, 39, 4399–4405.

Hillier, W. and Wydrzynski, T. (2004). Substrate water interactions within the photosystem II oxygen evolving complex, *Phys. Chem. Chem. Phys.*, 6, 4882–4889.

Hwang, H.J., Dilbeck, P., Debus, R.J. and Burnap, R.L. (2007). Mutation of arginine 357 of the CP43 protein of photosystem II severely impairs the catalytic S-state cycle of the H_2O oxidation complex, *Biochemistry*, 46, 11987–11997.

Kim, C.J., Bao, H., Burnap, R.L. and Debus, R.J. (2018). Impact of D1-V185 on the water molecules that facilitate O_2 formation by the catalytic Mn_4CaO_5 cluster in photosystem II, *Biochemistry*, 57, 4299–4311.

Kok, B., Forbush, B. and McGloin, M. (1970). Cooperation of charges in photosynthetic O_2 evolution-I. A linear four step mechanism, *Photochem. Photobiol.*, 11, 457–475.

Loll, B., Kern, J., Saenger, W., Zouni, A. and Biesiadka, J. (2005). Towards complete cofactor arrangement in the 3.0 angstrom resolution structure of photosystem II, *Nature*, 438, 1040–1044.

Luber, S., Rivalta, I., Umena, Y., Kawakami, K., Shen, J.R., Kamiya, N., Brudvig, G.W. and Batista, V.S. (2011). S_1-state model of the O_2-evolving complex of photosystem II, *Biochemistry*, 50, 6308–6311.

Mandal, M., Askerka, M., Banerjee, G., Amin, M., Brudvig, G.W., Batista, V.S. and Gunner, M. (2017). Characterization of ammonia binding to the second coordination shell of the oxygen-evolving complex of photosystem II, *Dalton Trans.*, 46, 16089–16095.

McEvoy, J.P. and Brudvig, G.W. (2006). Water-splitting chemistry of photosystem II, *Chem. Rev.*, 106, 4455–4483.

Nöring, B., Shevela, D., Renger, G. and Messinger, J. (2008). Effects of methanol on the S_i-state transitions in photosynthetic water-splitting, *Photosynth. Res.*, 98, 251–260.

Oyala, P.H., Stich, T.A., Debus, R.J. and Britt, R.D. (2015). Ammonia binds to the dangler manganese of the photosystem II oxygen-evolving complex, *J. Am. Chem. Soc.*, 137, 8829–8837.

Oyala, P.H., Stich, T.A., Stull, J.A., Yu, F., Pecoraro, V.L. and Britt, R.D. (2014). Pulse electron paramagnetic resonance studies of the interaction of methanol with the S_2 state of the Mn_4O_5Ca cluster of photosystem II, *Biochemistry*, 53, 7914–7928.

Pal, R., Negre, C.F.A., Vogt, L., Pokhrel, R., Ertem, M.Z., Brudvig, G.W. and Batista, V.S. (2013). S_0-state model of the oxygen-evolving complex of photosystem II, *Biochemistry*, 52, 7703–7706.

Pantazis, D.A. (2018). Missing pieces in the puzzle of biological water oxidation, *ACS Catal.*, 8, 9477–9507.

Pantazis, D.A., Ames, W., Cox, N., Lubitz, W. and Neese, F. (2012). Two interconvertible structures that explain the spectroscopic properties of the oxygen-evolving complex of photosystem II in the S_2 state, *Angew. Chem. Int. Ed.*, 51, 9935–9940.

Pokhrel, R. and Brudvig, G.W. (2014). Oxygen-evolving complex of photosystem II: Correlating structure with spectroscopy, *Phys. Chem. Chem. Phys.*, 16, 11812–11821.

Pokhrel, R., Debus, R.J. and Brudvig, G.W. (2015). Probing the effect of mutations of asparagine 181 in the D1 subunit of photosystem II, *Biochemistry*, 54, 1663–1672.

Pokhrel, R., McConnell, I.L. and Brudvig, G.W. (2011). Chloride regulation of enzyme turnover: Application to the role of chloride in photosystem II, *Biochemistry*, 50, 2725–2734.

Pokhrel, R., Service, R.J., Debus, R.J. and Brudvig, G.W. (2013). Mutation of lysine 317 in the D2 subunit of photosystem II alters chloride binding and proton transport, *Biochemistry*, 52, 4758–4773.

Rapatskiy, L., Cox, N., Savitsky, A., Ames, W.M., Sander, J., Nowaczyk, M.M., Rögner, M., Boussac, A., Neese, F., Messinger, J. and Lubitz, W. (2012). Detection of the water-binding sites of the oxygen-evolving complex of photosystem II using W-band ^{17}O electron–electron double resonance-detected NMR spectroscopy, *J. Am. Chem. Soc.*, 134, 16619–16634.

Retegan, M. and Pantazis, D.A. (2016). Interaction of methanol with the oxygen-evolving complex: atomistic models, channel identification, species dependence, and mechanistic implications, *Chem. Sci.*, 7, 6463–6476.

Retegan, M. and Pantazis, D.A. (2017). Differences in the active site of water oxidation among photosynthetic organisms, *J. Am. Chem. Soc.*, 139, 14340–14343.

Sarma, R., Angeles-Boza, A.M., Brinkley, D.W. and Roth, J.P. (2012). Studies of the di-iron(VI) intermediate in ferrate-dependent oxygen evolution from water, *J. Am. Chem. Soc.*, 134, 15371–15386.

Sauter, N.K., Echols, N., Adams, P.D., Zwart, P.H., Kern, J., Brewster, A.S., Koroidov, S., Alonso-Mori, R., Zouni, A. Messinger, J., Bergmann, U., Yano, J. and Yachandra, V.K. (2016). No observable conformational changes in PSII, *Nature*, 533, E1–E2.

Schuth, N., Liang, Z., Schonborn, M., Kussicke, A., Assuncao, R., Zaharieva, I., Zilliges, Y. and Dau, H. (2017). Inhibitory and non-inhibitory NH_3 binding at the water-oxidizing manganese complex of photosystem II suggests possible sites and a rearrangement mode of substrate water molecules, *Biochemistry*, 56, 6240–6256.

Siegbahn, P.E. (2013). Water oxidation mechanism in photosystem II, including oxidations, proton release pathways, O–O bond formation and O_2 release, *Biochim. Biophys. Acta-Bioenergetics*, 1827, 1003–1019.

Su, J.-H., Cox, N., Ames, W., Pantazis, D.A., Rapatskiy, L., Lohmiller, T., Kulik, L.V., Dorlet, P., Rutherford, A.W., Neese, F., Boussac, A., Lubitz, W. and Messinger, J. (2011). The electronic structures of the S_2 states of the oxygen-evolving complexes of photosystem II in plants and cyanobacteria in the presence and absence of methanol, *Biochim. Biophys. Acta-Bioenergetics*, 1807, 829–840.

Suga, M., Akita, F., Hirata, K., Ueno, G., Murakami, H., Nakajima, Y., Shimizu, T., Yamashita, K., Yamamoto, M. Ago, H. and Shen, J.-R. (2015). Native structure of photosystem II at 1.95 Å resolution viewed by femtosecond X-ray pulses, *Nature*, 517, 99–103.

Suga, M., Akita, F., Sugahara, M., Kubo, M., Nakajima, Y., Nakane, T., Yamashita, K., Umena, Y., Nakabayashi, M., Yamane, T., Nakano, T., Suzuki, M., Masuda, T., Inoue, S., Kimura, T., Nomura, T., Yonekura, S., Yu, L.-J., Sakamoto, T., Motomura, T., Chen, J.-H., Kato, Y., Noguchi, T., Tono, K., Joti, Y., Kameshima, T., Hatsui, T., Nango, E., Tanaka, R., Naitow, H., Matsuura, Y., Yamashita, A., Yamamoto, M., Nureki, O., Yabashi, M., Ishikawa, T., Iwata, S. and Shen, J.-R. (2017). Light-induced structural changes and the site of O=O bond formation in PSII caught by XFEL, *Nature*, 543, 131–135.

Umena, Y., Kawakami, K., Shen, J.-R. and Kamiya, N. (2011). Crystal structure of oxygen-evolving photosystem II at a resolution of 1.9 angstrom, *Nature*, 473, 55–60.

Velthuys, B. (1975). Binding of the inhibitor NH_3 to the oxygen-evolving apparatus of spinach chloroplasts, *Biochim. Biophys. Acta-Bioenergetics*, 396, 392–401.

Vinyard, D.J., Ananyev, G.M. and Dismukes, G.C. (2013). Photosystem II: The reaction center of oxygenic photosynthesis, *Annu. Rev. Biochem.*, 82, 577–606.

Vinyard, D.J. and Brudvig, G.W. (2015). Insights into substrate binding to the oxygen-evolving complex of photosystem II from ammonia inhibition studies, *Biochemistry*, 54, 622–628.

Vinyard, D.J., Khan, S. and Brudvig, G.W. (2015). Photosynthetic water oxidation: binding and activation of substrate waters for O-O bond formation, *Faraday Discuss.*, 185, 37–50.

Vogt, L., Vinyard, D.J., Khan, S. and Brudvig, G.W. (2015). Oxygen-evolving complex of photosystem II: An analysis of second-shell residues and hydrogen-bonding networks, *Curr. Opin. Chem. Biol.*, 25, 152–158.

Young, I.D., Ibrahim, M., Chatterjee, R., Gul, S., Fuller, F.D., Koroidov, S., Brewster, A.S., Tran, R., Alonso-Mori, R., Kroll, T., Michels-Clark, T., Laksmono, H., Sierra, R.G., Stan, C.A., Hussein, R., Zhang, M., Douthit, L., Kubin, M., de Lichtenberg, C., Pham, L.V., Nilsson, H., Cheah, M.H., Shevela, D., Saracini, C., Bean, M.A., Seuffert, I., Sokaras, D., Weng, T.-C., Pastor, E., Weninger, C., Fransson, T., Lassalle, L., Brauer, P., Aller, P., Docker, P.T., Andi, B., Orville, A.M., Glownia, J.M., Nelson, S., Sikorski, M., Zhu, D.L., Hunter, M.S., Lane, T.J., Aquila, A., Koglin, J.E., Robinson, J., Liang, M.N., Boutet, S., Lyubimov, A.Y., Uervirojnangkoorn, M., Moriarty, N.W., Liebschner, D., Afonine, P.V., Waterman, D.G., Evans, G., Wernet, P., Dobbek, H., Weis, W.I., Brunger, A.T., Zwart, P.H., Adams, P.D., Zouni, A., Messinger, J., Bergmann, U., Sauter, N.K., Kern, J., Yachandra, V.K. and Yano, J. (2016). Structure of photosystem II and substrate binding at room temperature, *Nature*, 540, 453–457.

Zahariou, G., Chrysina, M., Petrouleas, V. and Ioannidis, N. (2014). Can we trap the S_3Y_Z metalloradical intermediate during the S-state transitions of photosystem II? An EPR investigation, *FEBS Lett.*, 588, 1827–1831.

Chapter 15

Structure and Mechanism of Ni- and Fe-containing Carbon Monoxide Dehydrogenases

Holger Dobbek

Institut für Biologie, Humboldt-Universität zu Berlin
Unter den Linden 6, 10099 Berlin, Germany
holger.dobbek@hu-berlin.de

Carbon monoxide dehydrogenases catalyze the reversible oxidation of CO with water to CO_2, two protons and two electrons. Microorganisms may use one of two different types of carbon monoxide dehydrogenases (CODHs): the aerobic CODHs containing Cu and Mo or the anaerobic Ni and Fe-containing CODHs. The latter allow bacteria to use CO as a sole source of electrons or to transform CO_2 to acetyl-CoA by reducing it to CO. The active site of Ni,Fe-containing CODHs contains a unique $[NiFe_4S_4]$-cluster, the only known cofactor where Ni is integrated into a heterocubane structure. Ni binds the substrates CO and CO_2, while Fe in *exo* of the cluster activates water for nucleophilic attack on CO. The arrangement of Ni and Fe shows striking similarities to the $Mn_4Ca^{2+}O_5$-cluster of photosystem II. The review summarizes our present understanding of the structure of CODHs and its constituting metal centers.

1. Introduction

Carbon monoxide (CO) is a colorless, odorless and flammable gas, toxic for many respiring organisms due to its tendency to bind to the dioxygen-binding site of essential hemoproteins, such as hemoglobin, myoglobin and cytochrome *c* oxidase.

CO is an abundant molecule, produced by incomplete combustion and volcanic emissions, but its concentration is kept at trace gas levels as it is used by several microorganisms as a carbon and energy source [Khalil *et al.*, 1999; King, 1999; Lu and Khalil, 1993; Ragsdale, 2004]. The former process incorporates the carbon atom of CO into the organic material, while in the latter process, catalyzed by carbon monoxide dehydrogenases (CODHs), it is oxidized to CO_2 [Ragsdale, 2004]. The ability of CODHs to catalyze CO oxidation/CO_2 reduction attracted attention as CODHs catalyze the reduction with no overpotential: the midpoint potential of the catalytically relevant oxidation states of the active site cluster matches that of the redox couple CO_2/CO [Appel *et al.,* 2013].

Depending on physiological function and source organism, CODHs act either alone or in complex with other enzymes [Appel *et al.*, 2013; Ragsdale, 2007]. Monofunctional CODHs, consisting only of a CODH subunit catalyze CO oxidation, which is coupled to reductive reactions in other enzymes, ultimately generating a proton-motive force. The monofunctional CODHs of *Rhodospirillum rubrum* and *Carboxydothermus hydrogenoformans* for example support hydrogenogenic growth, where the organisms exploit the small redox midpoint potential difference between the oxidation of CO to CO_2 and the reduction of protons to dihydrogen to generate a proton gradient across the membrane, fuelling ATP synthesis [Kerby *et al.*, 1995; Svetlichny *et al.*, 1991; Svetlitchnyi *et al.*, 2001]. Bifunctional CODHs consist of a complex between a CODH and another metalloenzyme called acetyl-CoA synthase (ACS; condensing CO, CH_3^+ and CoA to acetyl-CoA) [Appel *et al.,* 2013]. The task of the CODH subunit is here to reduce CO_2, typically using electrons obtained from dihydrogen oxidation. Multifunctional CODHs catalyze the decarbonylation of acetyl-CoA or the synthesis of acetyl-CoA from two reducing equivalents, CO_2, CoA and a methylated pterin. The enzyme complex comprises five different subunits (α, β, γ, δ and ε), in which CODH is complexed with ACS, a corrinoid-iron/sulfur protein and methyltransferase [Abbanat and Ferry, 1991]. Multifunctional CODHs mostly occur in methanogenic archaea [Adam *et al.*, 2018; Grahame, 2003].

In this short review on CODHs, I will focus on the structure of the active site metal center and its role in supporting the rapid oxidation of CO.

2. Structure of CODHs

Crystal structures of several CODHs have been determined. We have highly resolved structures for the monofunctional CODHs from *R. rubrum* ($CODH_{Rr}$) [Drennan *et al.*, 2001] and *C. hydrogenoformans* (CODH-II_{Ch} and CODH-IV_{Ch}) [Dobbek *et al.*, 2001; Domnik *et al.*, 2017], the bifunctional CODH/ACS complex of *Moorella thermoacetica* (CODH/ACS_{Mt}) [Darnault *et al.*, 2003; Doukov *et al.*,

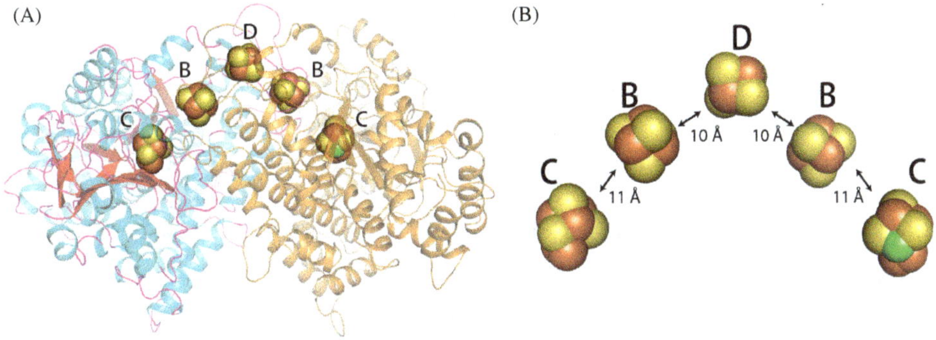

Figure 1. Overall structure of CODHs. (A) Shown is the bacterial type Ni,Fe-CODH from *C. hydrogenoformans* (CODH-II_{Ch}) in cartoon presentation. One subunit of the homodimer is color-coded according to secondary structure, with α-helices in light blue and β-sheets in red, the second subunit is colored in orange. The five Fe/S-clusters are shown as van-der-Waals spheres and labeled corresponding to the associated names. (B) The five Fe/S-clusters form a V-shaped electron transfer chain with short distances between the metal clusters. At the tips of the inverted V, we find the active site Ni,Fe-cluster (cluster C).

2002], and the CODH component of the multifunctional ACDS complex from *Methanosarcina barkeri* ($CODH_{Mb}$) [Gong *et al.*, 2008].

The CODH component is a mushroom-shaped homodimer with five metal clusters, three of which are cubane-type [4Fe4S]-clusters with standard coordination by four cysteine residues (two clusters B, one cluster D) (Figure 1). The CODH component of the ACDS complex has two additional [4Fe4S]-clusters (clusters E and F). Here I will use the monofunctional CODH-II_{Ch} as an example, as its structure was studied in different catalytic states and its active site metal cluster has been determined with highest resolution.

One of the central roles of the protein matrix is to guide substrates and products of CO oxidation, including not only CO and CO_2, but also water, protons and electrons, each appearing to have a defined way to reach or egress from the active site. Most visible is the chain of [4Fe4S]-clusters serving as a conduit for electrons (Figure 1B). All clusters are in a short distance, with shortest Fe–Fe distances of 10–12 Å facilitating rapid electron transfer [Page *et al.*, 1999].

Despite their small size and rapid diffusion, even CO and CO_2 appear to be guided in the protein through gas channels. The longest channel is running from one end to the other through the complete dimer, branching off at the active site cluster C [Dobbek *et al.,* 2001; Jeoung and Dobbek, 2012]. Probably not the entire length of the channel is used for gas transport as the middle part of the channel is hydrophilic and filled with water molecules. This channel of monofunctional CODHs partly coincides with the gas channel that connects the active sites of CODH (cluster C) and acetyl-CoA synthase (cluster A) in bifunctional

CODH/ACS complexes [Darnault *et al.*, 2003; Doukov *et al.*, 2002]. Furthermore, a dynamically formed gas channel, discovered by molecular dynamics and density functional theory calculations may allow CO_2 to diffuse from the solvent to cluster C in the bifunctional CODH/ACS_{Mt} complex [Wang *et al.*, 2013]. Finally, a set of conserved histidines forms a pathway for protons from the active site to the surface of the protein matrix. As a chain of water molecules runs near the proposed proton pathway, it may be linked to the transport of water molecules [Kim *et al.*, 2004].

All these pathways are enclosed in the three domains structuring monofunctional CODHs. The N-terminal domain (residues 3–237) is predominantly α-helical and accommodates the binding regions for cluster B and D. The middle (238–406) and C-terminal (407–636) domain have a Rossman-fold topology between which the active site cluster C is confined.

3. Active Site Cluster C

The active site harbors cluster C, where "C" stands for CO oxidation [Anderson *et al.*, 1993]. Cluster C is a [$NiFe_4S_4$]-cluster featuring a distorted $NiFe_3S_4$ heterocubane with μ_3-sulfido ligands connecting a Fe(II) ion in *exo* with the heterocubane cluster [Dobbek *et al.*, 2001; Drennan *et al.*, 2001]. This Fe(II) ion (Fe1) is also known as *ferrous component II* (FCII), early identified to hold a water or hydroxyl-ligand [DeRose *et al.*, 1998]. Each of the five metals (4 × Fe, 1 × Ni) is coordinated by a cysteine residue and the Fe(II) ion in *exo* is additionally coordinated by a histidine residue (Figure 2A).

This arrangement is very similar to the $Mn_4Ca^{2+}O_5$-cluster of photosystem II (PSII) (Figures 2B and 2C) [Barber, 2017]. The $Mn_4Ca^{2+}O_5$-cluster of PSII is the only other heterocubane cluster found as cofactor of natural enzymes, formed from $CaMn_3$, bridged by four oxygen atoms, with a fourth Mn ion placed in *exo*, called the "dangler" Mn [Ferreira *et al.*, 2004]. Based on these similarities and parallels in the catalyzed reaction, the possibility that both catalyze a surface type reaction between a nucleophilic and an electrophilic reactant seems plausible, an intriguing link between cluster organization and catalytic mechanism [Barber, 2017].

Cluster C may adopt at least three different oxidation states, termed C_{ox}, C_{red1} and C_{red2} [Appel *et al.*, 2013]. The most oxidized state, C_{ox}, is formed above –200 mV, shows no paramagnetic signal and is catalytically inactive [Feng and Lindahl, 2004; Lindahl, 2002]. One electron reduction forms C_{red1}, which is paramagnetic (S = 1/2 spins) and has the proposed oxidation state {[Ni^{2+} Fe^{2+}]:[Fe_3 S_4]$^-$} [Spangler *et al.*, 1996]. At a midpoint potential of E°' = –530 mV, coinciding with the midpoint potential of the CO_2/CO couple (E°' = –558 mV), C_{red2} forms [Lindahl *et al.*, 1990]. C_{red2} shows only minor changes compared to C_{red1} in the

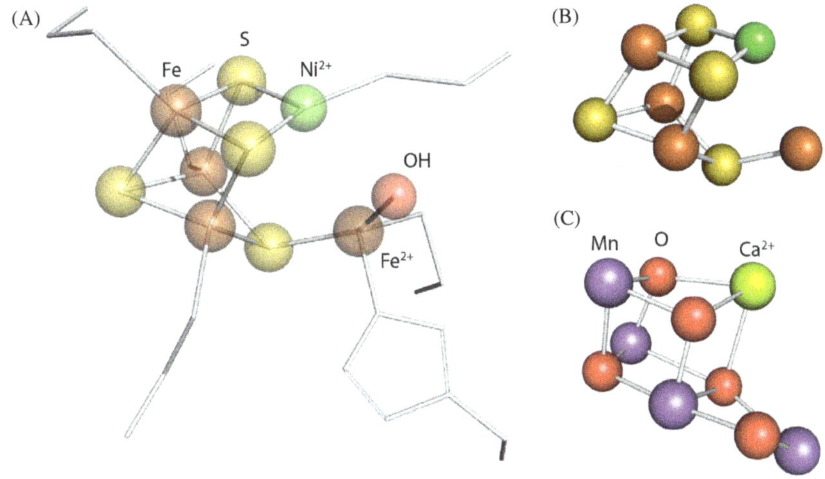

Figure 2. Structure of cluster C. (A) Ball-and-stick representation of the Ni,Fe,S-cluster with green Ni, yellow S, orange Fe and red O atoms. Coordinating amino acids are shown as white sticks. (B, C) Cluster C (B) and the $Mn_4Ca^{2+}O_5$-cluster (C) in comparable orientations with the heterocubane-cluster on the left side and the additional ion in *exo* on the right side. Same color-code as in A and purple Mn and light green Ca ions.

paramagnetic resonance spectrum and its Mössbauer parameters (reflecting the state and structure of the Fe core), making it unlikely that the Fe/S moiety of cluster C took up the two electrons, needed to reach C_{red2} from C_{red1} [Lindahl *et al.*, 1990a, 1990b]. Several alternatives for the place of the two additional electrons in the C_{red2} have been considered, including a zero-valent nickel ion (Ni^0), a Ni–Fe bond (dative metal–metal bond) and a hydride-bound Ni^{2+} [Amara *et al.*, 2011; Lindahl, 2012]. One-electron oxidation of C_{red2} results in C_{int}, which is devoid of a paramagnetic signal [Anderson and Lindahl, 1996].

4. Mechanism of CO Oxidation at Cluster C

The following mechanism appears to be consistent with the spectroscopic and structural investigations of CODHs. CO oxidation starts from the oxidation state C_{red1}, in which CO binds to cluster C (Figures 3 and 4). CO is rapidly oxidized to CO_2 whereby some CODHs, such as CODH-IV_{Ch} appear to be "perfect enzymes" (k_{cat}/K_M of 3×10^8 to 10^9 M^{-1} s^{-1} over the complete temperature range), being only limited by the diffusion of CO [Bar-Even *et al.*, 2011; Domnik *et al.*, 2017]. Unfortunately, the high turnover number makes it difficult to trap a CO-bound state. Although CO bound complexes have been posited, evidence is ambiguous and hampered in the reported cases because a comparison before and after treatment with CO is lacking [Drennan *et al.*, 2001; Gong *et al.*, 2008].

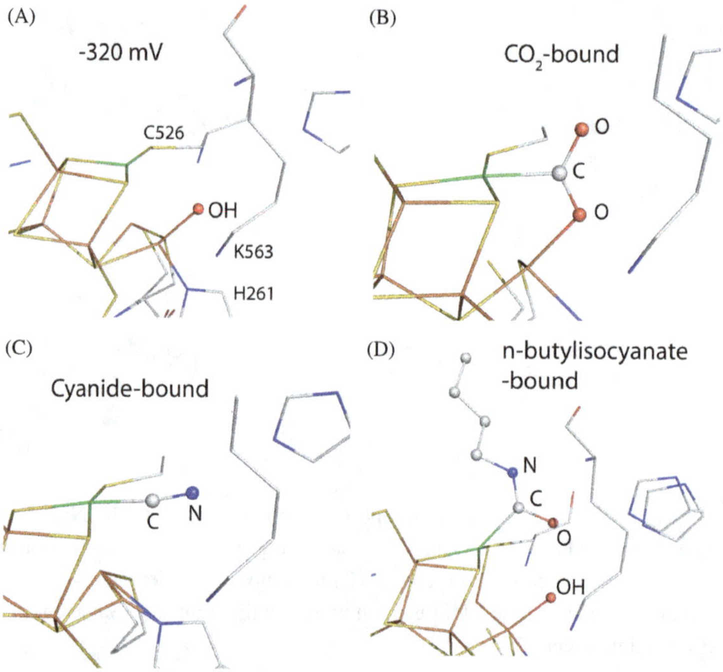

Figure 3. Ligand-bound states of cluster C. (A) The oxidized cluster is ready to react with CO and has a hydroxyl-ligand bound to the Fe^{2+}-ion in *exo*. (B) CO_2 is bound between Ni (in green) and the *exo*-Fe. CO_2 is activated and reduced, resulting in the smaller O–C–O angle. (C) Cyanide is binding to the open square-planar coordination site of Ni with a slight tilt. (D) *N*-butylisocyanide is oxidized by CODH to *n*-butylisocyanate, which was captured in the active site.

A more stable state, amenable to spectroscopic and structural investigation can be reached with cyanide. Cyanide is isoelectronic to CO [Crabtree, 2005] and is a slow-binding, competitive inhibitor of CO inhibition [Ensign *et al.*, 1989; Ha *et al.*, 2007]. Cyanide binds to CODHs with high affinity comparable to the dissociation constants estimated for CO (K_i for cyanide of 21,7 μM, K_M for CO of 18 μM for CODH-II_{Ch}) [Ha *et al.*, 2007; Svetlitchnyi *et al.*, 2001] — making it a model for the binding of CO. IR spectroscopy can follow cyanide binding by a specific band at 2110 cm^{-1} [Ciaccafava *et al.*, 2016], while EPR spectroscopy shows that the signal of the C_{red1} state changes [DeRose *et al.*, 1998].

The structure of CODH-II_{Ch} in the presence of cyanide shows cyanide binding to an open coordination site of Ni with a tilted Ni–C–N axis [Ciaccafava *et al.*, 2016; Jeoung and Dobbek, 2009; Kung *et al.*, 2009] (Figure 3C). Hydrogen bonds with Lys563 and His93 stabilize the cyanide-ligand. Fe1 looses its OH-ligand and shifts from the "A" position to an alternative position "B" closer to the Ni-ion. A combined crystallographic, IR-spectroscopic and computational study shows that an exchange of a nearby lysine residue (Lys563) against alanine hardly influences the bound cyanide and that Lys563 is deprotonated, probably from protonating the OH-ligand at Fe1 upon its dissociation as water [Ciaccafava *et al.*, 2016].

Cluster C has at least one open coordination site at Ni and the hydroxy-ligand bound to Fe1 appears perfectly aligned for a nucleophilic attack on CO bound to the open coordination site of Ni (Figure 3A, 4 top). When CO binds to a metal, it changes its properties and reverses its polarity [Crabtree, 2005]. CO binds preferentially with its C atom to metals where its HOMO is centered, donating σ-electrons to the metal. Stable binding to the metal is accomplished by back-bonding in which the metal accepts $d\pi$-electrons from CO into its π^*-orbital. This additional shift of charge polarizes and activates CO, facilitating a nucleophilic attack at the electrophilic carbon atom. The nucleophile in CODHs is the hydroxyl-ligand at Fe1. Binding to a metal with low π-basicity, such as Ni^{2+} leaves the carbon atom strongly polarized, increasing its electrophilic tendency. A Ni^{2+}-ion is an atypical partner to form a metal carbonyl [Macgregor *et al.*, 1994] and it is likely the presence of the sulfido-ligands acting as π-donor ligands that allows transient metal-CO complexes to form.

The steps following the nucleophilic attack of CO by the hydroxyl-ligand have become clearer from investigating CO_2-reduction starting from the C_{red2} state. CO_2 reduction is slower than CO oxidation with kinetics strongly depending on the source of electrons [Ensign, 1995]. With monofunctional $CODH_{Rr}$ CO_2 reduction is catalyzed with a k_{cat} of 45 s^{-1} and a K_M for CO_2 of 0.18 mM when methylviologen

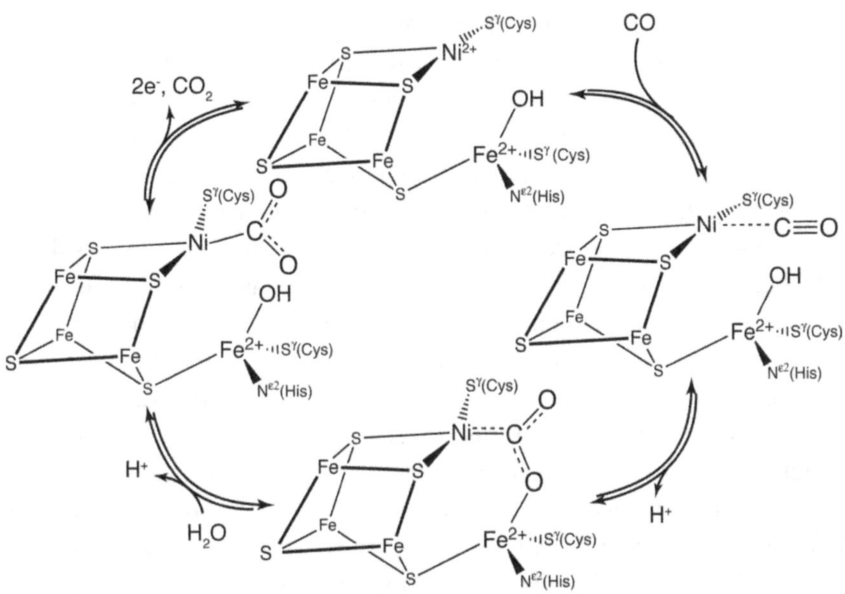

Figure 4. Structure-based mechanism of CO oxidation at the Ni,Fe-cluster of CODHs. See text for details.

is present as an electron mediator [Ensign, 1995]. CO_2 needs to be activated to be reduced and the reaction occurs in a first step by an attack of a nucleophile. The LUMO of CO_2 is localized at the carbon atom, rendering it susceptible to nucleophilic attack at this position [Appel *et al.*, 2013; Crabtree, 2005]. Activation is facilitated when an electrophilic center of the catalyst interacts with the HOMOs of CO_2 centered at the oxygens. Both result in a decrease of the C–O bond order and a decrease of the O–C–O bond angle. The carbon-centered LUMO is stabilized by the bending, facilitating the two-electron reduction, yielding either CO or formate. Which of the two products is generated, is most likely determined by the initial catalyst-CO_2 complex formed, with a η^1-CO_2 complex preferentially producing CO [Mondal *et al.*, 2015; Song *et al.*, 2014].

Crystal structures provided direct insights into CO_2 activation from a CO_2–bound structure. When CODH-II_{Ch} crystals are incubated in a solution containing high concentrations of a strong reducing agent (Ti(III)-citrate) and carbonate as a source for CO_2, we generate turnover conditions in the crystal as the substrate and reducing agent can diffuse through the pores of the protein crystal, allowing electrons and CO_2 to reach the active site cluster [Jeoung and Dobbek, 2007]. If we let the mixture react for several minutes, we expect to accumulate the state before the rate-limiting step in the crystal. When we cryo-cool the crystal by plunging in liquid nitrogen, we can capture and preserve the state throughout the diffraction experiment, which is carried out at 100 K. This approach allowed us to observe a $[NiFe_4S_4(CO_2)]$-cluster, where the substrate CO_2 binds as carboxylate ligand bridging Ni and Fe1 acting as a μ_2-η^2-ligand [Jeoung and Dobbek, 2007] (Figure 3B). CO_2 completes the square-planar coordination of Ni (η^1-CO_2 mode), as expected for preferential CO production and replaces the water ligand at Fe1 (η^1-OCO mode) (Figure 4, bottom). Hydrogen-bonds of the bound CO_2 to a lysine and a histidine residue, stabilize the accumulating electrons at the oxygen atoms. To gain further insight into CO_2 activation, it was crucial to push the crystallographic resolution from the initial 1.5 Å to a "true atomic" resolution of 1.03 Å. This allowed defining the individual positions of light atoms such as carbon and oxygen independently, which would have no clearly defined maxima at the lower resolution. The high resolution allowed us to determine the O–C–O-bending angle to 117° and the C–O bond length (1.30 Å and 1.32 Å) implying that CO_2 is reduced by two electrons [Fesseler *et al.*, 2015]. Also unobserved at the initial lower resolution was a short Ni–C bond of 1.8 Å, exposing a substantial π-backbonding between the carboxylate and Ni, resembling a Ni-carbon.

Looking from the angle of CO oxidation, the next step to release CO_2 from the observed Ni,Fe-bound carboxylate, would require a rearrangement of CO_2 and Ni. This step is likely supported by an incoming water molecule at Fe1, replacing the oxygen atom of CO_2, probably relying on a switch of the coordination geometry of Ni from square-planar to distorted tetrahedral (Figure 4, left). This would bring the product closer to the gas channel, forming a state similar to that observed with *n*-butylisocyanide-treated crystals [Jeoung and Dobbek, 2012]. *N*-butylisocyanide is one of several alternative substrates, including COS, N_2O and NH_2OH, which are reduced by CODHs and are in terms of their reactivity more comparable to CO_2. In contrast, *n*-butylisocyanide is a substrate analogous to CO, which acts as a rapidly binding competitive inhibitor and slow-turnover substrate of Ni,Fe-CODHs [Kumar and Ragsdale, 1995]. A crystal of CODH-II_{Ch} incubated with *n*-butyl isocyanide (nBIC) oxidized the isocyanide to *n*-butyl isocyanate, which remained bound in the crystal, where it is stabilized by hydrogen-bonds with the hydroxyl-ligand of Fe1 and other side chains in active site (Figure 3D). An analogous arrangement for CO_2 would support its facile release, removing one product from the active site. Two other products of CO oxidation, protons and electrons, leave at an as yet unspecified late step of the catalytic cycle. The two electrons should be transferred one-by-one to cluster B and possibly to cluster D from where they would be relayed to external electron acceptors. In case of monofunctional CODHs the electron acceptor is likely the ferredoxin CooF [Ensign and Ludden, 1991].

Not all steps in this reaction sequence are firmly established, requiring further experiments. The main goals for the future are therefore, (I) to capture elusive steps in the crystal and (II) combine spectroscopic and structural investigations at single crystals to link the rich spectroscopic data to structures.

Acknowledgment

Research on CODHs in my laboratory was financed by the Deutsche Forschungsgemeinschaft (DO 785) and the German excellence initiative (EXC 314; "Unifying concepts in catalysis — UniCat").

References and Citations

Abbanat, D.R. and Ferry, J.G. (1991). Resolution of component proteins in an enzyme complex from *Methanosarcina thermophila* catalyzing the synthesis or cleavage of acetyl-CoA, *Proc. Natl. Acad. Sci. USA*, 88, 3272–3276.

Adam, P.S., Borrel, G. and Gribaldo, S. (2018). Evolutionary history of carbon monoxide dehydrogenase/acetyl-CoA synthase, one of the oldest enzymatic complexes, *Proc. Natl. Acad. Sci. USA*, 115, 1166–1173.

Amara, P., Mouesca, J.M., Volbeda, A. and Fontecilla-Camps, J.C. (2011). Carbon monoxide dehydrogenase reaction mechanism: a likely case of abnormal CO_2 insertion to a Ni–H(−) bond, *Inorg. Chem.*, 50, 1868–1878.

Anderson, M.E. and Lindahl, P.A. (1996). Spectroscopic states of the CO oxidation/CO_2 reduction active site of carbon monoxide dehydrogenase and mechanistic implications, *Biochemistry*, 35, 8371–8380.

Anderson, M.E., Lindahl, P.A., Derose, V.J. and Hoffman, B.M. (1993). Identification of a Cyanide Binding Site in CO Dehydrogenase from *Clostridium thermoaceticum* Using EPR and ENDOR Spectroscopies, *J. Am. Chem. Soc.*, 115, 2204–12205.

Appel, A.M., Bercaw, J.E., Bocarsly, A.B., Dobbek, H., Dubois, D.L., Dupuis, M., Ferry, J.G., Fujita, E., Hille, R., Kenis, P.J.A., Kerfeld, C.A., Morris, R.H., Peden, C.H.F., Portis, A.R., Ragsdale, S.W., Rauchfuss, T.B., Reek, J.N.H., Seefeldt, L.C., Thauer, R.K. and Waldrop, G.L. (2013). Frontiers, Opportunities, and Challenges in Biochemical and Chemical Catalysis of CO_2 Fixation, *Chem. Rev.*, 113, 6621–6658.

Bar-Even, A., Noor, E., Savir, Y., Liebermeister, W., Davidi, D., Tawfik, D.S. and Milo, R. (2011). The moderately efficient enzyme: evolutionary and physicochemical trends shaping enzyme parameters, *Biochemistry*, 50, 4402–4410.

Barber, J. (2017). A mechanism for water splitting and oxygen production in photosynthesis, *Nat. Plants*, 3, 17041.

Barber, J. (2017). Photosynthetic water splitting by the $Mn_4Ca^{2+}O_X$ catalyst of photosystem II: Its structure, robustness and mechanism, *Q. Rev. Biophys.*, 50, e13.

Ciaccafava, A., Tombolelli, D., Domnik, L., Fesseler, J., Jeoung, J.H., Dobbek, H., Mroginski, M.A., Zebger, I. and Hildebrandt, P. (2016). When the inhibitor tells more than the substrate: the cyanide-bound state of a carbon monoxide dehydrogenase, *Chem. Sci.*, 7, 3162–3171.

Crabtree, R.H. (2005). *The Organometallic Chemistry of the Transition Metals*. John Wiley & Sons, Inc., New York.

Darnault, C., Volbeda, A., Kim, E.J., Legrand, P., Vernede, X., Lindahl, P.A. and Fontecilla-Camps, J.C. (2003). Ni–Zn–[Fe4S4] and Ni–Ni–[Fe4S4] clusters in closed and open subunits of acetyl-CoA synthase/carbon monoxide dehydrogenase, *Nat. Struct. Biol.*, 10, 271–279.

Derose, V.J., Telser, J., Anderson, M.E., Lindahl, P.A. and Hoffman, B.M. (1998). A multinuclear ENDOR study of the C-cluster in CO dehydrogenase from *Clostridium thermoaceticum*: evidence for H(x)O and histidine coordination to the [Fe4S4] center, *J. Am. Chem. Soc.*, 120, 8767–8776.

Dobbek, H., Svetlitchnyi, V., Gremer, L., Huber, R. and Meyer, O. (2001). Crystal structure of a carbon monoxide dehydrogenase reveals a [Ni–4Fe-–S] cluster, *Science*, 293, 1281–1285.

Domnik, L., Merrouch, M., Goetzl, S., Jeoung, J.H., Leger, C., Dementin, S., Fourmond, V. and Dobbek, H. (2017). CODH-IV: A high-efficiency CO-scavenging CO dehydrogenase with resistance to O_2, *Angew. Chem. Int. Ed. Engl.*, 56, 15466–15469.

Doukov, T.I., Iverson, T.M., Seravalli, J., Ragsdale, S.W. and Drennan, C.L. (2002). A Ni–Fe–Cu center in a bifunctional carbon monoxide dehydrogenase/acetyl-CoA synthase, *Science*, 298, 567–572.

Drennan, C.L., Heo, J., Sintchak, M.D., Schreiter, E. and Ludden, P.W. (2001). Life on carbon monoxide: X-ray structure of *Rhodospirillum rubrum* Ni–Fe–S carbon monoxide dehydrogenase, *Proc. Natl. Acad. Sci. USA,* 98, 11973–11978.

Ensign, S.A. (1995). Reactivity of carbon-monoxide dehydrogenase from *Rhodospirillum rubrum* with carbon-dioxide, carbonyl sulfide, and carbon-disulfide, *Biochemistry*, 34, 5372–5381.

Ensign, S.A., Hyman, M.R. and Ludden, P.W. (1989). Nickel-specific, slow-binding inhibition of carbon monoxide dehydrogenase from *Rhodospirillum rubrum* by cyanide, *Biochemistry*, 28, 4973–4979.

Ensign, S.A. and Ludden, P.W. (1991). Characterization of the CO oxidation/H_2 evolution system of *Rhodospirillum rubrum*: Role of a 22-kDa iron-sulfur protein in mediating electron transfer between carbon monoxide dehydrogenase and hydrogenase, *J. Biol. Chem.*, 266, 18395–18403.

Feng, J. and Lindahl, P.A. (2004). Carbon monoxide dehydrogenase from *Rhodospirillum rubrum*: effect of redox potential on catalysis, *Biochemistry*, 43, 1552–1559.

Ferreira, K.N., Iverson, T.M., Maghlaoui, K., Barber, J. and Iwata, S. (2004). Architecture of the photosynthetic oxygen-evolving center, *Science*, 303, 1831–1838.

Fesseler, J., Jeoung, J.H. and Dobbek, H. (2015). How the [NiFe4S4] cluster of CO dehydrogenase activates CO_2 and NCO, *Angew. Chem. Int. Ed. Engl.*, 54, 8560–8564.

Gong, W., Hao, B., Wei, Z., Ferguson, D.J., Jr., Tallant, T., Krzycki, J.A. and Chan, M.K. (2008). Structure of the alpha2epsilon2 Ni-dependent CO dehydrogenase component of the *Methanosarcina barkeri* acetyl-CoA decarbonylase/synthase complex, *Proc. Natl. Acad. Sci. USA,* 105, 9558–9563.

Grahame, D.A. (2003). Acetate C–C bond formation and decomposition in the anaerobic world: The structure of a central enzyme and its key active-site metal cluster, *Trends Biochem. Sci.*, 28, 221–224.

Ha, S.W., Korbas, M., Klepsch, M., Meyer-Klaucke, W., Meyer, O. and Svetlitchnyi, V. (2007). Interaction of potassium cyanide with the [Ni–4Fe–5S] active site cluster of CO dehydrogenase from *Carboxydothermus hydrogenoformans*, *J. Biol. Chem.*, 282, 10639–10646.

Jeoung, J.H. and Dobbek, H. (2007). Carbon dioxide activation at the Ni,Fe-cluster of anaerobic carbon monoxide dehydrogenase, *Science*, 318, 1461–1464.

Jeoung, J.H. and Dobbek, H. (2009). Structural basis of cyanide inhibition of Ni, Fe-containing carbon monoxide dehydrogenase, *J. Am. Chem. Soc.*, 131, 9922–9923.

Jeoung, J.H. and Dobbek, H. (2012). *n*-Butyl isocyanide oxidation at the [NiFe4S4OH(x)] cluster of CO dehydrogenase, *J. Biol. Inorg. Chem.*, 17, 167–173.

Kerby, R.L., Ludden, P.W. and Roberts, G.P. (1995). Carbon monoxide-dependent growth of *Rhodospirillum rubrum*, *J. Bacteriol.*, 177, 2241–2244.

Khalil, M.a.K., Pinto, J.P. and Shearer, M.J. (1999). Atmospheric carbon monoxide, *Chemosphere: Global Change Sci.*, 1, xi–xiii.

Kim, E.J., Feng, J., Bramlett, M.R. and Lindahl, P.A. (2004). Evidence for a proton transfer network and a required persulfide-bond-forming cysteine residue in Ni-containing carbon monoxide dehydrogenases, *Biochemistry*, 43, 5728–5734.

King, G.M. (1999). Characteristics and significance of atmospheric carbon monoxide consumption by soils *Chemosphere: Global Change Sci.*, 1, 53–63.

Kumar, M. and Ragsdale, S.W. (1995). *N*-Butyl isocyanide — A structural and functional analog of carbon-monoxide for carbon-monoxide dehydrogenase from *Clostridium thermoaceticum*, *J. Am. Chem. Soc.*, 117, 11604–11605.

Kung, Y., Doukov, T.I., Seravalli, J., Ragsdale, S.W. and Drennan, C.L. (2009). Crystallographic snapshots of cyanide- and water-bound C-clusters from bifunctional carbon monoxide dehydrogenase/acetyl-CoA synthase, *Biochemistry*, 48, 7432–7440.

Lindahl, P.A. (2002). The Ni-containing carbon monoxide dehydrogenase family: Light at the end of the tunnel?, *Biochemistry*, 41, 2097–2105.

Lindahl, P.A. (2012). Metal-metal bonds in biology, *J. Inorg. Biochem.*, 106, 172–178.

Lindahl, P.A., Munck, E. and Ragsdale, S.W. (1990a). CO Dehydrogenase from *Clostridium thermoaceticum* — Epr and electrochemical studies in CO_2 and argon atmospheres, *J. Biol. Chem.*, 265, 3873–3879.

Lindahl, P.A., Ragsdale, S.W. and Munck, E. (1990b). Mossbauer study of CO dehydrogenase from *Clostridium thermoaceticum*, *J. Biol. Chem.*, 265, 3880–3888.

Lu, Y. and Khalil, M.a.K. (1993). Methane and carbon-monoxide in O_h chemistry — the effects of feedbacks and reservoirs generated by the reactive products, *Chemosphere*, 26, 641–655.

Macgregor, S.A., Lu, Z., Eisenstein, O. and Crabtree, R.H. (1994). Why nickel(II) binds CO best in trigonal bipyramidal and square pyramidal geometries and possible consequences for CO dehydrogenase, *Inorg. Chem.*, 33, 3616–3618.

Mondal, B., Song, J., Neese, F. and Ye, S. (2015). Bio-inspired mechanistic insights into CO_2 reduction, *Curr. Op. Chem. Biol.*, 25, 103–109.

Page, C.C., Moser, C.C., Chen, X. and Dutton, P.L. (1999). Natural engineering principles of electron tunnelling in biological oxidation-reduction, *Nature*, 402, 47–52.

Ragsdale, S.W. (2004). Life with carbon monoxide, *Crit. Rev. Biochem. Mol. Biol.*, 39, 165–195.

Ragsdale, S.W. (2007). Nickel and the carbon cycle, *J. Inorg. Biochem.*, 101, 1657–1666.

Song, J., Klein, E.L., Neese, F. and Ye, S. (2014). The mechanism of homogeneous CO_2 reduction by Ni(cyclam): Product selectivity, concerted proton-electron transfer and C–O bond cleavage, *Inorg. Chem.*, 53, 7500–7507.

Spangler, N.J., Lindahl, P.A., Bandarian, V. and Ludden, P.W. (1996). Spectroelectrochemical characterization of the metal centers in carbon monoxide dehydrogenase (CODH) and nickel-deficient CODH from *Rhodospirillum rubrum*, *J. Biol. Chem.*, 271, 7973–7977.

Svetlichny, V.A., Sokolova, T.G., Gerhardt, M., Ringpfeil, M., Kostrikina, N.A. and Zavarzin, G.A. (1991). *Carboxydothermus hydrogenoformans* Gen-Nov, Sp-Nov, a CO-utilizing thermophilic anaerobic bacterium from hydrothermal environments of Kunashir-Island, *Syst. Appl. Microbiol.*, 14, 254–260.

Svetlitchnyi, V., Peschel, C., Acker, G. and Meyer, O. (2001). Two membrane-associated NiFeS-carbon monoxide dehydrogenases from the anaerobic carbon-monoxide-utilizing eubacterium *Carboxydothermus hydrogenoformans*, *J. Bacteriol.*, 183, 5134–5144.

Wang, P.H., Bruschi, M., De Gioia, L. and Blumberger, J. (2013). Uncovering a dynamically formed substrate access tunnel in carbon monoxide dehydrogenase/acetyl-CoA synthase, *J. Am. Chem. Soc.*, 135, 9493–9502.

Chapter 16

Selective Replacement of the Damaged D1 Reaction Center Subunit During the Repair of the Oxygen-Evolving Photosystem II Complex

Shengxi Shao*, Jianfeng Yu* and Peter J. Nixon*,†,‡

**Department of Life Sciences, Imperial College London, South Kensington Campus, London, SW7 2AZ, UK*

†School of Biological Sciences, Nanyang Technological University, Singapore

‡p.nixon@imperial.ac.uk

The multi-subunit photosystem II (PSII) pigment-protein complex found in plants, algae and cyanobacteria is nature's biological catalyst for producing oxygen from water. PSII needs sunlight to drive water oxidation but too much light can cause irreversible damage to pigments and proteins within PSII and loss of enzyme activity. Damaged PSII complexes can, however, be repaired in a highly selective process involving the replacement of damaged components by newly synthesized copies and the recycling of undamaged protein subunits and co-factors. Although substantial progress has been made to identify the enzymes and accessory factors involved in repair, many fundamental questions remain unanswered. In this chapter we discuss recent ideas on how the damaged D1 reaction center subunit, which is the subunit most prone to damage in PSII, is specifically recognized for replacement. Detachment of CP43 allowing access by FtsH proteases to the N-terminal tail of D1 seems to underpin selective degradation.

1. Introduction

PSII, the light-driven water:plastoquinone oxidoreductase of oxygenic photosynthesis, has long been considered a weak link in the photosynthetic electron transport chain due to its susceptibility to irreversible oxidative damage by light [Powles, 1984; Vass, 2012]. This effect is physiologically extremely important as loss of PSII activity has been linked to the decline in overall maximum rate of photosynthesis displayed by plants, algae and cyanobacteria when exposed to high intensities of light, in an effect termed "chronic photoinhibition" or 'photoinhibition' [Adir *et al.*, 2003]. Preventing or delaying chronic photoinhibition by improving the various photoprotective mechanisms found in plants has been advocated as one way to improve the efficiency of photosynthesis and so improve biomass and crop yield and a variety of efforts are being directed in this direction [Cardona *et al.*, 2018]. Often overlooked is the fact that the photoprotective mechanisms operating in plants are not infallible and PSII is irreversibly damaged at all light intensities, even at levels that are sub-saturating for photosynthesis [Park *et al.*, 1996; Tyystjärvi and Aro, 1996]. Repair of damaged PSII complexes is therefore an important photoprotective process in its own right and also a potential target for crop improvement [Bečková *et al.*, 2017; Cardona *et al.*, 2018].

Early radiolabeling studies revealed that mature plants and chloroplasts exposed to light predominantly synthesized a 32-kDa membrane protein [Eaglesham and Ellis, 1974; Hoffman-Falk *et al.*, 1982]. In the pioneering work by Eaglesham and Ellis (1974) it was suggested that this protein, labelled peak D in their early study, was likely to represent a rapidly-turning-over minor protein of the thylakoid membrane. Subsequent work, notably by Itzhak Ohad and colleagues, linked this rapid turnover to the repair of PSII that had been damaged by light [Ohad *et al.*, 1984]. We now know that the 32-kDa rapidly-turning-over protein is identical to the D1 reaction center (RC) protein of PSII, which has roles in binding the Mn_4CaO_5 cluster involved in water oxidation, the chlorophyll *a* and pheophytin *a* molecules involved in primary charge separation across the membrane and the secondary plastoquinone electron acceptor, Q_B, which is reduced and protonated by PSII and released into the thylakoid membrane as plastoquinol. The rapid turnover of the D1 protein is now accepted to reflect the selective degradation and replacement of a D1 subunit in a PSII complex that has been irreversibly damaged in the light [Tyystjärvi and Aro, 1996]. The half-life of D1 has been estimated to be as short as 40 minutes in *Pisum sativum* leaves [Aro *et al.*, 1993a] which makes D1 one of the most unstable proteins in plants. But this still means that PSII can complete 1–10 million catalytic turnovers before it is damaged and needs to be repaired [Park *et al.*, 1996].

The crystal structures of the dimeric PSII holoenzyme isolated from thermophilic cyanobacteria [Ferreira *et al.*, 2004; Umena *et al.*, 2011] have confirmed in striking detail that D1 lies at the heart of the PSII complex pinned in by the CP43 chlorophyll-binding subunit on one side and the D2 RC subunit and CP47 chlorophyll-binding subunit on the other and surrounded by several low-molecular-mass subunits (Figure 1). This structural information, reinforced by recent structural data on PSII supercomplexes in chloroplasts [Wei *et al.*, 2016], provides dramatic visual evidence that accessing and selectively replacing damaged D1 by a newly synthesized copy is a far from straightforward task.

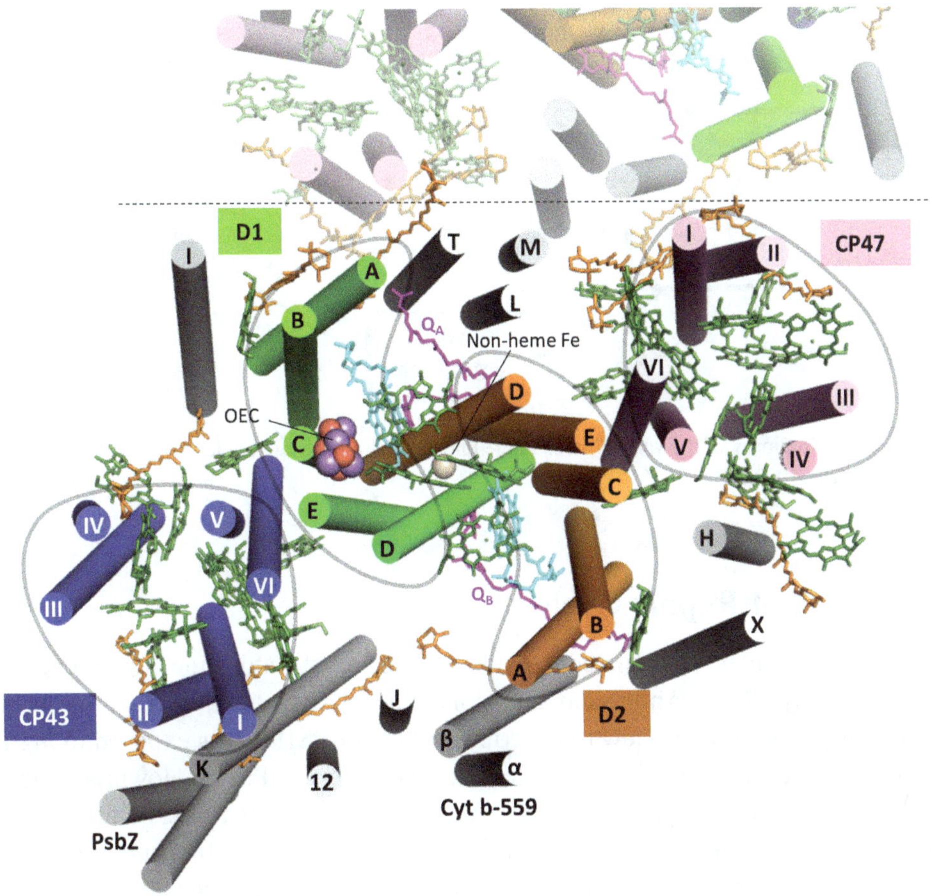

Figure 1. Subunit organization of the isolated PSII complex from the cyanobacterium *Thermosynechococcus vulcanus* (PDB: 3WU2) viewed from the lumenal surface. For clarity the lumenal extrinsic proteins are not shown and only one of the two complexes in the PSII dimer is shown in detail. The transmembrane α helices are shown as cylinders and color-coded according to the subunit: D1 in green (helices A–E), D2 in orange (helices A–E), CP43 in blue (helices I–VI), CP47 in pink (helices I–VI) and the 12 low-molecular-mass subunits in grey (label on helix indicates subunit, e.g. H is PsbH and 12 is Ycf12). Also shown are the chlorin rings of the chlorophylls (green) and pheophytins (cyan), the β-carotenes (orange), the plastoquinone electron acceptors Q_A and Q_B (purple), the non-heme iron (Fe) and the Mn_4O_5Ca cluster of the oxygen-evolving complex (OEC).

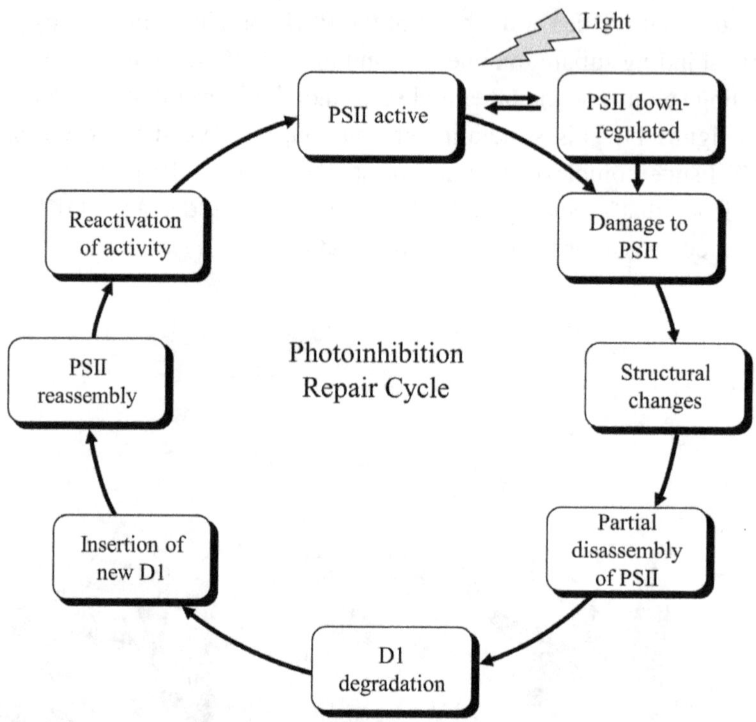

Figure 2. Summary of the various steps in the Photoinhibition Repair Cycle, based on the scheme in [Aro *et al.*, 1993b].

2. The PSII Repair Cycle

Figure 2 provides an updated version of the "photoinhibition-repair cycle" presented by Bertil Andersson and colleagues [Aro *et al.*, 1993b] in their extremely influential review to describe the fate of PSII that is exposed to high light. Upon an increase in light intensity, excitation of PSII is downregulated and PSII activity protected by a variety of mechanisms. If PSII is damaged, a PSII repair cycle operates in which the relevant damaged protein subunit, typically D1, is recognized, degraded by proteases and replaced by a newly synthesized subunit, and the PSII complex reactivated to form the holoenzyme [Nixon *et al.*, 2010]. Over the past 25 years many of the molecular details have now been fleshed out, especially the photoprotective mechanisms used to protect PSII activity, which are discussed in detail elsewhere [Derks *et al.*, 2015]. Here we will focus on the mechanisms of damage to D1, the sites of damage, and discuss the consequence of this for how damaged D1 might be recognized by proteases and selectively degraded.

2.1. *Production of reactive species in PSII by visible light*

Historically, PSII photoinhibition has been classified as either acceptor-side or donor-side depending on the nature of the perturbation to electron flow that leads to damage [Aro *et al.*, 1993a; Barber and Andersson, 1992; Vass, 2012]. The main species involved in acceptor-side photoinhibition is considered to be singlet oxygen ($^1O_2$) produced from triplet oxygen ($^3O_2$) by the transfer of excitation energy from a chlorophyll triplet state, which in the case of the PSII RC is usually generated by recombination of $^3(P680^+Pheo_{D1}^-)$ formed by spin conversion of the $^1(P680^+Pheo_{D1}^-)$ radical pair [Telfer, 2014], where P680 is the primary electron donor of PSII and $Pheo_{D1}$ the primary electron acceptor [Diner and Rappaport, 2002]. This process is prevalent at high irradiances when the plastoquinone pool in the thylakoid membrane and the plastoquinone electron acceptors, Q_A and Q_B, within PSII are reduced and forward electron transfer is limited to primary charge separation between P680 and $Pheo_{D1}$, but it has also been suggested to occur at low-light intensities during charge recombination between long-lived Q_B^- and the donor side [Keren *et al.*, 1997]. In addition, superoxide can be formed by direct reduction of oxygen by reduced electron transfer components on the acceptor side [Pospíšil, 2012; Vass, 2012] and the hydroxyl radical by reactions involving the non-heme iron [Pospíšil *et al.*, 2004].

In contrast, donor-side photoinhibition is considered to be caused by impaired electron donation to the PSII RC, leading to accumulation of long-lived $P680^{+\bullet}$ and tyrosine $Y_z^\bullet$ radicals, which then oxidize lipid and protein side chains in an oxygen independent process. In the presence of oxygen, reactive hydroperoxides and singlet oxygen can also be formed [Pathak *et al.*, 2017]. Donor-side photoinhibition is thought to be promoted by inactivation of the Mn_4O_5Ca cluster through excitation by blue light in a still poorly understood process [Hakala *et al.*, 2005; Ohnishi *et al.*, 2005]. Additionally, incomplete oxidation of water to peroxide by the Mn_4O_5Ca cluster is thought to lead to the formation of superoxide by PSII-mediated oxidation of hydrogen peroxide and hydroxyl radicals through Fenton chemistry involving Mn^{2+} released from damaged Mn_4O_5Ca clusters [Pospíšil, 2012].

The production of reactive oxygen species (ROS) on both the donor side and acceptor side of the complex means that PSII can in principle be damaged at multiple sites within the complex with the type and extent of damage likely to be dependent on the prevailing light quantity (intensity) and light quality (spectral profile) and on the physiological status of the organism [Pospíšil, 2016]. Enzymatic removal of damaged protein from PSII therefore needs to be sufficiently flexible to recognize and cope with the different types of damage that can occur. Knowledge of the PSII structure also explains why the D1 subunit is damaged and replaced more often than the other PSII proteins — the D1 subunit is most intimately involved in binding the co-factors in PSII that produce damaging species and so is the closest protein and the one most likely to be damaged. This situation has probably not arisen by chance as replacement of just a single subunit in the complex is clearly an evolutionary advantage as it minimizes the overall energetic cost involved in protein synthesis and protein degradation [Murata and Nishiyama, 2018; Rutherford and Faller, 2003].

2.2. *Sites of damage in D1*

Exposing isolated PSII complexes to visible light is known to cause the destruction of chlorophyll and carotenoid [Telfer, 2014] and induce extensive damage in proteins through crosslinking, protein fragmentation and oxidation of side chains [Lupinkova and Komenda, 2004]. Mass spectrometry has been used successfully to explore the identity of the amino-acid residues that are oxidized in D1 and other PSII subunits [Sharma *et al.*, 1997]. The most recent data suggest that in the case of D1, amino-acid residues lining water channels leading to the Mn_4O_5Ca cluster are the first to be oxidized in the light [Weisz *et al.*, 2017], followed by residues around $Pheo_{D1}$, Q_A and the non-heme iron [Kale *et al.*, 2017]. These results are consistent with the known sites of production of ROS in PSII as described in Section 2.1. However, work has so far been limited to experiments on isolated thylakoid membranes or isolated PSII complexes and so there is still uncertainty whether these sites are actually damaged in nature.

2.3. *Main proteases involved in degrading D1*

Early biochemical studies by Bertil Andersson and colleagues [Haußühl *et al.*, 2001; Lindahl *et al.*, 1996, 2000; Spetea *et al.*, 1999] implicated the involvement of two classes of protease in the degradation of damaged D1 in chloroplasts: FtsH proteases, which are membrane-embedded hexameric metalloprotease complexes that couple protein proteolysis to the ATP-driven translocation of membrane proteins from the membrane [Kato and Sakamoto, 2018], and Deg proteases, which are a type of serine protease [Cheregi *et al.*, 2016]. Initially it was suggested that damaged D1 was first cleaved by the DegP2 protease (now renamed Deg2) close to the Q_B-binding pocket of D1 [Haußühl *et al.*, 2001] with the resulting N-terminal D1 fragment removed by FtsH [Haußühl *et al.*, 2001; Lindahl *et al.*, 2000]. Chloroplasts and the evolutionary related cyanobacteria encode multiple FtsH and Deg subunits and work has since been directed at understanding the physiological relevance of these and other proteases in PSII repair, typically by studying PSII repair in defined mutants [Komenda *et al.*, 2012b]. Somewhat surprisingly, the analysis of D1 degradation in *ftsH* knock-out mutants, initially in cyanobacteria [Silva *et al.*, 2003; Silva and Nixon, 2001] and later in plants [Bailey *et al.*, 2002; Kato *et al.*, 2009], revealed the importance of FtsH for removing full-length damaged D1 not just D1 fragments [Nixon *et al.*, 2005]. This led to a revised model in which FtsH alone was able to degrade D1 [Nixon *et al.*, 2005; Silva *et al.*, 2003]. Consistent with this, the three Deg proteases found in cyanobacteria are not required for D1 degradation [Barker *et al.*, 2006], in line with later data showing that they are more involved in quality control outside the thylakoid membrane system [Cheregi *et al.*, 2015]. In the case of chloroplasts, Deg proteases are found on both the stromal and luminal sides of the thylakoid membrane and are now suggested to play a supporting role in D1 degradation [Kato *et al.*, 2012].

2.4. *Prompt selective repair versus emergency repair*

PSII repair is a multilayered response dependent on the rate of damage and the capacity for repair. Based on the analysis of cyanobacterial mutants, the FtsH proteases have been proposed to act as the "first-line of defence" to enable prompt degradation of damaged D1 so that the rate of repair balances the rate of damage and there is no net loss of PSII activity [Nixon *et al.*, 2005]. However, if there is an imbalance between the rate of damage and the capacity of the FtsH pathway to repair PSII, such as in PSII donor-side mutants that are more susceptible to damage [Komenda *et al.*, 2010], back-up or emergency pathways, such as those involving the Deg proteases in chloroplasts [Kato *et al.*, 2012], become more important, as the degree of damage becomes more extensive and additional PSII proteins such as D2 and CP43 are damaged and need replacing [Yao *et al.*, 2012]. Under these extreme, but physiologically relevant conditions, ROS can induce extensive damage in PSII including protein cleavage and crosslinking of subunits [Lupinkova and Komenda, 2004; Mizusawa *et al.*, 2003], and damage to lipid and pigments [Mattila *et al.*, 2015]. Prompt and selective PSII repair is therefore important not only for maintaining PSII activity but also for preventing the wholescale production of ROS that could lead to extensive damage to the thylakoid membrane plus the repair machinery itself [Nishiyama *et al.*, 2001].

3. Overview of FtsH Proteases and Mechanism

FtsH proteases are distributed throughout nature except for the Archaea [Shao *et al.*, 2018] with each subunit consisting of a transmembrane domain consisting of 1–2 transmembrane helices linked to a soluble fragment containing an AAA+ module (ATPase associated with diverse activities) and a zinc metalloprotease domain. The high resolution structure of the intact hexameric FtsH protease complex, be it homo or hetero-oligomeric, is still not known, although progress has been made understanding the structure of the soluble fragment of FtsH. Most recently, cryo-electron microscopy (cryo-EM) was used to determine the structure of a soluble engineered version of the yeast YME1 AAA+ protease found in the inner mitochondrial membrane with what seems to be a bound peptide substrate trapped in the process of translocation (Puchades *et al.*, 2017). The protease is formed of two stacked rings described by the authors as forming "an asymmetric spiral staircase of ATPase domains atop a planar protease ring" [Puchades *et al.*, 2017]. ATP hydrolysis seems to lead to dramatic conformational changes in the AAA+ modules that cause translocation of the unfolded target protein, usually from the free N- or C-terminus, through a hydrophobic pore in the AAA+ module, into the protease domain where it is cleaved into smaller peptides. In the case of membrane protein substrates, it is possible that the transmembrane domain of the FtsH complex interacts with the transmembrane hydrophobic segments of the target protein to facilitate the unfolding and translocation steps needed to pull the target from the membrane [Yang *et al.*, 2018].

Recent phylogenetic analyses have revealed that the FtsH proteases involved in PSII repair form a distinct clade that emerged early in evolution, possibly in response to the need to repair ancestral RCs that were prone to photoinhibition [Shao *et al.*, 2018]. Most cyanobacteria encode 4 FtsH protease subunits (FtsH1-4) [Shao *et al.*, 2018]. Two of these, FtsH2 and FtsH3, are known to play a major role in D1 degradation and to form a hexameric FtsH complex within the thylakoid membrane composed of alternating subunits [Boehm *et al.*, 2012a]. Besides a role in PSII repair in visible light, the FtsH2/FtsH3 complex is also involved in maintaining PSII activity in cells exposed to ultra-violet light [Cheregi *et al.*, 2007] or heat stress [Kamata *et al.*, 2005] and plays a house-keeping role in removing unassembled PSII proteins [Komenda *et al.*, 2006] and soluble proteins [Stirnberg *et al.*, 2007]. Biochemical data also suggest the presence of a FtsH1/FtsH3 hetero-complex in the cytoplasmic membrane and an FtsH4 homocomplex in the thylakoid membrane [Krynická *et al.*, 2014; Sacharz *et al.*, 2015]. It is still unclear what determines the substrate specificities of the various FtsH complexes but their location in different regions of the membrane is likely to be one important factor. D1 degradation in chloroplasts also involves a related set of FtsH complexes [Kato and Sakamoto, 2018; Shao *et al.*, 2018].

Thylakoid FtsH subunits are known to interact with other proteins in the membrane, including prohibitin [Boehm *et al.*, 2012a] and Psb29/Thf1 [Bečková *et al.*, 2017] in cyanobacteria and EngA in chloroplasts [Kato *et al.*, 2018]. The role of these accessory proteins in FtsH function is currently unclear although Psb29/Thf1 and EngA are thought to play a role in biogenesis [Bečková *et al.*, 2017; Kato *et al.*, 2018] possibly to counteract damage to FtsH [Kato *et al.*, 2018].

4. Recognition of Damaged D1 by FtsH

4.1. *Importance of the N-terminal tail*

Figure 3 summarizes our current view of the FtsH pathway for prompt D1 degradation, based mainly on experimental data obtained from analysis of cyanobacterial mutants, although evolutionary considerations and the available experimental data suggest that FtsH proteases function similarly in chloroplasts [Komenda *et al.*, 2012b]. It should be noted that this model was proposed on the basis of mutant studies — whether FtsH alone can degrade damaged D1 awaits the development of a reconstituted *in vitro* system, along the lines of one recently developed for *E. coli* FtsH [Yang *et al.*, 2018].

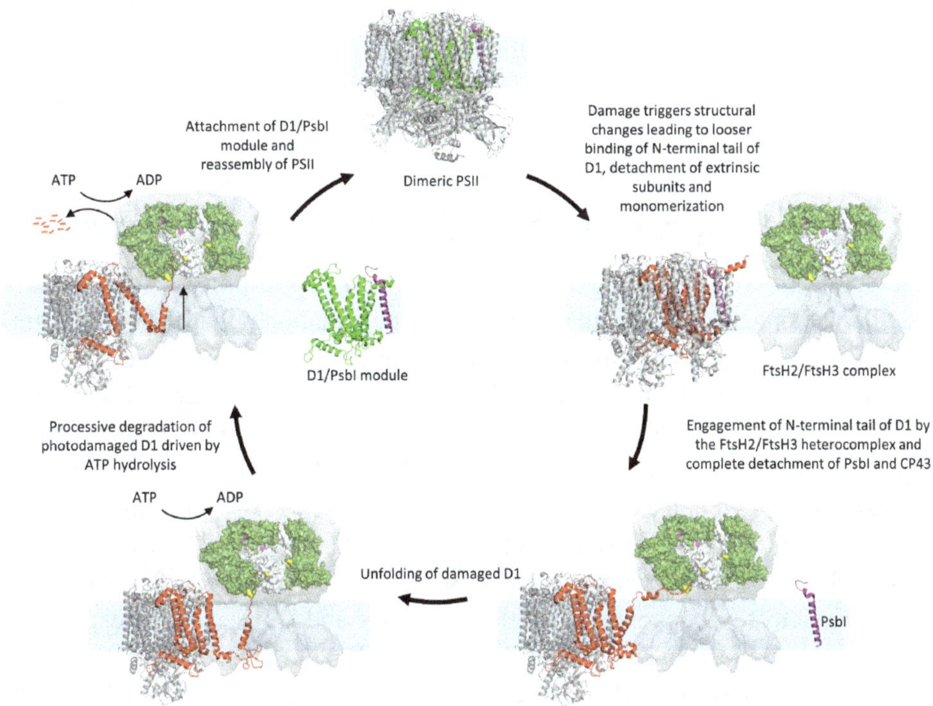

Figure 3. FtsH pathway for the degradation of damaged D1 during PSII repair. PSII components are drawn based on the PSII structure from *T. vulcanus* (PDB: 3WU2). The FtsH hetero-hexamer model is based on the cryo-EM structure of a yeast *m*-AAA protease (EMDB: EMD-1712) and the crystal structure of the cytosolic region of FtsH from *Thermus thermophilus* (PDB: 2DHR). For clarity accessory proteins and the SCPs involved in binding chlorophyll are not shown. A post-translational model for replacement of damaged D1 is shown involving a D1/PsbI module containing either a precursor form of D1 or mature D1.

In line with studies of FtsH in other systems [Chiba *et al.*, 2002], the N-terminal tail of D1 exposed to the cytoplasm has been shown to be vitally important for efficient D1 degradation in the cyanobacterium *Synechocystis* sp. PCC 6803 (hereafter *Synechocystis* 6803), since truncation by 20 amino-acid residues specifically blocks FtsH-mediated D1 degradation [Komenda *et al.*, 2007]. This experiment provided important evidence in support of an earlier proposal suggesting N-terminal mediated degradation of D1 [Nixon *et al.*, 2005; Silva *et al.*, 2003]. A similar D1 truncation mutant has also been made in tobacco, but interpreting the effect on D1 degradation is complicated by reduced accumulation of PSII and the presence of back-up pathways for degrading D1 [Michoux *et al.*, 2016]. Undamaged chlorophyll/chlorophyllide released during D1 degradation in cyanobacteria is thought to be captured by small Cab-like proteins (SCPs), which are members of the chlorophyll *a*/*b*-binding superfamily of proteins, to help prevent additional damage before being recycled and reincorporated into PSII [Komenda and Sobotka, 2016; Vavilin and Vermaas, 2007]. The related one-helix proteins (OHPs) are likely to perform an equivalent role in chloroplasts [Hey and Grimm, 2018].

4.2. *Importance of accessibility for selective D1 degradation*

Recent ideas have emphasized the importance of accessibility in determining whether a damaged D1 subunit is selected for degradation rather than an undamaged one. Evidence comes from experiments on D1 turnover in a cyanobacterial mutant lacking the CP43 subunit, which assembles a PSII sub-complex, termed RC47, inactive for oxygen-evolution, but still capable of light-induced charge separation between Y_Z and Q_A [Boehm *et al.*, 2012b]. D1 was selectively removed by FtsH from this complex both in the light and, unlike the WT and a mutant lacking PsbO, also *in the dark*. In a mutant lacking both CP43 and CP47, which assembles a PSII RC assembly complex composed of D1, D2 and the Ycf39 and Ycf48 assembly factors [Knoppová *et al.*, 2014], both D1 and D2 were now degraded in the dark. In contrast unassembled CP47 and CP43 complexes were much more stable [Krynická *et al.*, 2015]. These observations suggest that the D1 and D2 subunits are intrinsically more prone to proteolysis by FtsH in the dark than CP47 and CP43 and that binding of CP43 and CP47 protects undamaged D1 and D2 from FtsH-mediated degradation. This has led to the proposal that damage to PSII, causing partial or complete detachment of CP43, allowing access of FtsH to the N-terminal tail of D1, is a key event that determines specific degradation of damaged D1 [Krynická *et al.*, 2015] (Figure 3). Likewise, selective degradation of D2 would require detachment of CP47. The FtsH2/FtsH3 complex is predicted to contain a short lumenally exposed region on the opposite side of the membrane to the N-terminal tail, so FtsH complexes might also interact with lumenal parts of D1 during the course of degradation as well as transmembrane regions (Figure 3).

4.3. *What triggers detachment of CP43?*

It is still unclear the nature of the damage that causes detachment of CP43 to promote D1 degradation. It is plausible that there is no single event or "trigger" but rather it is the accumulation of multiple damaging events. Given that CP43 is intimately associated with D1, damage by ROS produced on both the donor and acceptor side of PSII has the potential to weaken the binding of CP43 to D1. From analysis of the crystal structure of cyanobacterial PSII, there are a number of potential binding interactions between CP43 and D1 that could be affected by oxidative damage. Besides amino-acid residues, targets include lipids at the interface of CP43 and D1 [Loll *et al.*, 2007], chlorophyll and carotenoid pigments and the Mn_4O_5Ca cluster which is ligated not only by amino-acid ligands in D1 but also by CP43-Glu354 [Ferreira *et al.*, 2004]. Following damage, movement of lipid molecules into the interface between CP43 and D1 will also help weaken the interactions between the transmembrane helices and promote the further peeling away of CP43. Disassembly of dimeric PSII complexes to form monomeric complexes has been suggested to involve detachment of the extrinsic proteins from the luminal surface of PSII, especially the PsbO extrinsic protein, which is known to stabilize dimeric complexes [Komenda *et al.*, 2010]. Detachment of CP43 could also be aided by binding of accessory factors to the lumenal side of PSII, such as Psb27 to CP43 [Komenda *et al.*, 2012a] and Ycf48 to D1 (Yu *et al*, 2018) and the unfoldase activity of the FtsH complex itself could contribute to the disassembly process once it has latched onto the N-terminus of D1 (Figure 3).

One dramatic feature of the PSII crystal structures is how the C-terminus of CP43 folds back to pin down the N-terminal tail of D1 (Figure 4). We speculate that altered binding of CP43 to D1 in this region would act to release the N-terminal tail of D1 to allow the initial engagement with the FtsH complex which might then couple further detachment of CP43 with D1 degradation (Figure 3). Also present close to the N-terminal tail of D1 is the PsbI subunit which is known to be important for binding of CP43 to PSII [Dobáková *et al.*, 2007]. PsbI binds to the surface helix of D1 and interacts with the first transmembrane helix of D1 as well as the second transmembrane helix close to the binding site of chlorophyll $ChlZ_{D1}$ which has been suggested to be oxidized in damaged PSII complexes [Stewart *et al.*, 1998; Tracewell and Brudvig, 2008; Vasil'ev *et al.*, 2003]. PsbI is not degraded during D1 turnover but is recycled [Dobáková *et al.*, 2007]. Hence detachment of PsbI, possibly related to oxidation of $ChlZ_{D1}$, might be involved in the initial disassembly processes in PSII.

4.4. *Incorporation of newly synthesized D1*

It is possible that there are multiple pathways for incorporating D1 into PSII during repair. In the case of *Synechocystis* 6803, newly synthesized D1, which is synthesized with a C-terminal extension that must be processed to permit assembly of the Mn_4O_5Ca cluster [Nixon *et al.*, 1992], can be detected in the membrane in the form of a sub-complex or module containing precursor forms of D1, PsbI and possibly Ycf48, and it is therefore possible that a post-translational mechanism might operate [Dobáková *et al.*, 2007] (Figure 3). Upon light stress, D1 present in PSII RC intermediates, which are normally used for the assembly of PSII, have been suggested to be redirected into the repair pathway to provide D1 and D2 if necessary [Knoppová *et al.*, 2014]. In addition, early studies using isolated chloroplasts suggested a co-translational mechanism for the insertion of D1 into PSII [Zhang *et al.*, 1999]. Such a mechanism has the advantage that it would allow synchronization between synthesis and degradation of damaged D1 [Komenda and Barber, 1995] and indeed there is increasing experimental evidence in cyanobacteria to support the presence of specific repair zones in the membrane [Sacharz *et al.*, 2015] containing the FtsH2/FtsH3 complex to degrade D1, the YidC insertase to insert D1 into the membrane, and chlorophyll synthase/SCP complexes to synthesize and deliver chlorophyll to D1 [Chidgey *et al.*, 2014; Knoppová *et al.*, 2014; Yu *et al.*, 2018].

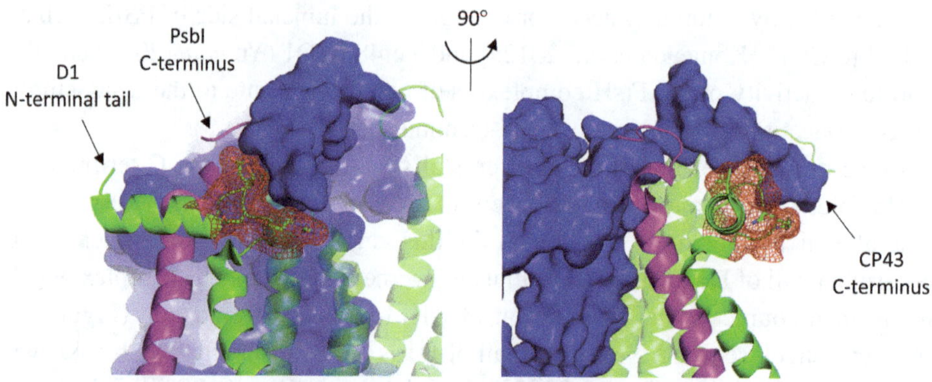

Figure 4. View of the local structure around the N-terminal tail of D1 (green) in PSII from *T. vulcanus* (PDB: 3WU2). The C-terminal region of CP43 (blue) is in close proximity to residues 24–27 of D1 (shown in stick and mesh) so that mobility of this region of D1 is constrained. The C-terminal region of PsbI subunit (in purple) makes contacts with both the surface helix of D1 and CP43. The first 10 residues of the N-terminal tail of D1 are absent from the crystal structure and are not shown.

5. Evolutionary Implications

There is widespread agreement from structural and phylogenetic analyses that photosynthetic RCs in all domains of life originated from a single ancestral RC [Cardona, 2015; Orf *et al.*, 2018]. According to this scenario, the last common ancestor evolved into type I RCs containing FeS clusters on the acceptor side or type II RCs containing quinone terminal electron acceptors. The homodimeric RC from Heliobacteria [Gisriel *et al.*, 2017] and heterodimeric PSI complex are type I complexes, whereas PSII and the RC from purple bacteria are type II RCs. There is an on-going debate about the precise sequence of events that led to the diversity of RCs found in nature [Hohmann-Marriott and Blankenship, 2011], but it is clear that PSII represents a "split" version of PSI in the sense that the 11 transmembrane helices seen in each of the two large PSI RC subunits, PsaA and PsaB, are each split in PSII into a RC subunit of five transmembrane helices (D1 and D2) and an antenna core complex of six transmembrane helices (CP47 and CP43), to form a CP43+D1 pair and a CP47+D2 pair (Figure 5). Our model highlighting the importance of detaching CP43 for prompt D1 degradation offers an explanation as to why PSII has retained this split RC structure rather than adopt the two fused RC subunits seen in PSI. We have recently tested the effect of creating a PSI-like PSII complex in which we have made a mutant of *Synechocystis* 6803 that expresses CP43–D1 and CP47–D2 fusion proteins. The resulting strain still assembles a functional oxygen-evolving PSII complex, but is unable to repair damaged PSII efficiently (Yu, Shao, Komenda and Nixon, unpublished data). Overall these recent data support the idea that the need to repair PSII promptly and efficiently has been an important constraint in the evolution of PSII and led to retention of a split RC.

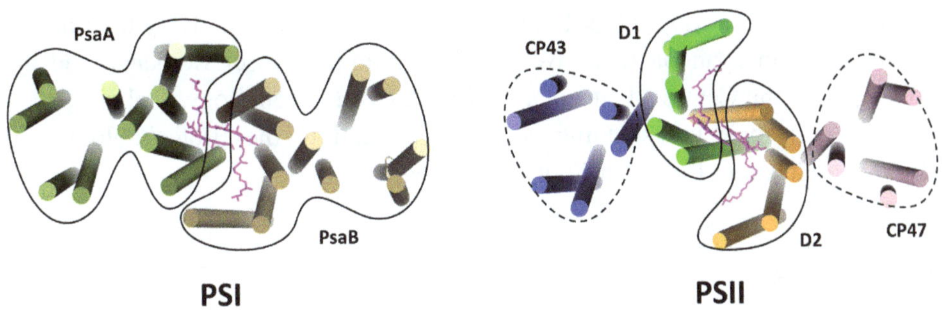

Figure 5. Comparison of the organization of the transmembrane helices found in PSI and PSII. The cylindrical models are based on the structure of PSI from *Thermosynechococcus elongatus* (PDB: 1JB0) and PSII from *T. vulcanus* (PDB:3WU2).

6. Conclusions and Outlook

PSII repair is a physiological important process for maintenance of photosynthesis in plants exposed to excessive light and, in principle, a detailed understanding of its mechanism might allow the generation of crop plants with enhanced tolerance to photoinhibition [Komenda *et al.*, 2012b]. In the case of chloroplasts, PSII complexes are located in the appressed membranes of the granal stacks and, when damaged, PSII supercomplexes must disassemble and migrate to the margins to engage with the repair apparatus [Herbstova *et al.*, 2012]. In contrast to the situation in cyanobacteria, the repair process in chloroplasts is regulated by the phosphorylation and dephosphorylation of threonine residues at the N-terminus of the D1, D2, CP43 and PsbH core proteins of PSII and in associated LHC subunits, but as yet the precise role of phosphorylation is unclear [Rochaix *et al.*, 2012]. Manipulating the levels of the kinases, phosphatases and proteases as well as membrane architecture might have a positive impact on the efficiency of PSII repair [Bečková *et al.*, 2017; Cardona *et al.*, 2018]. Accessory factors involved in the assembly and repair of PSII, such as the Ycf48 protein, involved in stabilizing newly synthesized D1 destined for PSII assembly or repair [Yu *et al.*, 2018], are also promising targets.

Acknowledgements

We dedicate this chapter to Bertil Andersson on the occasion of his 70th birthday and his retirement as President of Nanyang Technological University, Singapore. Bertil and his colleagues made many important contributions to our understanding of the PSII repair cycle. In particular the identification of the FtsH and Deg proteases as the likely proteases involved in degrading D1 was a major step forward in the area and the "trigger" for P.J.N. to become involved in studying their physiological role in cyanobacteria. We are also grateful to Josef Komenda and his colleagues in the Centre Algatech at the Institute of Microbiology, Třeboň, Czech Republic for their longstanding collaboration and to the Biotechnology and Biological Sciences Research Council for funding.

References

Adir, N., Zer, H., Shochat, S. and Ohad, I. (2003). Photoinhibition — A historical perspective. *Photosynth. Res.*, 76, 343–370.

Aro, E.M., McCaffery, S. and Anderson, J.M. (1993a). Photoinhibition and D1 protein degradation in peas acclimated to different growth irradiances. *Plant Physiol.*, 103, 835–843.

Aro, E.M., Virgin, I. and Andersson, B. (1993b). Photoinhibition of photosystem II. Inactivation, protein damage and turnover. *Biochim. Biophys. Acta — Bioenerg.*, 1143, 113–134.

Bailey, S., Thompson, E., Nixon, P.J., Horton, P., Mullineaux, C.W., Robinson, C. *et al.* (2002). A critical role for the Var2 FtsH homologue of Arabidopsis thaliana in the photosystem II repair cycle *in vivo*. *J. Biol. Chem.*, 277, 2006–2011.

Barber, J., and Andersson, B. (1992). Too much of a good thing: light can be bad for photosynthesis. *Trends Biochem. Sci.*, 17, 61–66.

Barker, M., de Vries, R., Nield, J., Komenda, J. and Nixon, P.J. (2006). The Deg proteases protect *Synechocystis* sp. PCC 6803 during heat and light stresses but are not essential for removal of damaged D1 protein during the photosystem two repair cycle. *J. Biol. Chem.*, 281, 30347–30355.

Bečková, M., Yu, J., Krynická, V., Kozlo, A., Shao, S., Koník, P. *et al.* (2017). Structure of Psb29/Thf1 and its association with the FtsH protease complex involved in photosystem II repair in cyanobacteria. *Philos. Trans. R. Soc. B Biol. Sci.*, 372, 20160394.

Boehm, M., Yu, J., Krynická, V., Barker, M., Tichý, M., Komenda, J. *et al.* (2012a). Subunit organization of a *Synechocystis* hetero-oligomeric thylakoid FtsH complex involved in Photosystem II repair. *Plant Cell*, 24, 3669–3683.

Boehm, M., Yu, J., Reisinger, V., Bečková, M., Eichacker, L.A., Schlodder, E. *et al.* (2012b). Subunit composition of CP43-less photosystem II complexes of *Synechocystis* sp. PCC 6803: implications for the assembly and repair of photosystem II. *Philos. Trans. R. Soc. Lond. B. Biol. Sci.*, 367, 3444–3454.

Cardona, T. (2015). A fresh look at the evolution and diversification of photochemical reaction centers. *Photosynth. Res.*, 126, 111–134.

Cardona, T., Shao, S. and Nixon, P.J. (2018). Enhancing photosynthesis in plants: The light reactions. *Essays Biochem.*, 62, 85–94.

Cheregi, O., Miranda, H., Gröbner, G. and Funk, C. (2015). Inactivation of the Deg protease family in the cyanobacterium *Synechocystis* sp. PCC 6803 has impact on the outer cell layers. *J. Photochem. Photobiol. B Biol.*, 152, 383–394.

Cheregi, O., Sicora, C., Kós, P.B., Barker, M., Nixon, P.J. and Vass, I. (2007). The role of the FtsH and Deg proteases in the repair of UV-B radiation-damaged Photosystem II in the cyanobacterium *Synechocystis* PCC 6803. *Biochim. Biophys. Acta,* 1767, 820–828.

Cheregi, O., Wagner, R. and Funk, C. (2016). Insights into the cyanobacterial Deg/HtrA proteases. *Front. Plant Sci.*, 7, 1–14.

Chiba, S., Akiyama, Y. and Ito, K. (2002). Membrane protein degradation by FtsH can be initiated from either end. *J. Bacteriol.*, 184, 4775–4782.

Chidgey, J.W., Linhartová, M., Komenda, J., Jackson, P.J., Dickman, M.J., Canniffe, D.P. *et al.* (2014). A cyanobacterial chlorophyll synthase-HliD complex associates with the Ycf39 protein and the YidC/Alb3 insertase. *Plant Cell,* 26, 1267–1279.

Derks, A., Schaven, K. and Bruce, D. (2015). Diverse mechanisms for photoprotection in photosynthesis. Dynamic regulation of photosystem II excitation in response to rapid environmental change. *Biochim. Biophys. Acta — Bioenerg.*, 1847, 468–485.

Diner, B.A. and Rappaport, F. (2002). Structure, dynamics, and energetics of the primary photochemistry of photosystem II of oxygenic photosynthesis. *Annu. Rev. Plant Biol.*, 53, 551–580.

Dobáková, M., Tichý, M. and Komenda, J. (2007). Role of the PsbI protein in photosystem II assembly and repair in the cyanobacterium *Synechocystis* sp. PCC 6803. *Plant Physiol.*, 145, 1681–1691.

Eaglesham, A.R.J. and Ellis, R.J. (1974). Protein synthesis in chloroplasts. *Biochim. Biophys. Acta - Nucleic Acids Protein Synth.*, 335, 396–407.

Ferreira, K.N., Iverson, T.M., Maghlaoui, K., Barber, J. and Iwata, S. (2004). Architecture of the photosynthetic oxygen-evolving center. *Science,* 303, 1831–1838.

Gisriel, C., Sarrou, I., Ferlez, B., Golbeck, J.H., Redding, K.E. and Fromme, R. (2017). Structure of a symmetric photosynthetic reaction center–photosystem. *Science,* 357, 1021–1025.

Hakala, M., Tuominen, I., Keränen, M., Tyystjärvi, T. and Tyystjärvi, E. (2005). Evidence for the role of the oxygen-evolving manganese complex in photoinhibition of Photosystem II. *Biochim. Biophys. Acta – Bioenerg.*, 1706, 68–80.

Haußühl, K., Andersson, B. and Adamska, I. (2001). A chloroplast DegP2 protease performs the primary cleavage of the photodamaged D1 protein in plant photosystem II. *EMBO J.*, 20, 713–722.

Herbstova, M., Tietz, S., Kinzel, C., Turkina, M.V. and Kirchhoff, H. (2012). Architectural switch in plant photosynthetic membranes induced by light stress. *Proc. Natl. Acad. Sci. USA,* 109, 20130–20135.

Hey, D. and Grimm, B. (2018). ONE-HELIX PROTEIN2 (OHP2) is required for the stability of OHP1 and assembly factor HCF244 and is functionally linked to PSII biogenesis. *Plant Physiol.*, 177, 1453–1472.

Hoffman-Falk, H., Mattoo, A.K., Marder, J.B., Edelman, M. and Ellis, R.J. (1982). General occurrence and structural similarity of the rapidly synthesized, 32,000-dalton protein of the chloroplast membrane. *J. Biol. Chem.*, 257, 4583–4587.

Hohmann-Marriott, M.F. and Blankenship, R.E. (2011). Evolution of photosynthesis. *Annu. Rev. Plant Biol,.* 62, 515–548.

Kale, R., Hebert, A.E., Frankel, L.K., Sallans, L., Bricker, T.M. and Pospíšil, P. (2017). Amino acid oxidation of the D1 and D2 proteins by oxygen radicals during photoinhibition of Photosystem II. *Proc. Natl. Acad. Sci.*, 114, 2988–2993.

Kamata, T., Hiramoto, H., Morita, N., Shen, J.R., Mann, N.H. and Yamamoto, Y. (2005). Quality control of photosystem II: An FtsH protease plays an essential role in the turnover of the reaction center D1 protein in *Synechocystis* PCC 6803 under heat stress as well as light stress conditions. *Photochem. Photobiol. Sci.*, 4, 983–990.

Kato, Y., Hyodo, K., and Sakamoto, W. (2018). The Photosystem II repair cycle requires FtsH turnover through the EngA GTPase. *Plant Physiol.*, 178, 596–611.

Kato, Y., Miura, E., Ido, K., Ifuku, K. and Sakamoto, W. (2009). The variegated mutants lacking chloroplastic FtsHs are defective in D1 degradation and accumulate reactive oxygen species. *Plant Physiol.*, 151, 1790–1801.

Kato, Y. and Sakamoto, W. (2018). FtsH protease in the thylakoid membrane: Physiological functions and the regulation of protease activity. *Front. Plant Sci.*, 9, 1–8.

Kato, Y., Sun, X., Zhang, L. and Sakamoto, W. (2012). Cooperative D1 degradation in the Photosystem II repair mediated by chloroplastic proteases in Arabidopsis. *Plant Physiol.*, 159, 1428–1439.

Keren, N., Berg, A., van Kan, P.J.M., Levanon, H. and Ohad, I. (1997). Mechanism of photosystem II photoinactivation and D1 protein degradation at low light: the role of back electron flow. *Proc. Natl. Acad. Sci. USA,* 94, 1579–1584.

Knoppová, J., Sobotka, R., Tichý, M., Yu, J., Konĭk, P., Halada, P. *et al.* (2014). Discovery of a chlorophyll binding protein complex involved in the early steps of Photosystem II assembly in *Synechocystis*. *Plant Cell,* 26, 1200–1212.

Komenda, J. and Barber, J. (1995). Comparison of *psbO* and *psbH* deletion mutants of *Synechocystis* PCC 6803 indicates that degradation of D1 protein is regulated by the Q_B site and dependent on protein synthesis. *Biochemistry,* 34, 9625–9631.

Komenda, J., Barker, M., Kuviková, S., de Vries, R., Mullineaux, C.W., Tichý, M. *et al.* (2006). The FtsH protease Slr0228 is important for quality control of photosystem II in the thylakoid membrane of *Synechocystis* sp. PCC 6803. *J. Biol. Chem.*, 281, 1145–1151.

Komenda, J., Knoppová, J., Krynická, V., Nixon, P.J. and Tichý, M. (2010). Role of FtsH2 in the repair of Photosystem II in mutants of the cyanobacterium *Synechocystis* PCC 6803 with impaired assembly or stability of the $CaMn_4$ cluster. *Biochim. Biophys. Acta - Bioenerg.*, 1797, 566–575.

Komenda, J., Knoppová, J., Sobotka, R., Halada, P., Yu, J., Nickelsen, J. *et al.* (2012a). The Psb27 assembly factor binds to the CP43 complex of Photosystem II in the cyanobacterium *Synechocystis* sp. PCC 6803. *Plant Physiol.*, 158, 476–486.

Komenda, J., and Sobotka, R. (2016). Cyanobacterial high-light-inducible proteins — Protectors of chlorophyll-protein synthesis and assembly. *Biochim. Biophys. Acta — Bioenerg.*, 1857, 288–295.

Komenda, J., Sobotka, R. and Nixon, P.J. (2012b). Assembling and maintaining the Photosystem II complex in chloroplasts and cyanobacteria. *Curr. Opin. Plant Biol.*, 15, 245–251.

Komenda, J., Tichy, M., Prasil, O., Knoppova, J., Kuvikova, S., de Vries, R. *et al.* (2007). The exposed N-terminal tail of the D1 subunit is required for rapid D1 degradation during Photosystem II repair in *Synechocystis* sp. PCC 6803. *Plant Cell,* 19, 2839–2854.

Krynická, V., Shao, S., Nixon, P. J. and Komenda, J. (2015). Accessibility controls selective degradation of photosystem II subunits by FtsH protease. *Nat. Plants* 1, 15168.

Krynická, V., Tichý, M., Krafl, J., Yu, J., Kaňa, R., Boehm, M. *et al.* (2014). Two essential FtsH proteases control the level of the Fur repressor during iron deficiency in the cyanobacterium *Synechocystis* sp. PCC 6803. *Mol. Microbiol.*, 94, 609–624.

Lindahl, M., Spetea, C., Hundal, T., Oppenheim, A.B., Adam, Z. and Andersson, B. (2000). The thylakoid FtsH protease plays a role in the light-induced turnover of the photosystem II D1 protein. *Plant Cell,* 12, 419–431.

Lindahl, M., Tabak, S., Cseke, L., Pichersky, E., Andersson, B. and Adam, Z. (1996). Identification, characterization, and molecular cloning of a homologue of the bacterial FtsH protease in chloroplasts of higher plants. *J. Biol. Chem.*, 271, 29329–29334.

Loll, B., Kern, J., Saenger, W., Zouni, A. and Biesiadka, J. (2007). Lipids in photosystem II: interactions with protein and cofactors. *Biochim. Biophys. Acta,* 1767, 509–519.

Lupinkova, L. and Komenda, J. (2004). Oxidative modifications of the Photosystem II D1 protein by reactive oxygen species : From isolated protein to cyanobacterial cells. *Photochem. Photobiol.*, 79, 152–162.

Mattila, H., Khorobrykh, S., Havurinne, V. and Tyystjärvi, E. (2015). Reactive oxygen species: Reactions and detection from photosynthetic tissues. *J. Photochem. Photobiol. B Biol.* 152, 176–214.

Michoux, F., Ahmad, N., Wei, Z.-Y., Belgio, E., Ruban, A.V. and Nixon, P.J. (2016). Testing the role of the N-terminal tail of D1 in the maintenance of photosystem II in tobacco chloroplasts. *Front. Plant Sci.*, 7, 844.

Mizusawa, N., Tomo, T., Satoh, K. and Miyao, M. (2003). Degradation of the D1 protein of photosystem II under illumination *in vivo*: Two different pathways involving cleavage or intermolecular cross-linking. *Biochemistry,* 42, 10034–10044.

Murata, N. and Nishiyama, Y. (2018). ATP is a driving force in the repair of photosystem II during photoinhibition. *Plant Cell. Environ.*, 41, 285–299.

Nishiyama, Y., Yamamoto, H., Allakhverdiev, S.I., Inaba, M., Yokota, A. and Murata, N. (2001). Oxidative stress inhibits the repair of photodamage to the photosynthetic machinery. *EMBO J.*, 20, 5587–5594.

Nixon, P.J., Barker, M., Boehm, M., de Vries, R. and Komenda, J. (2005). FtsH-mediated repair of the photosystem II complex in response to light stress. *J. Exp. Bot.*, 56, 357–363.

Nixon, P.J., Michoux, F., Yu, J., Boehm, M. and Komenda, J. (2010). Recent advances in understanding the assembly and repair of photosystem II. *Ann. Bot.*, 106, 1–16.

Nixon, P.J., Trost, J.T. and Diner, B.A. (1992). Role of the carboxy-terminus of polypeptide D1 in the assembly of a functional water-oxidizing manganese cluster in photosystem II of the cyanobacterium *Synechocystis* sp. PCC 6803:assembly requires a free carboxyl group at C-terminal position 344. *Biochemistry,* 31, 10859–10871.

Ohad, I., Kyle, D.J. and Arntzen, C.J. (1984). Membrane protein damage and repair: removal and replacement of inactivated 32-kilodalton polypeptides in chloroplast membranes. *J. Cell Biol.* 99, 481–485.

Ohnishi, N., Allakhverdiev, S.I., Takahashi, S., Higashi, S., Watanabe, M., Nishiyama, Y. *et al.* (2005). Two-step mechanism of photodamage to photosystem II: Step 1 occurs at the oxygen-evolving complex and step 2 occurs at the photochemical reaction center. *Biochemistry,* 44, 8494–8499.

Orf, G. S., Gisriel, C. and Redding, K.E. (2018). Evolution of photosynthetic reaction centers: insights from the structure of the heliobacterial reaction center. *Photosynth. Res.*, 138, 1–27.

Park, Y.-I., Anderson, J.M. and Chow, W. (1996). Photoinactivation of functional photosystem II and D1-protein synthesis *in vivo* are independent of the modulation of the photosynthetic apparatus by growth irradiance. *Planta,* 198, 300–309.

Pathak, V., Prasad, A. and Pospíšil, P. (2017). Formation of singlet oxygen by decomposition of protein hydroperoxide in photosystem II. *PLoS One,* 12, 1–17.

Pospíšil, P. (2012). Molecular mechanisms of production and scavenging of reactive oxygen species by photosystem II. *Biochim. Biophys. Acta - Bioenerg*., 1817, 218–231.

Pospíšil, P. (2016). Production of reactive oxygen species by Photosystem II as a response to light and temperature stress. *Front. Plant Sci.* 7, 1–12.

Pospíšil, P., Arató, A., Krieger-Liszkay, A. and Rutherford, A.W. (2004). Hydroxyl radical generation by photosystem II. *Biochemistry,* 43, 6783–6792.

Powles, S.B. (1984). Photoinhibition of photosynthesis induced by visible light. *Annu. Rev. Plant Physiol*., 35, 15–44.

Puchades, C., Rampello, A.J., Shin, M., Giuliano, C.J., Wiseman, RL., Glynn, S.E. *et al.* (2017). Structure of the mitochondrial inner membrane AAA+ protease YME1 gives insight into substrate processing. *Science,* 358, eaao0464.

Rochaix, J.D., Lemeille, S., Shapiguzov, A., Samol, I., Fucile, G., Willig, A. *et al.* (2012). Protein kinases and phosphatases involved in the acclimation of the photosynthetic apparatus to a changing light environment. *Philos. Trans. R. Soc. B Biol. Sci*., 367, 3466–3474.

Rutherford, A.W. and Faller, P. (2003). Photosystem II: Evolutionary perspectives. *Philos. Trans. R. Soc. B Biol. Sci*., 358, 245–253.

Sacharz, J., Bryan, S.J., Yu, J., Burroughs, N.J., Spence, E.M., Nixon, P.J. *et al.* (2015). Sub-cellular location of FtsH proteases in the cyanobacterium *Synechocystis* sp: PCC 6803 suggests localised PSII repair zones in the thylakoid membranes. *Mol. Microbiol*., 96, 448–462.

Shao, S., Cardona, T. and Nixon, P.J. (2018). Early emergence of the FtsH proteases involved in photosystem II repair. *Photosynthetica,* 56, 163–177.

Sharma, J., Panico, M., Shipton, C.A., Nilsson, F., Morris, H.R. and Barber, J. (1997). Primary structure characterization of the photosystem II D1 and D2 subunits. *J. Biol. Chem*., 272, 33158–33166.

Silva, P. and Nixon, P.J. (2001). Identification of possible assembly and repair factors in photosystem two preparations of *Synechocystis* sp . PCC 6803 : A new model for D1 turnover. In *PS2001 Proceedings 12th International Congress of Photosynthesis* (Canberra, Australia: CSIRO).

Silva, P., Thompson, E., Bailey, S., Kruse, O., Mullineaux, C.W., Robinson, C. *et al.* (2003). FtsH is involved in the early stages of repair of photosystem II in *Synechocystis* sp. PCC 6803. *Plant Cell,* 15, 2152–2164.

Spetea, C., Hundal, T., Lohmann, F. and Andersson, B. (1999). GTP bound to chloroplast thylakoid membranes is required for light-induced, multienzyme degradation of the photosystem II D1 protein. *Proc. Natl. Acad. Sci. USA,* 96, 6547–6552.

Stewart, D.H., Cua, A., Chisholm, D.A., Diner, B.A., Bocian, D.F. and Brudvig, G.W. (1998). Identification of histidine 118 in the D1 polypeptide of photosystem II as the axial ligand to chlorophyll Z. *Biochemistry,* 37, 10040–10046.

Stirnberg, M., Fulda, S., Huckauf, J., Hagemann, M., Krämer, R. and Marin, K. (2007). A membrane-bound FtsH protease is involved in osmoregulation in *Synechocystis* sp.

PCC 6803: the compatible solute synthesizing enzyme GgpS is one of the targets for proteolysis. *Mol. Microbiol.*, 63, 86–102.

Telfer, A. (2014). Singlet oxygen production by PSII under light stress: Mechanism, detection and the protective role of β-carotene. *Plant Cell Physiol,*. 55, 1216–1223.

Tracewell, C.A. and Brudvig, G.W. (2008). Multiple redox-active chlorophylls in the secondary electron-transfer pathways of oxygen-evolving photosystem II. *Biochemistry,* 47, 11559–11572.

Tyystjärvi, E. and Aro, E.M. (1996). The rate constant of photoinhibition, measured in lincomycin-treated leaves, is directly proportional to light intensity. *Proc. Natl. Acad. Sci. USA,* 93, 2213–2218.

Umena, Y., Kawakami, K., Shen, J.R. and Kamiya, N. (2011). Crystal structure of oxygen-evolving photosystem II at a resolution of 1.9Å. *Nature,* 473, 55–60.

Vasil'ev, S., Brudvig, G.W. and Bruce, D. (2003). The X-ray structure of photosystem II reveals a novel electron transport pathway between P680, cytochrome b559 and the energy-quenching cation, Chl_Z^+. *FEBS Lett.*, 543, 159–163.

Vass, I. (2012). Molecular mechanisms of photodamage in the Photosystem II complex. *Biochim. Biophys. Acta - Bioenerg.*, 1817, 209–217.

Vavilin, D. and Vermaas, W. (2007). Continuous chlorophyll degradation accompanied by chlorophyllide and phytol reutilization for chlorophyll synthesis in *Synechocystis* sp. PCC 6803. *Biochim. Biophys. Acta,* 1767, 920–9.

Wei, X., Su, X., Cao, P., Liu, X., Chang, W., Li, M. *et al.* (2016). Structure of spinach photosystem II-LHCII supercomplex at 3.2 Å resolution. *Nature,* 534, 69–74.

Weisz, D.A., Gross, M.L. and Pakrasi, H.B. (2017). Reactive oxygen species leave a damage trail that reveals water channels in Photosystem II. *Sci. Adv.*, 3, eaao3013.

Yang, Y., Guo, R., Gaffney, K., Kim, M., Muhammednazaar, S., Tian, W. *et al.* (2018). Folding-degradation relationship of a membrane protein mediated by the universally conserved ATP-dependent protease FtsH. *J. Am. Chem. Soc.*, 140, 4656–4665.

Yao, D.C.I., Brune, D.C., and Vermaas W.F.J. (2012). Lifetimes of photosystem I and II proteins in the cyanobacterium *Synechocystis* sp. PCC 6803. *FEBS Lett.*, 586, 169–173.

Yu, J., Knoppová, J., Michoux, F., Bialek, W., Cota, E., Shukla, M.K. *et al.* (2018). Ycf48 involved in the biogenesis of the oxygen-evolving photosystem II complex is a seven-bladed beta-propeller protein. *Proc. Natl. Acad. Sci. USA*, 201800609.

Zhang, L., Paakkarinen, V., van Wijk, K.J. and Aro, E. M. (1999). Co-translational assembly of the D1 protein into photosystem II. *J. Biol. Chem.*, 274, 16062–16067.

Chapter 17

DFT Investigations on the Reaction Mechanisms in Photosynthetic Oxygen Evolution

Yu Guo
KTH Royal Institute of Technology
SE-100 44 Stockholm, Sweden
yg2@kth.se

Photosynthetic water oxidation and oxygen evolution are fundamental biological processes in nature, with significant and challenging scientific issues centering on the oxygen-evolving complex, as well as some conundrums and controversies in this field. Theoretical and computational approaches are very important to elucidate the inherent mechanisms, which should also be conducive for the development of the artificial functional materials applied to water splitting. Based on the cutting-edge progress and using density functional theory as the tool, systematic investigations were carried out from the aspects of interaction between ammonia and the Mn cluster, and the mechanism of O–O bond formation. The related achievements make sense to the mechanism of photosynthetic oxygen evolution.

1. Introduction

Nature endows the oxygen-evolving complex (OEC) with the exceptional talent of water splitting under illumination, which constitutes the primary biochemical events of photosynthesis and prepares for the metabolism of substance and energy in the following carbon fixation [Barber, 2009, 2016a, 2016b; Cox *et al.*, 2013;

Dau *et al.*, 2012; Mcevoy and Brudvig, 2006]. The biological process is attached with molecular oxygen release to the atmosphere, a function of great significance to life activities on the earth. Embedded in photosystem II (PSII), the natural catalyst is actually a metalloenzyme centered by Mn_4CaO_5 cluster undergoing the "S_i-state cycle" (i = 0–4) to fulfill water oxidation and oxygen evolution, completed through the storage of incremental oxidizing equivalents driven by light [Kok *et al.*, 1970]. Humans have been improving the X-ray resolution of PSII through decades, trying to obtain a clearer structure, as well as clarification on the mechanism of water splitting [Ferreira *et al.*, 2004; Guskov *et al.*, 2009; Kamiya and Shen, 2003; Suga *et al.*, 2017; Umena *et al.*, 2011; Young *et al.*, 2016; Zouni *et al.*, 2001].

NH_3 and H_2O are isoelectronic species and structural analogues, leading to similar electronic characteristics, nucleophilicity, and accessibility to the OEC. It has been proven that there are two individual sites for NH_3 binding: one of them near Cl^- and S-state independent, while the other binding to Mn only upon S_2 state formation but to be cast off subsequently between the following S_3 and S_1 states [Andréasson *et al.*, 1988; Beck and Brudvig, 1986; Beck *et al.*, 1986; Boussac *et al.*, 1990; Boussac *et al.*, 2000; Britt *et al.*, 1989; Chu *et al.*, 2004; Dau *et al.*, 1995; Evans *et al.*, 2005; Kim *et al.*, 1990; Pokhrel and Brudvig, 2014]. The key difference is that unlike the Cl^- site, NH_3 binding to Mn does not inhibit O_2 production activity, and the exchange kinetics of substrate water molecules in the S_2 state is almost immune to NH_3 involvement. In other words, the position occupied by NH_3 definitely does not originate from a substrate water molecule. Consequently, studies on NH_3 binding to Mn site have important implications on the identification of substrate water molecules and O_2 creation style.

Recently, the S_3-state OEC has been successfully deciphered as four Mn in 4+ formal oxidation state and local octahedral configuration [Cox *et al.*, 2014], which is a powerful clue to uncover the mystery of subsequent oxygen creation [Britt and Oyala, 2014]. However, subject to the instantaneity of the S_4 state, it is difficult to be captured by any traditional experimental technique, in particular the fugacious transition state (TS) of dioxygen formation. To date, there have been a variety of experimental conjectures about O–O bond formation, yet stemming indirectly from the suppositions from other S-states [Nilsson *et al.*, 2014b, 2012] Fortunately, quantum chemical methods can be very helpful to probe into the vital step in a straightforward way [Vogiatzis *et al.*, 2018].

In this chapter, I present my mechanistic explorations by density functional theory (DFT) calculations on NH_3 binding to the Mn site in the S_2 state [Guo *et al.*, 2016], and comprehensive evaluations on O–O bond formation in the S_4 state [Guo *et al.*, 2017a], from which significance of the structural flexibility of Mn cluster could be reflected [Guo *et al.*, 2017b].

2. Ammonia Binding to the OEC in the S_2 State

Controversies still exist with regard to the interaction way between NH_3 and OEC in the S_2 state, for NH_3-altered g≈2.0 electron paramagnetic resonance (EPR) multiline signal (MLS) [Lohmiller *et al.*, 2014; Pokhrel and Brudvig, 2014; Vinyard *et al.*, 2015]. So far, there have been three kinds of proposals from different experimental techniques. For O5 substitution scheme, NH_3 would become $NH_2^-/NH^{2-}/N^{3-}$ in place of O5, which was initially inferred from electron spin echo envelope modulation (ESEEM) [Britt *et al.*, 1989], extended X-ray-absorption fine structure (EXAFS) [Dau *et al.*, 1995], Fourier transform infrared spectroscopy (FTIR) [Chu, 2013; Chu *et al.*, 2004; Hou *et al.*, 2011], thermoluminescence [Ono and Inoue, 1988] and flash-induced O_2 measurement [Vinyard and Brudvig, 2015]. However, this view has been questioned in recent years and W1 substitution scheme was proposed by EPR & time-resolved membrane inlet mass spectrometry (TR-MIMS) & Q-band ^{1}H-electron nuclear double resonance (ENDOR) [Navarro *et al.*, 2013], ^{14}N nuclear quadrupole coupling tensors [Schraut and Kaupp, 2014], and the recent compelling D1-Asp61 mutation experiment [Oyala *et al.*, 2015]. Afterwards, Mn4 addition scheme appeared by quantum mechanics/molecular mechanics (QM/MM) and EXAFS simulations, where NH_3 replaces neither, but binds as an additional terminal ligand to Mn4 of the g≥4 structure and triggers the "redox switch mechanism" to NH_3-altered g≈2.0 MLS [Askerka *et al.*, 2015; Vinyard *et al.*, 2015].

It has been reported that two almost isoenergetic tautomers in the S_2 state that account for the g≈2.0 MLS and g≥4.1 EPR signals stem from the mobile μ-oxo bridge O5, corresponding to two interconvertible open (S_2-A) and closed-cubane (S_2-B) structures (Figure 1) [Chatterjee *et al.*, 2016; Pantazis *et al.*, 2012]. This fact of multiformity necessarily diversifies avenues of the interaction between NH_3 and the OEC in the S_2 state. Therefore, taking both S_2-A and S_2-B structures as starting points (except for the Mn4 addition scheme), I have investigated the three prevalent proposals for NH_3 binding to the OEC based on the designed reaction mechanisms, with the common ground that all the ultimate products belonging to the open-cubane forms with the formal oxidation state Mn1–Mn4 (abbreviation for Mn1, Mn2, Mn3, Mn4) (III, IV, IV, IV) in accordance with the experimentally observed NH_3-bound g≈2.0 modified MLS. Note that the selectivity of EPR signal is definitely the oxidation state/spin coupling pattern, rather than the geometry, however, the requirement of the open-cubane structure here is based on my finding that the oxidation state Mn1–Mn4 (III, IV, IV, IV) cannot exist in any closed-cubane structure in the S_2 state, including the NH_3 addition form as the sixth ligand of Mn4 (but Mn1–Mn4 (IV, IV, IV, III) could exist in an open-cubane with an extra hydroxide binding to Mn1(IV), see discussions in Mn4 addition scheme). This is

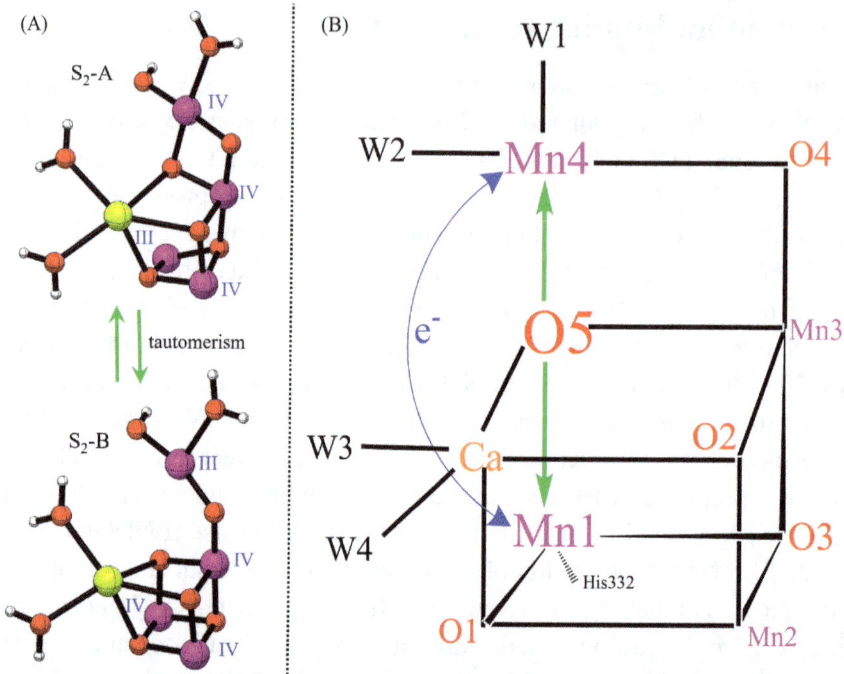

Figure 1. Two tautomers A and B rooted in the interconversion of Mn_4CaO_5 cluster in the S_2 state with its atomic labels used in this paper. All amino acid ligands are not shown for clarity (the same below).

the reason why I take the open-cubane structure as the restriction of NH_3-altered $g \approx 2.0$ MLS.

2.1. *O5 substitution scheme*

Both open and closed-cubane forms of the S_2 state contain a pentacoordinate Mn(III), which makes it possible for NH_3 binding to its vacant coordination site as the sixth ligand for further transformations. This has prompted me to look for logical and suitable routes for NH_3 converting to the nitrogen bridge between Mn3 and Mn4 in the altered $g \approx 2.0$ MLS structure with both open and closed-cubane structures as starting points.

2.1.1. *Closed-cubane structure as the reactant*

From a series of experimental observations, NH_3-altered $g \approx 2.0$ MLS in the S_2 state is most likely to come from $g \geq 4.1$ signal in pace with the increase of temperature and NH_3 concentration, which corresponds to the closed-cubane structure

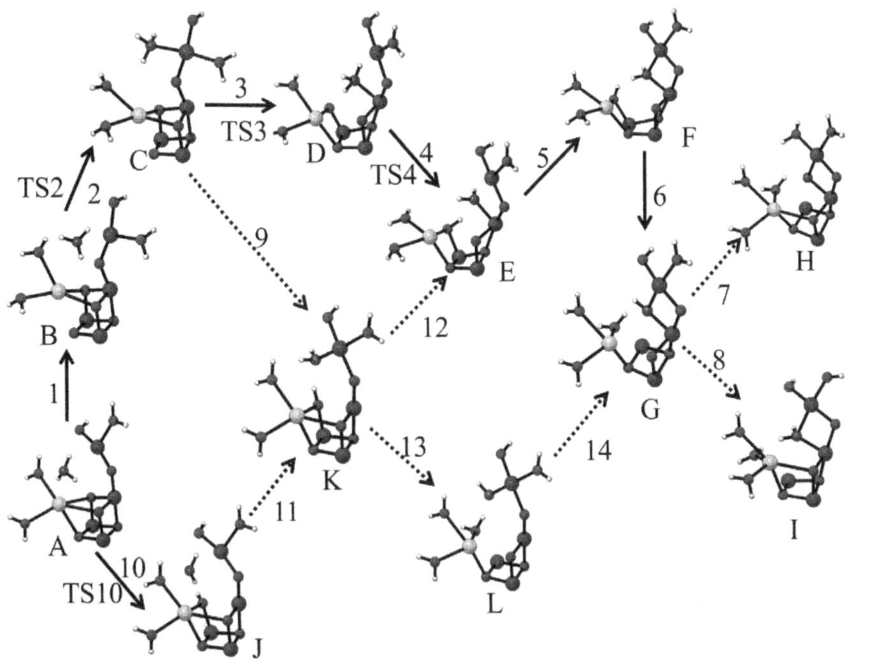

Scheme 1. Diagrammatic presentations for the proposed conversion steps from the g≥4.1 EPR signal to NH_3-altered g≈2.0 MLS structure in the O5 substitution mechanism.

as the reactant [Pokhrel and Brudvig, 2014]. I have constructed three conceivable pathways that should be adequate for this situation (Scheme 1) and studied some key steps to judge its feasibility.

Figure 2 shows the outline of Gibbs free energy tendency chart for this sequence of steps. Once binding to Mn4, NH_3 holds a relatively advantageous position approaching Mn3 in order to break the Mn3–O5 bond (step 3). This Mn3 ligand exchange event is essential for O5 substitution mechanism because it is a direct implementation of NH_3 entering the interior of the OEC through edging O5 out and this bond destruction is necessary for NH_3 becoming the nitrogen bridge afterwards. However, the activation free energy of this step is computed to be more than 30 kcal/mol, in addition to its endothermal change ~20 kcal/mol. Subsequently, NH_3 easily deprotonates to O5 and becomes NH_2^- group. For the sake of reinstitution of bonding interaction between Mn4 and NH_3, I did a relaxed potential energy surface (PES) scan and detected a TS with a negligible activation free energy barrier (about 1 kcal/mol), and it is connected with two almost isoenergetic stationary points E and F. Thus F with amido bridge has been produced, but it does not belong to the g≈2.0 signal because of its unchanged valence state distribution. Then I tried to optimize G with imido bridge linking Mn3 and Mn4 and $O5H_2$

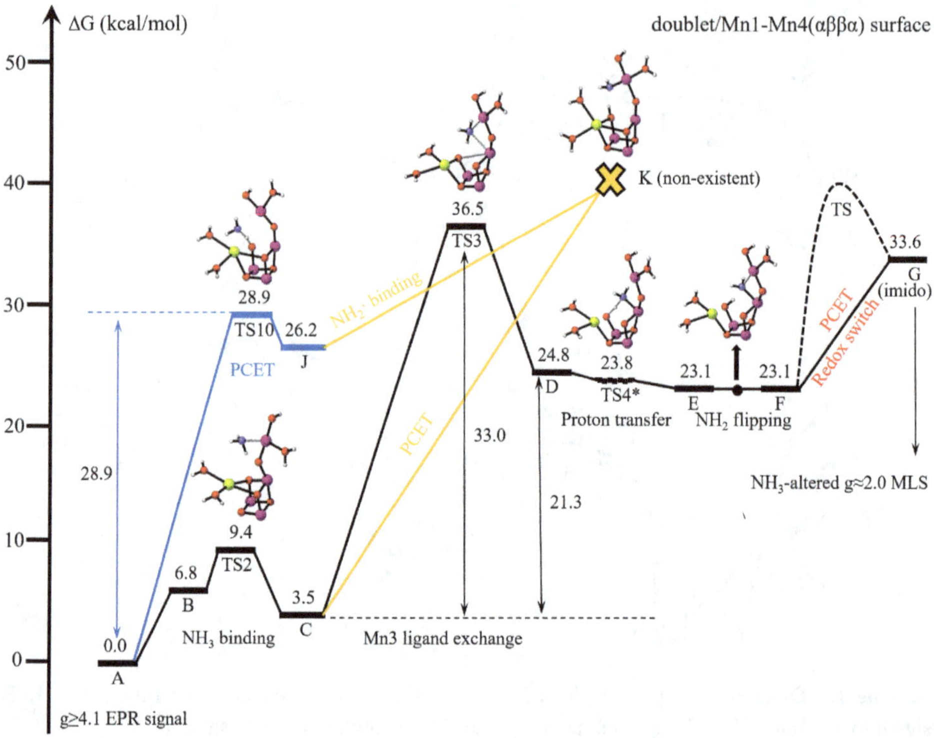

Figure 2. Calculated relative Gibbs free energy profile (in kcal/mol) for the substance transformation from g ≥ 4.1 EPR signal to NH_3-altered g ≈ 2.0 MLS in the O5 substitution mechanism ($\alpha\beta\beta\alpha$ denotes the spin coupling way of Mn1–Mn4. Only one spin surface is typically shown because of the likeness among all the cases, similarly hereinafter.)

ligand to Ca, finding valence rearrangement from Mn1–Mn4 (IV, IV, IV, III) to (III, IV, IV, IV). Hence reaction F to G (step 6) can be interpreted as a proton-coupled electron transfer (PCET) mechanism with proton transfer (PT) from NH_2^- to $O5H^-$ and electron transfer (ET) from Mn4 to Mn1, acting as the redox switch of spin-allowed *d–d* transition between Mn(III) and Mn(IV). Thus, product G with imido bridge and valence states Mn1–Mn4 (III, IV, IV, IV) is in line with the experimental observation in terms of NH_3-altered g ≈ 2.0 MLS. Further, the imido group may convert to nitrido (H) or amido (I) group through deprotonation (step 7) or protonation (step 8) with the nearby W2 and W1 ligands, reconciling with some suppositions from experiments [Pokhrel and Brudvig, 2014].

Focusing back on the intermediate C, PCET of PT from NH_3 to O5 accompanied by ET from Mn4 to Mn3 (step 9) may also be the first means weakening the Mn3–O5 bond, which is followed by Mn3 ligand exchange between NH_2^- and $O5H^-$ (step 12), producing E or directly F. However, any attempts to locate the TS of step 9 failed, so I resorted to a relaxed PES scan monitoring the PT process. Surprisingly, the energy diagram presents an upward trend throughout with the increasing N–H distance even when the proton gets into the bonding area of O5 (beyond 25 kcal/mol than C). During the scan period, the root mean square (RMS) gradient norm never approaches close enough to zero and tentative optimization of the hypothetical D will automatically turn back to C. This fully proves that the imaginary intermediate K is extremely unstable and perceived to be non-existent. Besides, protonation of O5 might happen (step 10) before NH_3 binding to Mn4 and was found to be a PCET reaction. Nevertheless, it is a hard reaction in virtue of its high barrier close to 30 kcal/mol and the thermodynamic disadvantage for the product J. Since routes to break Mn3–O5 bond in the closed-cubane structure by PT and ET are blocked, and the pathway of steps 1 ~ 6 suffers from the prohibitively high barriers of step 3, the possibility of the closed-cubane structure as the reactant to NH_3-altered $g \approx 2.0$ MLS is ruled out.

2.1.2. *Open-cubane structure as the reactant*

Although spectroscopic evidence indicates NH_3 binding does not modify D1-His332 signal which does not enable NH_3 binding to Mn1 [Navarro *et al.*, 2013], I take the possibility into account that NH_3 (or its deprotonated derivative) binding to Mn1 exists only in the intermediates which experience instant conversion thus avoiding exerting a detectable influence on D1-His332 signal. Accordingly, I have found solutions for transformations from the wild-type to nitrogen bridged open-cubane structure (Scheme 2).

This situation shares some commonalities with the last one while the most distinguishing characteristic lies in the extra step 7 or 11 as the second redox switch mechanism. It is an intriguing movement for NH^{2-} or NH_2^- (but failure for N^{3-}) which forms imido or amido bridge in the open-cubane structure accompanied by ET from Mn4 to Mn1, resembling the action of shuttling O5 in concomitance with charge transfer in the wild-type Mn_4CaO_5 cluster [Pantazis *et al.*, 2012]. This kind of reactions can only take place when Mn1 and Mn4 are ferromagnetically coupled (FC). The main point is that NH^{2-} or NH_2^- shifting brings out relatively evident inclination of thermodynamic stability within the range of 5.3 ~ 9.3 and 11.3 ~ 12.2 kcal/mol respectively, preferring the closed-cubane to open-cubane structure, in contrast to ~1 kcal/mol energy gap for the wild-type OEC. In the meanwhile, the calculated $\Delta G^{\neq}$ for NH^{2-} and NH_2^- transfer to Mn4 lies in 8.3 ~ 16.3 and 19.4 ~ 20.5 kcal/mol, respectively, compared with ~10 kcal/mol in O5 shifting case. The medium free energy barriers reserve the likelihood that deprotonated ammonia might turn into the bridged group replacing O5 of the $g \approx 2.0$ MLS structure by means of initial NH_3 binding to Mn1.

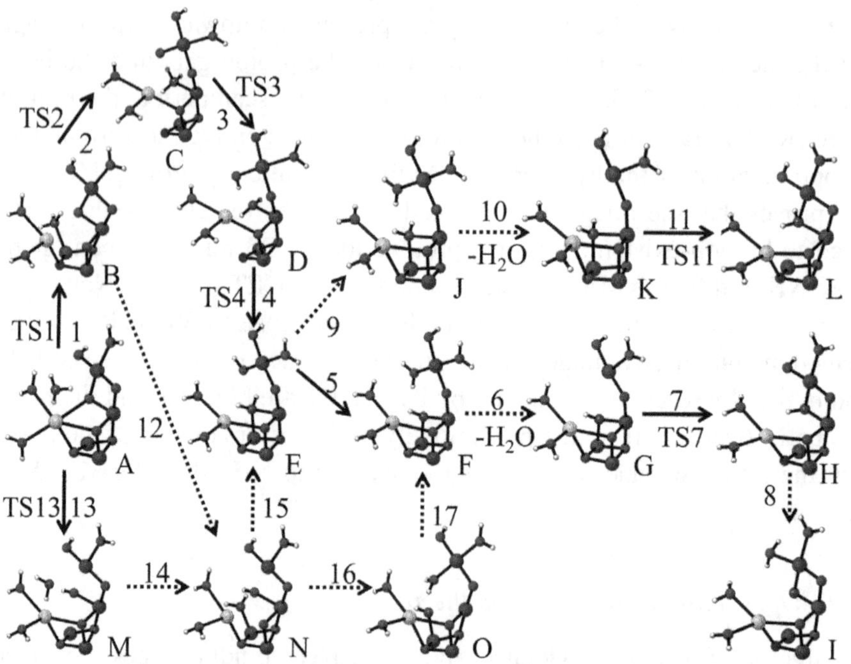

Scheme 2. Diagrammatic presentations for the proposed conversion steps from the wild-type to NH_3-altered g ≈ 2.0 MLS structure in the O5 substitution mechanism.

I have also sketched the Gibbs free energy profiles for these series of steps (Figure 3). Similar to Scheme 1, crucial points may also rest with the extent of the difficulty for NH_3 inserting into the OEC and synchronously squeezing O5 out (step 2), and indeed the free energy barriers are over 30 kcal/mol, although it is endothermic less than 10 kcal/mol. The following two steps of NH_3 deprotonation to O5 (step 3) and NH_2^- flipping (step 4) to bind on Mn1 manifest different energetic landscapes. Specifically, PT from NH_3 to O5 needs a driving force of 5.2 kcal/mol, and NH_2^- flipping is required to overcome a barrier ~10 kcal/mol. The free energy shows a trend of escalation in succession from C to E production. The effective prerequisite to remove O5 from Mn4 and make preparation for the subsequent nitrogen bridge shifting is a PCET reaction, which serves as the first redox switch altering the valence state distributions from the former Mn1–Mn4 (III, IV, IV, IV) to the transient (IV, IV, IV, III) prior to the second redox switch. After NH^{2-} or NH_2^- shifting to Mn4, the open-cubane structure with an imido (H) or amido (L) bridge in agreement with the experimental NH_3-altered g ≈ 2.0 MLS has been synthetized theoretically, and a nitrido bridged substance (I) may form through NH^{2-} deprotonation to W2 (step 8).

Likewise, I have examined the probability of the cleavage of the Mn3/Mn4–O5 bond resulting from PCET with NH_3. For one thing, after NH_3 binding to Mn1, I did not see any stationary point in the relaxed scan curves of PT from NH_3 to O5

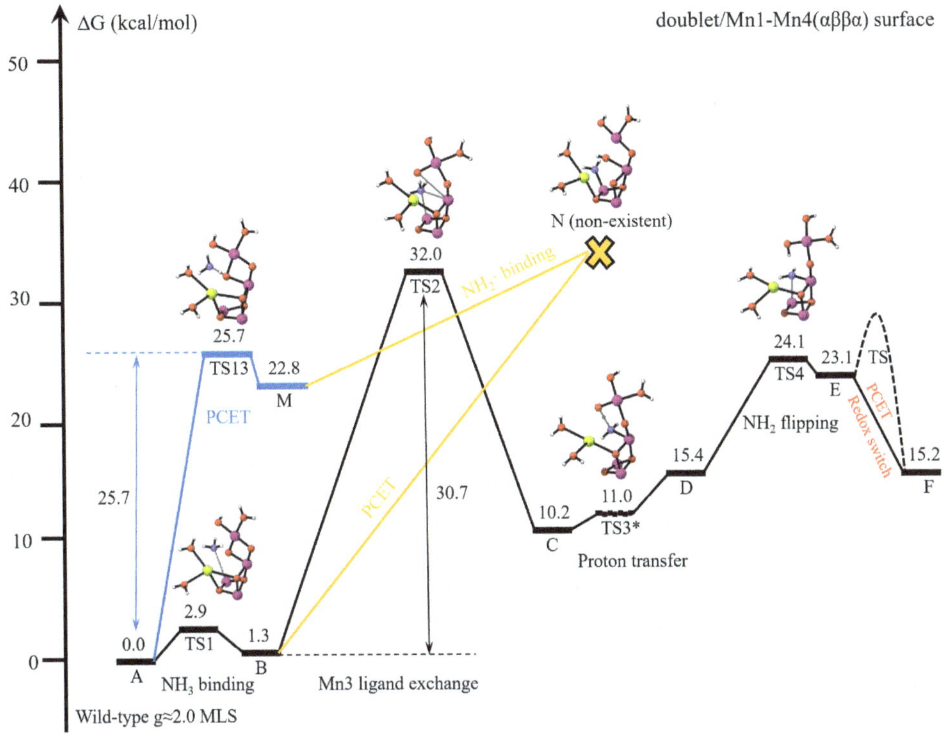

Figure 3. Calculated relative Gibbs free energy profile (in kcal/mol) for the substance transformation from wild-type g ≈ 2.0 MLS to NH_3-altered g ≈ 2.0 MLS in the O5 substitution mechanism.

(step 12) with the dramatically straight rising energy; for another, TS13 of PCET (step 13) ahead of NH_3 binding was exactly found, yet with a disadvantageous reactivity (positive ΔG over 20 kcal/mol and $\Delta G^{\neq}$ between 25 ~ 30 kcal/mol). Accordingly, B is incapable of transferring proton from NH_3 to O5 and N turns out to be non-existent. Similar to Scheme 1, initial PCET ways to disrupt the O5 bridge for NH_3 addition in the open-cubane structure are deemed to be forbidden, and the reaction channel passing by NH_3 direct coordination to Mn3 is banned, I also exclude the possibility of the open-cubane structure as the reactant to NH_3-altered g ≈ 2.0 MLS. Besides, the severe geometric distortion of the cluster in the nitrogen-bridged OEC and the apparent difficulty for ammonia to leave subsequently during the cycle also make the O5 substitution scheme too suspectable to be accepted. Thus, explorations for other alternative mechanisms are necessary.

2.2. *W1 substitution scheme*

Up-to-date spectrum analyses affirmed that NH_3 replaces the terminal ligand W1 rather than the μ-oxo bridge O5 [Navarro *et al.*, 2013; Oyala *et al.*, 2015]. In view

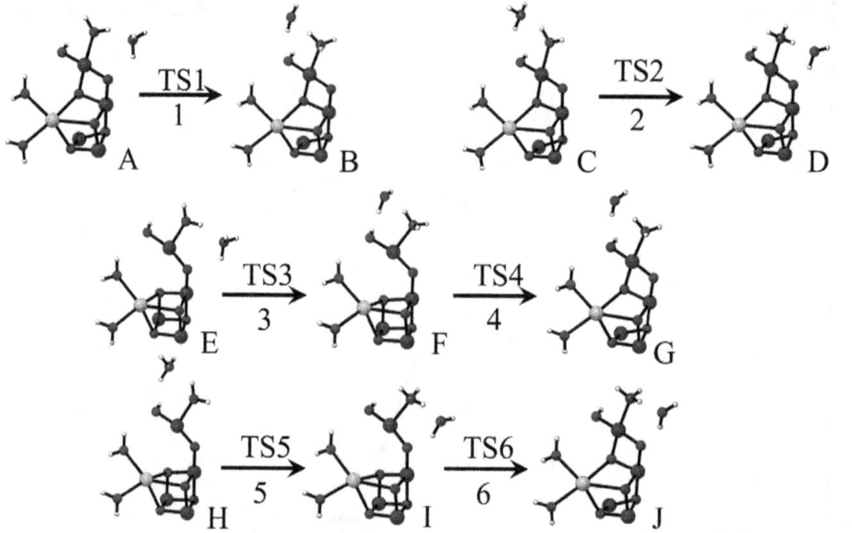

Scheme 3. Graphical expressions for the W1 substitution mechanism both on the open and closed-cubane structures from O4 and W2 sides in the S_2 state.

of the tautomerism in the S_2 state, I performed TS searches for W1 substitution by NH_3 both on the open and closed-cubane structures, the latter of which demands O5 shifting to Mn4 as the redox switch for NH_3-altered $g \approx 2.0$ MLS. At the same time, NH_3 attacking from both O4 and W2 sides were studied (see Scheme 3).

All possible TSs for different spin states have been located and Gibbs free energy changes are expressed in the diagrams (Figures 4 and 5). For W1 substitution on the open-cubane structure, NH_3 attacking processes from both O4 (step 1) and W2 (step 2) sides have similar free energy barriers for all spin states, *i.e.* 7.1 ~ 7.9 and 5.4 ~ 8.0 kcal/mol, respectively. For cases on the closed-cubane structure, a minor distinction of energy ranges emerges concerning NH_3 attacking direction. Concretely, the reaction barriers for O4 side cases (step 3) are 10.4 ~ 11.4 kcal/mol while the range declines to 4.0 ~ 7.9 kcal/mol for W2 side cases (step 5). Once W1 replacement by NH_3 is achieved on the closed-cubane structure, O5 transfer is needed to isomerizate to the almost isoenergetic open-cubane structure for $g \approx 2.0$ MLS, where Mn1 and Mn4 must be FC. Free energy barriers for steps 4 and 6 are calculated to be within the scope of 8.8 ~ 11.2 and 8.6 ~ 9.6 kcal/mol, respectively.

W1 substitution can be easily realized, but in consideration of the O4 channel as the access of substrate analogues [Retegan and Pantazis, 2016], I am more inclined to W1 replacement by NH_3 from O4 side. Given the incessant and efficient

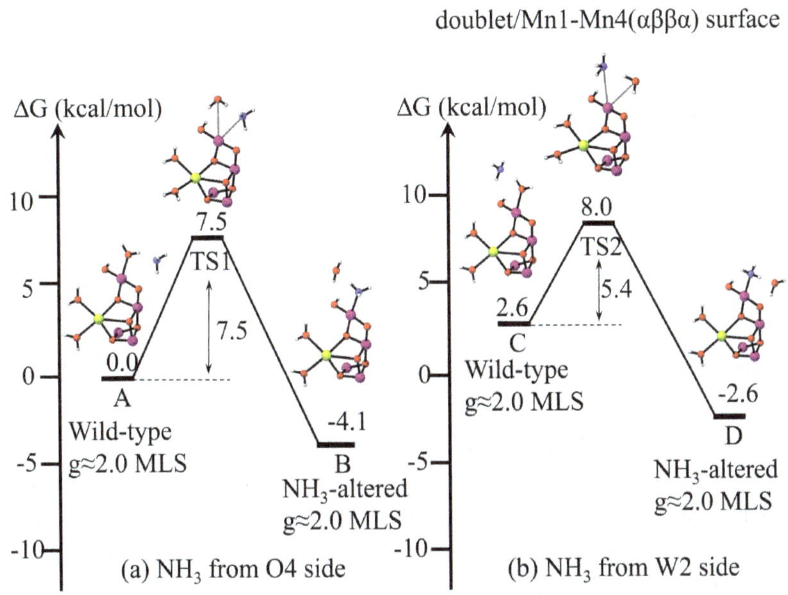

Figure 4. Calculated relative Gibbs free energy profile (in kcal/mol) of the W1 substitution mechanism for NH_3 attacking from O4 (left) and W2 (right) sides on the open-cubane structure from the wild-type g ≈ 2.0 MLS to NH_3-altered g ≈ 2.0 MLS.

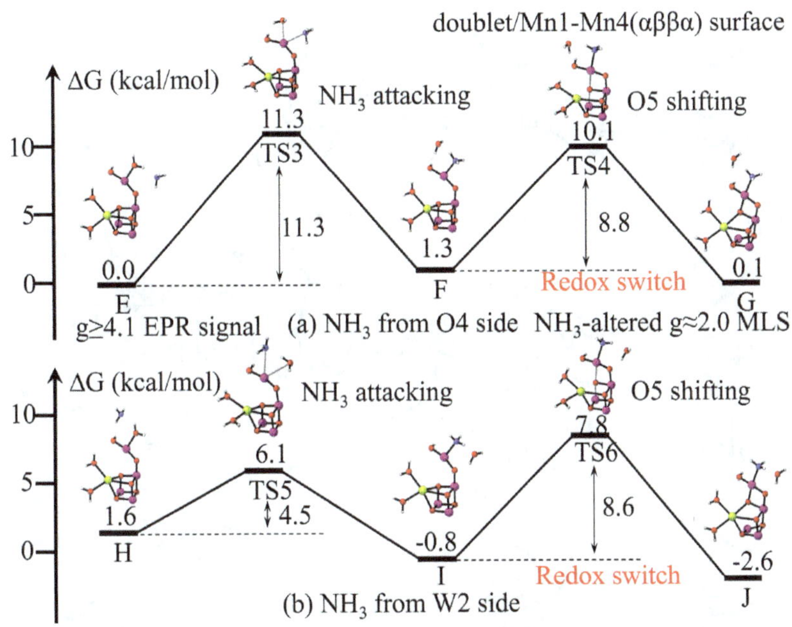

Figure 5. Calculated relative Gibbs free energy profile (in kcal/mol) of the W1 substitution mechanism for NH_3 attacking from O4 (upper) and W2 (lower) sides on the closed-cubane structure from the g ≥ 4.1 EPR signal to NH_3-altered g ≈ 2.0 MLS.

mutual transformation of the isomers before or after NH_3 binding, I recommend equivalent reaction probabilities on different isomers. Evidently, the W1 substitution mechanism manifests great superiority over the O5 substitution mechanism because of the large energy difference, so the latter should be ruled out. This conclusion computationally confirms the recent EPR finding [Oyala *et al.*, 2015] and serves as the theoretical guarantee for NH_3 terminally binding to Mn4 in place of W1.

2.3. *Mn4 addition scheme*

For the sake of coherence with the recent pulse EPR discovery [Oyala *et al.*, 2015], Mn4 addition scheme was proposed by QM/MM calculations and theoretical simulation of EXAFS spectra [Askerka *et al.*, 2015]. It means that instead of substituting W1, NH_3 binds to Mn4 as an additional terminal ligand on the closed-cubane structure, simultaneously with the carousel rearrangement of W1 and W2 around Mn4. Key points also involve PT from W2 to O5 accompanied by stabilization of the formal oxidation states from Mn1–Mn4 (IV, IV, IV, III) to (III, IV, IV, IV) in accordance with NH_3-altered $g \approx 2.0$ MLS. I have probed into its feasibility and investigated other possible routes to the experimental result based on the Mn4 addition rationale (Scheme 4).

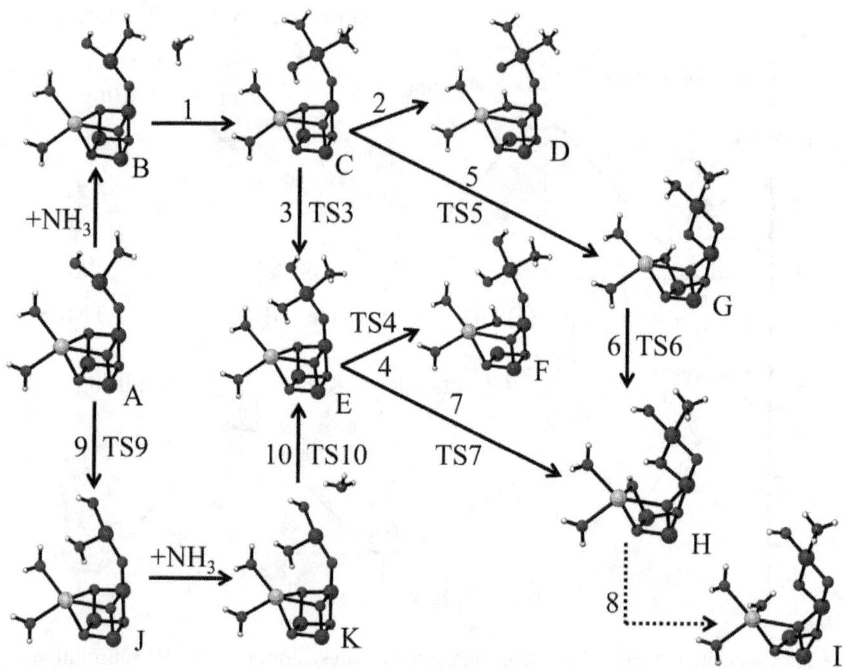

Scheme 4. Diagrammatic presentations for the proposed conversion steps from the $g \geq 4.1$ EPR signal to NH_3-altered $g \approx 2.0$ MLS structure in the Mn4 addition mechanism.

I did attempt to achieve the so-called "redox switch mechanism" but the spin density distributions of Mn could hardly change throughout the processes of NH_3 binding to Mn4 and NH_3 deprotonation to O5, so none of C/D/E/F could be eligible for NH_3-altered g≈2.0 MLS. Actually, the formal oxidation states of Mn1–Mn4 (III, IV, IV, IV) never prevail in a closed-cubane in the S_2 state. In this case, I managed to pursue an effective way for NH_3-altered g≈2.0 MLS through switching to an open-cubane (Figure 6). Step 5 renders an interesting path to actualize transformation between a closed and an open-cubane ($\Delta G^{\neq}$ = 22.3~23.7 kcal/mol). It is actually Mn3 ligand exchange between O5 and W2 (OH^-) assisted by PT from W2 to O5. But G does not satisfy g≈2.0 MLS because spin density distributions are basically consistent with C (nearly 3, 3, 3, 4 for Mn1, Mn2, Mn3, Mn4) of g ≥ 4.1 signal, which is due to the existence of Mn1-bonded $O5H^-$ stabilizing Mn1(IV). Noticing the double protonated form of W1, I am conscious of the possibility of $O5H^-$ changing into $O5H_2$ through two-step PT from W1 to $O5H^-$ *via* W2 (steps 6 and 8). In parallel with step 5, H could also be generated with W2 in the form of water by step 7 ($\Delta G^{\neq}$ = 26.7~27.7 kcal/mol). The final product I with oxidation states Mn1–Mn4 (III, IV, IV, IV) can only be obtained when Mn1

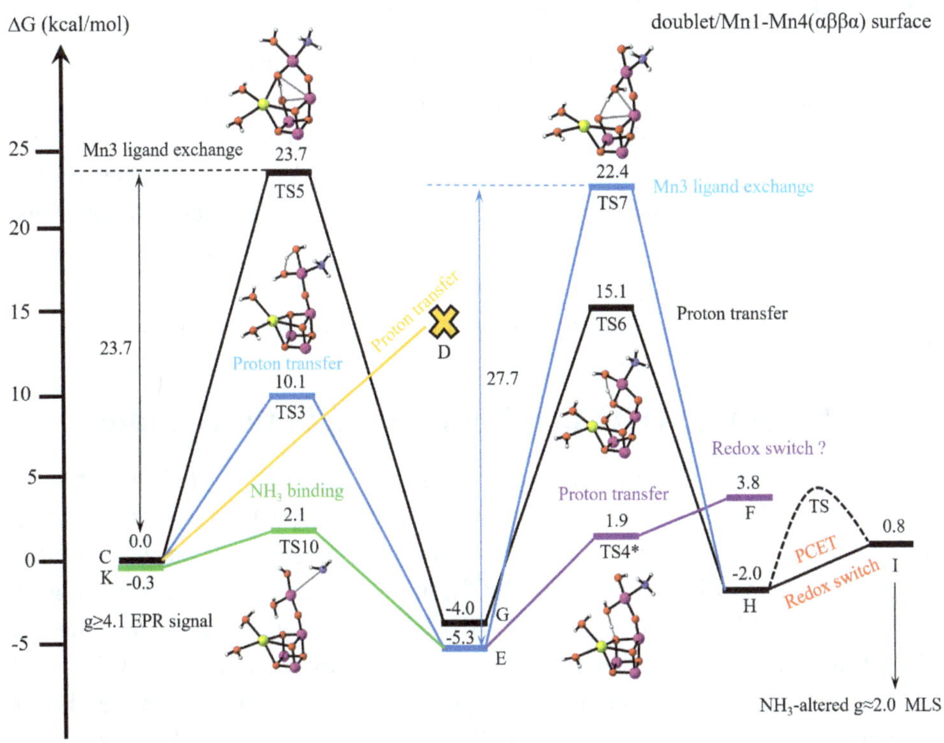

Figure 6. Calculated relative Gibbs free energy profile (in kcal/mol) for the substance transformation from g ≥ 4.1 EPR signal to NH_3-altered g≈2.0 MLS in Mn4 addition mechanism.

and Mn4 are FC. This reflects ET from Mn4 to Mn1 coupled with PT from W2 to $O5H^-$, precisely the ligands of Mn4 and Mn1, conforming with the PCET identity for step 8.

In brief, the experimental NH_3-altered $g \approx 2.0$ MLS could be realized by NH_3 terminal addition to Mn4 of the S_2-B structure, with the necessity of formation of the open-cubane structure through Mn3 ligand exchange between W2 and O5 assisted by proton transfer, and a subsequent PCET as the redox switch mechanism for valence rearrangement. However, compared with the W1 substitution mechanism, the much larger barrier heights (22.3 ~ 27.7 kcal/mol) of the antecedent conversion from the closed to open-cubane form makes the Mn4 addition scheme highly unfavorable. Besides energetics, actually, no direct computational work supports Mn4 addition scheme with respect to the spin state and EPR properties so far [Pantazis, 2018]. Afterwards its infeasibility was experimentally reinforced by Marchiori *et al.*'s ESEEM and ENDOR study through structural-spectrum analysis [Marchiori *et al.*, 2018]. All these demonstrations obviously prefer W1 substitution to NH_3 addition.

2.4. *Implications for oxygen evolution style*

In combination with the recent pulse EPR finding and my theoretical support, I strongly favor W1 substitution by NH_3 in the S_2 state and thus W1 is excluded from the candidates of substrates. This immediately rules out two ways for O_2 creation involving W1 in the S_4 state, *i.e.* W1–W2 and W1–O4 coupling mechanisms. The determination of W1 substitution (rather than O5) retains the possibility of oxo-oxyl mechanism involving O5 [Siegbahn, 2006]. However, this cannot excludes W3–W2 nucleophilic attack mechanism [Barber *et al.*, 2004]. Other possible versions of O–O bond formation are discussed in the next section.

3. Mechanism of O–O Bond Formation in the S_4 State

There is a striking ambiguity that what manner is actually selected by nature to make O_2 in the S_4 state. Generally, the most popular perspectives can be classified as two categories: Siegbahn's radical coupling mechanism between μ-oxo bridge and Mn1(IV)-oxyl [Siegbahn, 2006, 2008], and nucleophilic attack of Ca-bound water (W3) on Mn4(IV)-oxyl/Mn4(V)-oxo advocated by Barber *et al.* [Barber, 2009; Barber *et al.*, 2004; Ferreira *et al.*, 2004; Sproviero *et al.*, 2008; Vinyard *et al.*, 2015]. Siegbahn insists on oxo-oxyl coupling happening inside the cluster by energy considerations from DFT calculations, with the special restriction of spin alignment rule among Mn1-oxyl–oxo-Mn4 [Siegbahn, 2006]. Barber *et al.* early put forward the nucleophilic attack mechanism occurring on the cubane surface based on the structural analysis on the oxygen-evolving center of his significant discovery of PSII architecture in 2004 (for the first time they found Ca^{2+} is ligated to the cubane and may be a binding site for a substrate water molecule as a nucleophile) [Barber *et al.*, 2004; Ferreira *et al.*, 2004]. Furthermore, the recently found striking structural similarity between the Mn_4CaO_5 cluster of OEC and the Fe_4NiS_5 cluster of Fe–Ni carbon monoxide dehydrogenase (CODH) may suggest some mechanistic commonality in both enzyme, *i.e.* nucleophilic attack of Ca^{2+}-bound hydroxyl/water onto highly electrophilic Mn4-bound oxo/oxyl for O–O bond formation in PSII, compared to nucleophilic attack of Fe4-bound hydroxyl/water onto Ni^{2+}-bound CO group for CO_2 evolution in Fe–Ni CODH [Barber, 2017].

Apart from these two contradictory versions as the mainstream, there have been many alternative schemes, some of which match certain experimental speculations [Cox and Messinger, 2013; Kusunoki, 2007; Nilsson *et al.*, 2014a; Saito *et al.*, 2012; Yamaguchi *et al.*, 2012; Yamanaka *et al.*, 2011; Young *et al.*, 2016]. Figure 7 is basically a capsule of all the potentialities of dioxygen formation. For an open-cubane, effective ways are hidden in W_x-O5, W3-W2, W3-W_x, W1-W2, O2-O4, W1-O4, W2-O5 and W3-O5; while for a closed-cubane, interactive substrates may be W_x-O5, W3-W_x, W1-W2, O2-O4, W1-O4, W2-W_x or W3-O5 (Figure 8).

3.1. *Potential energy surface investigations*

The S_4 state was set up based on the proposed oxidation state distributions of Mn1–Mn4 (IV, IV, IV, V) or (IV, IV, IV, IV)-oxyl radical, as a result of the potential uncertainty of metal or ligand-centered oxidation during S_3–S_4 [Britt and Oyala, 2014; Britt *et al.*, 2015; Cox *et al.*, 2013; Rapatskiy *et al.*, 2012]. Multistate reactivity was also considered owing to spin crossover phenomenon. For both open and closed-cubane structures, possible spin topologies are confined to $\alpha\alpha\beta\beta\alpha$, $\alpha\beta\alpha\beta\alpha$, $\alpha\beta\beta\alpha\alpha$, $\beta\alpha\alpha\beta\alpha$, $\beta\alpha\beta\alpha\alpha$, $\beta\beta\alpha\alpha\alpha$ on the doublet surface, $\alpha\alpha\alpha\beta\beta$, $\alpha\alpha\beta\alpha\beta$, $\alpha\beta\alpha\alpha\beta$, $\beta\alpha\alpha\alpha\beta$ on the sextet surface, $\alpha\alpha\alpha\beta\alpha$, $\alpha\alpha\beta\alpha\alpha$, $\alpha\beta\alpha\alpha\alpha$, $\beta\alpha\alpha\alpha\alpha$ on the octet surface, $\alpha\alpha\alpha\alpha\beta$ on the 12-et surface and $\alpha\alpha\alpha\alpha\alpha$ on the 14-et surface. Given single proton release in the S_3–S_4 transition [Haumann *et al.*, 2005], all the possible protonation states were tested for both the reactive groups and the titratable ligands. There are over 500 potential energy surfaces surveyed.

3.2. *The open-cubane oxo-oxyl coupling as the energetically optimal mechanism*

I do not describe the reaction details for all the mechanisms, but state that my statistics show the open-cubane oxo-oxyl coupling has the best thermodynamic superiority over the others. Although hydroxyl is a better nucleophile than water,

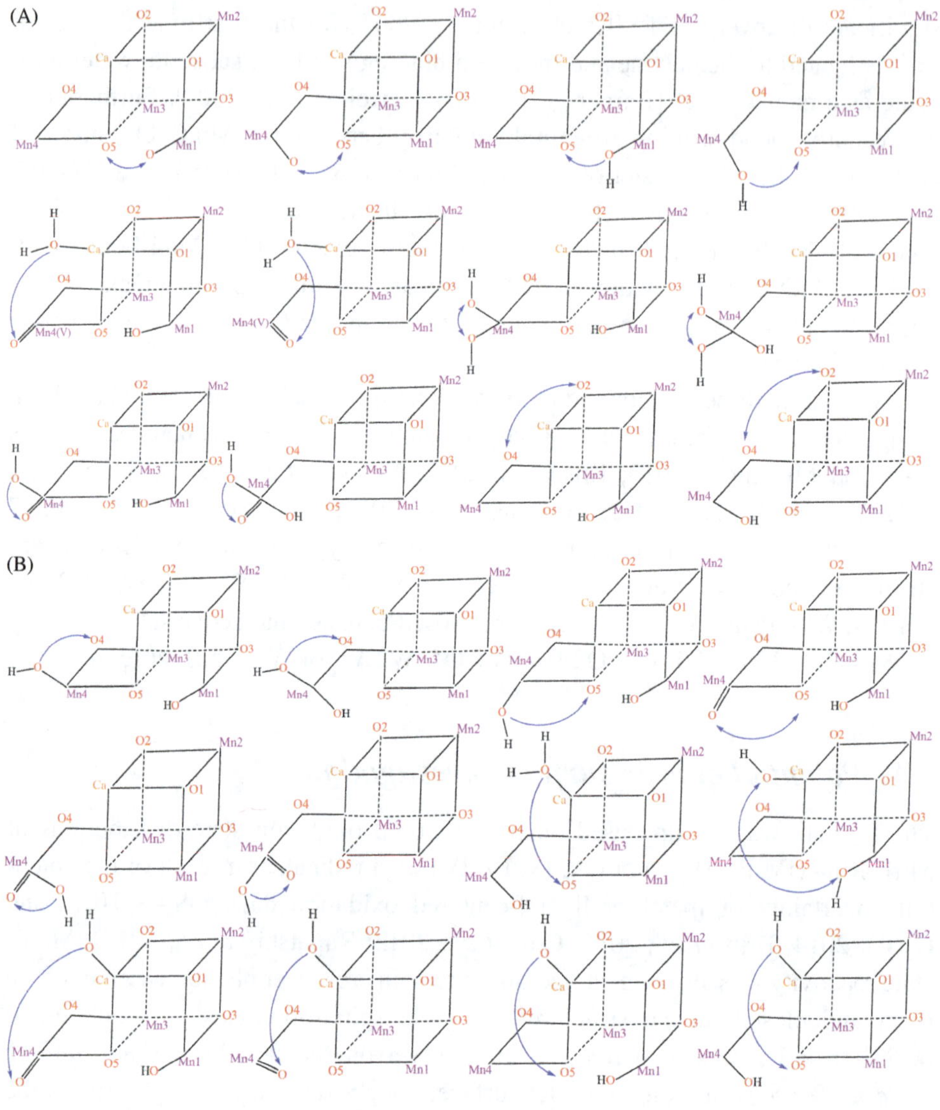

Figure 7. Diagrams of all the possible schemes for O–O bond formation in both open (A) and closed-cubane (B) structures in the S_4 state.

the cases of Ca-bound hydroxyl as the reactant are forbidden because the deprotonation of W3 is rather difficult.

The key species is O10 with obvious radical character (nearly integer 1.0 spins), transferring its unpaired electron to Mn4(IV) through a spin-allowed way and inducing coupling of O10 and O5 to form peroxide product Pro-O_2^{2-}. Therefore, FC between O10 and Mn4 is energetically favorable, which escapes from producing paired electrons in the *d* orbital of Mn4(III) with exorbitant energy. In addition, antiferromagnetically coupled (AFC) Mn1/O10 in a

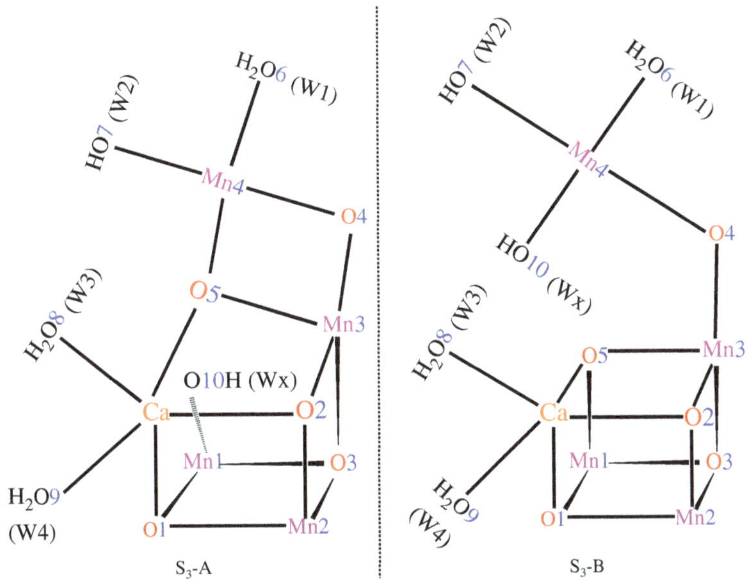

Figure 8. A sketch map for atomic notations for the open (left) and closed-cubane (right) structures in the S_3 state (uniformly in the S_4 state).

low-energy level decides the preferred spin states of reactants. In this way four spin topologies $\beta\alpha\beta\alpha\alpha$, $\beta\beta\alpha\alpha\alpha$, $\alpha\alpha\alpha\beta\beta$ and $\beta\alpha\alpha\alpha\alpha$ stand out as valid candidates, which coincide with Siegbahn's spin alignment rule [Siegbahn, 2006]. Next, the peroxide group experiences two-step ET processes, initially from O10–O5 moiety to Mn1(IV) creating superoxide radical intermediate Int-$O_2^{\cdot-}$ (about 1.0 spins for O10–O5 collectively) and then to Mn3(IV), releasing molecular O_2 and generating Mn1–Mn4 (III, IV, III, III) valence state arrangement for the most reduced S_0 state. Furthermore, $\beta\beta\alpha\alpha\alpha$ and $\beta\alpha\alpha\alpha\alpha$ must be excluded because of singlet $^1O_2$ formation, rather than the normal and more stable triplet $^3O_2$ arising from FC between Mn1 and Mn3 which ought to be highlighted as an effective supplement. It follows that only doublet/$\beta\alpha\beta\alpha\alpha$ and sextet/$\alpha\alpha\alpha\beta\beta$ meet all the qualifications for smooth $^3O_2$ release, with equivalent opportunities due to the undistinguishable energetic priority. Given the ground state septet/$\alpha\alpha\alpha\beta$ for the previous S_3 state and doublet/$\beta\alpha\beta\alpha$ for the later S_0 state [Krewald *et al.*, 2015], it can be deduced that a preceding or subsequent spin crossing is required for O–O bonding on doublet/$\beta\alpha\beta\alpha\alpha$ or sextet/$\alpha\alpha\alpha\beta\beta$ surface, respectively. Low barriers of 5.9 and 5.3 kcal mol^{-1} and driving forces of –15.0 and –14.9 kcal mol^{-1} were obtained for the two spin states, which commendably demonstrates its reactive feasibility (Figure 9).

The oxyl radical can stably bind to Mn1 antiferromagnetically, and transfers its unpaired electron to Mn4 ferromagnetically in the open-cubane structure, which are essential for the low-barrier reactivity (Figure 10, left). On the contrary,

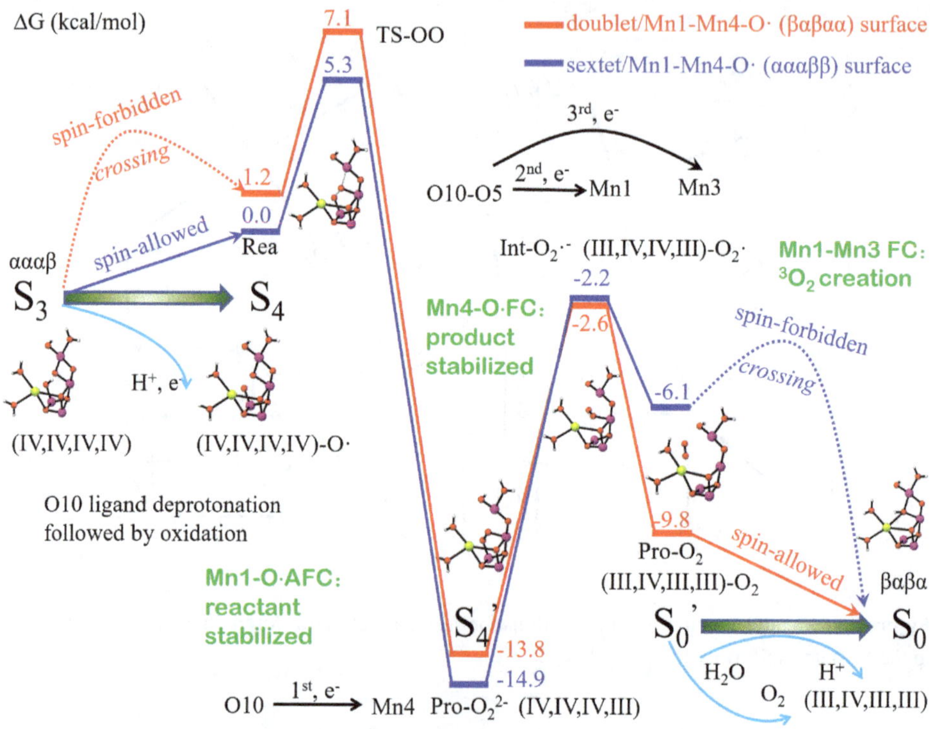

Figure 9. Reaction energetics of O_2 evolution and detailed information for the S_3–S_0 transition under the open-cubane oxo-oxyl coupling mechanism.

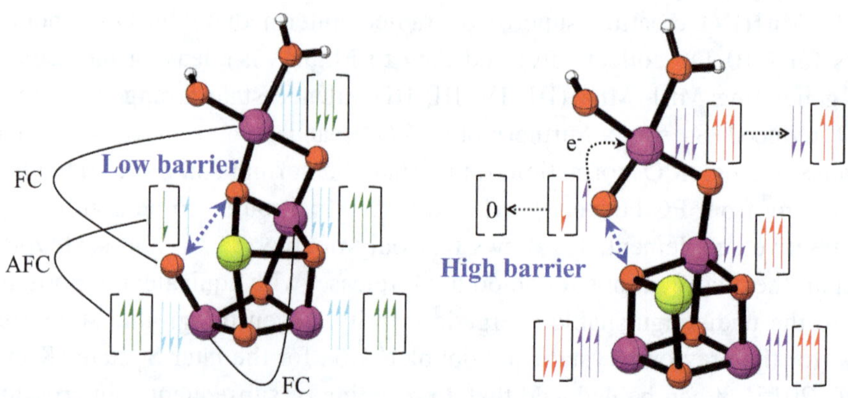

Figure 10. A diagram of magnetic coupling ways of the stable spin states for the oxo-oxyl coupling mechanism in the open (left) and closed-cubane (right) structures.

however, an awkward electronic situation exists in the closed-cubane oxo-oxyl coupling mechanism. A prerequisite for low-barrier reactivity is the radical character of Mn4-bound oxyl, but it can only be FC to Mn4 and this causes severe instability of the reactants. Alternatively, AFC Mn4 and oxyl, although much lower in energy, loses its radical character due to a local ET from oxyl radical to Mn4 and this increases the reaction barrier dramatically (Figure 10, right). Similarly, the situation is also found in the nucleophilic attack mechanism of Ca-bound water (W3) onto Mn4(IV)-bound oxyl (W2), and the electronic effect is the key negative factor to be disfavor. This could be another factor against W3–W2 nucleophilic attack for O–O bond formation besides the effect of high-energy protonated peroxide product [Siegbahn, 2017].

3.3. *Correlations to recent experimental revelations*

Recently, structural changes from the X-ray free electron laser (XFEL) experiment clearly show an oxygen insertion on Mn1 close to O5, with other displacements of waters and ligands during the S_1–S_3 transition [Suga *et al.*, 2017]. This provides strong evidence in agreement with my computational support for the open-cubane oxo-oxyl coupling mechanism in the S_4 state. In addition, the Fourier map also indicates that the crystal water W665 disappears from its original position upon 2-flash illumination, accompanying with W567 moving towards O4 during S_1–S_3. Particularly, in the light of the pivot mechanism for water binding to Mn4 in S'_3[Retegan *et al.*, 2015], I speculate that W567 is actually the new water binding to Mn4 and occupying W1 site after the carrousel rearrangement of water ligands, serving as one of substrates in the next cycle. Cleavage of the hydrogen bond between W567 and O4, which is attributed to the decrease of pK_a of O4 upon oxidation of Mn4 from (III) to (IV) [Vinyard *et al.*, 2017], could be a necessary factor facilitating W567 coordination to the vacant sphere of Mn4. Simultaneously, W665, hydrogen-bonded to W567, is pulled closer to the cofactor by the movement of W567 and locates near the preceding site of W567. The hydrogen bond between W665 and W542 (connected to Asn87) was broken early and W665 forms a short hydrogen bond with O4 due to PT to a nearby Brønsted base, possibly the titratable guanidine group of CP43–Arg357 (Figure 11).

Accordingly, the two substrates in S_4 are descended from W2 and O5 as the fast and slow exchanging substrate water molecules (W_f and W_s) in S_2. This accounts for the source of superficial W665 displacement and W567 shifting to O4 reflected from XFEL in S_3, *i.e.* W665 "becoming" W567 in its deprotonated form, while W567 binding to Mn4 and 'becoming' W1 with rotation of its water ligands. Moreover, the S-cycle advancement, achieved through the structural flexibility and polymorphism of the Mn_4CaO_5 cluster, conforms with the TR-MIMS findings on substrate exchanging kinetics [Cox and Messinger, 2013; Hendry and Wydrzynski, 2002; Nilsson *et al.*, 2014a], which indicate W_f and W_s are already bound in S_2 and they are in similar chemical environment between the S_2 and S_3 states as the exchange rates are not widely divergent towards S_3.

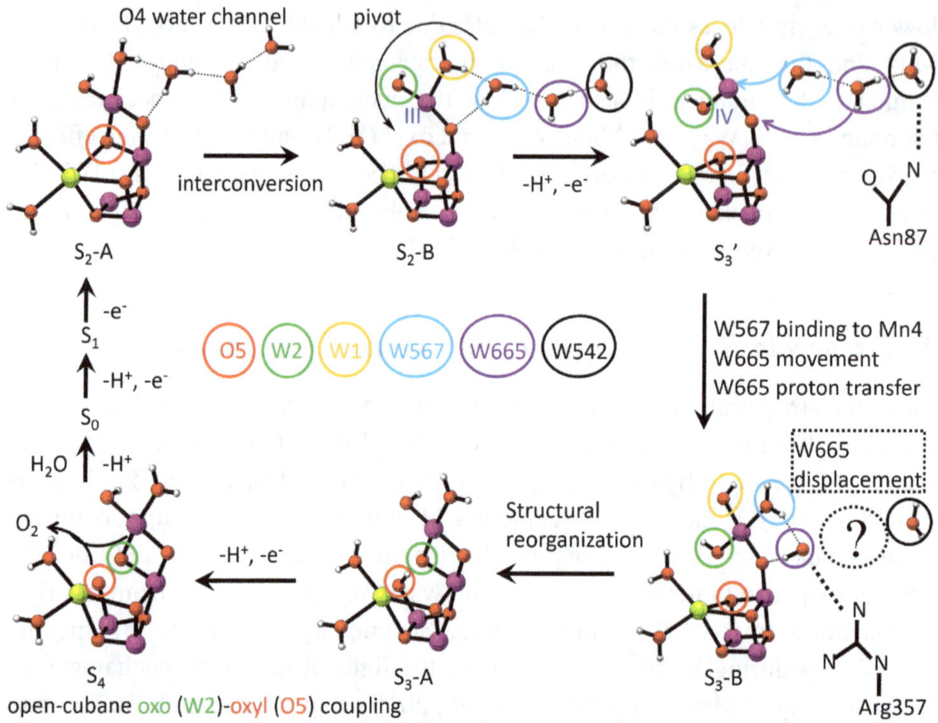

Figure 11. A graphic interpretation for the theoretical ratiocination on the observed structural changes of the light-induced XFEL experiment for the S_1–S_3 transition.

My current calculations are in line with Siegbahn's idea of O–O bond formation [Siegbahn, 2006], Suga *et al.*'s XFEL observation [Suga *et al.*, 2017] and Kern *et al.*'s most recent discovery [Kern *et al.*, 2018]. However, it should be noted that beyond energetic views from DFT studies, the nucleophilic attack mechanism suggested by Barber *et al.* has also gained wide support by other experimental and theoretical investigations [Gao *et al.*, 2009; Gupta *et al.*, 2015; Sproviero *et al.*, 2008; Vinyard *et al.*, 2015], and very recently there appear some new schemes regarding O_2 evolution in PSII [Corry and O'malley, 2018; Kawashima *et al.*, 2018; Pushkar *et al.*, 2018; Shoji *et al.*, 2018a; Vaddypally *et al.*, 2017; Zhang *et al.*, 2018; Zhang and Sun, 2018]. Let's keep an open mind on all these proposals and wish to see a definite conclusion in the near future.

4. Appendix

All the DFT calculations are based on B3LYP*/LANL2DZ(Mn,Ca), 6-31G*(C,H,O,N) for geometry optimizations and frequencies calculations and

B3LYP*/LANL2TZ(f)(Mn), LANL08(Ca), cc-pvtz(-f)(C,H,O,N), SMD(eps = 6.0), DFT-D3(BJ-damping) for single point energies.

Annotations: B3LYP*: a minor revision of B3LYP (3-parameter hybrid Becke exchange/Lee-Yang-Parr correlation functional) with Hartree-Fock (HF) exact exchange reduced from 20% to 15%. LANL2DZ: an effective core potential (ECP) basis set called Los Alamos National Laboratory 2 double-ζ. 6-31G*: Pople's split-valence double-zeta basis with added *d* polarization functions on non-hydrogen atoms. LANL2TZ(f): an ECP basis set higher than LanL2DZ with triple-ζ quality and polarization *f* functions. LANL08: a completely uncontracted basis set derived from the original Hay-Wadt set corresponding to triple-ζ valence. cc-pvtz(-f): correlation consistent polarized valence triple-ζ without polarization f functions. SMD: (Solvation, Minnesota, Density) a universal continuum solvation model based on the quantum mechanical charge density of a solute molecule interacting with a continuum description of the solvent. Eps = 6.0: dielectric constant set to 6.0 mimicking the protein matrix. DFT-D3(BJ-damping): Grimme's third-generation DFT dispersion correction with Becke-Johnson damping. For more computational details, please refer to the papers [Guo *et al.*, 2016, 2017b].

Acknowledgments

I dedicate this chapter to the memory of my PhD supervisor, Professor Zhong-Zhi Yang of Liaoning Normal University. I would like to acknowledge distinguished Professor James Barber FRS of Imperial College London for a short-term academic partnership, and for providing me the great opportunity to write this chapter.

References and Citations

Andréasson, L.-E., Hansson, Ö. and Schenck, K.V. (1988). The interaction of ammonia with the photosynthetic oxygen-evolving system. *Biochim. Biophys. Acta Bioenerg.*, 936, 351–360.

Askerka, M., Vinyard, D.J., Brudvig, G.W. and Batista, V.S. (2015). NH_3 binding to the S_2 state of the O_2-evolving complex of photosystem II: analogue to H_2O binding during the $S_2 \rightarrow S_3$ transition. *Biochemistry*, 54, 5783–5786.

Barber, J. (2006). Photosystem II: an enzyme of global significance. *Biochem. Soc. Trans.*, 34, 619–631.

Barber, J. (2009). Photosynthetic energy conversion: natural and artificial. *Chem. Soc. Rev.*, 38, 185–196.

Barber, J. (2016a). Mn_4Ca cluster of photosynthetic oxygen-evolving center: structure, function and evolution. *Biochemistry*, 55, 5901–5906.

Barber, J. (2016b). Photosystem II: the water splitting enzyme of photosynthesis and the origin of oxygen in our atmosphere. *Q. Rev. Biophys.*, 49, 1–21.

Barber, J. (2017). A mechanism for water splitting and oxygen production in photosynthesis. *Nat. Plants*, 3, 17041.

Barber, J., Ferreira, K., Maghlaouia, K. and Iwata, S. (2004). Structural model of the oxygen-evolving centre of photosystem II with mechanistic implications. *Phys. Chem. Chem. Phys.*, 6, 4737–4742.

Beck, W.F. and Brudvig, G.W. (1986). Binding of amines to the O_2-evolving center of photosystem II. *Biochemistry*, 25, 6479–6486.

Beck, W.F., Depaula, J.C. and Brudvig, G.W. (1986). Ammonia binds to the manganese site of the O_2-evolving complex of photosystem II in the S_2 state. *J. Am. Chem. Soc.*, 108, 4018–4022.

Boussac, A., Rutherford, W. and Styring, S. (1990). Interaction of ammonia with the water splitting enzyme of photosystem II. *Biochemistry*, 29, 24–32.

Boussac, A., Sugiura, M., Inoue, Y. and Rutherford, A.W. (2000). EPR study of the oxygen evolving complex in His-tagged photosystem II from the cyanobacterium *Synechococcus elongatus*. *Biochemistry*, 39, 13788–13799.

Britt, R.D., and Oyala, P.H. (2014). One step closer to O_2. *Science*, 345, 736.

Britt, R.D., Suess, D.L.M. and Stich, T.A. (2015). An Mn(V)-oxo role in splitting water? *Proc. Natl. Acad. Sci. USA*, 112, 5265–5266.

Britt, R.D., Zimmermann, J.-L., Sauer, K. and Klein, M.P. (1989). Ammonia binds to the catalytic Mn of the oxygen-evolving complex of photosystem II: evidence by electron spin-echo envelope modulation spectroscopy. *J. Am. Chem. Soc.*, 111, 3522–3532.

Chatterjee, R., Han, G., Kern, J., Gul, S., Fuller, F.D., Htchenko, A.G., Young, I.D., Weng, T.-C., Nordlund, D., Alonso-Mori, R., Bergmann, U., Sokaras, D., Hatakeyama, M., Yachandra, V.K. and Yano, J. (2016). Structural changes correlated with magnetic spin state isomorphism in the S_2 state of the Mn_4CaO_5 cluster in the oxygen-evolving complex of photosystem II. *Chem. Sci.*, 7, 5236–5248

Chu, H.-A. (2013). Fourier transform infrared difference spectroscopy for studying the molecular mechanism of photosynthetic water oxidation. *Front. Plant Sci.*, 4, 146.

Chu, H.-A., Feng, Y.-W., Wang, C.-M., Chiang, K.-A. and Ke, S.-C. (2004). Ammonia-induced structural changes of the oxygen-evolving complex in photosystem II as revealed by light-induced FTIR difference spectroscopy. *Biochemistry*, 43, 10877–10885.

Corry, T.A. and O'malley, P.J. (2018). Evidence of O–O bond formation in the final metastable S_3 state of nature's water oxidizing complex implying a novel mechanism of water oxidation. *J. Phys. Chem. Lett.*, 9, 6269–6274.

Cox, N. and Messinger, J. (2013). Reflections on substrate water and dioxygen formation. *Biochim. Biophys. Acta Bioenerg.*, 1827, 1020–1030.

Cox, N., Pantazis, D.A., Neese, F. and Lubitz, W. (2013). Biological water oxidation. *Acc. Chem. Res.*, 46, 1588–1596.

Cox, N., Retegan, M., Neese, F., Pantazis, D.A., Boussac, A. and Lubitz, W. (2014). Electronic structure of the oxygen-evolving complex in photosystem II prior to O–O bond formation. *Science*, 345, 804–808.

Dau, H., Andrews, J.C., Roelofs, T.A., Latimer, J.M.J., Liang, W., Yachandra, V.K., Sauer, K. and Klein, M.P. (1995). Structural consequences of ammonia binding to the manganese center of the photosynthetic oxygen-evolving complex: an X-ray absorption spectroscopy study of isotropic and oriented photosystem II particles. *Biochemistry*, 34, 5274–5287.

Dau, H., Zaharieva, I. and Haumann, M. (2012). Recent developments in research on water oxidation by photosystem II. *Curr. Opin. Chem. Biol.*, 16, 3–10.

Evans, M.C.W., Ball, R.J. and Nugent, J.H.A. (2005). Ammonia displaces methanol bound to the water oxidizing complex of photosystem II in the S_2 state. *FEBS Lett.*, 579, 3081–3084.

Ferreira, K.N., Iverson, T.M., Maghlaoui, K., Barber, J. and Iwata, S. (2004). Architecture of the photosynthetic oxygen-evolving center. *Science*, 303, 1831–1838.

Gao, Y., Åkermark, T., Liu, J., Sun, L. and Åkermark, B. (2009). Nucleophilic attack of hydroxide on a MnV oxo complex: a model of the O–O bond formation in the oxygen evolving complex of photosystem II. *J. Am. Chem. Soc.*, 131, 8726–8727.

Guo, Y., He, L.-L., Zhao, D.-X., Gong, L.-D., Liu, C. and Yang, Z.-Z. (2016). How does ammonia bind to the oxygen-evolving complex in the S_2 state of photosynthetic water oxidation? Theoretical support and implications for the W1 substitution mechanism. *Phys. Chem. Chem. Phys.*, 18, 31551–31565.

Guo, Y., Li, H., He, L.-L., Zhao, D.-X., Gong, L.-D. and Yang, Z.-Z. (2017a). The open-cubane oxo-oxyl coupling mechanism dominates photosynthetic oxygen evolution: a comprehensive DFT investigation on O–O bond formation in the S_4 state. *Phys. Chem. Chem. Phys.*, 19, 13909–13923.

Guo, Y., Li, H., He, L.-L., Zhao, D.-X., Gong, L.-D. and Yang, Z.-Z. (2017b). Theoretical reflections on the structural polymorphism of the oxygen-evolving complex in the S_2 state and the correlations to substrate water exchange and water oxidation mechanism in photosynthesis. *Biochim. Biophys. Acta Bioenerg.*, 1858, 833–846.

Gupta, R., Taguchi, T., Lassalle-Kaiser, B., Bominaar, E.L., Yano, J., Hendrich, M.P. and Borovik, A.S. (2015). High-spin Mn-oxo complexes and their relevance to the oxygen-evolving complex within photosystem II. *Proc. Natl. Acad. Sci. USA*, 112, 5319–5324.

Guskov, A., Kern, J., Gabdulkhakov, A., Broser, M., Zouni, A. and Saenger, W. (2009). Cyanobacterial photosystem II at 2.9-Å resolution and the role of quinones, lipids, channels and chloride. *Nature Struct. Mol. Biol.*, 16, 334–342.

Haumann, M., Liebisch, P., Müller, C., Barra, M., Grabolle, M. and Dau, H. (2005). Photosynthetic O_2 formation tracked by time-resolved X-ray experiments. *Science*, 310, 1019–1021.

Hendry, G. and Wydrzynski, T. (2002). The two substrate-water molecules are already bound to the oxygen-evolving complex in the S_2 state of photosystem II. *Biochemistry*, 41, 13328–13334.

Hou, L.-H., Wu, C.-M., Huang, H.-H. and Chu, H.-A. (2011). Effects of ammonia on the structure of the oxygen-evolving complex in photosystem II as revealed by light-induced FTIR difference spectroscopy. *Biochemistry*, 50, 9248–9254.

Kamiya, N. and Shen, J.-R. (2003). Crystal structure of oxygen-evolving photosystem II from Thermosynechococcus vulcanus at 3.7-Å resolution. *Proc. Natl. Acad. Sci. US.A*, 100, 98–103.

Kawashima, K., Takaoka, T., Kimura, H., Saito, K. and Ishikita, H. (2018). O_2 evolution and recovery of the water-oxidizing enzyme. *Nat. Commun.*, 9, 1247.

Kern, J., Chatterjee, R., Young, I.D., Fuller, F.D., Lassalle, L., Ibrahim, M., Gul, S., Fransson, T., Brewster, A.S., Alonso-Mori, R., Hussein, R., Zhang, M., Douthit, L., Lichtenberg, C.D., Cheah, M.H., Shevela, D., Wersig, J., Seuffert, I., Sokaras, D., Pastor, E., Weninger, C., Kroll, T., Sierra, R.G., Aller, P., Butryn, A., Orville, A.M., Liang, M., Batyuk, A., Koglin, J.E., Carbajo, S., Boutet, S., Moriarty, N.W., Holton, J.M., Dobbek, H., Adams, P.D., Bergmann, U., Sauter, N.K., Zouni, A., Messinger, J., Yano, J. and Yachandra, V.K. (2018). Structures of the intermediates of Kok's photosynthetic water oxidation clock. *Nature*, 563, 421–425.

Kim, D.H., Britt, R.D., Klein, M.P. and Sauer, K. (1990). The g = 4.1 EPR signal of the S_2 state of the photosynthetic oxygen-evolving complex arises from a multinuclear Mn cluster. *J. Am. Chem. Soc.*, 112, 9389–9391.

Kok, B., Forbush, B. and Mcgloin, M. (1970). Cooperation of charges in photosynthetic O_2 evolution. I. a linear four-step mechanism. *Photochem. Photobiol.*, 11, 457–475.

Krewald, V., Retegan, M., Cox, N., Messinger, J., Lubitz, W., Debeer, S., Neese, F. and Pantazis, D.A. (2015). Metal oxidation states in biological water splitting. *Chem. Sci.*, 6, 1676–1695.

Kusunoki, M. (2007). Mono-manganese mechanism of the photosytem II water splitting reaction by a unique Mn_4Ca cluster. Biochim. *Biophys. Acta Bioenerg.*, 1767, 484–492.

Lohmiller, T., Krewald, V., Navarro, M.P., Retegan, M., Rapatskiy, L., Nowaczyk, M.M., Boussac, A., Neese, F., Lubitz, W., Pantazis, D.A. and Cox, N. (2014). Structure, ligands and substrate coordination of the oxygen-evolving complex of photosystem II in the S_2 state: A combined EPR and DFT study. *Phys. Chem. Chem. Phys.*, 16, 11877–11892.

Marchiori, D.A., Oyala, P.H., Debus, R.J., Stich, T.A. and Britt, R.D. (2018). Structural effects of ammonia binding to the Mn4CaO5 cluster of photosystem II. *J. Phys. Chem. B*, 122, 1588–1599.

Mcevoy, J.P. and Brudvig, G.W. (2006). Water-splitting chemistry of photosystem II. *Chem. Rev.*, 106, 4455–4483.

Navarro, M.P., Ames, W.M., Nilsson, H., Lohmiller, T., Pantazis, D.A., Rapatskiy, L., Nowaczyk, M.M., Neese, F., Boussac, A., Messinger, J., Lubitz, W. and Cox, N. (2013). Ammonia binding to the oxygen-evolving complex of photosystem II identifies the solvent-exchangeable oxygen bridge (μ-oxo) of the manganese tetramer. *Proc. Natl. Acad. Sci.* USA, 110, 15561–15566.

Nilsson, H., Krupnik, T., Kargul, J. and Messinger, J. (2014a). Substrate water exchange in photosystem II core complexes of the extremophilic red alga cyanidioschyzon merolae. *Biochim. Biophys. Acta Bioenerg.*, 1837, 1257–1262.

Nilsson, H., Rappaport, F., Boussac, A. and Messinger, J. (2014b). Substrate-water exchange in photosystem II is arrested before dioxygen formation. *Nat. Commun.*, 5, 4305.

Ono, T.-A. and Inoue, Y. (1988). Abnormal S-state turnovers in NH_3-binding Mn centers of photosynthetic O_2 evolving system. *Arch. Biochem. Biophys.*, 264, 82–92.

Oyala, P.H., Stich, T.A., Debus, R.J. and Britt, R.D. (2015). Ammonia binds to the dangler manganese of the photosystem II oxygen-evolving complex. *J. Am. Chem. Soc.*, 137, 8829–8837.

Pantazis, D.A. (2018). Missing pieces in the puzzle of biological water oxidation. *ACS Catal.*, 8, 9477–9507.

Pantazis, D.A., Ames, W., Cox, N., Lubitz, W. and Neese, F. (2012). Two interconvertible structures that explain the spectroscopic properties of the oxygen-evolving complex of photosystem II in the S_2 state. *Angew. Chem. Int. Ed.*, 51, 9935–9940.

Pokhrel, R. and Brudvig, G.W. (2014). Oxygen-evolving complex of photosystem II: Correlating structure with spectroscopy. *Phys. Chem. Chem. Phys.*, 16, 11812–11821.

Pushkar, Y., Davis, K.M. and Palenik, M.C. (2018). Model of the oxygen evolving complex which is highly predisposed to O–O bond formation. *J. Phys. Chem. Lett.*, 9, 3525–3531.

Rapatskiy, L., Cox, N., Savitsky, A., Ames, W.M., Sander, J., Nowaczyk, M.M., RöGner, M., Boussac, A., Neese, F., Messinger, J. and Lubitz, W. (2012). Detection of the water-binding sites of the oxygen-evolving complex of photosystem II using W-band ^{17}O electron-electron double resonance-detected NMR spectroscopy. *J. Am. Chem. Soc.*, 134, 16619–16634.

Retegan, M., Krewald, V., Mamedov, F., Neese, F., Lubitz, W., Cox, N. and Pantazis, D.A. (2015). A five-coordinate Mn(IV) intermediate in biological water oxidation: spectroscopic signature and a pivot mechanism for water binding. *Chem. Sci.*, 7, 72–84.

Retegan, M. and Pantazis, D.A. (2016). Interaction of methanol with the oxygen-evolving complex: atomistic models, channel identification, species dependence, and mechanistic implications. *Chem. Sci.*, 7, 6463–6476.

Saito, T., Shoji, M., Kanda, K., Isobe, H., Yamanaka, S., Kitagawa, Y., Yamada, S., Kawakami, T., Okumura, M. and Yamaguchi, K. (2012). Theory of chemical bonds in metalloenzymes. XVII. symmetry breaking in manganese cluster structures and chameleonic mechanisms for the O-O bond formation of water splitting reaction. *Int. J. Quantum Chem.*, 112, 121–135.

Schraut, J. and Kaupp, M. (2014). On ammonia binding to the oxygen-evolving complex of photosystem II: a quantum chemical study. *Chem. Eur. J.*, 20, 7300–7308.

Shoji, M., Isobe, H., Shigeta, Y., Nakajima, T. and Yamaguchi, K. (2018a). Concerted mechanism of water insertion and O_2 release during the S_4 to S_0 transition of the oxygen-evolving complex in photosystem II. *J. Phys. Chem. B*, 122, 6491–6502.

Shoji, M., Isobe, H., Shigeta, Y., Nakajima, T. and Yamaguchi, K. (2018b). Nonadiabatic one-electron transfer mechanism for the O–O bond formation in the oxygen-evolving complex of photosystem II. *Chem. Phys. Lett.*, 698, 138–146.

Siegbahn, P.E.M. (2006). O–O bond formation in the S_4 state of the oxygen-evolving complex in photosystem II. *Chem. Eur. J.*, 12, 9217–9227.

Siegbahn, P.E.M. (2008). Theoretical studies of O–O bond formation in photosystem II. *Inorg. Chem.*, 47, 1779–1786.

Siegbahn, P.E.M. (2009). Structures and energetics for O_2 formation in photosystem II. *Acc. Chem. Res.*, 42, 1871–1880.

Siegbahn, P.E.M. (2017). Nucleophilic water attack is not a possible mechanism for O–O bond formation in photosystem II. *Proc. Natl. Acad. Sci. USA.*, 114, 4966–4968.

Sproviero, E.M., Gascón, J.A., Mcevoy, J.P., Brudvig, G.W. and Batista, V.S. (2008). Quantum mechanics/molecular mechanics study of the catalytic cycle of water splitting in photosystem II. *J. Am. Chem. Soc.*, 130, 3428–3442.

Suga, M., Akita, F., Sugahara, M., Kubo, M., Nakajima, Y., Nakane, T., Yamashita, K., Umena, Y., Nakabayashi, M., Yamane, T., Nakano, T., Suzuki, M., Masuda, T., Inoue, S., Kimura, T., Nomura, T., Yonekura, S., Yu, L.-J., Sakamoto, T., Motomura, T., Chen, J.-H., Kato, Y., Noguchi, T., Tono, K., Joti, Y., Kameshima, T., Hatsui, T., Nango, E., Tanaka, R., Naitow, H., Matsuura, Y., Yamashita, A., Yamamoto, M., Nureki, O., Yabashi, M., Ishikawa, T., Iwata, S. and Shen, J.-R. (2017). Light-induced structural changes and the site of O=O bond formation in PSII caught by XFEL. *Nature*, 543, 131–135.

Umena, Y., Kawakami, K., Shen, J.-R. and Kamiya, N. (2011). Crystal structure of oxygen-evolving photosystem II at a resolution of 1.9A. *Nature*, 473, 55–60.

Vaddypally, S., Kondaveeti, S.K., Karki, S., Vliet, M.M.V., Levis, R.J. and Zdilla, M.J. (2017). Reactive pendant Mn=O in a synthetic structural model of a proposed S_4 state in the photosynthetic oxygen evolving complex. *J. Am. Chem. Soc.*, 139, 4675–4681.

Vinyard, D.J. and Brudvig, G.W. (2015). Insights into substrate binding to the oxygen-evolving complex of photosystem II from ammonia inhibition studies. *Biochemistry*, 54, 622–628.

Vinyard, D.J., Khan, S., Askerka, M., Batista, V.S. and Brudvig, G.W. (2017). Energetics of the S_2 State Spin Isomers of the Oxygen-Evolving Complex of Photosystem II. *J. Phys. Chem. B*, 121, 1020–1025.

Vinyard, D.J., Khan, S. and Brudvig, G.W. (2015). Photosynthetic water oxidation: binding and activation of substrate waters for O–O bond formation. *Faraday Discuss.*, 185, 37–50.

Vogiatzis, K.D., Polynski, M.V., Kirkland, J.K., Townsend, J., Hashemi, A., Liu, C. and Pidko, E.A. (2018). Computational approach to molecular catalysis by 3d transition metals: challenges and opportunities. *Chem. Rev.*, 2019, 119(4), 2453–2523.

Yamaguchi, K., Yamanaka, S., Isobe, H., Tanaka, K. and Ueyama, N. (2012). Spin hamiltonian models for artificial and native water splitting systems revealed by hybrid DFT calculations. oxygen activation by high-valent Mn and Ru ions. *Int. J. Quantum Chem.*, 112, 3849–3866.

Yamanaka, S., Isobe, H., Kanda, K., Saito, T., Umena, Y., Kawakami, K., Shen, J.-R., Kamiya, N., Okumura, M., Nakamura, H. and Yamaguchi, K. (2011). Possible mechanisms for the O–O bond formation in oxygen evolution reaction at the

$CaMn_4O_5(H_2O)_4$ cluster of PSII refined to 1.9Å X-ray resolution. *Chem. Phys. Lett.*, 511, 138–145.

Young, I.D., Ibrahim, M., Chatterjee1, R., Gul, S., Fuller, F.D., Koroidov, S., Brewster, A.S., Tran, R., Alonso-Mori, R., Kroll, T., Michels-Clark, T., Laksmono, H., Sierra, R.G., Stan, C.A., Hussein, R., Zhang, M., Douthit, L., Kubin, M., Lichtenberg, C.D., Pham, L.V., Nilsson, H., Cheah, M.H., Shevela, D., Saracini, C., Bean, M A., Seuffert, I., Sokaras, D., Weng, T.-C., Pastor, E., Weninger, C., Fransson, T., Lassalle, L., Bräuer, P., Aller, P., Docker, P.T., Andi, B., Orville, A.M., Glownia, J.M., Nelson, S., Sikorski, M., Zhu, D., Hunter, M.S., Lane, T.J., Aquila, A., Koglin, J.E., Robinson, J., Liang, M., Boutet, S., Lyubimov, A.Y., Uervirojnangkoorn, M., Moriarty, N.W., Liebschner, D., Afonine, P. V., Waterman, D.G., Evans, G., Wernet, P., Dobbek, H., Weis, W.I., Brunger, A.T., Zwart, P.H., Adams, P.D., Zouni, A., Messinger, J., Bergmann, U., Sauter, N.K., Kern, J., Yachandra, V.K. and Yano, J. (2016). Structure of photosystem II and substrate binding at room temperature. *Nature*, 540, 453–457.

Zhang, B., Daniel, Q., Fan, L., Liu, T., Meng, Q. and Sun, L. (2018). Identifying Mn^{VII}-oxo species during electrochemical water oxidation by manganese oxide. *iScience*, 4, 144–152.

Zhang, B. and Sun, L. (2018). Why nature chose the Mn_4CaO_5 cluster as water-splitting catalyst in photosystem II: a new hypothesis for the mechanism of O–O bond formation. *Dalton Trans.*, 47, 14381–14387

Zouni, A., Witt, H.-T., Kern, J., Fromme, P., Krauss, N., Saenger, W. and Orth, P. (2001). Crystal structure of photosystem II from Synechococcus elongatus at 3.8 Å resolution. *Nature*, 409, 739–743.

Chapter 18

Local Cycle of Photosynthesis and Quasi-Aerobic Respiration Facilitated by Manganese Oxides — A Hypothesis on the Evolution of Phototrophy

Holger Dau[*,‡], Dennis J. Nürnberg[*,§] and Robert L. Burnap[†,¶]

**Freie Universität Berlin, Department of Physics,*
Arnimallee 14, 14195 Berlin, Germany
[†]Oklahoma State University, Department of Microbiology and Molecular Genetics, Stillwater, Oklahoma, USA
[‡]holger.dau@fu-berlin.de
[§]dennis.nuernberg@fu-berlin.de
[¶]rob.burnap@okstate.edu

Understanding the origin of oxygenic photosynthesis remains one of the grand challenges of science. It connects with important contemporary problems ranging from the development of alternative energy sources to understanding planetary evolution and attendant questions on the origin of life. The catalytic center of the photosynthetic water-splitting reaction is a manganese-calcium oxide coordinated within the protein matrix of the photosystem II complex. This metal cluster bears striking resemblance to certain poorly crystalline manganese oxides, which also share the capacity to catalyze the electrochemical splitting of water. The assembly of the metal cluster within the protein matrix is an oxidative process driven by photochemical charge separation of the photosystem II complex. This feature of natural assembly demonstrates what may be a primordial capacity to oxidize soluble Mn^{2+} as a source of electrons for cellular metabolism, and hence, presents a logical evolutionary intermediate on the path from anoxygenic to oxygenic photosynthesis as first proposed by Zubay over twenty years ago. This hypothesis is now considered in the framework of recent findings on the structure and biogenic production of manganese oxides, how extant bacteria metabolize metals in the environment, and how contemporaneous metal-based photosynthesis and quasi-aerobic respiration may have presaged the current oxygen cycle.

1. The World of Manganese Oxides

1.1. *Structure and formation of (inorganic) manganese oxides*

Oxides are often conceived as a class of crystalline materials with a limited variety of crystalline structures. For manganese in the 2+ oxidation state, this is indeed an appropriate perspective. There is Mn(II)O, which is "the" Mn(II) oxide. It is a simple crystal with a rock salt structure, like NaCl. In conjunction with other cations, Mn(II) oxides with more complex structures can form, like the manganese-calcium oxide denoted as marokite, but also this mineral is found as a well ordered, classical crystal [Najafpour *et al.*, 2010]. However, for high-valent Mn(III/IV) oxides the situation is different. There exists an almost unlimited variety of often poorly crystalline oxides, which mostly contain additional cations and water molecules [Post, 1999]. For us, these manganese oxides are of special interest. They are the prevalent form of manganese oxides found, *e.g.*, as manganese nodules on the seafloor [Post, 1999]. It has been hypothesized that these manganese oxides are often of biogenic origin [Spiro *et al.*, 2010]. Synthetic oxides of this type catalyze water oxidation (see [Zaharieva *et al.*, 2011, 2012] and references therein). The manganese-calcium cluster of photosynthesis shares structural motifs and further properties with these oxides [Zaharieva *et al.*, 2016].

How does this interesting class of oxides appear? A description of their structure may best start with minerals denoted as birnesites [Post, 1999]. The prevalent structural feature of a birnesite is a layered form of MnO_2, where coordination bonds connect each manganese ion to six oxygen ligands; each oxygen is connected to three Mn ions. The six oxygen ligands form octahedra around the central Mn ion; the MnO_2 layer is a planar layer of edge-sharing MnO_6 octahedra. Using a molecular terminology to describe the oxide layers, neighboring Mn ions are interconnected by di-μ-oxo bridges (two O^{2-} connecting the Mn ion; also denoted di-μ-oxido bridges); each oxygen bridges between three Mn ions and thus is denoted as μ_3-O. Birnesites consists of stacks of the described MnO_2 layers with layers of water molecules between the MnO_2 layers resulting in an alternating arrangement of oxide and water layers.

If a birnesite were composed exclusively of oxide and water layers as described above, the formal manganese oxidation state would be an integer number, namely +4.0, as illustrated by the following assignment of oxidations states and neutral overall charge:

$$\text{Cation-free birnesite} \Leftrightarrow [Mn^{4+}(O^{2-})_2(H_2O)^0_n]^0.$$

Such manganese-only birnesites are atypical, most birnesites contain additional cations within the water layer [Post and Veblen, 1990; Luo and Suib, 1997; Post, 1999; Lee *et al.*, 2014]. Essentially all mono-atomic cations can be found in birnesites. Prominent examples are Na^+, K^+, and Ca^{2+} resulting in sodium birnesite, potassium birnesite and calcium birnesite. The cation also can be a proton, either in form of an H_3O^+ ion or bound to the oxide by protonation of a di-μ-oxo bridge (transforming the μ-oxo in a μ-hydroxo bridge). The cation even may be a low valent Mn ion (Mn^{2+} and Mn^{3+}). We note that especially for Mn^{3+} ions as well as for protonation of bridging oxygens, the distinction between an interlayer cation and defects of the oxide layer itself becomes blurry. Additional cations also may bind to the MnO_2 layer in a way that cubane type structures are formed, *e.g.* a distorted $CaMn_3O_8$ cubane, in close analogy to a structural motif present in the biological water oxidation catalyst. In any event, the presence of additional cations lowers the mean Mn oxidation state resulting in mixed $Mn^{3+/4+}$ oxide:

$$\text{Cation-containing birnesite} \Leftrightarrow [Mn^{4+}{}_{1-m}Mn^{3+}{}_{m}(O^{2-})_2\,Cat^{+}{}_{m}(H_2O)^0_n]^0$$

(where m is a non-integer number smaller than unity).

Buserites are a manganese oxide modification that is especially closely related to birnesites. Instead of a single water/cation layer between two MnO_2 layers as found in birnesites, there are two water/cation layers in buserites [Golden *et al.*, 1987]. Like birnesites, buserites belong to the class of layered manganese oxides, which is distinguished from tunnel-type manganese oxides. The tunnel-type oxides share crucial features with birnesites and buserites. They are also composed of edge-sharing MnO_6 octahedra; they also may contain additional water molecules and cations. They differ from layered manganese oxides by an arrangement of the MnO_6 octahedra in form of linear tunnels with rectangular cross-sections of various dimensions. For example, a 1 × 2 tunnel formed by one row of MnO_6 octahedra at one side of the cross-section rectangle and two rows at the other side is denoted as ramsdelite. The smallest tunnel size is encountered in pyrolusite (1 × 1) tunnel; the tunnel size is especially large in todorokite (3 × 3). For understanding the structural relation between layered and tunnel-type oxides, one also may view tunnel-type oxides as layered oxides that are cross-linked by rows of MnO_6 octahedra.

The layers and tunnels in manganese oxides are filled by water molecules and cations. It is likely that also anions like carbonate and phosphate can be constituents of manganese oxides, possibly as ligands of Mn ions, but the structural role of anions in manganese oxides is largely still unknown. The knowledge on the structure of manganese oxides mostly comes from powder diffraction experiments, which provide insight only in sufficiently well-ordered oxides. Specifically, the arrangement of water molecules and cations in the interlayer or inner-tunnel space often remains hypothetical. Even worse regarding scarcity of structural information, are the poorly diffracting or non-diffracting manganese oxides, where X-ray absorption spectroscopy can provide insight in prevalent local structural motifs. Although less well studied, these amorphous manganese oxides are by no means rare exceptions. For amorphous manganese oxides, extended X-ray absorption fine-structure (EXAFS) spectra suggest that defect-rich MnO_2 layers or layer fragments may form complex structures with interlayer cations, *e.g.* a $CaMn_3O_6$ heterocubane motif. [Zaharieva *et al.*, 2011, 2012] It is conceivable that this cubane and further structural motifs that are assumed to be mechanistically decisive in photosynthetic water oxidation (at the protein-bound $CaMn_4O_5$ cluster of photosystem II) are also present in some manganese oxides [Zaharieva *et al.*, 2016].

Naturally occurring manganese oxides may be largely or even completely of biogenic origin as discussed further below. Synthetic $Mn^{3+/4+}$ oxides can be synthesized by various routes, typically starting with (i) oxidation of dissolved Mn^{2+} ions or (ii) reduction of permanganate ($[MnO_4]^-$) ions (or oxidation of Mn^{2+} ions coupled to $[MnO_4]^-$ reduction), mostly followed by specific thermal protocols that eventually result in powders of manganese oxides. Both routes involve strong oxidants and often extreme pH, which are unlikely to be present in a natural environment. Especially informative are manganese oxides synthesized by electrodeposition on inert substrate electrodes. By application of electric potentials around 1 V_{NHE} (*versus* NHE, at pH 7), dissolved Mn^{2+} ions are oxidized and high-valent $Mn^{3+/4+}$ oxides are spontaneously formed at appropriate concentrations of Mn^{2+} ions and together with the cations and anions of a suitable pH buffer (*e.g.* phosphate buffer). In seminal work by Zaharieva, various manganese oxide films were electrodeposited on the substrate electrode and investigated by electrochemical methods in conjunction with X-ray absorption spectroscopy (XAS). The XAS experiments revealed that the electrodeposited manganese oxides are of the birnesite type. They were either well ordered (but still non-crystalline) or they contained more complex structural motifs, depending on the deposition protocol [Zaharieva *et al.*, 2012]. In the electrodeposited manganese oxides, additional cations were present, specifically potassium or calcium. The electrodeposited oxides were redox-active meaning that the variation of the electric potential caused changes from an average oxidation state often well below +3.5 to a level close to

+4.0 at 1.2 V_{NHE} (at pH 7) [Zaharieva *et al.*, 2016]. Regarding range and spectral characteristics of the oxidation state changes, a close resemblance with the redox-cycling of the photosynthetic manganese complex was observed. Whereas redox-activity was observed in all manganese oxides, only the imperfectly ordered oxides containing more complex structural motifs were able to oxidize water efficiently [Zaharieva *et al.*, 2012; Zaharieva *et al.*, 2016].

Starting with a pioneering study on synthetic marokite [Najafpour *et al.*, 2010], water oxidation by manganese oxide particles has been extensively investigated by Kurz and coworkers. Structurally, the electrodeposited oxides are found to be similar to the marokite precursor and various other synthetic manganese oxides [Zaharieva *et al.*, 2011, 2012]. Even though the redox state changes of the synthetic oxide particles in aqueous solution have not been characterized, it is reasonable to assume that their electrochemical and catalytic properties correspond to the properties of the electrodeposited manganese oxides investigated by Zaharieva and coworkers.

In summary, $Mn^{3+/4+}$ oxides of the birnesite type are readily formed by oxidation of dissolved Mn^{2+} ions at potentials of around 1 V *versus* NHE, which is close to the H_2O/O_2 equilibrium potential of about 0.8 V *versus* NHE (at pH 7 and atmospheric O_2-partial pressure). The formation of the $Mn^{3+/4+}$ oxides may be described as oxidative self-assembly. The formation of biogenic manganese oxides and also of the $CaMn_4$ cluster of oxygenic photosynthesis likely involves similar redox potentials and related modes of oxidative self-assembly. Depending on the potential of an oxidizing or reducing electrode or chemical group, they can change their oxidation state from about +3.3 to almost +4.0. At lower redox potentials, dissolution of the oxide films is often observed, which likely results from reduction of Mn ions to the Mn^{2+} level (unpublished results). Some specific variants of the birnesite-type oxides are efficient in oxidizing water oxidation and associated O_2 formation, others are not; calcium ions favor water oxidation, but are not essential.

1.2. *Biogenic manganese oxides and bacterial metal oxide metabolism*

1.2.1. *Formation of manganese deposits*

Microorganisms, primarily bacteria and fungi, are thought to be the main source of manganese oxides in the environment and have been shown to accelerate oxide formation by several orders of magnitude over the abiotic processes [Tebo *et al.*, 2005; Spiro *et al.*, 2010; Kuhn *et al.*, 2017]. Since the first description of biological manganese oxide formation in soils [Mann and Quastel, 1946], the presence of biogenic manganese oxides has been confirmed in various marine and terrestrial habitats [Tebo *et al.*, 2005]. Manganese oxides are subject to redox cycling due to a variety of abiotic and biotic reduction and re-oxidation reactions occurring rapidly in sediments and soils. The accumulation of metal oxides in geological formations is typically the consequence of specific patterns of oxidation and precipitation involving both biological metabolism and abiotic processes. Importantly for the present discussion, interpretation of geologic deposits containing manganese as evidence of the ancient formation of manganese oxides, paradoxically, derives from the identification of reduced manganese minerals in rocks [Johnson *et al.*, 2013; Fischer *et al.*, 2015]. Manganese oxides produced by abiotic and biotic processes are prone to additional secondary and tertiary redox and chemical transformations *in situ*, typically involving the formation of a (Mn^{2+})-carbonate or (Mn^{2+})-phosphate as the intermediate or terminal mineral phase. Accordingly, a careful understanding of the formation of the minerals and the redox chemistry is an active area of research.

1.2.2. *Manganese oxide reduction in bacterial respiration*

It is well-recognized that bacteria are capable of mediating both metal-reduction as well as metal-oxidation and some species are capable of both, depending upon the environmental conditions [Spiro *et al.*, 2010; Toyoda and Tebo, 2013; Wright *et al.*, 2016]. The aforementioned reductive formation of Mn(II) carbonates and phosphates can occur chemically by reactions with ferrous iron and sulphide. However, biological reduction of metals is an ecologically widespread and highly active extant process that occurs by bacterial respiration of organic compounds in the absence of oxygen. Given that these are likely ancient metabolic capabilities, this is a process that is especially relevant for the consideration of the evolution of photosynthesis leading to the development of oxygenic photosynthesis. Biologically mediated metal redox processes are generally integrally woven into various metabolic pathways, involving a large variety of organic compounds depending upon the bacterial species, the ecological niche, and the prevailing conditions. For example, anaerobic catabolism of organic acids, such as lactic acid, occurs in certain bacteria as a form of anoxic respiration utilizing an oxidized metal as the terminal electron acceptor instead of oxygen. Through pathways such as this, the coupling of inorganic and biochemical processes governs the cycling of minerals and organic compounds in a variety of natural environments. Accordingly, understanding these processes is important for interpreting existing geological formations and provides a framework for understanding the early evolution of metabolism including the evolution of photosynthesis and the advent of oxygenic photosynthesis.

An understanding of the biogenic formation of manganese oxides remains fragmentary, although considerable progress has been made recently. It is worth first considering the process of manganese reduction, however. Members of the extant bacterial phyla *Shewanella* and *Geobacter* are recognized to be key bacterial species involved in the reduction of metals, primarily Fe^{3+} and $Mn^{3+/4+}$[Spiro *et al.*, 2010; Toyoda and Tebo, 2013; Wright *et al.*, 2016]. Metabolically diverse pathways have evolved to catabolize organic compounds *via* respiratory pathways using a variety of alternative oxidants besides oxygen. Metal-reducing bacteria, such as the well-studied *Shewanella oneidensis*, utilize oxidized metals as sinks for metabolically-derived electrons as mentioned above. Although, it is highly probable that the transfer of negative potential catabolic reductants to high potential metal oxidants is coupled to the production of high energy intermediates, such as ATP. The details of this biochemistry remain poorly understood and most studies have been confined to taking stock of the genetic potential estimated from the bioinformatic analysis of genomes and physiological experiments looking at the overall features of this type of metabolism. It is important to note in this regard that the utilization of reduced metals as sinks for metabolic reductant is, by itself, providing a metabolic function in that it allows the recycling of reduced cofactors that are produced during fermentive catabolism in much the same way that dissimilatory release of ethanol provides an advantage during fermentation. Here the energy dividend of ATP from sugar breakdown *via* substrate level phosphorylation, albeit modest, is sustainably enabled by dissipating excess reductant in the form of ethanol. Without ethanol dissimilation, the reduced cofactor of the reaction, NADH, cannot be recycled and the entire process would grind to a halt. Analogously, metal reduction enables anaerobic fermentation by providing a sink for catabolically-derived electrons that would otherwise accumulate as reduced cofactors in the cell.

One of the best studied metal-reducing bacterial species, *Shewanella oneidensis*, has remarkable metabolic flexibility. It is a facultative anaerobe capable of fermentation, yet it is also capable of performing respiration of different carbon compounds using a variety of different terminal electron acceptors, including Fe^{3+}, $Mn^{3+/4+}$, and oxygen, depending on the environmental conditions. Indeed, its ability to couple the oxidation of soluble organic compounds, such as lactate, to various soluble and insoluble electron acceptors reflects an impressive genetic potential to deploy a variety of different metabolic activities that utilize different reduced growth substrates and different electron acceptors. A feature that reflects the diversity of different modes of redox metabolism, is the finding that over 42 *c*-type cytochromes have been identified in the genome. These findings are relatively recent, and the rich biochemistry of this organism remains an area of active research with many key questions remaining to be resolved. Generally, it is known that the catabolism of various organic growth substrates typically produces a flow of electrons leading to the reduction of the menaquinone pool of the cytoplasmic (inner) membrane. Under aerobic conditions, this pool is reoxidized using the canonical, O_2-dependent respiratory electron transfer chain involving the cytochrome b-c_1 complex, a soluble *c*-type cytochrome, and, finally, cytochrome oxidase mediating the transfer of electrons to molecular oxygen. Under anaerobic conditions, non-O_2-dependent respiratory pathways are mobilized. In this case, the menoquinol pool is reoxidized by a membrane-bound tetra-heme cytochrome complex, CymA, which serves as a redox hub for a variety of electron shuttles on the periplasmic side of the membrane and which serve to transport electrons to specific terminal electron acceptors in the environment (*e.g.* metal oxides) and maintain the redox poise of the quinone pool. However, the mechanism for the electron transfer to terminal acceptors either involves electrically conductive type IV pili or tubular extracellular extensions of the outer membrane as in case of *Shewanella oneidensis* [Pirbadian *et al.*, 2014] (also known as "bacterial nanowires") or, perhaps a diffusible redox mediator. Importantly, it appears that respiration of metals may only involve Complex I, the NADH dehydrogenase with its attendant proton pumping activity, but not the downstream portion of the pathway since the menaquinone pool is being oxidized by CymA. CymA has four low-spin *c*-hemes with midpoint reduction potentials of +10, −108, −136, and −229 mV [Paquete and Louro, 2010]. As this is likely the terminal electron acceptor in the electron transport chain, it suggests that metal-reducing bacteria using this system cannot take full advantage of the large potentials available for metals such as high valent manganese oxides. Thus, a considerable amount of the available free energy potential in the remaining redox span between the reduced quinone and the manganese oxides is essentially wasted. In any event, this anaerobic respiratory mechanism allows *Shewanella oneidensis* and its relatives to thrive in diverse, redox-stratified environments and participate in the overall cycling of metals and organic matter by the microbial communities inhabiting these environments. Although the precise mechanism of electron transfer to metal oxides is still being worked out, it is clear that this is a major environmental mechanism to mobilize soluble Mn^{2+} ions for oxide precipitates and is a process that figures prominently in reaction schemes for understanding manganese mineral formations, including those with mixed valence species.

Modern cyanobacteria such as *Synechocystis* sp. PCC 6803 have access to manganese oxides *via* type IV pili, in addition to their specific uptake system for Mn^{2+} [Lamb and Hohmann-Marriott, 2017]. This could be an additional mechanism to obtain Mn during transient limitations in addition to the proposed storage in the periplasm [Keren *et al.*, 2002] but could be also an important mechanism to prevent the accumulation of toxic concentrations of Mn in the cytoplasm and reuse them when necessary. The export activity of the Mn transporter Mnx was recently confirmed to transport Mn from the cytoplasm into the thylakoid lumen

[Brandenburg *et al.*, 2017]. However, it is unknown whether Mn excretion is possible. Studies on the green algae *Chlorella* and the cyanobacterium *Microcystis* however have shown that manganese oxides are formed in dense cultures [Richardson *et al.*, 1988].

1.2.3. *Bacterial manganese oxide formation*

Strikingly, a number of bacteria, including *Shewanella oneidensis*, have been demonstrated to have even greater metabolic flexibility: in addition to reducing metals as part of anaerobic respiration, a number of bacterial species are also capable of oxidizing metals under aerobic growth conditions [Spiro *et al.*, 2010]. For example, both *Shewanella loihica and Shewanella oneidensis* are capable of performing anaerobic respiration using manganese oxides as a terminal electron acceptor, but when transferred to aerobic conditions, both species oxidize soluble Mn^{2+} to form finely grained, poorly crystalline oxide nanoparticles at Mn oxidation states of 3.86 that extensively accumulate in the immediate environment of the cells [Wright *et al.*, 2016]. In fact, numerous bacterial species have been identified as having the capability of oxidizing Mn^{2+} [Spiro *et al.*, 2010]. In all cases examined, the Mn^{2+} oxidation and concomitant oxide nanoparticle formation depends upon multicopper oxidase (also referred to as a laccase) [Dick *et al.*, 2008; Soldatova *et al.*, 2017; Soldatova *et al.*, 2017]. It is known that fungi are also capable of manganese oxidation and this activity also strictly depends of the action of multicopper oxidases [Dick *et al.*, 2008; Spiro *et al.*, 2010]. One of the most geologically significant Mn^{2+}-oxidizing bacteria is the marine *Bacillus* sp. strain SG-1, which accumulates Mn^{4+} on its spore coat and is one of the bacterial species crucial for the formation of manganese nodules on the ocean floor. In the context of understanding the relationships between manganese oxide formation in these bacteria and the evolution and assembly of the water oxidizing complex of photosynthesis, it is interesting to note that the formation of manganese oxides by *Bacillus* is greatly accelerated by the presence of Ca^{2+}, but not other cations [Toyoda and Tebo, 2013]. This may reflect non-enzymatic oxide deposition processes since recent studies show that the purified enzyme does not require Ca^{2+} [Soldatova *et al.*, 2017; Soldatova *et al.*, 2017]. Perhaps more interestingly, the enzyme relies upon activation of the electron transfer to active site Cu by formation of hydrolysis of a coordinating H_2O to form Mn-OH^- species followed by formation of a bridge with a second Mn^{2+} and triggering electron transfer, *via* the Cu and then to molecular oxygen. The activation of electron transfer by deprotonation is a crucial aspect of both the assembly and the catalytic activity of the oxide core of oxygenic photosynthesis. It parallels the redox leveling function in PSII that proton release facilitates electron transfer by preventing the accumulation of positive charge with each successive electron withdrawal from the Mn_4CaO_5 during the catalytic cycle. The MnO_2 nanoparticles formed by bacteria *via* multicopper oxidase enzymatic activity consist of stacked hexagonal planar layers of edge-sharing MnO_6 octahedra resembling birnesites. The sheets are intercalated with water and cations, almost invariably Ca^{2+}, thus forming Ca^{2+} birnesites as discussed above. Moreover, the biogenic particles are extremely small and poorly crystalline, appearing as amorphous material surrounding the bacteria. Importantly, the molecular structure of these oxides displays numerous structural defects in the hexagonal arrays which provide open coordination sites for various ligands to the metal atoms. This accessibility of the $Mn^{3+/4+}$ components explains the characteristic rapid re-reduction of biogenic manganese oxides in the presence of metal respiring bacteria exploiting the oxides as terminal electron acceptors. In contrast, highly crystalline oxide arrays act as kinetically poor sinks for metabolic electrons underlining the importance of the structural defects in the biogenic oxides in promoting their reactivity.

While the bioenergetic basis of the metal reduction metabolism in bacteria such as *Shewanella* and *Geobacter* is understood, at least in broad terms, the physiological advantage of Mn oxidation and the selective evolutionary pressures leading to its evolutionary development of enzymes such as bacterial lactases, are less clear. Perhaps the most likely explanation is that manganese oxidation is an adaptive metabolic feature that allows *Shewanella* to accumulate oxidant during aerobic conditions that can then be used during periods of anaerobicity. This scenario fits the characteristics of the redox stratified and changing environments that *Shewanella* thrives in. This possibility would also be consistent with the observation that bacterial reduction of highly crystalline manganese oxides is considerably slower than highly crystalline manganese oxides. The fine poorly crystalline nanoparticles deposited by bacteria under aerobic conditions are, on the other hand, rapidly reduced by the same bacteria under anaerobic conditions.

1.3. *The manganese complex of photosynthetic water oxidation*

1.3.1. *Structure of the photosystem II (PSII) manganese complex*

In oxygenic photosynthesis, water is oxidized by a manganese complex bound to the proteins of PSII, which is often denoted as the oxygen-evolving complex (OEC). First clear evidence for a functional role of a polynuclear manganese complex at the electron-donor side of PSII came from detection of an EPR multiline signal, which suggested that minimally two electronically coupled manganese ions are involved [Dismukes and Siderer, 1981]. A few years later [Goodin *et al.*, 1984;

Yachandra *et al.*, 1986], pioneering developments by Sauer, Klein, Yachandra and their coworkers in Berkeley promoted X-ray absorption spectroscopy (XAS) to a prime experimental approach for (i) investigating redox state changes of the PSII manganese complex and (ii) gaining information on the structure of the manganese complex, as summarized elsewhere with focus on the functional relation to the crucial electron and proton transfer steps of biological water oxidation [Dau and Haumann, 2008]. Already early XAS investigations revealed a structure of the metal complex with di-μ-oxo bridging between high-valent $Mn^{3+/4+}$ ions resulting in a model of two "dimers" of di-μ-oxo-bridged $Mn^{3+/4+}$ ions mutually interconnected by a mono-μ-oxo bridge and connected by an additional μ-oxo bridge to a Ca^{2+} ion [Yachandra *et al.*, 1993]. This dimer-of-dimers model later was modified, based on advanced EPR spectroscopy, to a monomer-trimer model (also denoted as "dangler model") [Peloquin *et al.*, 2000; Dau and Haumann, 2008]. Various specific (and mechanistically important) details of these structures did not withstand the test of time, but the prevalence of structural motifs prevailing also in inorganic manganese oxides has been confirmed.

In 2001, Zouni and her Berlin coworkers published the first crystal structure of PSII containing its manganese complex [Zouni *et al.*, 2001], thereby starting the era of increasingly detailed crystallographic models of the PSII manganese complex. At the resolution of 3.8 Å, individual manganese and oxygen atoms were not resolved, but the highly compact electron density suggested an even more compact arrangement of the four metal ions (4 × Mn, 1 × Ca) than predicted by the dimers-of-dimers model. A next milestone had been the crystallographic model of Barber and his coworkers in London [Ferreira *et al.*, 2004], who proposed a Mn_3CaO_4 cubane structure connected by a mono-μ-oxo bridge to a "dangling" fourth Mn ion. Later chemical feasibility of this structure was verified by synthesis of a stunningly close molecular complex [Zhang *et al.*, 2015]. In 2008, a mechanistically relevant modification of the London model was suggested by Siegbahn based on computational chemistry. At the same time, based on XAS data and a refined crystallographic model of the ligating protein sidechains [Loll *et al.*, 2005], Dau and coworkers predicted a structure of the PSII manganese complex and its protein environment, which inter alia shares with the Siegbahn proposal a tight connection of the dangling Mn ion *via* di-μ-oxo bridges to a distorted Mn_3CaO_4, with a five-coordinated Mn^{3+} ion in the lower oxidations states of the metal complex [Dau *et al.*, 2008]. Albeit using very different approaches, both models were similar to the crystallographic model reported in 2010 at the international photosynthesis congress in Beijing (and published 2011 [Umena *et al.*, 2011]) by Shen and his coworkers in Japan, namely a structural model of the PSII manganese complex at 1.9 Å resolution (see Figure 1). Shen's structural model was a stunning breakthrough in PSII structure elucidation, but comparison with XAS data revealed a modification of the metal-oxo core

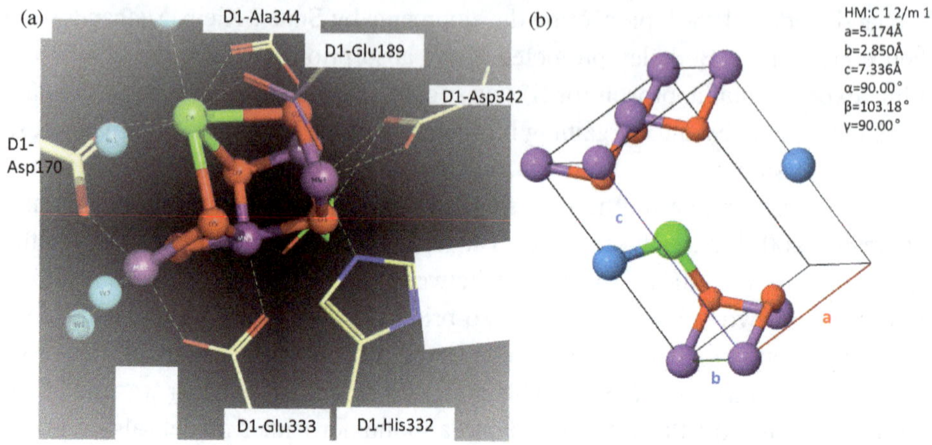

Figure 1. (a) Catalytic center of water oxidation of the photosystem II complex. Manganese atoms (purple), calcium (green), water molecules (cyan), H-bonding (yellow dashes) and oxo-bridge atoms (red) of the Mn_4CaO_5 cluster. Key amino acid ligands are shown to indicate the role of the protein in coordinating the metal cluster structure. (Based upon PDB entry 4UB6). (b) Birnessite-like structure (unit cell) with manganese atoms (purple), calcium (green), water molecules (blue) and oxo-bridge atoms (red). One manganese atom the (in unit cell corner "a") and corresponding oxos were omitted from the image to highlight the resemblance of structural motifs to the PSII Mn_4CaO_5 cluster. Adapted from American Mineralogist Crystal Structure Database (http://rruff.geo.arizona.edu/).

caused by X-ray photoreduction of the manganese ions [Yano *et al.*, 2005; Grabolle *et al.*, 2006; Grundmeier and Dau, 2012]. However, the free electron laser (FEL) facilities providing strong femtosecond X-ray pulses facilitated the determination of radiation-damage-free PSII structures [Suga *et al.*, 2015], which confirmed all central aspects of Shen's initial structural model. Today FEL crystallography on PSII is on its way towards time-resolved structure determination for various intermediate states [Young *et al.*, 2016] of the complete electron-proton reaction cycle of PSII water oxidation previously concluded from time-resolved spectroscopic experiments [Dau and Haumann, 2007; Klauss *et al.*, 2012].

1.3.2. *Assembly (photoactivation) of the PSII manganese complex*

Unlike many metalloprotein assembly processes in nature, the assembly of the catalytic manganese cluster of photosynthetic water oxidation is an active process that not only requires the diffusion of metal ions into the coordination environment of the protein, but also requires an energy input, which in this case is the strong oxidant provided by light-driven charge separation at the photochemical reaction center. Essentially the same photochemical mechanism that drives the catalytic oxidation of water is first utilized to oxidatively assemble the four manganese and one calcium ions into the active Mn_4CaO_5 core. The oxidative assembly, known as photoactivation, forms an oxo bridged structure that shares fundamental similarities with the inorganic oxides discussed above. Importantly, the process of natural assembly demonstrates what may be a primordial capacity to oxidize soluble Mn^{2+} as a source of electrons for metabolism [Zubay, 1996; Johnson *et al.*, 2013; Fischer *et al.*, 2015]. The ability to oxidize soluble Mn^{2+} is hypothesized to represent an ancient phototropic ecological strategy to exploit metals as a source of reductant for biosynthetic metabolism. This is distinct for the oxidation of Mn^{2+} by bacteria as discussed above since in this case metal oxidation has a clear physiological advantage: the source of metabolic reductant. On the other hand, the Mn^{2+} oxidation by *Shewanella* and other genera of bacteria discussed above has no immediate physiological role, but instead appears to be an ecological strategy for preparing for shifts in the redox state of the environment, as discussed above. Moreover, the hypothesized mode of Mn^{2+} phototrophy [Zubay, 1996; Johnson *et al.*, 2013; Fischer *et al.*, 2015] is further postulated to have led to the evolution of the Mn_4CaO_5 core and its attendant ability to utilize water as the source of reductant (*vide infra*); and it is in this context that the mechanism of photoactivation is especially intriguing since both the assembly and the hypothesized precursor of oxygenic phototrophy, Mn^{2+}-based phototrophy, share soluble Mn^{2+} oxidation in common.

1.3.3. *Variation of PSII core subunit D1 (PsbA)*

The OEC of PSII is mainly coordinated by amino acids of the D1 protein (Figure 1A), encoded by the *psbA* gene. A detailed study by Cardona, Murray and Rutherford on the diversity of D1 proteins identified several variants of D1 which form distinct phylogenetic clades that might have evolved before the origin of water oxidation [Cardona *et al.*, 2015]. All of these ancient forms are characterized by dramatically different Mn_4CaO_5 binding sites. The authors defined three groups of atypical D1 forms. The most ancient D1 variant was found in the cyanobacterium *Gloeobacter kilaueensis* JS-1 which has no thylakoid membranes and grows only in epilithic biofilms in a lava cave in Kīlauea Caldera, Hawai'i [Saw *et al.*, 2013]. The so-called G0–D1 lacks all ligands to the Mn_4CaO_5 core with the exception of Glu189.

A second group of D1 variants, named G1–D1 is more widely distributed but is also lacking most of the ligands needed for the assembly of the Mn_4CaO_5 core. The G1–D1 (previously known as "super-rogue" D1 [Murray, 2012]) is found in all cyanobacteria that synthesize chlorophyll *f* when exposed to near-infrared light [Gan *et al.*, 2014]. Chlorophyll *f* is the most red-shifted chlorophyll that still allows oxygenic photosynthesis [Chen *et al.*, 2010; Gan *et al.*, 2014; Nürnberg *et al.*, 2018]. A recent study by Bryant and coworkers showed that G1–D1 has obtained a new function by synthesizing chlorophyll *f* and was named accordingly chlorophyll *f* synthase ChlF [Ho *et al.*, 2016].

A third group of atypical D1, G2–D1 (formerly named "rogue" D1 [Murray, 2012] or "sentinel" D1 [Wegener *et al.*, 2015]) has also been shown to be unable to oxidize water efficiently and restore phototrophic growth of a D1-less mutant of the cyanobacterium *Synechocystis* sp. PCC 6803 [Wegener *et al.*, 2015]. Pakrasi and coworkers suggested that G2–D1 is incorporated into PSII to allow oxygen-sensitive processes such as nitrogen fixation to occur in cyanobacterial cells [Wegener *et al.*, 2015].

Presently it is not known whether atypical D1 forms have further functions. In particular, it is unknown whether they can assemble alternative metal clusters or promote light-driven manganese oxide formation from dissolved Mn^{2+} ions.

1.3.4. *Location of PSII in the photosynthetic cell*

The PSII is a membrane-integral protein complex. In most organisms, there is a specialized membrane system carrying the PSII and other photosynthetic protein complexes, the thylakoid membrane. In higher plants, the PSII is strongly enriched in specific regions of the thylakoid membrane system, which are denoted as grana stacks. This organizational principle was discovered in the early 1980s, when the PhD student Bertil Andersson ventured into this area, eventually resulting in an exceptionally influential review jointly written with Jane Anderson [Anderson and Andersson, 1982]. The plant PSII thus is concentrated at an innermost location of the plant cell. It is first segregated from the stroma space of the chloroplast, second separated by the chloroplast envelope membranes from the cytosol (and by mitochondrial membranes even further detached from the respiratory enzymes) and third separated by the cell membrane (cytosolic membrane) from the environment of the plant cell. In cyanobacteria, this separation from the environment is less stringent. The same membrane system carries the photosynthetic and respiratory protein complexes, but it still is a cell-internal membrane system separated from the environment by a separate cytosolic membrane. However, there also is *Gloeobacter*, a cyanobacterium commonly found in rocks [Mares *et al.*, 2013]. This organism is believed to be an evolutionary relic, which is closer to ancient cyanobacteria than any other of the so far described cyanobacterial species. In this ancient organism, the photosynthetic and respiratory enzymes are located in the plasma membrane in specialized domains [Rexroth *et al.*, 2011]. The orientation of PSII is such that the electron-donor side of PSII with its manganese-binding residues are no longer inside of the cellular space enclosed by the cytoplasmic membrane, but in direct contact with the periplasmatic space, which is considered to be the main storage site for manganese ions in cyanobacterial cells [Keren *et al.*, 2002]. The localization of PSII in the plasma membrane has also significant influence on the extrinsic subunits of PSII, PsbO, PsbU and PsbV, which are only poorly conserved in *Gloeobacter* [Nakamura, 2003; De Las Rivas and Barber, 2004; Koyama *et al.*, 2008]. The direct contact of the oxidizing side of the photosystem in an ancient cyanobacterium is of high relevance for the hypotheses discussed in the following section.

2. Hypothesis — Manganese Oxides Facilitated Quasi-Aerobic Photosynthesis and Respiration

2.1. *Has the biological catalyst initially been created from external $Mn^{3+/4+}$ oxide particles? No, the other way around, an ancient photosystem was a manganese oxide producer*

Intriguingly, the metal-oxo core of the photosynthetic manganese complex resembles a tiny oxide particle. The striking similarity of structural motifs motivated a stimulating hypothesis by Sauer and Yachandra that addressed evolution of the protein-bound Mn-oxo cluster of PSII [Sauer and Yachandra, 2002]. They suggested that small oxide particles had been formed outside of PSII and subsequently were incorporated into PSII by binding to specific protein groups. However, there are arguments against this hypothesis:

- When oxygenic photosynthesis evolved, the dioxygen content of the atmosphere and in aqueous environments was negligibly small so that spontaneous formation of manganese oxides essentially can be excluded.
- Stable manganese oxide particles consisting only of a handful of metal ions have not been reported so far. Such small manganese oxide particles likely would decay rapidly in aqueous solution. If they existed, they would have a comparably high fraction of terminal water ligands so that stabilization of the metal-oxo cluster by bridging oxides between metal ions would be insufficient. Could complex formation involving appropriate monodentate or bidentate ligands solve the stability problem? Indeed, synthetic chemists have achieved multinuclear oxide-like manganese complexes where the small metal-oxo core is stabilized by ligand groups, *e.g.* carboxylate groups. However, these complexes are stable only in organic solvents or in the form of crystalline powders. In the presence of a few percent of water, they typically degrade rapidly.

- Assembly of the manganese-calcium complex of today's PSII does not involve any pre-assembled metal complexes that serve as precursors. The light-driven assembly starts directly by binding and oxidation of dissolved Mn^{2+} ions. It would be highly surprising if early in evolution, the starting point had been the incorporation of externally formed oxide particles and was thus very different. There is no obvious evolutionary path that leads from binding of an externally formed oxide particle to the oxidative self-assembly from dissolved Mn^{2+} ions.

In our opinion, the incorporation of extrinsic manganese oxide particles into PSII has not been an early step in evolution of the biological catalyst. Ruling out this option, we nonetheless see a close relation between manganese oxides and the catalytic site of photosynthetic water oxidation. Our working hypothesis turns the suggestion of Sauer and Yachandra upside down: *manganese oxide particles are formed by an early evolutionary PSII precursor as the product of Mn^{2+} oxidation to provide metabolic reductant.*

This hypothesis is, *inter alia*, motivated by the recent experiments on electrodeposited manganese oxides, which are readily formed by oxidative self-assembly [Zaharieva *et al.*, 2012; Gonzalez-Flores *et al.*, 2016; Zaharieva *et al.*, 2016], as discussed further above. Aside from demonstrating the facile formation of manganese oxides, these studies also teach a further lesson. Provided sufficiently positive oxidation potentials are available, the formation of $Mn^{3+/4+}$ oxides may be easily achieved, but the formation of an oxygen-evolving manganese oxide requires the presence of specific structural motifs. The need for specific structural motifs is in line with the current understanding of photosynthetic water oxidation: the protein environment in PSII has evolved such that precisely the one-and-only biological catalyst of water oxidation is formed by oxidative self-assembly [Bao and Burnap, 2016; Zhang *et al.*, 2017]. Moreover, this line of argument is consistent with the proposal that prior to oxygenic photosynthesis, certain phototrophs evolved photochemical reaction centers that had the ability to utilize metal ions, probably Fe^{2+} at first and later Mn^{2+}, as the source of reductant to supply cellular metabolism as first proposed by Zubay [Zubay, 1996], and discussed in more detail in the following.

2.2. *Local cycle of photosynthetic production of manganese oxides and their use as oxidants by (i) heterotrophic bacteria or (ii) the photosynthetic organisms*

Further above we described the structural properties of high-valent manganese oxides, recent insight on their formation and redox activity obtained by electrochemical approaches, and their formation *via* the activity of biological organisms.

We noted that the formation of high-valent manganese oxides is currently observed in numerous species of bacteria and, though not completely understood, it appears to be an adaptation to fluctuations in the redox poise of the environmental niche occupied by these bacteria. In contrast, the proposal for the use of Mn^{2+} as a metabolic reductant for phototrophy is fundamentally different and would have represented a turning point for the evolution of life on Earth since on the one hand, it would have been a highly advantageous metabolic innovation, and on the other, it set the stage for the most momentous innovation in the early biosphere: the evolution of oxygenic photosynthesis. The idea of metal-based phototrophy was first proposed by Zubay [Zubay, 1996] and later developed into more detailed hypotheses by Fisher and coworkers [Johnson *et al.*, 2013; Fischer *et al.*, 2015] and Martin and coworkers [Martin *et al.*, 2018]. Manganese phototrophy is proposed to have followed the evolutionary development of hydrogen, sulfur compound, and iron-based anoxic phototrophy in a succession according to the midpoint potentials of each substrate starting with hydrogen and ending with manganese [Martin *et al.*, 2018]. As these electron donors successively became increasingly limited in the environment, progressive evolutionary alterations in the oxidizing power of the photosynthetic reaction center enabled access to the increasingly more difficult to oxidize substrates. This is a most reasonable scenario, in our opinion.

We extend these ideas to develop the concept of an early ecological cycle of broad-scale reduction and oxidation analogous to the oxygenic photosynthesis/heterotrophic respiration cycle of the present biosphere but involving redox cycling of manganese driven by Mn-phototrophy. Two related variants of this hypothesis are proposed.

In the first scenario, the evolutionary development of manganese phototrophy in anoxygenic bacteria produced manganese oxides that could be used by other bacterial species in the same ecosystem. The oxides produced by manganese phototrophy may have been structurally similar to the poorly crystalline, amorphous microparticles produced by extant Mn-oxidizing bacteria, but instead of multicopper oxidases, the ancient photosystem would produce them as a photosynthetic waste product. These would enable a syntrophic relationship with metal respiring bacteria that could use the manganese oxides as sinks for respiratory electrons similar to the known physiology of *Shewanella* and other extant metal-reducing bacteria. This would be a syntrophic relationship because the phototroph and the metal-reducing bacteria would utilize one another's metabolic products.

The second scenario, is that this photosynthesis-respiratory capacity eventually evolved in single organisms, in analogy to cyanobacteria which, rather than manganese redox couples, utilized the H_2O/O_2 redox couple to mediate diurnal cycle of photosynthesis (day)/respiration (night) with, as detailed in the next section.

In both scenarios, a reaction center, an early PSII precursor, produced $Mn^{3+/4+}$ oxides by light-driven oxidation of dissolved Mn^{2+} ions, in a mechanism similar to the process of photoactivation, but without forming the compact oxide core that is the catalytic site for water oxidation. Assuming that upon illumination manganese oxides were more or less continuously formed, the conversion of dissolved Mn^{2+} ions into a high-valent $Mn^{3+/4+}$ oxide led to the deposition of manganese oxides in the environment of the phototroph, possibly also within the organism (*e.g.* in the periplasmatic space). Instead of exploiting water as a source of electrons and protons for photosynthetic CO_2 reduction, the PSII precursor employed the $Mn(H_2O)_6^{2+}$ hexaquo complex as electron and proton source and, instead of gaseous O_2, solid MnO_2-type oxides were formed. Such manganese oxides would also be subject to the biotic and abiotic transformations leading to the geologic deposits now taken as evidence for the evolution of manganese phototrophy as a precursor to oxygenic photosynthesis [Johnson *et al.*, 2013]. As recently proposed, the advent of this capacity may have relied upon the mutational conversion of anoxygenic reaction centers that had utilized a diffusible cytochrome as the secondary donor to reaction centers that could no longer bind cytochrome, but instead began to utilize soluble metal ions as the source of reductant [Fischer *et al.*, 2016].

In the first scenario, with the capacity to oxidize Mn^{2+} by the new phototrophs and the resultant formation of solid MnO_2-type oxides as their waste product, a new ecological niche was opened to heterotrophic bacteria that were the evolutionary precursors to *Shewanella* and other organisms that would evolve to utilize the manganese oxides as a terminal electron acceptor. In the second scenario, photosynthetic formation of MnO_2-type oxides facilitated a form of quasi-aerobic respiration based on manganese oxides as a terminal electron acceptor in the respiratory chain of the photosynthetic organism. With the evolutionary inception of water oxidation and the eventual accumulation of molecular oxygen in the atmosphere, later today's aerobic respiration would have emerged in response to the availability of this even more advantageous, albeit more dangerous, terminal electron acceptor that set the stage for the expansion of aerobic life on the planet.

2.3. *Local cycle of photosynthesis and quasi-aerobic respiration of the photosynthetic organisms based on the Mn^{2+}/MnO_2 redox couple*

The oxidation of water to dioxygen or alternatively the oxidation of dissolved Mn^{2+} ions to high-valent manganese oxides requires two photosystems operating in a complex chain of photosynthetic light reactions. The essence of the reaction center series operation motif is the cooperative action of a quinone-reducing type

II reaction center and a ferredoxin-reducing type I reaction center stepping up the reducing power of electrons transferred from a relatively high potential donor to a very negative potential acceptor capable of reducing CO_2. While type II-containing anoxygenic phototrophs can readily reduce CO_2, they do so using the electrochemical potential of the proton gradient to drive reverse electron flow through the NAD(P)H-dehydrogenase (Complex I) to produce NAD(P)H and reduced ferredoxin. Having both type I and type II reaction centers in the same organism may have been the consequence of gene duplication/diversification with the advantage of allowing alternative routes for utilizing reduced sulfur compounds as electron donors [Martin *et al.*, 2018]. This may have led to the further diversification of the type II reaction centers to utilize new sources of electrons, eventually including soluble Mn^{2+}. In comparison to CO_2 reduction in photosynthetic organisms with a single photosystem only, the number of photons needed for CO_2 reduction is higher (roughly doubled) and the biosynthesis costs for the complex protein machinery is increased. The latter cost issue is aggravated by the need for a sophisticated machinery to cope with changing light intensities and damage to photosystems operated at highly positive redox potentials. We assume that under conditions of energy limitation, it is likely that ancient photosynthetic organisms would have benefitted from greater energy efficiencies to accommodate high biosynthesis costs (but under some circumstances energetically inefficient pathways still can be advantageous [van Dijken *et al.*, 1993; Burnap, 2015]). Even if this assumption were incorrect (see, *e.g.*, [Molenaar *et al.*, 2009]), the energetic benefits of aerobic respiration would have been a major evolutionary advantage in the development of early photosynthetic organisms.

The following example illustrates the dramatic difference in energetic efficiency: The anaerobic glycolysis resulting in lactate formation facilitates formation of only about two ATP molecules per glucose molecule, whereas (aerobic) oxidative phosphorylation could produce up to 38 ATP molecules. The low efficiency of typical anaerobic respiration and fermentation roots in the lack of a high-potential electron acceptor, which in today's aerobic respiration is dioxygen. Dioxygen is a highly advantageous terminal electron acceptor in respiration because of its redox potential for reduction of O_2 to water of about +0.8 V (*versus* NHE at pH 7). The redox potential of $Mn^{3+/4+}$ oxides depends on the oxide type and average Mn oxidation state but typically it is in vicinity of the O_2/H_2O redox potential, especially at slightly acid pH values which may have prevailed at the time when oxygenic photosynthesis evolved (see Figure 3). Therefore, from an energetic viewpoint, high-valent manganese oxides could serve as a terminal electron acceptor of a respiratory electron transfer chain facilitating equally favorable energetics compared with O_2 molecules. This motivates our hypothesis of a quasi-aerobic respiration based on $Mn^{3+/4+}$ oxides serving as terminal electron acceptor.

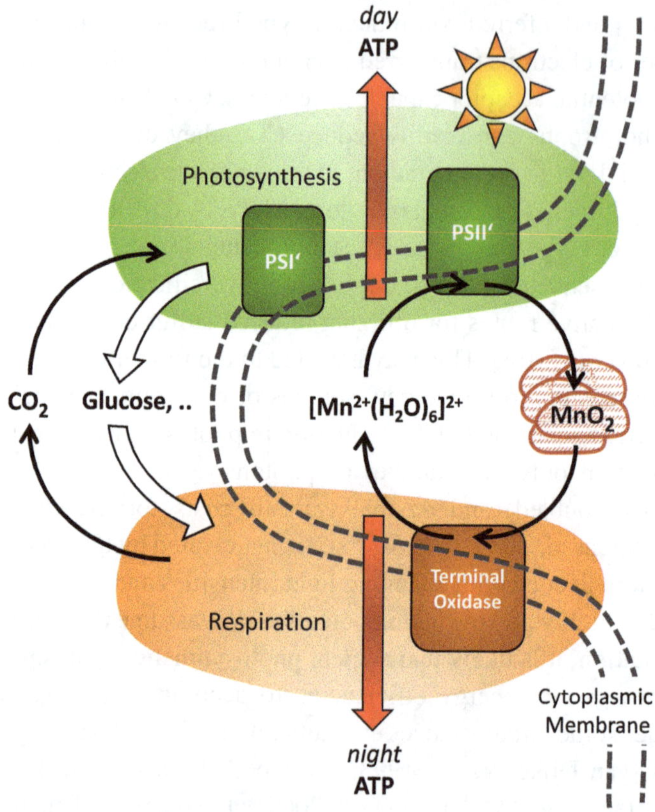

Figure 2. Proposed local cycle of photosynthesis and quasi-aerobic respiration facilitated by manganese oxides. The photosystem II (PSII) as well as the chain of respiratory enzymes are located in the plasma membrane. The manganese-oxidizing, electron-donor side of PSII is at the outer side of the cytoplasmic membrane, as found in the evolutionary ancient cyanobacterium *Gloeobacter* [Rexroth *et al.*, 2011].

For formation of the manganese oxide particles by early photosystems and their use in respiration, we envision a local cycle of photosynthesis and respiration as schematically shown in Figure 2. In comparison to the role of photosynthesis and respiration in today's global carbon cycle, water oxidation and O_2 evolution is replaced by Mn oxidation and formation of high-valent manganese oxides; O_2-reduction is replaced by reduction of high-valent manganese oxides.

Why manganese oxides? A similarly energy efficient local photosynthesis-respiration cycle cannot be established based on Fe oxides even though iron likely has been highly abundant in aquatic environments at the time when oxygenic photosynthesis evolved (and according to recent proposals, the Fe-phototrophy may have preceded Mn-phototrophy [Martin *et al.*, 2018]). The redox potential for oxidation of dissolved Fe^{2+} ions towards Fe oxide (Fe_2O_3) is close to 0 V (*versus* NHE at pH 7) and thus by about 0.6 V less positive than the potential for

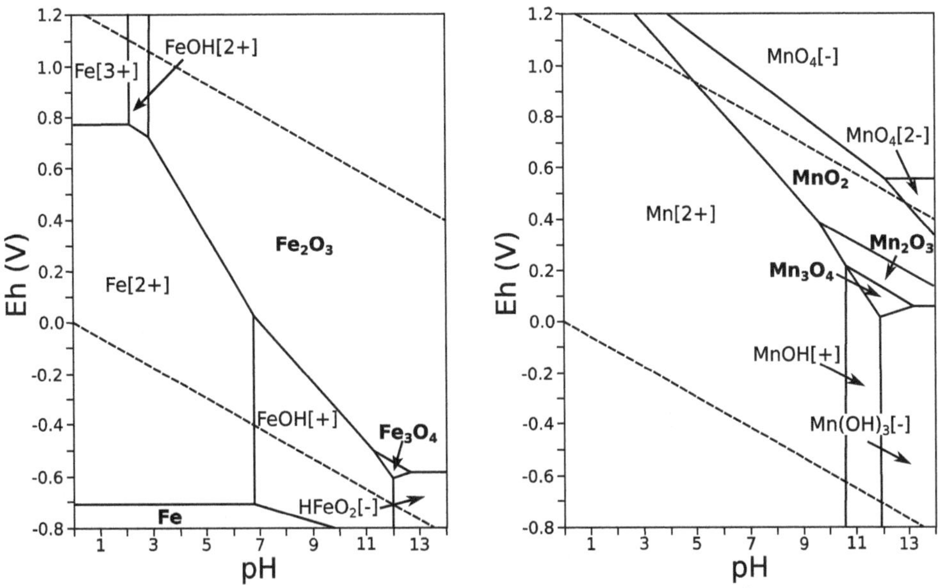

Figure 3. Pourbaix diagrams of iron (a) and manganese (b). The dotted lines indicate the equilibrium potentials of the H_2O/O_2 (upper dashed lines) and H^+/H_2 (lower dashed lines) reactions. The H_2O/O_2 and H^+/H_2 potentials are shown for a standard gas partial pressure of 1 bar. For a O_2 partial pressure of 0.21 bar (corresponding to today's atmospheric O_2 concentration of 21%), the H_2O/O_2 potential is by about 40 mV lower than the indicated values; at a partial pressure of 0.01 bar corresponding to an atmospheric O_2 concentration of 1% (*e.g.*), the H_2O/O_2 potential is by about 120 mV less positive (Pourbaix diagrams based on: Atlas of Eh-pH diagrams, Geological Survey of Japan Open File Report No. 419, National Institute of Advanced Industrial Science and Technology, Research Center for Deep Geological Environments, Naoto Takeno, May 2005).

manganese oxide formation (Figure 3). Using a rough battery picture illustrates the energetic disadvantage. The carbohydrate–MnO_2 battery may provide a voltage of about 1.0 V, assuming a CO_2/[CH_2O] potential of about –0.4 V and Mn^{2+}/MnO_2 potential of about 0.6 V. The carbohydrate–Fe_2O_3 battery provides a voltage of 0.4 V only (assuming a Fe^{2+}/Fe_2O_3 potential of about 0.0 V) and thus is energetically clearly inferior to the carbohydrate–MnO_2 battery. Other oxides based on, *e.g.*, Co or Ni could be energetically similarly favorable than manganese oxides. However, these are comparably scarce elements and their concentration in seawater most likely had been too low for light-driven oxide formation by an ancient photosystem. The seawater concentration of vanadium is comparably high, but the intricate redox chemistry of vanadium, with highly oxidized vanadium in soluble form (vanadates), cannot support a favorable photosynthesis–respiration cycle. Because of availability and suitable redox properties, "mighty manganese" [Schatz, 2003] may have been the only element that could have started the evolution of oxygenic photosynthesis by facilitating an efficient local redox cycle of photosynthesis and efficient respiration.

3. Verification of the Proposed Photosynthetic-Respiratory Manganese Oxide Cycle?

According to Karl Popper, a hypothesis (or theory) can never be verified, but any good hypothesis is accessible to empirical falsification. For hypotheses relating to biological evolution (or natural history in general), Popper's theory of science is not easily applicable. The impossibility of verification (proof) is especially obvious, but unless basic physical, chemical or biological rules are violated, also the options for stringent falsification are limited. Therefore (i) general plausibility and (ii) traces left by the hypothesized evolutionary paths become the central criteria in evaluation of the hypothesis. Plausibility includes agreement with the body of established scientific knowledge. Traces can be found in geological deposits, DNA sequences, and at the functional (physiological) level of still existing organisms. In the first section, we have attempted an overview of the truly fascinating world of manganese oxides, addressing their structural variety and modes of formation, the formation and utilization of manganese oxides by today's bacteria, and the protein-bound manganese complex of oxygenic photosynthesis. Aiming at the specific hypothesis presented in the second section, this overview has been focused on avenues of plausibility and possible traces of early evolutionary stages in the functioning of today's bacteria in ecological niches. The hypothesis (or rather hypotheses) of the second section have been focused on energetic feasibility and benefits, but largely ignore the biochemical mechanism needed to realize what is energetically feasible. Crucial open questions regarding biochemical realization are discussed in the following.

Can photosystems produce manganese oxides by reduction of Mn^{2+} ions?

Biological formation of manganese oxide nanoparticles by oxidation of Mn^{2+} ions by multicopper oxidases has been demonstrated, but this process relies on dioxygen serving as an oxidant, as discussed in Section 2.1.2. In clear contrast, we propose light-driven oxidation of dissolved Mn^{2+} ions by an ancient photosystem. The employment of dissolved Mn^{2+} ions as electron donor in photosynthesis is by no means a trivial step. The dissolved Mn^{2+} ion cannot simply enter the photosystem, become oxidized and subsequently leave the photosystem as dissolved Mn^{3+} or Mn^{4+} ions. This is essentially impossible because isolated $Mn^{3+/4+}$ ions are not stable in the aqueous solvent environment outside of the photosystem. The stabilization of the high-valent manganese ions by formation of numerous di-μ-O(H) bridges (up to six per manganese ion) is essential. Consequently, we propose that Mn^{2+} oxidation involves formation of manganese oxide nanoparticles within the photosystem (or in a surface exposed protein cavity), which leave the photosystem once their dimensions ensure stability in the aqueous protein environment. Whether (and how) photosystems can achieve formation of manganese oxide nanoparticles still is a crucial open question. An alternative to the oxide nanoparticle scenario could be photosystems that operate directly in permanent contact (attachment) to a microscopic or macroscopic "stone", which in the light grows continuously by adding more and high valent metal ions to its surface oxide layer. Preliminary results we recently obtained suggest that photosystems indeed may be able to form manganese oxide nanoparticles (unpublished). Thus, we currently favor a scenario of manganese oxide nanoparticles formed by an ancient PSII precursor and released into the periplasmatic space as schematically shown in Figure 2.

Can manganese oxides be utilized highly efficiently in respiration?

In specific ecological niches, bacteria reduce manganese oxides. The metal oxides may serve as terminal electron acceptors in metabolic (respiratory) processes, but also other functions have been proposed (see Section 1.2.2). However, based on the presently available (and still limited) insight, it is doubtful that these bacteria can use high-valent manganese oxides as terminal electron acceptors as efficiently as aerobic organisms use dioxygen. The ability to use high-valent manganese oxides as a terminal electron acceptor appears be hampered, presumably by the inability of the oxides to diffuse as O_2 does, directly to the terminal oxidase. Despite the huge energetic advantages in fully utilizing the oxidizing power of manganese oxides, the need for diffusible mediators or less well-defined contact mechanisms seems not to have been overcome. The evolutionary persistence of the CymA mechanism (midpoint potential ~0.0 V, see above discussion on *Shewanella* in Section 1.2.2), suggests that mechanisms that could more fully exploit the large potential available in high-valent manganese oxides have not been possible and underlines the radical advantage of O_2 as the terminal acceptor. Nevertheless, despite these constraints, having metal-based respiration very likely produced huge syntrophic advantages.

Like formation of manganese oxide nanoparticles by an ancient photosystem, the biochemical realization of highly efficient use of the oxidizing power of manganese oxides is by no means trivial. Searching for evolutionary traces in todays manganese-reducing bacteria so far has failed to support the employment of manganese oxides as terminal electron acceptors with an energetic efficiency that could compare to aerobic respiration. Future research in this area is needed, which may or may not provide support for the possibility of highly efficient use of manganese oxides in respiratory pathways. Today's metal ion reducers typically can operate on both, manganese oxides as well as iron oxides, using the same or closely related metabolisms. This versatility could be one reason why the high-efficiency respiratory utilization of manganese oxides is not realized. Even more important could be the pronounced differences between today's manganese reducers and early phototrophs regarding the spatial organization of the photosynthetic and respiratory enzymes, which is schematically shown for the (hypothetical) phototroph in Figure 2 (see also Section 2.3). The location of both the manganese oxide producing photosystem and the terminal oxidase in the outer cytosolic membrane could have favored pathways of using manganese oxides that are clearly more efficient than those possible in today's organisms for which manganese oxide reduction has been investigated. Specifically, a direct contact between manganese oxides and terminal oxidases becomes a viable option. For example, manganese oxide nanoparticles (produced by a PSII ancestor) reached the ancient terminal oxidase directly by diffusion, where they were reduced by a terminal oxidase. The terminal oxidase was similar to today's cyt-*c* oxidase, but required additional or modified subunits (in comparison to today's terminal oxidases).

We are aware of the highly hypothetical character of the proposed local cycle of photosynthesis and quasi-aerobic respiration mediated by manganese oxides. Due to the huge benefits of the proposed scheme for individual photosynthetic organisms and local bacterial communities in pre-aerobic times, we see it as a possible or even likely intermediate stage in the development of modern photosynthetic organisms. Supportive evidence is clearly needed, especially regarding biochemical feasibility. In this regard, we see interesting options by searching for traces in the functioning of still existing organisms or genetic variants, the latter aiming at reverse engineering of ancient protein functions. In vitro reconstruction of individual steps of the proposed photosynthetic formation of manganese oxides and their respiratory utilization could support our hypothesis even if definitive verification cannot be achieved.

Acknowledgments

Financial support by the Deutsche Forschungsgemeinschaft (SFB 1078/ project A4-Dau) and the Einstein Foundation (founded by the State of Berlin; program "Einstein Visiting Fellow", awarded to RB) is gratefully acknowledged.

References and Citations

Anderson, J.M. and Andersson, B. (1982). The architecture of photosynthetic membranes: Lateral and transverse organization. *Trends Biochem. Sci.*, 7, 288–292.

Bao, H. and Burnap, R.L. (2016). Photoactivation: The light-driven assembly of the water oxidation complex of Photosystem II. *Front. Plant Sci.*, 7, 578.

Brandenburg, F., Schoffman, H., Kurz, S., Kramer, U., Keren, N., Weber, A.P., *et al.* (2017). The synechocystis manganese exporter Mnx Is essential for manganese homeostasis in cyanobacteria. *Plant Physiol.*, 173, 1798–1810.

Burnap, R.L. (2015). Systems and photosystems: Cellular limits of autotrophic productivity in cyanobacteria. *Front. Bioeng. Biotechnol.*, 3, 1.

Cardona, T., Murray, J.W. and Rutherford, A.W. (2015). Origin and evolution of water oxidation before the last common ancestor of the cyanobacteria. *Mol. Biol. Evol.*, 32, 1310–1328.

Chen, M., Schliep, M., Willows, R.D., Cai, Z.L., Neilan, B.A. and Scheer, H. (2010). A red-shifted chlorophyll. *Science*, 329, 1318–1319.

Dau, H., Grundmeier, A., Loja, P. and Haumann, M. (2008). On the structure of the manganese complex of photosystem II: extended-range EXAFS data and specific atomic-resolution models for four S-states. *Philos. Trans. R. Soc. Lond B. Biol. Sci.*, 363, 1237–1243.

Dau, H. and Haumann, M. (2007). Eight steps preceding O–O bond formation in oxygenic photosynthesis — A basic reaction cycle of the Photosystem II manganese complex. *Biochim. Biophys. Acta.*, 1767, 472–483.

Dau, H. and Haumann, M. (2008). The manganese complex of photosystem II in its reaction cycle-Basic framework and possible realization at the atomic level. *Coord. Chem. Rev.*, 252, 273–295.

De Las Rivas, J. and Barber, J. (2004). Analysis of the structure of the PsbO protein and its implications. *Photosynth. Res.*, 81, 329–343.

Dick, G.J., Torpey, J.W., Beveridge, T.J. and Tebo, B.M. (2008). Direct identification of a bacterial manganese(II) oxidase, the multicopper oxidase MnxG, from spores of several different marine *Bacillus* species. *Appl. Environ. Microbiol.*, 74, 1527–1534.

Dismukes, G.C. and Siderer, Y. (1981). Intermediates of a polynuclear manganese center involved in photosynthetic oxidation of water. *Proc. Natl. Acad. Sci. USA*, 78, 274–278.

Ferreira, K.N., Iverson, T.M., Maghlaoui, K., Barber, J. and Iwata, S. (2004). Architecture of the photosynthetic oxygen-evolving center. *Science*, 303, 1831–1838.

Fischer, W.W., Hemp, J. and Johnson, J.E. (2015). Manganese and the evolution of photosynthesis. *Orig. Life Evol. Biosph.*, 45, 351–357.

Fischer, W.W., Hemp, J. and Johnson, J.E. (2016). Evolution of oxygenic photosynthesis. *Annu. Rev. Earth Planet Sci.*, 44, 647–683.

Gan, F., Zhang, S., Rockwell, N.C., Martin, S.S., Lagarias, J.C. and Bryant, D.A. (2014). Extensive remodeling of a cyanobacterial photosynthetic apparatus in far-red light. *Science*, 345, 1312–1317.

Golden, D.C., Chen, C.C. and Dixon, J.B. (1987). Transformation of birnessite to buserite, todorokite, and manganite under mild hydrothermal treatment. *Clays Clay Miner.*, 35, 271–280.

Gonzalez-Flores, D., Zaharieva, I., Heidkamp, J., Chernev, P., Martinez-Moreno, E., Pasquini, C., *et al.* (2016). Electrosynthesis of biomimetic manganese-calcium oxides for water oxidation catalysis — Atomic structure and functionality. *Chem. Sus. Chem.*, 9, 379–387.

Goodin, D.B., Yachandra, V.K., Britt, R.D., Sauer, K. and Klein, M.P. (1984). The state of manganese in the photosynthetic apparatus. 3. Light-induced changes in X-ray absorption (K-edge) energies of manganese in photosynthetic membranes. *Biochim. Biophys. Acta*, 767, 209–216.

Grabolle, M., Haumann, M., Muller, C., Liebisch, P. and Dau, H. (2006). Rapid loss of structural motifs in the manganese complex of oxygenic photosynthesis by X-ray irradiation at 10–300 K. *J. Biol. Chem.*, 281, 4580–4588.

Grundmeier, A. and Dau, H. (2012). Structural models of the manganese complex of photosystem II and mechanistic implications. *Biochim. Biophys Acta*, 1817, 88–105.

Ho, M.Y., Shen, G., Canniffe, D.P., Zhao, C. and Bryant, D.A. (2016). Light-dependent chlorophyll f synthase is a highly divergent paralog of PsbA of photosystem II. *Science*, 353, aaf9178.

Johnson, J.E., Webb, S.M., Thomas, K., Ono, S., Kirschvink, J.L. and Fischer, W.W. (2013). Manganese-oxidizing photosynthesis before the rise of cyanobacteria. *Proc Natl. Acad. Sci. USA*, 110, 11238–11243.

Keren, N., Kidd, M.J., Penner-Hahn, J.E. and Pakrasi, H.B. (2002). A light-dependent mechanism for massive accumulation of manganese in the photosynthetic bacterium *Synechocystis* sp. PCC 6803. *Biochemistry*, 41, 15085–15092.

Klauss, A., Haumann, M. and Dau, H. (2012). Alternating electron and proton transfer steps in photosynthetic water oxidation. *Proc. Natl. Acad. Sci. USA*, 109, 16035–16040.

Koyama, K., Suzuki, H., Noguchi, T., Akimoto, S., Tsuchiya, T. and Mimuro, M. (2008). Oxygen evolution in the thylakoid-lacking cyanobacterium *Gloeobacter violaceus* PCC 7421. *Biochim. Biophys. Acta*, 1777, 369–378.

Kuhn, T., Wegorzewski, A., Rühlemann, C. and Vink, A. (2017). Composition, Formation, and Occurrence of Polymetallic Nodules. In: *Deep-Sea Mining: Resource Potential, Technical and Environmental Considerations*, R. Sharma (ed.). Cham: Springer International Publishing, pp. 23–63.

Lamb, J.J. and Hohmann-Marriott, M.F. (2017). Manganese acquisition is facilitated by PilA in the cyanobacterium *Synechocystis* sp. PCC 6803. *PLoS One*, 12, e0184685.

Lee, S.Y., Gonzalez-Flores, D., Ohms, J., Trost, T., Dau, H., Zaharieva, I., *et al.* (2014). Screen-printed calcium-birnessite electrodes for water oxidation at neutral pH and an "electrochemical harriman series". *Chem. Sus. Chem.*, 7, 3442–3451.

Loll, B., Kern, J., Saenger, W., Zouni, A. and Biesiadka, J. (2005). Towards complete cofactor arrangement in the 3.0 Å resolution structure of photosystem II. *Nature*, 438, 1040–1044.

Luo, J. and Suib, S.L. (1997). Preparative parameters, magnesium effects, and anion effects in the crystallization of birnessites. *J. Phys. Chem. B*, 101, 10403–10413.

Mann, P.J.G. and Quastel, J.H. (1946). Manganese metabolism in soils. *Nature*, 158, 154–156.

Mares, J., Hrouzek, P., Kana, R., Ventura, S., Strunecky, O. and Komarek, J. (2013). The primitive thylakoid-less cyanobacterium *Gloeobacter* is a common rock-dwelling organism. *PLoS One*, 8, e66323.

Martin, W.F., Bryant, D.A. and Beatty, J.T. (2018). A physiological perspective on the origin and evolution of photosynthesis. *FEMS Microbiol. Rev.*, 42, 205–231.

Molenaar, D., van Berlo, R., de Ridder, D. and Teusink, B. (2009). Shifts in growth strategies reflect tradeoffs in cellular economics. *Mol. Syst. Biol.*, 5, 323.

Murray, J.W. (2012). Sequence variation at the oxygen-evolving centre of photosystem II: a new class of 'rogue' cyanobacterial D1 proteins. *Photosynth Res.*, 110, 177–184.

Najafpour, M. M., Ehrenberg, T., Wiechen, M. and Kurz, P. (2010). Calcium manganese(III) oxides ($CaMn_2O_4xH_2O$) as biomimetic oxygen-evolving catalysts. *Angew. Chem. Int. Ed. Engl.*, 49, 2233–2237.

Nakamura, Y. (2003). Complete genome structure of *Gloeobacter violaceus* PCC 7421, a cyanobacterium that lacks thylakoids. *DNA Res.*, 10, 181–201.

Nürnberg, D.J., Morton, J., Santabarbara, S., Telfer, A., Joliot, P., Antonaru, L. A. *et al.* (2018). Photochemistry beyond the red limit in chlorophyll *f*-containing photosystems. *Science*, 360, 1210–1213.

Paquete, C.M. and Louro, R.O. (2010). Molecular details of multielectron transfer: the case of multiheme cytochromes from metal respiring organisms. *Dalton Trans.*, 39, 4259–4266.

Peloquin, J.M., Campbell, K.A., Randall, D.W., Evanchik, M.A., Pecoraro, V.L., Armstrong, W.H. *et al.* (2000). ^{55}Mn ENDOR of the S2-state multiline EPR signal of photosystem II: Implications on the structure of the tetranuclear Mn cluster. *J. Am. Chem. Soc.*, 122, 10926–10942.

Pirbadian, S., Barchinger, S.E., Leung, K.M., Byun, H.S., Jangir, Y., Bouhenni, R.A. *et al.* (2014). Shewanella oneidensis MR-1 nanowires are outer membrane and periplasmic extensions of the extracellular electron transport components. *Proc. Natl. Acad. Sci USA*, 111, 12883–12888.

Post, J.E. (1999). Manganese oxide minerals: Crystal structures and economic and environmental significance. *Proc. Natl. Acad. Sci. USA*, 96, 3447–3454.

Post, J.E. and Veblen, D.R. (1990). Crystal structure determinations of synthetic sodium, magnesium, and potassium birnessite using TEM and the Rietveld method. *Am. Mineral*, 75, 477–489.

Rexroth, S., Mullineaux, C.W., Ellinger, D., Sendtko, E., Rogner, M. and Koenig, F. (2011). The plasma membrane of the cyanobacterium *Gloeobacter violaceus* contains segregated bioenergetic domains. *Plant Cell*, 23, 2379–2390.

Richardson, L.L., Aguilar, C. and Nealson, K.H. (1988). Manganese oxidation in pH and O2microenvironments produced by phytoplankton. *Limnol. Oceanogr.*, 33, 352–363.

Sauer, K. and Yachandra, V.K. (2002). A possible evolutionary origin for the Mn_4 cluster of the photosynthetic water oxidation complex from natural MnO_2 precipitates in the early ocean. *Proc. Natl. Acad. Sci. USA*, 99, 8631–8636.

Saw, J.H., Schatz, M., Brown, M.V., Kunkel, D.D., Foster, J.S., Shick, H. *et al.* (2013). Cultivation and complete genome sequencing of *Gloeobacter kilaueensis* sp. nov., from a lava cave in Kilauea Caldera, Hawai'i. *PLoS One*, 8, e76376.

Schatz, G. (2003). Mighty manganese. *FEBS Lett.*, 551, 1–2.

Soldatova, A.V., Romano, C.A., Tao, L., Stich, T.A., Casey, W.H., Britt, R.D. *et al.* (2017). Mn(II) oxidation by the multicopper oxidase complex Mnx: A coordinated two-stage Mn(II)/(III) and Mn(III)/(IV) mechanism. *J. Am. Chem. Soc.*, 139, 11381–11391.

Soldatova, A.V., Tao, L., Romano, C.A., Stich, T.A., Casey, W.H., Britt, R.D. *et al.* (2017). Mn(II) oxidation by the multicopper oxidase complex Mnx: A binuclear activation mechanism. *J. Am. Chem. Soc.*, 139, 11369–11380.

Spiro, T.G., Bargar, J.R., Sposito, G. and Tebo, B.M. (2010). Bacteriogenic manganese oxides. *Acc. Chem. Res.*, 43, 2–9.

Suga, M., Akita, F., Hirata, K., Ueno, G., Murakami, H., Nakajima, Y. *et al.* (2015). Native structure of photosystem II at 1.95 Å resolution viewed by femtosecond X-ray pulses. *Nature*, 517, 99–103.

Tebo, B.M., Johnson, H.A., McCarthy, J.K. and Templeton, A.S. (2005). Geomicrobiology of manganese(II) oxidation. *Trends Microbiol.*, 13, 421–428.

Toyoda, K. and Tebo, B.M. (2013). The effect of Ca(2+) ions and ionic strength on Mn(II) oxidation by spores of the marine *Bacillus* sp. SG-1. *Geochim. Cosmochim. Acta*, 101, 1–11.

Umena, Y., Kawakami, K., Shen, J.R. and Kamiya, N. (2011). Crystal structure of oxygen-evolving photosystem II at a resolution of 1.9 Å *Nature*, 473, 55–60.

van Dijken, J.P., Weusthuis, R.A. and Pronk, J.T. (1993). Kinetics of growth and sugar consumption in yeasts. *Antonie Van Leeuwenhoek*, 63, 343–352.

Wegener, K.M., Nagarajan, A. and Pakrasi, H.B. (2015). An atypical *psbA* gene encodes a sentinel D1 protein to form a physiologically relevant inactive photosystem II complex in cyanobacteria. *J. Biol. Chem.*, 290, 3764–3774.

Wright, M.H., Farooqui, S.M., White, A.R. and Greene, A.C. (2016). Production of manganese oxide nanoparticles by *Shewanella* species. *Appl. Environ. Microbiol.*, 82, 5402–5409.

Yachandra, V.K., DeRose, V.J., Latimer, M.J., Mukerji, I., Sauer, K. and Klein, M.P. (1993). Where plants make oxygen: A structural model for the photosynthetic oxygen-evolving manganese cluster. *Science*, 260, 675–679.

Yachandra, V.K., Guiles, R.D., McDermott, A., Britt, R.D., Dexheimer, S.L., Sauer, K. *et al.* (1986). The state of manganese in the photosynthetic apparatus. *Biochim. Biophys. Acta*, 850, 324–332.

Yano, J., Kern, J., Irrgang, K.D., Latimer, M.J., Bergmann, U., Glatzel, P. *et al.* (2005). X-ray damage to the Mn_4Ca complex in single crystals of photosystem II: A case study for metalloprotein crystallography. *Proc. Natl. Acad. Sci. USA*, 102, 12047–12052.

Young, I.D., Ibrahim, M., Chatterjee, R., Gul, S., Fuller, F., Koroidov, S. *et al.* (2016). Structure of photosystem II and substrate binding at room temperature. *Nature*, 540, 453–457.

Zaharieva, I., Chernev, P., Risch, M., Klingan, K., Kohlhoff, M., Fischer, A. *et al.* (2012). Electrosynthesis, functional, and structural characterization of a water-oxidizing manganese oxide. *Energy Environ. Sci.*, 5, 7081–7089.

Zaharieva, I., González-Flores, D., Asfari, B., Pasquini, C., Mohammadi, M.R., Klingan, K. *et al.* (2016). Water oxidation catalysis — role of redox and structural dynamics in biological photosynthesis and inorganic manganese oxides. *Energy Environ. Sci.*, 9, 2433–2443.

Zaharieva, I., Najafpour, M.M., Wiechen, M., Haumann, M., Kurz, P. and Dau, H. (2011). Synthetic manganese–calcium oxides mimic the water-oxidizing complex of photosynthesis functionally and structurally. *Energy Environ. Sci.*, 4, 2400–2408.

Zhang, C., Chen, C., Dong, H., Shen, J.R., Dau, H. and Zhao, J. (2015). Inorganic chemistry. A synthetic Mn(4)Ca-cluster mimicking the oxygen-evolving center of photosynthesis. *Science*, 348, 690–693.

Zhang, M., Bommer, M., Chatterjee, R., Hussein, R., Yano, J., Dau, H. *et al.* (2017). Structural insights into the light-driven auto-assembly process of the water-oxidizing Mn_4CaO_5-cluster in photosystem II. *Elife*, 6.

Zouni, A., Witt, H. T., Kern, J., Fromme, P., Krauss, N., Saenger, W. *et al.* (2001). Crystal structure of photosystem II from *Synechococcus elongatus* at 3.8 Å resolution. *Nature*, 409, 739–743.

Zubay, G. (1996). *Orgins of Life on Earth and in the Cosmos*. Academic Press.

Chapter 19

A Perspective on a Paradigm Shift in Plant Photosynthesis: Designing a Novel Type of Photosynthesis in Sorghum by Combining C_3 and C_4 Metabolism

Christer Jansson[*,¶¶], Todd Mockler[†], John P. Vogel[‡], Henrique DePaoli[‡], Samuel P. Hazen[§], Venkat Srinivasan[¶], Asaph Cousins[**], Peggy Lemaux[††], Jeff Dahlberg[‡‡] and Tom Brutnell[§§]

[*]Pacific Northwest National Laboratory (PNNL), The Environmental Molecular Science Laboratory (EMSL), Richland, WA, USA

[†]Donald Danforth Plant Science Center (DDPSC), St. Louis, MO, USA

[‡]Lawrence Berkeley National Laboratory (LBNL), The Joint Genome Institute (JGI)

[§]University of Massachusetts, Amherst (UMass), Amherst, MA, USA

[¶]Indian Institute of Technology Madras (IITM), Chennai, India

[**]Washington State University (WSU), Pullman, WA, USA

[††]University of California, Berkeley (UC Berkeley), Berkeley, CA, USA

[‡‡]Kearny Agricultural Research & Extension Center (KARE), Parlier, CA, USA

[§§]Viridis Genomics Consulting, USA

[¶¶]christer.jansson@pnnl.gov

Plant biomass is produced by photosynthesis — be it lignocellulosic feedstock for fuels or forage, grain or tubers for food, or extracted fractions for commodity or high-value chemicals. Meeting the increasing demand for plant biomass going into food, feed and energy production without negative environmental and economic impacts will require a 50–85% increase in crop productivity in the coming decades, without the expansion of agricultural land area. The potential of enhanced photosynthetic efficiency to help achieve the sustainable yield increases required to meet future demands for food and energy has spurred intense research towards understanding, modeling and engineering photosynthesis. In this perspective article we advocate for a bold and novel approach to engineer the photosynthetic machinery in sorghum, an annual C_4 grass with desirable agronomic and genomic properties for biomass production. As a C_4 crop, sorghum is well adapted to high productivity in warm and arid climates. However, reliance on C_4 photosynthesis also makes sorghum prone to frequent chilling injury, low productivity, and inferior quantum yields compared with C_3 vegetation during diurnal and seasonal cycles of low temperatures. We propose that research efforts should be directed towards designing sorghum that can extend the utilization of its potential growing season by rewiring the biochemical machinery to operate either C_3 or C_4 photosynthesis in a temperature-dependent manner. Such a $C_{3,4}$ sorghum would represent a paradigm shift in plant photosynthesis by combining the benefits of both C_3 and C_4 photosynthesis, and dramatically enhance sorghum production on an annual basis.

1. Sorghum — A Versatile Super Crop

Sorghum (*Sorghum bicolor* [L.] Moench) is a high-yielding and drought-resilient annual C_4 crop used for grain, forage or lignocellulosic or saccharine bioenergy production. Sorghum's center of origin is northeastern Africa, from which it migrated to drier areas of Africa where rainfall patterns were unpredictable and often severe. With its adaptation to extreme environments, sorghum exhibits extensive genetic variation for drought tolerance [Blum, 1979; Doggett, 1988]. Due to its agronomic properties, diverse germplasm, relatively small (700–772 Mbp), diploid genome, annotated whole-genome sequences and developed platform for genome editing and engineering, sorghum is an ideal crop in engineering efforts to enhance food and bioenergy production under future climate scenarios.

2. The Benefits of C_4 Photosynthesis

C_4 plants like sorghum, maize and switchgrass exhibit higher photosynthetic energy conversion efficiency compared to C_3 plants like wheat and rice by employing a biochemical pump that increases the concentration of CO_2 in the vicinity of ribulose 1,5-bisphosphate carboxylase/oxygenase (Rubisco) enzyme, thereby avoiding

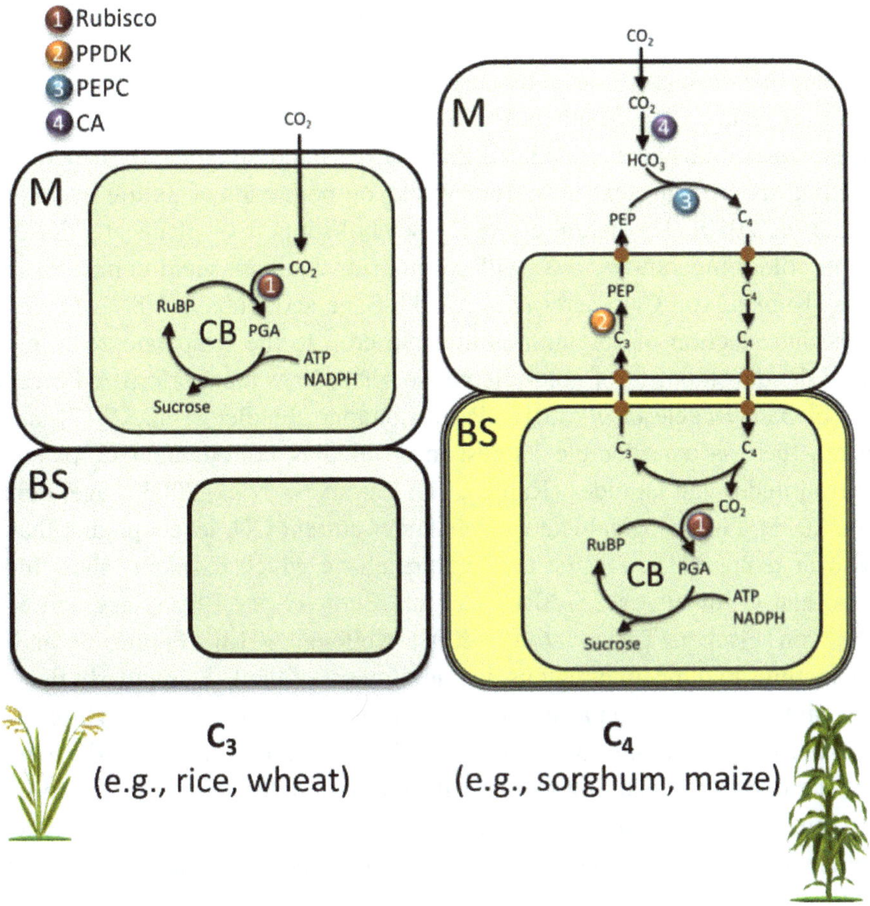

Figure 1. Salient features of C_3 vs. C_4 photosynthesis. BS, bundle sheath cell; C_3, C_3 acid (pyruvate); C_4, C_4 acid (malate/aspartate); CA, carbonic anhydrase; CB, Calvin-Benson cycle; M, mesophyll cell; PEP, phosphoenolpyruvate; PEPC, phosphoenolpyruvate carboxylase; PPDK, pyruvate phosphate dikinase; RuBP, ribuose 1,5-bisphosphate; Rubisco, ribulose 1,5-bisphosphate carboxylase/oxygenase.

energy and CO_2 losses due to photorespiration [Zhu, Long *et al.*, 2008; Zhu, Long *et al.*, 2010; Sage and Zhu, 2011; Slattery and Ort, 2015] (Figure 1). C_4 photosynthesis evolved 30–80 million years ago as an adaptation to declining atmospheric CO_2 concentrations that increased photorespiratory losses [Christin, Besnard *et al.*, 2008; Vicentini, Barber *et al.*, 2008]. Today, C_4 plants dominate grasslands and biomass production in the warmer climates of the tropics and subtropics [Edwards, Osborne *et al.*, 2010]. However, C_4 photosynthesis incurs an extra cost compared to C_3 photosynthesis (2 additional ATP for every mole of CO_2 fixed), and in temperate climates, C_3 photosynthesis often outperforms C_4 photosynthesis.

3. The Shortcomings of C_4 Photosynthesis

Sorghum like most crops does not currently use the full potential growing season [Zhu, Long *et al.*, 2010; Maulana and Tesso, 2013; Mullet, Morishige *et al.*, 2014]. For example, it has been concluded that maize would significantly benefit from extending its growing season by four weeks on both ends of its life cycle [Zhu, Long *et al.*, 2010]. The nature of the C_4 metabolism puts C_4 plants at a disadvantage in cold temperatures, *e.g.*, with an inferior quantum yield compared to C_3 plants [Kubien, von Cammerer *et al.*, 2003; Sage and Kubien, 2007]. Therefore, for the introduction of sorghum as a stable crop to the temperate zone, efforts toward the development of cold-tolerant sorghum lines are a prioritized breeding goal [Fiedler, Bekele *et al.*, 2016; Parra-Londono, Fiedler *et al.*, 2017], and in general, there is considerable interest in developing cold-tolerant C_4 plants for higher altitudes and latitudes [Kubien, von Cammerer *et al.*, 2003; Sage, Peixoto *et al.*, 2015]. Physiological leaf models under current CO_2 levels predict that the crossover temperature, *i.e.* the temperature above which C_4 plants show higher yields than C_3 plants, is 25–30°C [Collatz, Berry *et al.*, 1998; Sage and Sage, 2002; von Fischer, Tiezen *et al.*, 2008; Kellogg, 2013] (Figures 2 and 4). Temperatures in temperate regions, including most of the U.S., regularly fluctuate above and below 25°C, on a seasonal and diurnal basis, and most sorghum varieties under cultivation are vulnerable to cooler conditions during stand establishment at early season planting from April to May in majority of the U.S. sorghum belt [Burow, Burke *et al.*, 2011]. Stand establishment and early season vigor of sorghum is adversely affected by air and soil temperatures below 15°C during

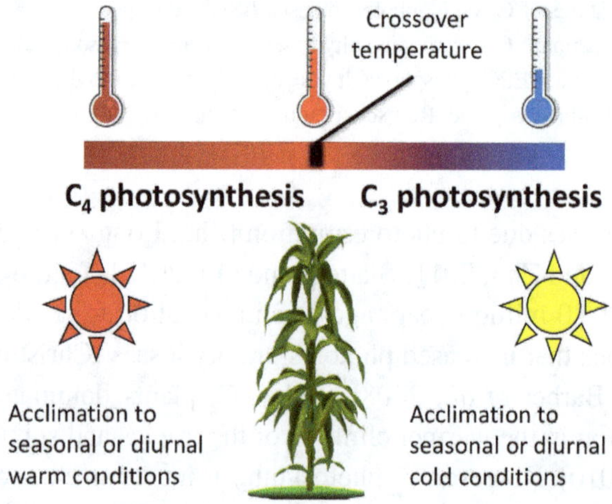

Figure 2. $C_{3,4}$ sorghum. Plants will shift between C_3 or C_4 photosynthesis depending on weather and season. Reproduced with permission from Jansson *et al.* (2018) *J. Exp. Bot.* 69, 3801–3809.

germination, emergence and early seedling growth, even in genotypes adapted to a temperate climate [Maulana, Weerasooriya *et al.*, 2017]. This cold sensitivity limits sorghum production both temporally and spatially by shortening the planting window in places where the crop is already under cultivation, and through hampering its spatial expansion to colder regions. Breeding efforts to extend its range or growing season have been limited [Maulana, Weerasooriya *et al.*, 2017].

Additionally, whole plant canopy models [Marshall-Colon, Long *et al.*, 2017] demonstrate that large fractions of leaf area in modern crop cultivars planted at high density experience shaded light conditions where they are energy limited [Drewry, Kumar *et al.*, 2014; Srinivasan, Kumar *et al.*, 2017]. Under these low-light conditions the competitive advantage of C_4 leaves decreases [Ubierna, Sun *et al.*, 2011; Wang, Guo *et al.*, 2012; Parra-Londono, Fiedler *et al.*, 2017] (Figure 3). Furthermore, whole canopy energy balance model simulations show that the temperature of shaded leaves is cooler than sunlit leaves [Drewry, Kumar *et al.*, 2010]. With the temperature optimum of C_4 photosynthesis higher than C_3 [Sage and Kubien, 2007], lower leaf temperatures in the shaded portion of the canopy decreases the advantage of C_4 leaves over C_3. Including the diurnal effect of variation of solar radiation where peak light levels are achieved only around mid-day and lower light levels are experienced during much of the mornings and evenings, the relative advantage of C_4 over the whole plant canopy further decreases. As

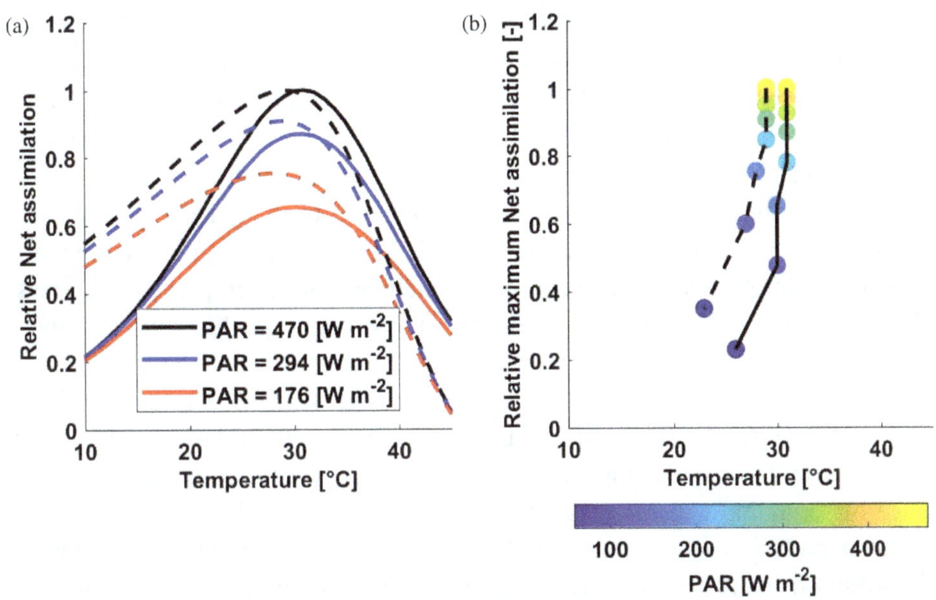

Figure 3. Physiological leaf-level model of gas exchange. Efficiency of C_3 *vs.* C_4 photosynthesis under different light intensities and leaf temperatures (a). Relative decrease in max CO_2 assimilation with decreasing light intensities (b). C_3, broken lines; C_4, solid lines.

breeding efforts in the grasses are focused on increasing stand density, it is likely that these constraints on C_4 photosynthesis will increase in the years ahead.

4. Designing Dual C_3-C_4 Photosynthesis

Considering the discussion above, a flexible, dual C_3-C_4 sorghum ($C_{3,4}$ sorghum) with a thermo-inducible switch would enable the implementation of C_3 or C_4 photo-synthesis depending on cyclic or episodic variations in temperatures across the entire potential growing season and a wide range of biogeographical regions. Such a sorghum plant would maximize yield under seasonal or episodic fluctuations in temperature by selecting the most favorable CO_2 assimilation route, and the advantage of the shift from C_4 to C_3 mode at lower temperatures would be accentuated in the shaded portion of the canopy. Importantly, a dual C_3-C_4 photosynthesis would generate an independent, broader and much faster response to temperature variations than the known 'temperature acclimation of photosynthesis' [Yamori, Hikosaka *et al.*, 2014], which gives plants only a limited ability to cope with extreme of temperatures on a non-diel and non-seasonal manner.

The yield gap for biomass sorghum and other C_4 grass energy crops amounts to 25–33% of the yield potential (Yp) [Mullet, Morishige *et al.*, 2014], pointing to the significant opportunity that exists to increase productivity in C_4 grasses, *e.g.*, by employing genetic improvements to extend season length (Sj) [Mullet, Morishige *et al.*, 2014]. Even for sorghum genotypes adapted to cold-weather conditions, mean seedling dry weight was 85% lower in early-season planting compared to normal-season planting in Kansas field trials, and seedling emergence and vigor were down 50% [Maulana, Weerasooriya *et al.*, 2017]. Another sorghum study in Kansas showed that cold stress at the flowering stage resulted in a 40-–60% loss in yield components both in field trials and under controlled greenhouse conditions [Maulana and Tesso, 2013]. As an analogous example, the 8-weeks longer growing season for the unusually cold-tolerant perennial C_4 grass Miscanthus resulted in 60% increase in cumulative intercepted radiation and 60% more biomass than for the typical C_4 grass maize, when grown in the US Midwest, where recorded yields of maize are among the highest in the world [Zhu, Long *et al.*, 2010]. Given the effects of cold temperature stress on sorghum establishment, growth, yield, lower leaf temperatures of the inner canopy, and limited geographic range of sorghum, it seems reasonable to assume that introducing a flexible $C_{3,4}$ sorghum cultivar has the potential to dramatically increase sorghum production.

Additionally, at atmospheric CO_2 levels above 600 ppm, C_4 photosynthesis is predicted to provide little advantage, a notion that is corroborated in open-air FACE experiments [Leakey, Uribelarrea *et al.*, 2006]. Consequently, with continued global change and rising atmospheric CO_2 levels, the ability for a C_4 crop plant

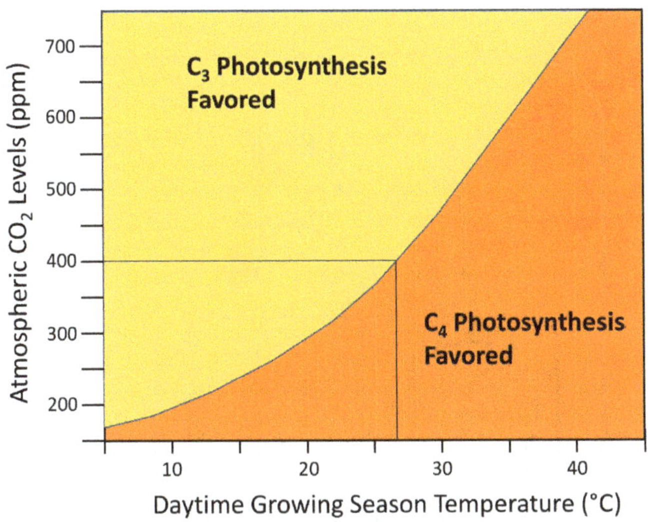

Figure 4. Conditions favoring C_3 or C_4 photosynthesis. Temperatures and CO_2 concentrations at which C_3 or C_4 plants are favored, based on the quantum yield of photosynthesis. Modified with permission from: Kellogg, E.A. (2013) *Curr. Bio.* 23, R594–R599.

to shift to C_3 mode becomes increasingly important even as global temperatures increase (Figure 4). Hence, the impact of $C_{3,4}$ sorghum will be further enhanced in future climate scenarios.

Importantly, $C_{3,4}$ sorghum would maintain all the vital genomic qualities of sorghum [Paterson, 2008; Brenton, Cooper *et al.*, 2016] that makes it amenable for continued breeding and engineering of the introduction of additional agronomic traits like pathogen resistance. As an annual crop, $C_{3,4}$ sorghum would offer the advantage to farmers who do not want to commit 10–15 years growing perennial bioenergy grasses like Miscanthus or switchgrass [Glithero, Wilson *et al.*, 2013], but instead prefer the flexibility of crop rotations and swift adjustments to market demands and prospects for financial returns.

As a C_4 grass, sorghum has evolved anatomical and biochemical adaptations to facilitate carbon exchange between the bundle sheath (BS) and mesophyll (M) cells (Figure 1). C_4 plants possess a carbon shuttle to move carbon initially fixed in the M cells into the BS through a C_4 acid. In comparison to C_3 grasses, C_4 grasses also have a higher vein density, increased plastid numbers and size in BS and often have a suberized BS layer.

For construction of $C_{3,4}$ sorghum, we recommend a two-pronged approach, each leveraging overlapping but independent features of C_4 metabolism, a "CB cycle installation approach" whereby Rubisco activity is installed in C_4 M cells (Figure 5) and a "CO_2 re-direction approach" that will bypass the initial CO_2-fixation step in M cells and direct the CO_2 to Rubisco in the BS cells (Figure 6).

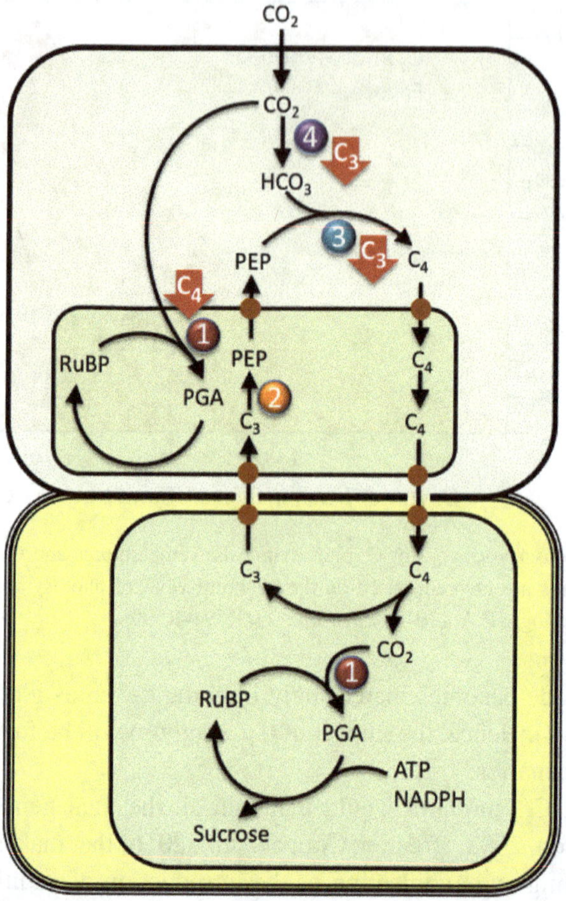

Figure 5. The "CB cycle installation" approach for developing $C_{3,4}$ sorghum. Downward arrows indicate enzymes / pathways suppressed in C_3 or C_4 mode. See Figure 1 for additional information.

From an evolutionary perspective, the CB cycle installation approach can be viewed as a re-installation of CB activity in M cells. As a case in point, in the rare C_3-C_4 intermediates, which are considered an evolutionary link between C_3 and C_4 plants [Stata, Sage *et al.*, 2016], Rubisco resides in both M and BS cells [Drincovich, Casati *et al.*, 1998, Ueno, 2011]. Suppressing the carbon shuttle in C_3 mode by temperature control of key enzymes (*e.g.*, PEPC and CA), would allow for CO_2 entering the M cell to be assimilated through the CB cycle (Figure 5).

BS cell-specific expression of Rubisco in C_4 plants is secured by an intricate network of control points, including transcriptional and post-transcriptional regulation [Patel and Berry, 2008; Wostrikoff, Clark *et al.*, 2012; Berry, Mure *et al.*, 2016] (Figure 7). Thus, the CB installation approach relies on the ability to modify

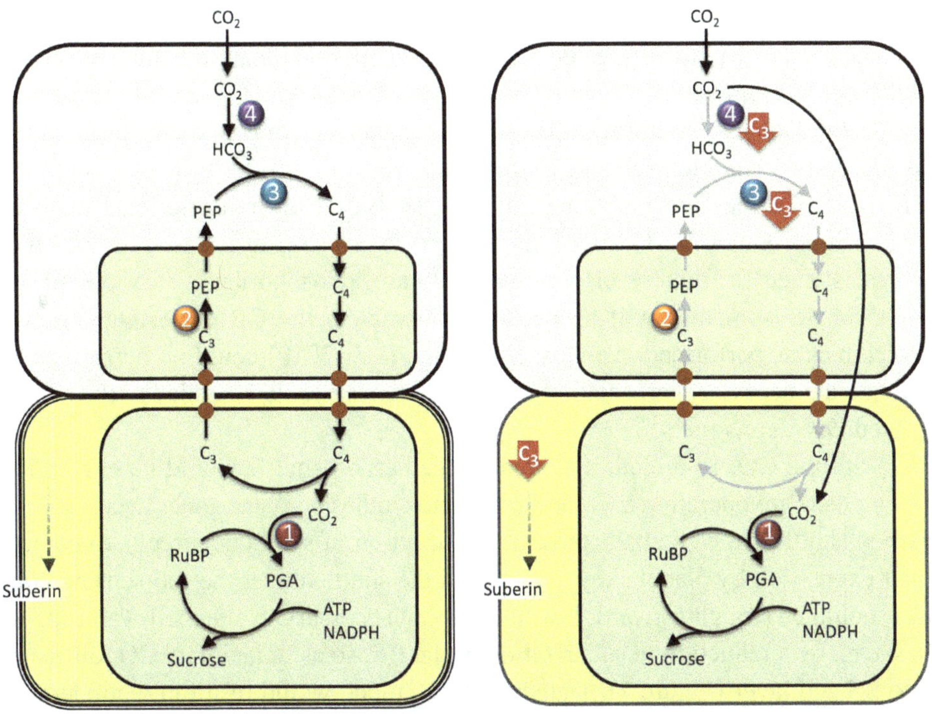

Figure 6. The "CO_2 re-direction" approach for developing $C_{3,4}$ sorghum. Downward arrows indicate enzymes/pathways suppressed in C_3 mode. See Figure 1 for additional information.

key regulatory steps to maintain CB cycle activity in the M cells. For example, Rubisco assembly factor 1 (Raf1), required for assembly of the large and small subunits (LSU and SSU, respectively) of Rubisco (Figure 6) is BS cell specific in C_4 plants [Berry, Mure *et al.*, 2016]. Another example is the RNA-binding protein RLSB that specifically docks to the mRNA of the Rubisco LSU, rbcL, and facilitates translation and/or promotes RNA stability [Berry, Mure *et al.*, 2016; Yerramsetty, Agar *et al.*, 2017]. Since rbcL is expressed in M cells of C_4 plants but not transcribed [Wostrikoff, Clark *et al.*, 2012], RLSB, which is also BS cell specific in C_4 plants, might be required for LSU synthesis. Thus, removing the BS-cell specificity for Raf1 and RLSB, or introducing M cell-specific Raf1and RLSB expression, are logical steps towards achieving Rubisco activity in C_4 M cells. This is akin to the genetic RLSB-manipulation approach proposed for shifting Rubisco to BS cells in C_3 plants [Li and Heckman, 2017]. Other regulatory proteins for BS-specific expression of Rubisco in C_4 BS cells have recently been implicated [Vitlin Gruber and Feiz, 2018]. Constructs for RLSB, Raf1, and other regulators, singly or in combination, will also include Rubisco SSU genes with M cell-specific and temperature-controlled expression, so that the CB cycle will be operational in C_3 but not in C_4 mode (Figure 5). Another option would be to insert nuclear copies of plastid-targeted SSU and LSU in combination behind strong M cell specific promoters.

The reductive phase of the CB cycle in C_4 plants is known to take place in the M cells via metabolite exchange [Gowik and Westhoff, 2011]. Thus, several of the CB cycle enzymes are functional in M cells. Based on transcriptional analysis [John, Smith-Unna *et al.*, 2014], augmented expression in the M cells of other enzymes, such as fructose bisphosphate adolase, phosphoribulose kinase, ribose 5-phosphate isomerase will be needed to complete the CB cycle. Quantitative proteomics as performed in maize [Majeran *et al.*, 2010] could be performed to provide the necessary information as to what genes should be added to the appropriated DNA constructs.

With the CO_2 re-direction approach, the carbon shuttle would be eliminated when plants are operating a C_3 mode of photosynthesis. As the genes necessary for carbon shuttle are well characterized, construction of synthetic circuits to control their expression by placing key enzymes in the shuttle under thermo-sensing control should be straightforward. It would also be necessary to alter cell-wall properties, *e.g.*, by a reduction in suberization of the BS, so as to facilitate CO_2 diffusion into BS cell layer (Figure 7). Transition to C_4 mode would result in the induction of genes that drive the carbon shuttle and the restoration of BS suberization. As early season growth will favor a C_3 mode of operation, the default state of the plants will be C_3 until induced by temperature to become C_4.

The making of a $C_{3,4}$ sorghum would represent a paradigm shift in plant photosynthesis and inspire similar experiments in other crops, *e.g.*, maize, as well as provide critical data and knowledge for a wide array of plant-engineering efforts. For example, deep insight into the regulatory machinery that confers BS cell-specific expression of Rubisco and the CB cycle will accelerate progress in the C_4-Rice project, and understanding the processes underpinning suberization of the BS cells will benefit projects aimed at increasing sink and source strength and biomass yield in C_3 crops [Ueno and Sentoku, 2006; Feldman, Murchie *et al.*, 2014].

At face value, construction of temperature-controlled dual C_3-C_4 photosynthesis may seem as a lofty but unattainable goal. Non-arguably, the approaches we suggest in this perspective represent major undertakings. However, as we have pointed out elsewhere [Jansson, Vogel *et al.*, 2018], neither dual C_3-C_4 photosynthesis nor temperature-inducible C_4 photosynthesis are unrecognized phenomena. Also, all C_4 enzymes are present in C_3 plants where they play a housekeeping role [Aubry, Brown *et al.*, 2011, Langdale, 2011], suggesting that designing dual C_3-C_4 photosynthesis may mostly be a matter of rewiring regulatory networks, rather than introducing novel functionalities. As to the nature of a molecular thermosensing system, we posit that

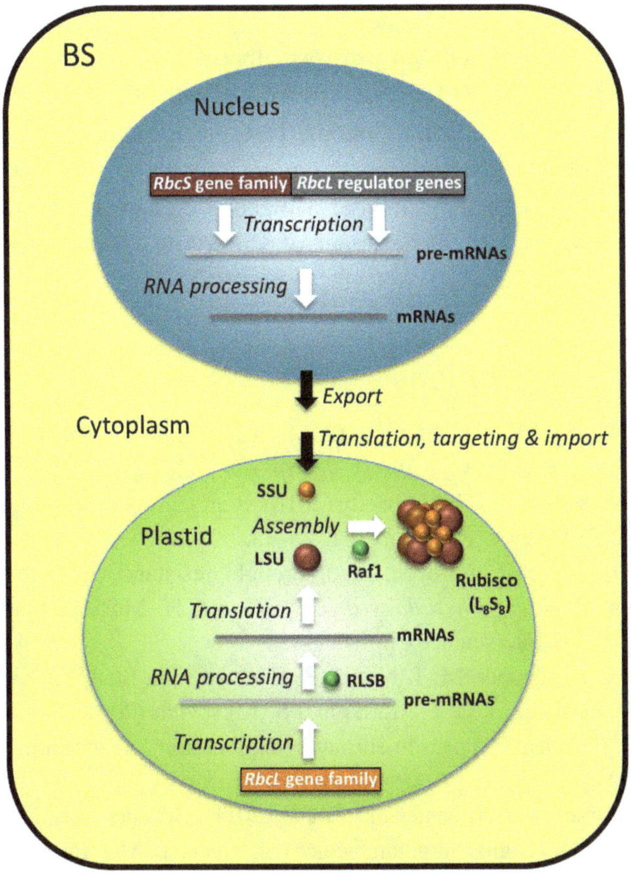

Figure 7. Mechanisms for bundle cell (BS)-specific expression of Rubisco activity in C_4 plants indicating a few identified key players.

attention should be paid to temperature-induced C_4 photosynthesis in certain aquatic systems [von Caemmerer, Edwards *et al.*, 2014], and to bacterial RNA thermometers [Kortmann and Narberhaus, 2012], and thermosensing DNA hairpins [Jonstrup, Fredsoe *et al.*, 2013]. To further enhance the impact of a dual C_3-C_4 photosynthesis, we suggest integrating temperature and light in the C_3-C_4 switch similar to what has been shown for phytochrome B in Arabidopsis [Legris, Klose *et al.*, 2016].

Acknowledgments

C. J. acknowledges the Pacific Northwest National Laboratory (PNNL) iPASS and PREMIS Initiatives, the DOE-ARPA-E CASP project, and DOE-BER EPICON project. The CASP project was supported by the US Department of Energy (DOE)

Advanced Research Projects Agency — Energy (ARPA-E) under award number 14/CJ000/09/02. The EPICON project was supported by US DOE Biological and Environmental Research (BER) under award number DE-SC0014081. The work conducted by the US DOE Joint Genome Institute is supported by the Office of Science of the US Department of Energy under Contract no. DE-AC02-05CH1123. Work performed at the Environmental Molecular Sciences Laboratory (EMSL) was funded in part by US Department of Energy (DOE) Contract DE-AC05-76RL01830 with PNNL.

References and Citations

Aubry, S., N. J. Brown and J. M. Hibberd (2011). The role of proteins in C_3 plants prior to their recruitment into the C_4 pathway. *J. Exp. Bot.,* 62, 3049–3059.

Berry, J. O., C. M. Mure and P. Yerramsetty (2016). Regulation of Rubisco gene expression in C_4 plants. *Curr. Opin. Plant Bio.,* 31, 23–28.

Blum, A. (1979). Genetic improvement of drought resistance in crop plants: A case for sorghum. In *Stress Physiology in Crop Plants*, H. Mussell, Staples, R.C., eds. (New York, Wiley Interscience), pp. 429–450.

Brenton, Z. W., E. A. Cooper, M. T. Myers, R. E. Boyles, N. Shakoor, K. J. Zielinski, B. L. Rauh, W. C. Bridges, G. P. Morris and S. Kresovich (2016). A genomic resource for the development, improvement, and exploitation of sorghum for bioenergy. *Genetics*, 204, 21–23.

Burow, G., J. J. Burke, Z. G. Xin and C. D. Franks (2011). Genetic dissection of early-season cold tolerance in sorghum (Sorghum bicolor (L.) Moench). *Mol. Breeding*, 28, 391–402.

Christin, P. A., G. Besnard, E. Samaritani, M. R. Duvall, T. R. Hodkinson, V. Savolainen and N. Salamin (2008). Oligocene CO_2 decline promoted C_4 photosynthesis in grasses. *Curr. Biol.*, 18, 37–43.

Collatz, G. J., J. A. Berry and J. S. Clark (1998). Effects of climate and atmospheric CO_2 partial pressure on the global distribution of C_4 grasses: present, past, and future. *Oecologia*, 114, 441–454.

Doggett, H. (1988). *Sorghum*, 2nd ed., Longman Scientific and Technical.

Drewry, D. T., P. Kumar and S. P. Long (2014). Simultaneous improvement in productivity, water use, and albedo through crop structural modification. *Global Change Biol.,* 20, 1955–1967.

Drewry, D. T., P. Kumar, S. P. Long, C. Bernacchi and X.-Z. Liang (2010). Ecohydrological responses of dense canopies to environmental variability: 1. Interplay between vertical structure and photosynthetic pathway. *J. Geophys. Res.*, 115, DOI:10.1029/2010JG001340.

Drincovich, M. F., P. Casati, C. S. Andreo, S. J. Chessin, V. R. Franceschi, G. E. Edwards and M. S. B. Ku (1998). Evolution of C_4 photosynthesis in Flaveria species — Isoforms of NADP-malic enzyme. *Plant Physiol.*, 117, 733–744.

Edwards, E. J., C. P. Osborne, C. A. E. Stromberg, S. A. Smith, W. J. Bond, P. A. Christin, A. B. Cousins, M. R. Duvall, D. L. Fox, R. P. Freckleton, O. Ghannoum, J. Hartwell,

Y. S. Huang, C. M. Janis, J. E. Keeley, E. A. Kellogg, A. K. Knapp, A. D. B. Leakey, D. M. Nelson, J. M. Saarela, R. F. Sage, O. E. Sala, N. Salamin, C. J. Still, B. Tipple and C. G. Consortium (2010). The origins of C_4 grasslands: Integrating evolutionary and ecosystem science. *Science,* 328, 587–591.

Feldman, A. B., E. H. Murchie, H. Leung, M. Baraoidan, R. Coe, S. M. Yu, S. F. Lo and W. P. Quick (2014). Increasing leaf vein density by mutagenesis: Laying the foundations for C_4 rice. *Plos One*, 9.

Fiedler, K., W. A. Bekele, C. Matschegewski, R. Snowdon, S. Wieckhorst, A. Zacharias and R. Uptmoor (2016). Cold tolerance during juvenile development in sorghum: a comparative analysis by genomewide association and linkage mapping. *Plant Breeding,* 135, 598–606.

Glithero, N. J., P. Wilson and S. J. Ramsden (2013). Prospects for arable farm uptake of short rotation coppice willow and miscanthus in England. *App. Energy,* 107, 209–218.

Gowik, U. and P. Westhoff (2011). The Path from C_3 to C_4 Photosynthesis 1. *Plant Physiol.*, 155, 56–63.

Jansson, C., J. Vogel, S. Hazen, T. Brutnell and T. Mockler (2018). Climate-smart crops with enhanced photosynthesis. *J.Exp. Bot.*, 69, 3801–3809.

John, C. R., R. D. Smith-Unna, H. Woodfield, S. Covshoff and J. M. Hibberd (2014). Evolutionary convergence of cell-specific gene expression in independent lineages of C-4 grasses. *Plant Physiol.,* 165, 62–75.

Jonstrup, A. T., J. Fredsoe and A. H. Andersen (2013). DNA hairpins as temperature switches, thermometers and ionic detectors. *Sensors*, 13, 5937–5944.

Kellogg, E. A. (2013). C_4 photosynthesis. *Curr. Biol.,* 23, R594–R599.

Kortmann, J. and F. Narberhaus (2012). Bacterial RNA thermometers: molecular zippers and switches. *Nat. Rev. Microbiol.,* 10, 255–265.

Kubien, D. S., S. Von Cammerer, R. T. Furbank and R. F. Sage (2003). C_4 photosynthesis at low temperature. A study using transgenic plants with reduced amounts of Rubisco. *Plant Physiol.*, 132, 1577–1585.

Langdale, J. A. (2011). C_4 cycles: Past, present, and future research on C_4 photosynthesis. *Plant Cell,* 23, 3879–3892.

Leakey, A. D. B., M. Uribelarrea, E. A. Ainsworth, S. L. Naidu, A. Rogers, D. R. Ort and S. P. Long (2006). Photosynthesis, productivity, and yield of maize are not affected by open-air elevation of CO2 concentration in the absence of drought. *Plant Physiol.*, 140, 779–790.

Legris, M., C. Klose, E. S. Burgle, C. Costigliolo, M. N'eme, M. Neme, A. Hiltbrunn, P. A. Wigge, E. Schäfer, R. D. Vierstra and J. J. Casal (2016). Phytochrome B integrates light and temperature signals in Arabidopsis. *Science,* 27, DOI: 10.1126/science.aaf5656.

Li, Y. and D. Heckman (2017). Combining genetic and evolutionary engineering to establish C_4 metabolism in C_3 plants. *J. Exp. Bot.,* 68.

Majeran, W., Frisco, G., Ponnala, L., Connolly, B., Hunag, M. S., Reiderl, E., Zhang, C. K., Asakura, Y., Bhulyan, N. H., Sun, Q., Turgeon, R. and van Wijk, K. J. (2010) Structural and metabolic transitions of C4 leaf development and differentiation defined by microscopy and quantitative proteomics in maize. *Plant Cell,* 22, 3509–3542.

Marshall-Colon, A., S. P. Long, D. K. Allen, G. Allen, D. A. Beard, B. Benes, S. von Caemmerer, A. J. Christensen, D. J. Cox, J. C. Hart, P. M. Hirst, K. Kannan, D. S. Katz, J. P. Lynch, A. J. Millar, B. Panneerselvam, N. D. Price, P. Prusinkiewicz, D. Raila, R. G. Shekar, S. Shrivastava, D. Shukla, V. Srinivasan, M. Stitt, M. J. Turk, E. O. Voit, Y. Wang, X. Y. Yin and X. G. Zhu (2017). Crops in silico: Generating virtual crops using an integrative and multi-scale modeling platform. *Front. Plant Sci.,* 8.

Maulana, F. and T. T. Tesso (2013). Cold temperature episode at seedling and flowering stages reduces growth and yield components in sorghum. *Crop Science,* 53, 564–574.

Maulana, F., D. Weerasooriya and T. Tesso (2017). Sorghum landrace collections from cooler regions of the world exhibit magnificent genetic differentiation and early season cold tolerance. *Front. Plant Sci.,* DOI: 10.3389/fpls.2017.00756.

Mullet, J., D. T. Morishige, R. F. McCormick, S. K. Truong, J. Hilley, B. McKinley, R. G. W. Anderson, S. N. Olson and W. Rooney (2014). Energy sorghum — a genetic model for the design of C_4 grass bioenergy crops. *J. Exp. Bot.,* 65, 3479–3489.

Parra-Londono, S., K. Fiedler, M. Kavka, B. Samans, S. Wieckhorst, A. Zacharias and R. Uptmoor (2017). Genetic dissection of early season cold tolerance in sorghum: genome wide association studies for seedling emergence and survival under field and controlled environment conditions. *Theor. Appl. Genet.,* https://doi.org/10.1007/s00122–017-3021–2.

Patel, M. and J. O. Berry (2008). Rubisco gene expression in C_4 plants. *J. Exp. Bot*., 59, 1625–1634.

Paterson, A. H. (2008). Genomics of sorghum. *Int. J. Plant Genomics,* DOI:10.1155/2008/362451.

Sage, R. F. and D. S. Kubien (2007). The temperature response of C_3 and C_4 photosynthesis. *Plant Cell Environ.,* 30, 1086–1106.

Sage, R. F., M. D. Peixoto, P. Friesen and B. Deen (2015). C_4 bioenergy crops for cool climates, with special emphasis on perennial C_4 grasses. *J. Exp. Bot.,* 66, 4195–4212.

Sage, R. F. and T. L. Sage (2002). Microsite characteristics of Muhlenbergia richardsonis (Trin.) Rydb., an alpine C(4) grass from the White Mountains, California. *Oecologia,* 132, 501–508.

Sage, R. F. and X. G. Zhu (2011). Exploiting the engine of C_4 photosynthesis. *J. Exp. Bot*., 62, 2989–3000.

Slattery, R. A. and D. R. Ort (2015). Photosynthetic energy conversion efficiency: Setting a baseline for gauging future improvements in important food and biofuel crops. *Plant Physiol.,* 168, 383–392.

Srinivasan, V., P. Kumar and S. P. Long (2017). Decreasing, not increasing, leaf area will raise crop yields under global atmospheric change. *Global Change Biol.,* 23, 1626–1635.

Stata, M., T. L. Sage, N. Hoffmann, S. Covshoff, G. K. S. Wong and R. F. Sage (2016). Mesophyll chloroplast investment in C-3, C-4 and C-2 species of the Genus Flaveria. *Plant Cell Physiol.,* 57, 904–918.

Ubierna, N., W. Sun and A. B. Cousins (2011). The efficiency of C_4 photosynthesis under low light conditions: assumptions and calculations with CO_2 isotope discrimination. *J. Exp. Bot.,* 62, 3119–3134.

Ueno, O. (2011). Structural and biochemical characterization of the C_3-C_4 intermediate *Brassica gravinae* and relatives, with particular reference to cellular distribution of Rubisco. *J. Exp. Bot.*, 62, 5347–5355.

Ueno, O. and N. Sentoku (2006). Comparison of leaf structure and photosynthetic characteristics of C-3 and C-4 Alloteropsis semialata subspecies. *Plant Cell Environ.*, 29, 257–268.

Vicentini, A., J. C. Barber, S. S. Aliscioni, L. M. Giussani and E. A. Kellogg (2008). The age of the grasses and clusters of origins of C_4 photosynthesis. *Global Change Biol.*, 14, 2963–2977.

Vitlin Gruber, A. and L. Feiz (2018). Rubisco Assembly in the Chloroplast. *Front. Mol. Biosci.*, DOI: 10.3389/fmolb.2018.00024.

von Caemmerer, S., G. E. Edwards, N. Koteyeva and A. B. Cousins (2014). Single cell C_4 photosynthesis in aquatic and terrestrial plants: A gas exchange perspective. *Aqutic Biol.*, 118.

von Fischer, J. C., L. L. Tiezen and D. S. Schimel (2008). Climate controls on C_3 *vs.* C_4 productivity in North American grasslands from carbon isotope composition of soil organic matter. *Global Change Biol.*, DOI: 10.1111/j.1365–2486.2008.01552 14, 1–15.

Wang, C. L., L. Y. Guo, Y. X. Li and Z. Wang (2012). Systematic comparison of C_3 and C_4 plants based on metabolic network analysis. *BMC Syst. Bio.*, 6.

Wostrikoff, K., A. Clark, S. Sato, T. Clemente and D. Stern (2012). Ectopic expression of Rubisco subunits in maize mesophyll cells does not overcome barriers to cell type-specific accumulation. *Plant Physiol.*, 160, 419–432.

Yamori, W., K. Hikosaka and D. A. Way (2014). Temperature response of photosynthesis in C_3, C_4, and CAM plants: Temperature acclimation and temperature adaptation. *Photosynth. Res.*, 119, 101–117.

Yerramsetty, P., E. M. Agar, W. C. Yim, J. C. Cushman and J. O. Berry (2017). An rbcL mRNA-binding protein is associated with C-3 to C-4 evolution and light-induced production of Rubisco in Flaveria. *J. Exp. Bot.*, 68, 4635–4649.

Zhu, X. G., S. P. Long and D. R. Ort (2008). What is the maximum efficiency with which photosynthesis can convert solar energy into biomass? *Curr. Opin. Biotechnol.*, 19, 153–159.

Zhu, X. G., S. P. Long and D. R. Ort (2010). Improving photosynthetic efficiency for greater yield. *Ann. Rev. Plant Biol.*, 61, 235–261.

Index

www.ingramcontent.com/pod-product-compliance
Lightning Source LLC
LaVergne TN
LVHW081931200926
848484LV00009B/2404
* 9 7 8 9 8 1 3 2 7 6 9 1 8 *